dvanced Derivatives Pricing and Risk Management

# ADVANCED DERIVATIVES PRICING AND RISK MANAGEMENT

## Theory, Tools and Hands-On Programming Application

Claudio Albanese and
Giuseppe Campolieti

ELSEVIER
ACADEMIC
PRESS

AMSTERDAM • BOSTON • HEIDELBERG • LONDON
NEW YORK • OXFORD • PARIS • SAN DIEGO
SAN FRANCISCO • SINGAPORE • SYDNEY • TOKYO

Elsevier Academic Press
30 Corporate Drive, Suite 400, Burlington, MA 01803, USA
525 B Street, Suite 1900, San Diego, California 92101-4495, USA
84 Theobald's Road, London WC1X 8RR, UK

This book is printed on acid-free paper. ♾



**Library of Congress Cataloging-in-Publication Data**
Application Submitted

**British Library Cataloguing in Publication Data**
A catalogue record for this book is available from the British Library

ISBN 13: 978-0-12-047682-4
ISBN 10: 0-12-047682-7



For all information on all Elsevier Academic Press publications visit our Web site at www.books.elsevier.com

Printed in the United States of America
05 06 07 08 09 10 9 8 7 6 5 4 3 2 1

# Contents

# Preface

This book originated in part from lecture notes we developed while teaching courses in financial mathematics in the Master of Mathematical Finance Program at the University of Toronto during the years from 1998 to 2003. We were confronted with the challenge of teaching a varied set of finance topics, ranging from derivative pricing to risk management, while developing the necessary notions in probability theory, stochastic calculus, statistics, and numerical analysis and while having the students acquire practical computer laboratory experience in the implementation of financial models. The amount of material to be covered spans a daunting number of topics. The leading motives are recent discoveries in derivatives research, whose comprehension requires an array of applied mathematical techniques traditionally taught in a variety of different graduate and senior undergraduate courses, often not included in the realm of traditional finance education. Our choice was to teach all the relevant topics *in the context* of financial engineering and mathematical finance while delegating more systematic treatments of the supporting disciplines, such as probability, statistics, numerical analysis, and financial markets and institutions, to parallel courses. Our project turned from a challenge into an interesting and rewarding teaching experience. We discovered that probability and stochastic calculus, when presented in the context of derivative pricing, are easier to teach than we had anticipated. Most students find financial concepts and situations helpful to develop an intuition and understanding of the mathematics. A formal course in probability running in parallel introduced the students to the mathematical theory of stochastic calculus, but only after they already had acquired the basic problem-solving skills. Computer laboratory projects were run in parallel and took students through the actual "hands-on" implementation of the theory through a series of financial models. Practical notions of information technology were introduced in the laboratory as well as the basics in applied statistics and numerical analysis.

This book is organized into two main parts: Part I consists of the main body of the theory and mathematical tools, and Part II covers a series of numerical implementation projects for laboratory instruction. The first part is organized into rather large chapters that span the main topics, which in turn consist of a series of related subtopics or sections. Chapter 1 introduces the basic notions of pricing theory together with probability and stochastic calculus. The relevant notions in probability and stochastic calculus are introduced in the finance context. Students learn about static and dynamic hedging strategies and develop an underlying framework for pricing various European-style contracts, including quanto and basket options. The martingale (or probabilistic) and Partial differential equation (PDE) formulations are presented as alternative approaches for derivatives pricing. The last part of Chapter 1 provides a theoretical framework for pricing American options. Chapter 2 is devoted to fixed-income derivatives. Numerical solution methods such as lattice models, model calibration, and Monte Carlo simulations are introduced within relevant projects in the second part of the book. Chapter 3 is devoted to more advanced mathematical topics in option pricing, covering some techniques for exact exotic option pricing within continuous-time state-dependent diffusion models. A substantial part of Chapter 3 is drawn partly from some of our recent research and hence covers derivations of new pricing formulas for complex state-dependent diffusion models for European-style contracts as well as barrier options. One focus of this chapter is to expose the reader to some of the more advanced, yet essential, mathematical tools for tackling derivative pricing problems that lie beyond the standard contracts and/or simpler models. Although the technical content in Chapter 3 may be relatively high, our goal has been to present the material in a comprehensive fashion. Chapter 4 reviews numerical methods and statistical estimation methodologies for value-at-risk and risk management.

Part II includes a dozen shorter "chapters," each one dedicated to a numerical laboratory project. The additional files distributed in the attached disk give the documentation and framework as they were developed for the students. We made an effort to cover a broad variety of information technology topics, to make sure that the students acquire the basic programming skills required by a professional financial engineer, such as the ability to design an interface for a pricing module, produce scenario-generation engines for pricing and risk management, and access a host of numerical library components, such as linear algebra routines. In keeping with the general approach of this book, students acquire these skills not in isolation but, rather, in the context of concrete implementation tasks for pricing and risk management models.

This book can presumably be read and used in a variety of ways. In the mathematical finance program, Chapters 1 and 2, and limited parts of Chapters 3 and 4 formed the core of the theory course. All the chapters (i.e., projects) in Part II were used in the parallel numerical laboratory course. Some of the material in Chapter 3 can be used as a basis for a separate graduate course in advanced topics in pricing theory. Since Chapter 4, on value-at-risk, is largely independent of the other ones, it may also possibly be covered in a parallel risk management course.

The laboratory material has been organized in a series of modules for classroom instruction we refer to as *projects* (i.e., numerical laboratory projects). These projects serve to provide the student or practitioner with an initial experience in actual quantitative implementations of pricing and risk management. Admittedly, the initial projects are quite far from being realistic financial engineering problems, for they were devised mostly for pedagogical reasons to make students familiar with the most basic concepts and the programming environment. We thought that a key feature of this book was to keep the prerequisites to a bare minimum and not assume that all students have advanced programming skills. As the student proceeds further, the exercises become more challenging and resemble realistic situations more closely. The projects were designed to cover a reasonable spectrum of some of the basic topics introduced in Part I so as to enhance and augment the student's knowledge in various basic topics. For example, students learn about static hedging strategies by studying problems with barrier options and variance swaps, learn how to design and calibrate lattice models and use them to price American and other exotics, learn how to back out a high-precision LIBOR zero-yield curve from swap and forward rates, learn how to set up and calibrate interest rate trees for pricing interest rate derivatives using a variety of one-factor short rate models, and learn about estimation and simulation methodologies for value-at-risk. As the assignments progress, relevant programming topics may be introduced in parallel. Our choice fell on the Microsoft technologies because they provide perhaps the easiest-to-learn-about rapid application development frameworks; however, the concepts that students learn also have analogues with other technologies. Students learn gradually how to design the interface for a pricing model using spreadsheets. Most importantly, they learn how to invoke and use numerical libraries, including LAPACK, the standard numerical linear algebra package, as well as a broad variety of random- and quasi-random-number generators, zero finders and optimizer routines, spline interpolations, etc. To a large extent, technologies can be replaced. We have chosen Microsoft Excel as a graphic user interface as well as a programming tool. This should give most PC users the opportunity to quickly gain familiarity with the code and to modify and experiment with it as desired. The Math Point libraries for visual basic (VB) and visual Basic for applications (VBA), which are used in our laboratory materials, were developed specifically for this teaching project, but an experienced programmer could still use this book and work in alternative frameworks, such as the Nag FORTRAN libraries under Linux and Java. The main motive of the book also applies in this case: We teach the relevant concepts in information technology, which are a necessary part of the professional toolkit of financial engineers, by following what according to our experience is the path of least resistance in the learning process.

Finally, we would like to add numerous acknowledgments to all those who made this project a successful experience. Special thanks go to the students who attended the Master of Mathematical Finance Program at the University of Toronto in the years from 1998 to 2003. They are the ones who made this project come to life in the first place. We thank Oliver Chen and Stephan Lawi for having taught the laboratory course in the fifth year of the program. We thank Petter Wiberg, who agreed to make the material in his Ph.D. thesis available to us for partial use in Chapter 4. We thank our coauthors in the research papers we wrote over the years, including Peter Carr, Oliver Chen, Ken Jackson, Alexei Kusnetzov, Pierre Hauvillier, Stephan Lawi, Alex Lipton, Roman Makarov, Smaranda Paun, Dmitri Rubisov, Alexei Tchernitser, Petter Wiberg, and Andrei Zavidonov.

PART • I

# Pricing Theory and Risk Management

CHAPTER • 1

# Pricing Theory

Pricing theory for derivative securities is a highly technical topic in finance; its foundations rest on trading practices and its theory relies on advanced methods from stochastic calculus and numerical analysis. This chapter summarizes the main concepts while presenting the essential theory and basic mathematical tools for which the modeling and pricing of financial derivatives can be achieved.

Financial assets are subdivided into several classes, some being quite basic while others are structured as complex contracts referring to more elementary assets. Examples of *elementary asset classes* include *stocks*, which are ownership rights to a corporate entity; *bonds*, which are promises by one party to make cash payments to another in the future; *commodities*, which are assets, such as wheat, metals, and oil that can be consumed; and *real estate* assets, which have a convenience yield deriving from their use. A more general example of an asset is that of a contractual *contingent claim* associated with the obligation of one party to enter a stream of more elementary financial transactions, such as cash payments or deliveries of shares, with another party at future dates. The value of an individual transaction is called a *pay-off* or *payout*. Mathematically, a pay-off can be modeled by means of a *payoff function* in terms of the prices of other, more elementary assets.

There are numerous examples of contingent claims. *Insurance policies*, for instance, are structured as contracts that envision a payment by the insurer to the insured in case a specific event happens, such as a car accident or an illness, and whose pay-off is typically linked to the damage suffered by the insured party. *Derivative assets* are claims that distinguish themselves by the property that the *payoff function* is expressed in terms of the price of an *underlying asset*. In finance jargon, one often refers to underlying assets simply as *underlyings*. To some extent, there is an overlap between insurance policies and derivative assets, except the nomenclature differs because the first are marketed by insurance companies while the latter are traded by banks.

A *trading strategy* consists of a set of rules indicating what positions to take in response to changing market conditions. For instance, a rule could say that one has to adjust the position in a given stock or bond on a daily basis to a level given by evaluating a certain function. The implementation of a trading strategy results in pay-offs that are typically random. A major difference that distinguishes derivative instruments from insurance contracts is that most traded derivatives are structured in such a way that it is possible to implement trading strategies in the underlying assets that generate streams of pay-offs that *replicate* the pay-offs of the derivative claim. In this sense, trading strategies are substitutes for derivative claims. One of the driving forces behind derivatives markets is that some market participants, such as market makers, have a competitive advantage in implementing replication strategies, while their clients are interested in taking certain complex risk exposures synthetically by entering into a single contract.

A key property of replicable derivatives is that the corresponding payoff functions depend only on prices of tradable assets, such as stocks and bonds, and are not affected by events, such as car accidents or individual health conditions that are not directly linked to an asset price. In the latter case, risk can be reduced only by diversification and reinsurance. A related concept is that of *portfolio immunization*, which is defined as a trade intended to offset the risk of a portfolio over at least a short time horizon. A perfect replication strategy for a given claim is one for which a position in the strategy combined with an offsetting position in the claim are perfectly immunized, i.e., risk free. The position in an asset that immunizes a given portfolio against a certain risk is traditionally called *hedge ratio*.[1] An immunizing trade is called a *hedge*. One distinguishes between *static* and *dynamic hedging*, depending on whether the hedge trades can be executed only once or instead are carried over time while making adjustments to respond to new information.

The assets traded to execute a replication strategy are called *hedging instruments*. A set of hedging instruments in a financial model is *complete* if all derivative assets can be replicated by means of a trading strategy involving only positions in that set. In the following, we shall define the mathematical notion of financial models by listing a set of hedging instruments and assuming that there are *no redundancies*, in the sense that no hedging instrument can be replicated by means of a strategy in the other ones. Another very common expression is that of *risk factor*: The risk factors underlying a given financial model with a complete basis of hedging instruments are given by the prices of the hedging instruments themselves or functions thereof; as these prices change, risk factor values also change and the prices of all other derivative assets change accordingly. The statistical analysis of risk factors allows one to assess the risk of financial holdings.

*Transaction costs* are impediments to the execution of replication strategies and correspond to costs associated with adjusting a position in the hedging instruments. The market for a given asset is *perfectly liquid* if unlimited amounts of the asset can be traded without affecting the asset price. An important notion in finance is that of *arbitrage*: If an asset is replicable by a trading strategy and if the price of the asset is different from that of the replicating strategy, the opportunity for riskless gains/profits arises. Practical limitations to the size of possible gains are, however, placed by the inaccuracy of replication strategies due to either market incompleteness or lack of liquidity. In such situations, either riskless replication strategies are not possible or prices move in response to posting large trades. For these reasons, arbitrage opportunities are typically short lived in real markets.

Most financial models in pricing theory account for finite liquidity indirectly, by postulating that prices are *arbitrage free*. Also, market incompleteness is accounted for indirectly and is reflected in corrections to the probability distributions in the price processes. In this stylized mathematical framework, each asset has a unique *price*.[2]

[1] Notice that the term *hedge ratio* is part of the finance jargon. As we shall see, in certain situations hedge ratios are computed as mathematical ratios or limits thereof, such as derivatives. In other cases, expressions are more complicated.

[2] To avoid the perception of a linguistic ambiguity, when in the following we state that a given asset is *worth* a certain amount, we mean that amount is the asset price.

Most financial models are built upon the *perfect-markets hypothesis*, according to which:

- There are no trading impediments such as transaction costs.
- The set of basic hedging instruments is complete.
- Liquidity is infinite.
- No arbitrage opportunities are present.

These hypotheses are robust in several ways. If liquidity is not perfect, then arbitrage opportunities are short lived because of the actions of arbitrageurs. The lack of completeness and the presence of transaction costs impacts prices in a way that is uniform across classes of derivative assets and can safely be accounted for implicitly by adjusting the process probabilities.

The existence of replication strategies, combined with the perfect-markets hypothesis, makes it possible to apply more sophisticated pricing methodologies to financial derivatives than is generally possible to devise for insurance claims and more basic assets, such as stocks. The key to finding derivative prices is to construct mathematical models for the underlying asset price processes and the replication strategies. Other sources of information, such as a country's domestic product or a takeover announcement, although possibly relevant to the underlying prices, affect derivative prices only indirectly.

This first chapter introduces the reader to the mathematical framework of pricing theory in parallel with the relevant notions of probability, stochastic calculus, and stochastic control theory. The dynamic evolution of the risk factors underlying derivative prices is *random*, i.e., not deterministic, and is subject to uncertainty. Mathematically, one uses *stochastic processes*, defined as random variables with probability distributions on sets of paths. Replicating and hedging strategies are formulated as sets of rules to be followed in response to changing price levels. The key principle of pricing theory is that if a given payoff stream can be replicated by means of a dynamic trading strategy, then the cost of executing the strategy must equal the price of a contractual claim to the payoff stream itself. Otherwise, arbitrage opportunities would ensue. Hence pricing can be reduced to a mathematical optimization problem: to replicate a certain payoff function while minimizing at the same time replication costs and replication risks. In perfect markets one can show that one can achieve perfect replication at a finite cost, while if there are imperfections one will have to find the right trade-off between risk and cost. The *fundamental theorem of asset pricing* is a far-reaching mathematical result that states;

- The solution of this optimization problem can be expressed in terms of a *discounted expectation of future pay-offs* under a pricing (or probability) measure.
- This representation is unique (with respect to a given discounting) as long as markets are complete.

Discounting can be achieved in various ways: using a bond, using the money market account, or in general using a reference *numeraire asset* whose price is positive. This is because pricing assets is a relative, as opposed to an absolute, concept: One values an asset by computing its worth as compared to that of another asset. A key point is that expectations used in pricing theory are computed under a probability measure tailored to the numeraire asset.

In this chapter, we start the discussion with a simple single-period model, where trades can be carried out only at one point in time and gains or losses are observed at a later time, a fixed date in the future. In this context, we discuss static hedging strategies. We then briefly review some of the relevant and most basic elements of probability theory in the context of multivariate continuous random variables. Brownian motion and martingales are then discussed as an introduction to stochastic processes. We then move on to further discuss continuous-time stochastic processes and review the basic framework of stochastic (Itô) calculus. Geometric Brownian motion is then presented, with some preliminary derivations of Black–Scholes formulas for single-asset and multiasset price models. We then proceed to introduce a more general mathematical framework for dynamic hedging and derive the fundamental theorem of asset pricing (FTAP) for continuous-state-space and continuous-time-diffusion processes. We then apply the FTAP to European-style options. Namely, by the use of change of numeraire and stochastic calculus techniques, we show how exact pricing formulas based on geometric Brownian motions for the underlying assets are obtained for a variety of situations, ranging from elementary stock options to foreign exchange and quanto options. The partial differential equation approach for option pricing is then presented. We then discuss pricing theory for early-exercise or American-style options.

## 1.1 Single-Period Finite Financial Models

The simplest framework in pricing theory is given by *single-period financial models*, in which calendar time $t$ is restricted to take only two values, current time $t = 0$ and a future date $t = T > 0$. Such models are appropriate for analyzing situations where trades can be made *only* at current time $t = 0$. Revenues (i.e., profits or losses) can be realized only at the later date $T$, while trades at intermediate times are not allowed.

In this section, we focus on the particular case in which only a finite number of *scenarios* $\omega_1, \ldots, \omega_m$ can occur. *Scenario* is a common term for an outcome or event. The *scenario set* $\Omega = \{\omega_1, \ldots, \omega_m\}$ is also called the *probability space*. A *probability measure* $P$ is given by a set of numbers $p_i$, $i = 1, \ldots, m$, in the interval $[0, 1]$ that sum up to 1; i.e.,

$$\sum_{i=1}^{m} p_i = 1, \qquad 0 \leq p_i \leq 1. \tag{1.1}$$

$p_i$ is the *probability* that scenario (event) $\omega_i$ occurs, i.e., that the $i$th state is attained. Scenario $\omega_i$ is *possible* if it can occur with strictly positive probability $p_i > 0$. Neglecting scenarios that cannot possibly occur, the probabilities $p_i$ will henceforth be assumed to be strictly positive; i.e., $p_i > 0$. A *random variable* is a function on the scenario set, $f : \Omega \to \mathbb{R}$, whose values $f(\omega_i)$ represent observables. As we discuss later in more detail, examples of random variables one encounters in finance include the price of an asset or an interest rate at some point in the future or the pay-off of a derivative contract. The *expectation* of the random variable $f$ is defined as the sum

$$E^P[f] = \sum_{i=1}^{m} p_i f(\omega_i). \tag{1.2}$$

Asset prices and other financial observables, such as interest rates, are modeled by *stochastic processes*. In a single-period model, a *stochastic process* is given by a value $f_0$ at current time $t = 0$ and by a random variable $f_T$ that models possible values at time $T$. In finance, probabilities are obtained with two basically different procedures: They can either be *inferred* from historical data by *estimating* a statistical model, or they can be *implied* from current asset valuations by *calibrating* a pricing model. The former are called *historical*, *statistical*, or, better, *real-world* probabilities. The latter are called *implied probabilities*. The calibration procedure involves using the fundamental theorem of asset pricing to represent prices as discounted expectations of future pay-offs and represents one of the central topics to be discussed in the rest of this chapter.

**Definition 1.1. Financial Model** *A finite, single-period financial model $\mathcal{M} = (\Omega, \mathcal{A})$ is given by a finite scenario set $\Omega = \{\omega_1, \ldots, \omega_m\}$ and n basic asset price processes for hedging instruments:*

$$\mathcal{A} = \{A_t^1, \ldots, A_t^n; t = 0, T\}. \tag{1.3}$$

*Here, $A_0^i$ models the current price of the ith asset at current (or initial) time $t = 0$ and $A_T^i$ is a random variable such that the price at time $T > 0$ of the ith asset in case scenario $\omega_j$ occurs is given by $A_T^i(\omega_j)$. The basic asset prices $A_t^i$, $i = 1, \ldots, n$, are assumed real and positive.*

**Definition 1.2. Portfolio and Asset** *Let $\mathcal{M} = (\Omega, \mathcal{A})$ be a financial model. A portfolio $\boldsymbol{\pi}$ is given by a vector with components $\pi_i \in \mathbb{R}$, $i = 1, \ldots, n$, representing the positions or holdings in the the family of basic assets with prices $A_t^1, \ldots, A_t^n$. The worth of the portfolio at terminal time T is given by $\sum_{i=1}^n \pi_i A_T^i(\omega)$ given the state or scenario $\omega$, whereas the current price is $\sum_{i=1}^n \pi_i A_0^i$. A portfolio is nonnegative if it gives rise to nonnegative pay-offs under all scenarios, i.e., $\sum_{i=1}^n \pi_i A_T^i(\omega_j) \geq 0, \ \forall j = 1, \ldots, m$. An asset price process $A_t = A_t(\omega)$ (a generic one, not necessarily that of a hedging instrument) is a process of the form*

$$A_t = \sum_{i=1}^n \pi_i A_t^i \tag{1.4}$$

*for some portfolio $\boldsymbol{\pi} \in \mathbb{R}^n$.*

The modeling assumption behind this definition is that market liquidity is infinite, meaning that asset prices don't vary as a consequence of agents trading them. As we discussed at the start of this chapter, this hypothesis is valid only in case trades are relatively small, for large trades cause market prices to change. In addition, a financial model with infinite liquidity is mathematically consistent only if there are no arbitrage opportunities.

**Definition 1.3. Arbitrage: Single-Period Discrete Case** *An arbitrage opportunity or arbitrage portfolio is a portfolio $\boldsymbol{\pi} = (\pi_1, \ldots, \pi_n)$ such that either of the following conditions holds:*
**A1.** *The current price of $\boldsymbol{\pi}$ is negative, $\sum_{i=1}^n \pi_i A_0^i < 0$, and the pay-off at terminal time T is nonnegative, i.e., $\sum_{i=1}^n \pi_i A_T^i(\omega_j) \geq 0$ for all j states.*
**A2.** *The current price of $\boldsymbol{\pi}$ is zero, i.e., $\sum_{i=1}^n \pi_i A_0^i = 0$, and the pay-off at terminal time T in at least one scenario $\omega_j$ is positive, i.e., $\sum_{i=1}^n \pi_i A_T^i(\omega_j) > 0$ for some jth state, and the pay-off at terminal time T is nonnegative.*

**Definition 1.4. Market Completeness** *The financial model $\mathcal{M} = (\Omega, \mathcal{A})$ is complete if for all random variables $f_t : \Omega \to \mathbb{R}$, where $f_t$ is a bounded payoff function, there exists an asset price process or portfolio $A_t$ in the basic assets contained in $\mathcal{A}$ such that $A_T(\omega) = f_T(\omega)$ for all scenarios $\omega \in \Omega$.*

This definition essentially states that any pay-off (or state-contingent claim) can be replicated, i.e., is attainable by means of a portfolio consisting of positions in the set of basic assets. If an arbitrage portfolio exists, one says there is *arbitrage*. The first form of arbitrage occurs whenever there exists a trade of negative initial cost at time $t = 0$ by means of which one can form a portfolio that under all scenarios at future time $t = T$ has a nonnegative pay-off. The second form of arbitrage occurs whenever one can perform a trade at zero cost at an initial time $t = 0$ and then be assured of a strictly positive payout at future time $T$ under at least one possible scenario, with no possible downside. In reality, in either case investors would want to perform arbitrage trades and take arbitrarily large positions in the arbitrage portfolios. The existence of these trades, however, infringes on the modeling assumption of infinite liquidity, because market prices would shift as a consequence of these large trades having been placed.

Let's start by considering the simplest case of a single-period economy consisting of only two hedging instruments (i.e., $n = 2$ basic assets) with price processes $A_t^1 = B_t$ and $A_t^2 = S_t$. The scenario set, or sample space, is assumed to consist of only two possible states of the world: $\Omega = \{\omega_+, \omega_-\}$. $S_t$ is the price of a risky asset, which can be thought of as a stock price. The riskless asset is a *zero-coupon bond*, defined as a process $B_t$ that is known to be worth the so-called *nominal* amount $B_T = N$ at time $T$ while at time $t = 0$ has worth

$$B_0 = (1 + rT)^{-1} N. \tag{1.5}$$

Here $r > 0$ is called the *interest rate*. As is discussed in more detail in Chapter 2, interest rates can be defined with a number of different *compounding rules*; the definition chosen here for $r$ corresponds to selecting $T$ itself as the compounding interval, with simple (or discrete) compounding assumed. At current time $t = 0$, the stock has known worth $S_0$. At a later time $t = T$, two scenarios are possible for the stock. If the scenario $\omega_+$ occurs, then there is an upward move and $S_T = S_T(\omega_+) \equiv S_+$; if the scenario $\omega_-$ occurs, there is a downward move and $S_T = S_T(\omega_-) \equiv S_-$, where $S_+ > S_-$. Since the bond is riskless we have $B_T(\omega_+) = B_T(\omega_-) = B_T$. Assume that the real-world probabilities that these events will occur are $p_+ = p \in (0, 1)$ and $p_- = (1 - p)$, respectively.

Figure 1.1 illustrates this simple economy. In this situation, the hypothesis of arbitrage freedom demands that the following strict inequality be satisfied:

$$\frac{S_-}{1 + rT} < S_0 < \frac{S_+}{1 + rT}. \tag{1.6}$$

In fact, if, for instance, one had $S_0 < \frac{S_-}{1+rT}$, then one could make unbounded riskless profits by initially borrowing an arbitrary amount of money and buying an arbitrary number of shares in the stock at price $S_0$ at time $t = 0$, followed by selling the stock at time $t = T$ at a higher return level than $r$. Inequality (1.6) is an example of a restriction resulting from the *condition of absence of arbitrage*, which is defined in more detail later.

A *derivative* asset, of worth $A_t$ at time $t$, is a claim whose pay-off is contingent on future values of risky underlying assets. In this simple economy the underlying asset is the stock. An example is a derivative that pays $f_+$ dollars if the stock is worth $S_+$, and $f_-$ otherwise, at final time $T$: $A_T = A_T(\omega_+) = f_+$ if $S_T = S_+$ and $A_T = A_T(\omega_-) = f_-$ if $S_T = S_-$. Assuming one can take fractional positions, this payout can be statically replicated by means of a portfolio

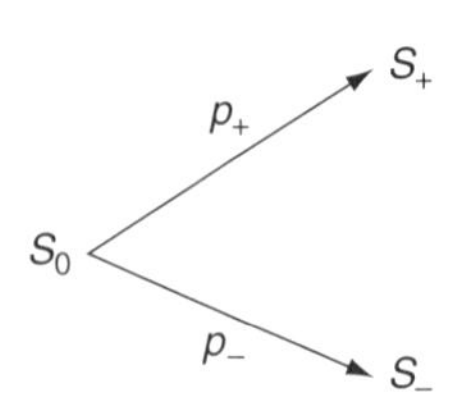


FIGURE 1.1 A single-period model with two possible future prices for an asset S.

consisting of $a$ shares of the stock and $b$ bonds such that the following replication conditions under the two scenarios are satisfied:

$$aS_- + bN = f_-, \tag{1.7}$$

$$aS_+ + bN = f_+. \tag{1.8}$$

The solution to this system is

$$a = \frac{f_+ - f_-}{S_+ - S_-}, \qquad b = \frac{f_- S_+ - f_+ S_-}{N(S_+ - S_-)}. \tag{1.9}$$

The price of the replicating portfolio, with pay-off identical to that of the derivative, must be the price of the derivative asset; otherwise there would be an arbitrage opportunity. That is, one could make unlimited riskless profits by buying (or selling) the derivative asset and, at the same time, taking a short (or long) position in the portfolio at time $t = 0$. At time $t = 0$, the arbitrage-free price of the derivative asset, $A_0$, is then

$$\begin{aligned} A_0 &= aS_0 + b(1+rT)^{-1}N \\ &= \left(\frac{S_0 - (1+rT)^{-1}S_-}{S_+ - S_-}\right)f_+ + \left(\frac{(1+rT)^{-1}S_+ - S_0}{S_+ - S_-}\right)f_-. \end{aligned} \tag{1.10}$$

Dimensional considerations are often useful to understand the structure of pricing formulas and detect errors. It is important to remember that prices at different moments in calendar time are not equivalent and that they are related by discount factors. The hedge ratios $a$ and $b$ in equation (1.9) are dimensionless because they are expressed in terms of ratios of prices at time $T$. In equation (1.10) the variables $f_\pm$ and $S_+ - S_-$ are measured in dollars at time $T$, so their ratio is dimensionless. Both $S_0$ and the discounted prices $(1+rT)^{-1}S_\pm$ are measured in dollars at time 0, as is also the derivative price $A_0$.

Rewriting this last equation as

$$A_0 = (1+rT)^{-1}\left[\left(\frac{(1+rT)S_0 - S_-}{S_+ - S_-}\right)f_+ + \left(\frac{S_+ - (1+rT)S_0}{S_+ - S_-}\right)f_-\right] \tag{1.11}$$

shows that price $A_0$ can be interpreted as the discounted expected pay-off. However, the probability measure is *not* the real-world one (i.e., not the physical measure $P$) with probabilities $p_\pm$ for up and down moves in the stock price. Rather, current price $A_0$ is the discounted expectation of future prices $A_T$, in the following sense:

$$A_0 = (1+rT)^{-1}\, E^Q[A_T] = (1+rT)^{-1}[q_+ A_T(\omega_+) + q_- A_T(\omega_-)] \tag{1.12}$$

under the measure $Q$ with probabilities (strictly between 0 and 1)

$$q_+ = \frac{(1+rT)S_0 - S_-}{S_+ - S_-}, \qquad q_- = \frac{S_+ - (1+rT)S_0}{S_+ - S_-}, \tag{1.13}$$

$q_+ + q_- = 1$. The measure $Q$ is called the *pricing measure*. Pricing measures also have other, more specific names. In the particular case at hand, since we are discounting with a constant interest rate within the time interval $[0, T]$, $Q$ is commonly named the *risk-neutral* or *risk-adjusted probability measure*, where $q_\pm$ are so-called risk-neutral (or risk-adjusted) probabilities. Later we shall see that this measure is also the *forward measure*, where the bond price $B_t$ is used as numeraire asset. In particular, by expressing all asset prices relative to (i.e., in units of) the bond price $A_t^i/B_t$, with $B_T = N$, regardless of the scenario and $B_0/B_T = (1+rT)^{-1}$, we can hence recast the foregoing expectation as: $A_0 = B_0\, E^Q[A_T/B_T]$. Hence $Q$ corresponds to the forward measure. We can also use as numeraire a discretely compounded *money-market account* having value $(1+rt)$ (or $(1+rt)N$). By expressing all asset prices relative to this quantity, it is trivially seen that the corresponding measure is the same as the forward measure in this simple model. As discussed later, the name *risk-neutral measure* shall, however, refer to the case in which the money-market account (to be defined more generally later in this chapter) is used as numeraire, and this measure generally differs from the forward measure for more complex financial models.

Later in this chapter, when we cover pricing in *continuous time*, we will be more specific in defining the terminology needed for pricing under general choices of numeraire asset. We will also see that what we just unveiled in this particularly simple case is a general and far-reaching property: Arbitrage-free prices can be expressed as discounted expectations of future pay-offs. More generally, we will demonstrate that asset prices can be expressed in terms of expectations of *relative asset price processes*. A pricing measure is then a *martingale measure*, under which all relative asset price processes (i.e., relative to a given choice of numeraire asset) are so-called *martingales*. Since our primary focus is on continuous-time pricing models, as introduced later in this chapter, we shall begin to explicitly cover some of the essential elements of martingales in the context of stochastic calculus and continuous-time pricing. For a more complete and elaborate mathematical construction of the martingale framework in the case of discrete-time finite financial models, however, we refer the reader to other literature (for example, see [Pli97, MM03]).

We now extend the pricing formula of equation (1.12) to the case of $n$ assets and $m$ possible scenarios.

**Definition 1.5. Pricing Measure** *A probability measure $Q = (q_1, \ldots, q_m)$, $0 < q_j < 1$, for the scenario set $\Omega = \{\omega_1, \ldots, \omega_m\}$ is a pricing measure if asset prices can be expressed as follows:*

$$A_0^i = \alpha\, E^Q[A_T^i] = \alpha \sum_{j=1}^{m} q_j A_T^i(\omega_j) \tag{1.14}$$

*for all $i = 1, \ldots, n$ and some real number $\alpha > 0$. The constant $\alpha$ is called the discount factor.*

**Theorem 1.1. Fundamental Theorem of Asset Pricing (Discrete, single-period case)** *Assume that all scenarios in $\Omega$ are possible. Then the following statements hold true:*

- *There is no arbitrage if and only if there is a pricing measure for which all scenarios are possible.*
- *The financial model is complete, with no arbitrage if and only if the pricing measure is unique.*

**Proof.** First, we prove that if a pricing measure $Q = (q_1, \ldots, q_m)$ exists and prices $A_0^i = \alpha\, E^Q[A_T^i]$ for all $i = 1, \ldots, n$, then there is no arbitrage. If $\sum_i \pi_i A_T^i(\omega_j) \geq 0$, for all $\omega_j \in \Omega$, then from equation (1.14) we must have $\sum_i \pi_i A_0^i \geq 0$. If $\sum_i \pi_i A_0^i = 0$, then from equation (1.14) we cannot satisfy the payoff conditions in (A2) of Definition 1.3. Hence there is no arbitrage, for any choice of portfolio $\boldsymbol{\pi} \in \mathbb{R}^n$.

On the other hand, assume that there is no arbitrage. The possible price-payoff $(m+1)$-tuples

$$\mathcal{P} = \left\{ \left( \sum_{i=1}^{n} \pi_i A_0^i, \sum_{i=1}^{n} \pi_i A_T^i(\omega_1), \ldots, \sum_{i=1}^{n} \pi_i A_T^i(\omega_m) \right), \qquad \boldsymbol{\pi} \in \mathbb{R}^n \right\} \tag{1.15}$$

make up a plane in $\mathbb{R} \times \mathbb{R}^m$. Since there is no arbitrage, the plane $\mathcal{P}$ intersects the octant $\mathbb{R}_+ \times \mathbb{R}^m_+$ made up of vectors of nonnegative coordinates only in the origin. Let $\mathcal{N}$ be the set of all vectors $(-\beta, \gamma_1, \ldots, \gamma_m)$ normal to the plane $\mathcal{P}$ and normalized so that $\beta > 0$. Vectors in $\mathcal{N}$ satisfy the normality condition

$$-\beta\left(\sum_{i=1}^{n} \pi_i A_0^i\right) + \sum_{j=1}^{m} \gamma_j \left(\sum_{i=1}^{n} \pi_i A_T^i(\omega_j)\right) = 0 \tag{1.16}$$

for all portfolios $\boldsymbol{\pi}$.

Next we obtain two Lemmas to complete the proof.

**Lemma 1.1.** *Suppose the financial model on the scenario set $\Omega$ and with instruments $(A^1, \ldots, A^n)$ is arbitrage free and let m be the dimension of the linear space $\mathcal{P}$. If the matrix rank* $\dim \mathcal{P} < m$, *then one can define $l = (m - \dim \mathcal{P})$ price-payoff tuples $(-B_0^k, B_T^k(\omega)), k = 1, \ldots, l$, so that the extended financial model with basic assets $(A^1, \ldots, A^n, B^1, \ldots, B^l)$ and scenario set $\Omega$ is complete and arbitrage free.*

**Proof.** The price-payoff tuples $(-B_0^k, B_T^k(\omega_1), \ldots, B_T^k(\omega_l))$ can be found iteratively. Suppose that $l = m - \dim \mathcal{P} > 0$. Then the complement to the linear space $\mathcal{P}$ has dimension $l + 1 \geq 2$. Let $X = (-X_0^k, X_T^k(\omega))$ and $Y = (-Y_0^k, Y_T^k(\omega))$ be two vectors orthogonal to each other and orthogonal to $\mathcal{P}$. Then there is an angle $\theta$ such that the vector $B^1 = \cos\theta X + \sin\theta Y$ has at least one strictly positive coordinate and one strictly negative coordinate, i.e., $B^1 \notin \mathbb{R} \times \mathbb{R}_+$. Hence the financial model with instruments $(A^1, \ldots, A^n, B^1)$ is arbitrage free. Iterating the argument, one can complete the market while retaining arbitrage freedom. □

**Lemma 1.2.** *If markets are complete, the space $\mathcal{N}$ orthogonal to $\mathcal{P}$ is spanned by a vector $(\beta, \gamma_1, \ldots, \gamma_m)$ lying in the main octant $\mathcal{B} = \mathbb{R}_+ \times \mathbb{R}^m_+$ of vectors with strictly positive coordinates.*

**Proof.** In fact if $\beta = 0$, then $\mathcal{P}$ contains the line $(x, 0, \ldots, 0)$ and all positive payouts would be possible, even for an empty portfolio, which is absurd. It is also absurd that $\gamma_j = 0, \forall j$. In fact, in this case, since markets are complete, there is an instrument paying one dollar in case the scenario $\omega_j$ occurs and zero otherwise, and since $\gamma_j = 0$, the price of this instrument at time $t = 0$ is zero, which is absurd. □

If markets are not complete, one can still conclude that the set $\mathcal{N}$ contains a vector $(\beta, \gamma_1, \ldots, \gamma_m)$ with strictly positive coordinates. In fact, thanks to Lemma 1.1, one can complete it while preserving arbitrage freedom by introducing auxiliary assets and the normal vector can be chosen to have positive coordinates. Hence, in all cases of $\pi_i$ values, according to equation (1.16) we have

$$A_0^i = \alpha\, E^Q[A_T^i] = \alpha \sum_{j=1}^{m} q_j A_T^i(\omega_j), \tag{1.17}$$

where $Q$ is the measure with probabilities

$$q_j = \frac{\gamma_j}{\sum_{j=1}^{m} \gamma_j} \tag{1.18}$$

and discount factor

$$\alpha = \beta^{-1} \sum_{j=1}^{m} \gamma_j. \tag{1.19}$$

□

The first project of Part II of this book is a study on single-period arbitrage. We refer the interested reader to that project for a more detailed and practical exposition of the foregoing theory. In particular, the project provides an explicit discussion of a numerical linear algebra implementation for detecting arbitrage in single-period, finite financial models.

## Problems

**Problem 1.** Consider the simple example in Figure 1.1 and assume the interest rate is $r$. Under what condition is there no arbitrage in the model?

**Problem 2.** Compute $E^Q[S_T]$ within the single-period two-state model. Explain your result.

**Problem 3.** Let $p_i^0$ denote the current price $A_0^i$ of the $i$th security and denote by $D_{ij} = A_T^i(\omega_j)$ the matrix elements of the $n \times m$ dividend matrix with $i = 1, \ldots, n$, $j = 1, \ldots, m$. Using equation (1.14) with $\alpha = (1 + rT)^{-1}$ show that the risk-neutral expected return on any security $A^i$ is given by the risk-free interest rate

$$E^Q\left[\frac{A_T^i - A_0^i}{A_0^i}\right] = \sum_{j=1}^{m} q_j\left(\frac{D_{ij}}{p_i^0} - 1\right) = rT, \tag{1.20}$$

where $q_j$ are the risk-neutral probabilities.

**Problem 4.** State the explicit matrix condition for market completeness in the single-period two-state model with the two basic assets as the riskless bond and the stock. Under what condition is this market complete?

**Problem 5.** *Arrow–Debreu securities* are claims with unit pay-offs in only one state of the world. Assuming a single-period two-state economy, these claims are denoted by $E_\pm$ and defined by

$$E_+(\omega) = \begin{cases} 1, & \text{if } \omega = \omega_+ \\ 0, & \text{if } \omega = \omega_- \end{cases}, \qquad E_-(\omega) = \begin{cases} 0, & \text{if } \omega = \omega_+ \\ 1, & \text{if } \omega = \omega_- \end{cases}.$$

(a) Find exact replicating portfolios $\boldsymbol{\pi}_+ = (a_+, b_+)$ and $\boldsymbol{\pi}_- = (a_-, b_-)$ for $E_+$ and $E_-$, respectively. The coefficients $a$ and $b$ are positions in the stock and the riskless bond, respectively.
(b) Letting $F_T$ represent an arbitrary pay-off, find the unique portfolio of Arrow–Debreu securities that replicates $F_T$.

## 1.2 Continuous State Spaces

This section, together with the next section, presents a review of basic elements of probability theory for random variables that can take on a continuum of values while emphasizing some of the financial interpretation of mathematical concepts.

Modern probability theory is based on measure theory. Referring the reader to textbook literature for more detailed and exhaustive formal treatments, we will just simply recall here that measure theory deals with the definition of *measurable sets* $D$, *probability measures* $\mu$, and *integrable functions* $f : D \to \mathbb{R}$ for which one can evaluate expectations as integrals

$$E[f] = \int_D f(\mathbf{x})\mu(d\mathbf{x}). \tag{1.21}$$

In finance, one typically deals with situations where the measurable set $D \subset \mathbb{R}^d$, with integer $d \geq 1$. Realizations of the vector variable $\mathbf{x} \in D$ correspond to scenarios for the risk factors or random variables in a financial model.

Future asset prices are real-valued functions of underlying risk factors $f(\mathbf{x})$ defined for $\mathbf{x} \in D$ and hence themselves define random variables. Probability measures $\mu(d\mathbf{x})$ are often defined as $\mu(d\mathbf{x}) = p(\mathbf{x})d\mathbf{x}$, where $p(\mathbf{x})$ is a real-valued continuous *probability distribution function* that is nonnegative and integrates to 1; i.e.,

$$p(\mathbf{x}) \geq 0, \qquad \int_D p(\mathbf{x})d\mathbf{x} = 1. \tag{1.22}$$

The expectation $E^P[f]$ of $f$ under the probability measure with $p$ as density is defined by the $d$-dimensional integral

$$E^P[f] = \int_D f(\mathbf{x})p(\mathbf{x})d\mathbf{x}. \tag{1.23}$$

The pair $(D, \mu(d\mathbf{x}))$ is called a *probability space*.

In particular, this formalism can also allow for the case of a finite scenario set of vectors $D = \{\mathbf{x}^{(1)}, \ldots, \mathbf{x}^{(N)}\}$, as was considered in the previous section. In this case the probability distribution is a sum of Dirac *delta functions*,

$$p(\mathbf{x}) = \sum_{i=1}^{N} p_i \delta(\mathbf{x} - \mathbf{x}^{(i)}). \tag{1.24}$$

As further discussed shortly, a delta function can be thought of as a singular function that is positive, integrates to 1 over all space, and corresponds to the infinite limiting case of a sequence of integrable functions with support only at the origin. Probabilistically, a distribution, such as equation (1.24), which is a sum of delta functions, corresponds to a situation where only the scenarios $\mathbf{x}^{(1)}, \ldots, \mathbf{x}^{(N)}$ can possibly occur, and they do with probabilities $p_1, \ldots, p_N$. These probabilities must be positive and add up to 1; i.e.,

$$\sum_{i=1}^{N} p_i = 1. \tag{1.25}$$

In the case of a finite scenario set (i.e., a finite set of possible events with finite integer $N$), the random variable $f = f(\mathbf{x})$ is a function defined on the set of scenarios $D$, and its expectation under the measure with $p$ as density is given by the finite sum

$$E^P[f] = \sum_{i=1}^{N} p_i f(\mathbf{x}^{(i)}). \tag{1.26}$$

For an infinitely countable set of scenarios, then, the preceding expressions must be considered in the limit $N \to \infty$. Hence in the case of a discrete set of scenarios (as opposed to a continuum) the probability density function collapses into the usual *probability mass function*, as occurs in standard probability theory of discrete-valued random variables.

The Dirac delta function is not an ordinary function in $\mathbb{R}^d$ but, rather, a so-called *distribution*. Mathematically, a distribution is defined through its value when integrated against a smooth function. One can regard $\delta(\mathbf{x} - \mathbf{x}')$, $\mathbf{x}, \mathbf{x}' \in \mathbb{R}^d$, as the limit of an infinitesimally narrow $d$-dimensional normal distribution:

$$\int_{\mathbb{R}^d} f(\mathbf{x})\delta(\mathbf{x} - \mathbf{x}')d\mathbf{x} = \lim_{\sigma \to 0} \frac{1}{(\sigma\sqrt{2\pi})^d} \int_{\mathbb{R}^d} f(\mathbf{x}) \exp\left(-\frac{|\mathbf{x} - \mathbf{x}'|^2}{2\sigma^2}\right) d\mathbf{x} = f(\mathbf{x}'). \tag{1.27}$$

For example, in one dimension a representation of the delta function is

$$\delta(x - x') = \lim_{\sigma \to 0} \frac{1}{\sigma\sqrt{2\pi}} e^{-(x-x')^2/2\sigma^2}. \tag{1.28}$$

*Events* are modeled as subsets $G \subset D$ for which one can compute the integral that gives the expectation $E^P[\mathbf{1}_G]$. The function $\mathbf{1}_G(\mathbf{x})$ denotes the random variable equal to 1 for $\mathbf{x} \in G$ and to zero otherwise; $\mathbf{1}_G(\mathbf{x})$ is called the *indicator function* of the set $G$. This expectation is interpreted as the *probability P(G)* that event $G \subset D$ will occur; i.e.,

$$P(G) = E^P[\mathbf{1}_G] = \int_{\mathbb{R}^d} \mathbf{1}_G(\mathbf{x}) p(\mathbf{x}) d\mathbf{x} = \int_G p(\mathbf{x}) d\mathbf{x}. \tag{1.29}$$

Examples of events are subsets, e.g., such as

$$G = \{\mathbf{x} \in D: \ a < f(\mathbf{x}) < b\}, \tag{1.30}$$

with $b > a$ and where $f$ is some function. An important concept associated with events is that of *conditional expectation*. Given a random variable $f$, the expectation of $f$ conditioned to knowing that event $G$ will occur is

$$E^P[f|G] = \frac{E^P[f \cdot \mathbf{1}_G]}{P(G)}. \tag{1.31}$$

Two probability measures $\tilde{\mu}(d\mathbf{x}) = \tilde{p}(\mathbf{x})d\mathbf{x}$ and $\mu(d\mathbf{x}) = p(\mathbf{x})d\mathbf{x}$ are said to be equivalent (or absolutely continuous with respect to one another) if they share the same sets of null probability; i.e., $\tilde{\mu} \sim \mu$ if the probability condition $P(G) > 0$ implies $\tilde{P}(G) > 0$, where

$$\tilde{P}(G) = E^{\tilde{P}}[\mathbf{1}_G] = \int_{\mathbb{R}^d} \mathbf{1}_G(\mathbf{x}) \tilde{p}(\mathbf{x}) d\mathbf{x} = \int_G \tilde{p}(\mathbf{x}) d\mathbf{x}, \tag{1.32}$$

with $E^{\tilde{P}}[\,]$ denoting the expectation with respect to the measure $\tilde{\mu}$. When computing the expectation of a real-valued random variable, say, of the general form of a function of a random vector (such functions are further defined in the next section), $f = f(\mathbf{X}) : \mathbb{R}^d \to \mathbb{R}$, it is sometimes useful to switch from one choice of probability measure to another, equivalent one. One can use the following *change of measure* (known as the *Radon–Nikodym theorem*) for computing expectations:

$$E^P[f] = \int_D f(\mathbf{x}) \mu(d\mathbf{x}) = \int_D f(\mathbf{x}) \frac{d\mu}{d\tilde{\mu}}(\mathbf{x}) \tilde{\mu}(d\mathbf{x}) = E^{\tilde{P}}\left[f \frac{d\mu}{d\tilde{\mu}}\right]. \tag{1.33}$$

The nonnegative random variable denoted by $\frac{d\mu}{d\tilde{\mu}}$ is called the *Radon–Nikodym derivative* of $\mu$ with respect to $\tilde{\mu}$ (or $P$ w.r.t. $\tilde{P}$). From this result it also follows that $\frac{d\mu}{d\tilde{\mu}} = \left(\frac{d\tilde{\mu}}{d\mu}\right)^{-1}$ and $E^{\tilde{P}}[\frac{d\mu}{d\tilde{\mu}}] = 1$. As will be seen later in the chapter, a more general adaptation of this result for computing certain types of conditional expectations involving martingales will turn out to form one of the basic tools for pricing financial derivatives using changes of numeraire. Another particular example of the use of this change-of-measure technique is in the Monte Carlo estimation of integrals by so-called importance-sampling methods, as described in Chapter 4.

Just as integrals are approximated with arbitrary accuracy by finite integral sums, continuous probability distributions can be approximated by discrete ones. For instance, let $D \subset \mathbb{R}^d$ be a bounded domain and $p(\mathbf{x})$ be a continuous probability density on $D$ and let $\{G_1, \ldots, G_m\}$

be a *partition of D* made up of a family of nonintersecting events $G_i \subset D$ whose union covers the entire state space $D$ and that have the shape of hypercubes. Let $p_i$ be the probability of event $G_i$ under the probability measure with density $p(x)$. Then an approximation for $p(\mathbf{x})$ is

$$p(\mathbf{x}) = \sum_{i=1}^{m} p_i \delta(\mathbf{x} - \mathbf{x}_i), \tag{1.34}$$

where $\mathbf{x}_i$ is the center of the hypercube corresponding to event $G_i$. Let $\delta$ be the volume of the largest hypercube among the cubes in the partition $\{G_1, \ldots, G_m\}$ and let $f(\mathbf{x})$ be a random variable on $D$. In the limit $\delta \to 0$, as the partition becomes finer and finer, the number of events $m(\delta)$ will diverge to $\infty$. In this limit, we find

$$E^P[f] = \lim_{\delta \to 0} \sum_{i=1}^{m(\delta)} p_i f(\mathbf{x}_i). \tag{1.35}$$

By using sums as approximations to expectations, which are essentially multidimensional Riemann integrals, one can extend the theorem in the previous section to the case of continuous probability distributions. Consider a single-period financial model with current (i.e., initial) time $t = 0$ and time horizon $t = T$ and with $n$ basic assets whose current prices are $A_0^i$, $i = 1, \ldots, n$. The prices of these basic assets at time $T$ are indexed by a continuous state space represented by the domain $\Omega \subset \mathbb{R}^d$, and the values of the basic assets are random variables $A_T^i(\mathbf{x})$, with $\mathbf{x} \in \Omega$. That is, the asset prices $A_t^i$ are random variables assumed to take on real positive values, i.e., $A_t^i : \Omega \to \mathbb{R}_+$. Let's denote by $p(\mathbf{x})d\mathbf{x}$ the real-world probability measure in $\Omega$ and assume that the measure of all open subsets of $\Omega$ is strictly positive. A portfolio is modeled by a vector $\boldsymbol{\pi}$ whose components denote positions or holdings $\pi_i$, $i = 1, \ldots, n$, in the basic assets. The definition of arbitrage extends as follows.

**Definition 1.6. Nonnegative Portfolio** *A portfolio is nonnegative if it gives rise to nonnegative expected pay-offs under almost all events $G \subset \Omega$ of nonzero probability, i.e., such that*

$$E^P\left[\sum_{i=1}^{n} \pi_i\, A_T^i(\mathbf{x}) \middle| \mathbf{x} \in G\right] \geq 0. \tag{1.36}$$

**Definition 1.7. Arbitrage: Single-Period Continuous Case** *The market admits arbitrage if either of the following conditions holds:*
**A1.** *There is a nonnegative portfolio $\boldsymbol{\pi}$ of negative initial price $\sum_{i=1}^{n} \pi_i A_0^i < 0$.*
**A2.** *There is a nonnegative portfolio of zero initial cost, $\sum_{i=1}^{n} \pi_i A_0^i = 0$, for which the expected payoff is strictly positive, i.e., $E^P\left[\sum_{i=1}^{n} \pi_i A_T^i\right] > 0$.*

**Definition 1.8. Pricing Measure: Single-Period Continuous Case**[3] *A probability measure $Q$ of density $q(\mathbf{x})d\mathbf{x}$ on $D$ is a pricing measure if all asset prices at current time $t = 0$ can be expressed as follows:*

$$A_0^i = \alpha E^Q[f_i] = \alpha \int_\Omega f_i(\mathbf{x}) q(\mathbf{x}) d\mathbf{x} \tag{1.37}$$

*for some real number $\alpha > 0$. The constant $\alpha$ is called the discount factor. The functions $f_i(x) = A_T^i(x)$ are payoff functions for a given state or scenario $\boldsymbol{x}$.*

[3]Later we relate such pricing measures to the case of arbitrary choices of numeraire asset wherein the pricing formula involves an expectation of asset prices relative to the chosen numeraire asset price. Changes in numeraire correspond to changes in the probability measure.

Market completeness is defined in a manner similar to that in the single-period discrete case of the previous section. From the foregoing definitions of arbitrage and pricing measure we then have the following result, whose proof is left as an exercise.

**Theorem 1.2. Fundamental Theorem of Asset Pricing (Continuous Single-Period Case)** *Assume that all scenarios in $\Omega$ are possible. Then the following statements hold true.*

- *There is no arbitrage if and only if there is a pricing measure for which all scenarios are possible.*
- *If the linear span of the set of basic instruments $A_T^i$, $i = 1, \ldots, n$, is complete and there is no arbitrage, then there is a unique pricing measure $Q$ consistent with the prices $A_0^i$ of the reference assets at current time $t = 0$.*

The single-period pricing formalism can also be extended to the case of a multiperiod discrete-time financial model, where trading is allowed to take place at a finite number of intermediate dates. This feature gives rise to dynamic trading strategies, with portfolios in the basic assets being rebalanced at discrete points in time. The foregoing definitions and notions of arbitrage and asset pricing must then be modified and extended substantially. Rather than present the theory for such discrete-time models, we shall instead introduce more important theoretical tools in the following sections that will allow us ultimately to consider continuous-time financial models. Multiperiod discrete-time (continuous-state-space) models can then be obtained, if desired, as special cases of the continuous models via a discretization of time. A further discretization of the state space leads to discrete-time multiperiod finite financial models.

## 1.3 Multivariate Continuous Distributions: Basic Tools

*Marginal probability distributions* arise, for instance, when one is computing expectations on some reduced subspace of random variables. Consider, for example, a set of continuous random variables that can be separated or grouped into two random vector spaces $\mathbf{X} = (X_1, \ldots, X_m)$ and $\mathbf{Y} = (Y_1, \ldots, Y_{n-m})$ that can take on values $\mathbf{x} = (x_1, \ldots, x_m) \in \mathbb{R}^m$ and $\mathbf{y} = (y_1, \ldots, y_{n-m}) \in \mathbb{R}^{n-m}$, respectively, with $1 \leq m < n$, $n \geq 2$. The function $p(\mathbf{x}, \mathbf{y})$ is the joint probability density or *probability distribution function (pdf)* in the product space $\mathbb{R}^n = \mathbb{R}^m \times \mathbb{R}^{n-m}$. The integral

$$p_y(\mathbf{y}) \equiv \int_{\mathbb{R}^m} p(\mathbf{x}, \mathbf{y}) d\mathbf{x} \tag{1.38}$$

defines a marginal density $p_y(\mathbf{y})$. This function describes a probability density in the subspace of random vectors $\mathbf{Y} \in \mathbb{R}^{n-m}$ and integrates to unity over $\mathbb{R}^{n-m}$. The *conditional density function*, denoted by $p(\mathbf{x}|\mathbf{Y} = \mathbf{y}) \equiv p(\mathbf{x}|\mathbf{y})$ for the random vector $\mathbf{X}$, is defined on the subspace of $\mathbb{R}^m$ (for a given vector value $\mathbf{Y} = \mathbf{y}$) and is defined by the ratio of the joint probability density function and the marginal density function for the random vector $\mathbf{Y}$ evaluated at $\mathbf{y}$:

$$p(\mathbf{x}|\mathbf{y}) = \frac{p(\mathbf{x}, \mathbf{y})}{p_y(\mathbf{y})}, \tag{1.39}$$

assuming $p_y(\mathbf{y}) \neq 0$. From the foregoing two relations it is simple to see that, for any given $\mathbf{y}$, the conditional density also integrates to unity over $\mathbf{x} \in \mathbb{R}^m$.

Conditional distributions play an important role in finance and pricing theory. As we will see later, derivative instruments can be priced by computing conditional expectations. Assuming a conditional distribution, the *conditional expectation* of a continuous random variable $g = g(\mathbf{X}, \mathbf{Y})$, given $\mathbf{Y} = \mathbf{y}$, is defined by

$$E[g|\mathbf{Y} = \mathbf{y}] = \int_{\mathbb{R}^m} g(\mathbf{x}, \mathbf{y}) p(\mathbf{x}|\mathbf{y}) d\mathbf{x}. \tag{1.40}$$

Given any two continuous random variables $X$ and $Y$, then $E[X|Y = y]$ is a number while $E[X|Y]$ is itself a random variable as $Y$ is random, i.e., has not been fixed. We then have the following property that relates unconditional and conditional expectations:

$$E[X] = E\big[E[X|Y]\big] = \int_{-\infty}^{\infty} E[X|Y = y] p_y(y) dy. \tag{1.41}$$

This property is useful for computing expectations by conditioning. More generally, for a random variable given by the function $g = g(\mathbf{X}, \mathbf{Y})$ we have the property

$$\begin{aligned} E[g] &= \int_{\mathbb{R}^{n-m}} \int_{\mathbb{R}^m} g(\mathbf{x}, \mathbf{y}) p(\mathbf{x}, \mathbf{y}) d\mathbf{x} d\mathbf{y} \\ &= \int_{\mathbb{R}^{n-m}} \left[ \int_{\mathbb{R}^m} g(\mathbf{x}, \mathbf{y}) p(\mathbf{x}|\mathbf{y}) d\mathbf{x} \right] p_y(\mathbf{y}) d\mathbf{y} \\ &= \int_{\mathbb{R}^{n-m}} E[g|\mathbf{Y} = \mathbf{y}] p_y(\mathbf{y}) d\mathbf{y} = E\big[E[g|\mathbf{Y}]\big]. \end{aligned} \tag{1.42}$$

Functions of random variables, such as $g(\mathbf{X}, \mathbf{Y})$, are of course also random variables. In general, the pdf of a random variable given by a mapping $f = f(\mathbf{X}) : \mathbb{R}^n \to \mathbb{R}$ is the function $p_f : \mathbb{R} \to \mathbb{R}$,

$$p_f(\xi) = \lim_{\delta\xi \to 0} \frac{\mathrm{P}\big(f(\mathbf{X}) \in [\xi, \xi + \delta\xi)\big)}{(\delta\xi)}, \tag{1.43}$$

defined on some open or closed interval between $a$ and $b$. This interval may be finite or infinite; some examples are $\xi \in [0, 1]$, $[0, \infty)$, and $(-\infty, \infty)$. The *cumulative distribution function (cdf)* $C_f$ for the random variable $f$ is defined as

$$C_f(z) = \int_a^z p_f(\xi) d\xi \tag{1.44}$$

and gives the probability $P(a \le f \le z)$, with $C_f(b) = 1$. Let us consider another independent real-valued random variable $g \in (c, d)$, where *(c,d)* is generally any other interval. We recall that any two random variables $f$ and $g$ are independent if the joint pdf (or cdf) of $f$ and $g$ is given by the product of the respective marginal pdfs (or cdfs). The sum of two independent random variables $f$ and $g$ is again a random variable $h = f + g$. The cumulative distribution function, denoted by $C_h$, for the random variable $h$ is given by the convolution integral

$$\begin{aligned} C_h(\zeta) &= \iint\limits_{\xi+\eta\le\zeta} p_f(\xi) p_g(\eta) d\xi d\eta \\ &= \int_a^b p_f(\xi) C_g(\zeta - \xi) d\xi = \int_c^d p_g(\eta) C_f(\zeta - \eta) d\eta, \end{aligned} \tag{1.45}$$

where $p_g$ and $C_g$ are the density and cumulative distribution functions, respectively, for the random variable $g$. By differentiating the cumulative distribution function we find the density function for the variable $h$:

$$p_h(\zeta) = \int_a^b p_f(\xi)p_g(\zeta - \xi)d\xi = \int_c^d p_g(\eta)p_f(\zeta - \eta)d\eta. \tag{1.46}$$

The preceding formulas are sometimes useful because they provide the cumulative (or density) functions for a sum of two independent random variables as convolution integrals of the separate density and cumulative functions.

The definition for cumulative distribution functions extends into the multivariate case in the obvious manner. Given a pdf $p : \mathbb{R}^n \to \mathbb{R}$ for $\mathbb{R}^n$-valued random vectors $\mathbf{X} = (X_1, \ldots, X_n)$, the corresponding cdf is the function $C_p : \mathbb{R}^n \to \mathbb{R}$ defined by the joint probability

$$C_p(\mathbf{x}) = P(X_1 \leq x_1, \ldots, X_n \leq x_n) = \int_{-\infty}^{x_n} \cdots \int_{-\infty}^{x_1} p(\mathbf{x}')d\mathbf{x}'. \tag{1.47}$$

We recall that any two random variables $X_i$ and $X_j$ $(i \neq j)$ are *independent* if the joint probability $P(X_i \leq a, X_j \leq b) = P(X_i \leq a)P(X_j \leq b)$ for all $a, b \in \mathbb{R}$, i.e., if the events $\{X_i \leq a\}$ and $\{X_j \leq b\}$ are independent. Hence, for two independent random variables the joint cdf and joint pdf are equal to the product of the marginal cdf and marginal pdf, respectively: $p(x_i, x_j) = p_i(x_i)p_j(x_j)$ and $C_p(x_i, x_j) = C_i(x_i)C_j(x_j)$.

Another useful formula for multivariate distributions is the relationship between probability densities (within the same probability measure, say, $\mu(d\mathbf{x})$) expressed on different variable spaces or coordinate variables. That is, if $p(\mathbf{x})$ and $p_{\tilde{\mathbf{X}}}(\tilde{\mathbf{x}})$ represent probability densities on $n$-dimensional real-valued vector spaces $\mathbf{x}$ and $\tilde{\mathbf{x}}$, respectively and the two spaces are related by a one-to-one continuously differentiable mapping $\tilde{\mathbf{x}} = \tilde{\mathbf{x}}(\mathbf{x})$, then

$$p(\mathbf{x}) = p_{\tilde{\mathbf{X}}}(\tilde{\mathbf{x}})\left|\frac{d\tilde{\mathbf{x}}}{d\mathbf{x}}\right|, \tag{1.48}$$

where $\frac{d\tilde{\mathbf{x}}}{d\mathbf{x}}$ is the Jacobian matrix of the invertible transformation $\mathbf{x} \to \tilde{\mathbf{x}}$. The notation $|\mathbf{M}|$ refers to the determinant of a matrix $\mathbf{M}$.

A probability distribution that plays a distinguished role is the $n$-dimensional *Gaussian* (or *normal*) distribution, with *mean* (or average) vector $\boldsymbol{\mu} = (\mu_1, \ldots, \mu_n)$, defined on $\mathbf{x} \in \mathbb{R}^n$ as follows:

$$p(\mathbf{x}; \boldsymbol{\mu}, \mathbf{C}) = \frac{1}{\sqrt{(2\pi)^n|\mathbf{C}|}} \exp\left(-\frac{1}{2}(\mathbf{x} - \boldsymbol{\mu}) \cdot \mathbf{C}^{-1} \cdot (\mathbf{x} - \boldsymbol{\mu})\right). \tag{1.49}$$

The shorthand notation $\mathbf{x} \sim N_n(\boldsymbol{\mu}, \mathbf{C})$ is also used to denote the values of an $n$-dimensional random vector with components $x_1, \ldots, x_n$ that are obtained by sampling with distribution $p(\mathbf{x}; \boldsymbol{\mu}, \mathbf{C})$. $\mathbf{C} = (C_{ij})$ is called *covariance matrix* and enjoys the property of being *positive definite*, i.e., is such that the inner product $(\mathbf{x}, \mathbf{C}\mathbf{x}) \equiv \mathbf{x} \cdot (\mathbf{C}\mathbf{x}) > 0$ for all real vectors $\mathbf{x}$, and $C_{ij} = C_{ji}$. It follows that the cdf of the $n$-dimensional multivariate normal random vector is defined by the $n$-dimensional Gaussian integral

$$\Phi_n(\mathbf{x}; \boldsymbol{\mu}, \mathbf{C}) = \int_{-\infty}^{x_n} \cdots \int_{-\infty}^{x_1} p(\mathbf{x}'; \boldsymbol{\mu}, \mathbf{C})d\mathbf{x}'. \tag{1.50}$$

A particularly important special case of equation (1.50) for $n = 1$ is the univariate standard normal cdf (i.e., $\Phi_1(x; 0, 1)$), defined by

$$N(x) \equiv \frac{1}{\sqrt{2\pi}} \int_{-\infty}^{x} e^{-y^2/2}dy. \tag{1.51}$$

The mean of a random vector $\mathbf{X}$ with given pdf $p(\mathbf{x})$, is defined by the components

$$\mu_i = E[X_i] = \int_{\mathbb{R}^n} x_i p(\mathbf{x}) d\mathbf{x} = \int_{\mathbb{R}} x p_i(x) dx, \tag{1.52}$$

and the covariance matrix elements are defined by the expectations

$$C_{ij} \equiv \mathrm{Cov}(X_i, X_j) = E[(X_i - \mu_i)(X_j - \mu_j)] = \int_{\mathbb{R}^n} (x_i - \mu_i)(x_j - \mu_j) p(\mathbf{x}) d\mathbf{x}, \tag{1.53}$$

for all $i, j = 1, \ldots, n$. The *standard deviation* of the random variable $X_i$ is defined as the square root of the variance:

$$\sigma_i \equiv \sqrt{\mathrm{Var}(X_i)} = \sqrt{E[(X_i - \mu_i)^2]}, \tag{1.54}$$

and the *correlation* between two random variables $X_i$ and $X_j$ is defined as follows:

$$\rho_{ij} \equiv \mathrm{Corr}(X_i, X_j) = \frac{C_{ij}}{\sigma_i \sigma_j}. \tag{1.55}$$

Since $\sqrt{C_{ii}} = \sigma_i$, the correlation matrix has a unit diagonal, i.e., $\rho_{ii} = 1$. As well, they obey the inequality $|\rho_{ij}| \leq 1$ (see Problem 1 of this section). For random variables that may be positively or negatively correlated (e.g., as is the case for different stock returns) it follows that

$$-1 \leq \rho_{ij} \leq 1. \tag{1.56}$$

In the particular case of a multivariate normal distribution with positive definite covariance matrix as in equation (1.49), the strict inequalities $-1 < \rho_{ij} < 1$ hold.

The main property of normal distributions is that the convolution of two normal distributions is also normal. A random variable that is a sum of random normal variables is, therefore, also normally distributed (see Problem 2). Because of this property, multivariate normal distributions can be regarded as affine transformations of standard normal distributions with $\boldsymbol{\mu} = \mathbf{0}_{n\times 1}$ and $\mathbf{C} = \mathbf{I}_{n\times n}$ (the identity matrix). Consider the vector $\boldsymbol{\xi} = (\xi_1, \ldots, \xi_n)$ of independent standard normal variables with zero mean and unit covariance, i.e., with probability density

$$p(\boldsymbol{\xi}) = \prod_{i=1}^{n} \frac{e^{-\xi_i^2/2}}{\sqrt{2\pi}}. \tag{1.57}$$

If $\mathbf{L} = (L_{ij})$, is an $n$-dimensional matrix, then the random vector $\mathbf{X} = \boldsymbol{\mu} + \mathbf{L}\boldsymbol{\xi}$ is normally distributed with mean $\boldsymbol{\mu}$ and covariance $\mathbf{C} = \mathbf{L}\mathbf{L}^\dagger$, $\dagger \equiv$ matrix transpose. Indeed, taking expectations over the components gives

$$E[X_i] = E\left[\mu_i + \sum_{j=1}^{n} L_{ij}\xi_j\right] = \mu_i, \tag{1.58}$$

and

$$E[(X_i - \mu_i)(X_j - \mu_j)] = E\left[\left(\sum_{k=1}^{n} L_{ik}\xi_k\right)\left(\sum_{l=1}^{n} L_{jl}\xi_l\right)\right] = \sum_{k=1}^{n} L_{ik}L_{jk} = C_{ij}. \tag{1.59}$$

Here we have used $E[\xi_i] = 0$ and $E[\xi_i \xi_j] = \delta_{ij}$, where $\delta_{ij}$ is Kronecker's delta, with value 1 if $i = j$ and zero otherwise.

Conversely, given a positive definite matrix $\mathbf{C}$, one can show that there is a lower triangular matrix $\mathbf{L} = (L_{ij})$ with $L_{ij} = 0$ if $j > i$, such that $\mathbf{C} = \mathbf{L}\mathbf{L}^{\dagger}$. The matrix $\mathbf{L}$ can be evaluated with a procedure known as *Cholesky factorization*. As discussed later in the book, this algorithm is at the basis of Monte Carlo methods for generating scenarios obeying a multivariate normal distribution with a given covariance matrix.

A special case of a multivariate normal is the bivariate distribution defined for $\mathbf{x} = (x_1, x_2) \in \mathbb{R}^2$:

$$p(x_1, x_2; \mu_1, \mu_2, \sigma_1, \sigma_2, \rho) = \frac{e^{-\frac{1}{2(1-\rho^2)}\left[\frac{(x_1-\mu_1)^2}{\sigma_1^2} + \frac{(x_2-\mu_2)^2}{\sigma_2^2} - 2\rho\frac{(x_1-\mu_1)}{\sigma_1}\frac{(x_2-\mu_2)}{\sigma_2}\right]}}{2\pi\sigma_1\sigma_2\sqrt{1-\rho^2}}.$$

The parameters $\mu_i$ and $\sigma_i > 0$ are the mean and the standard deviation of $X_i$, $i = 1, 2$, respectively, and $\rho$ $(-1 < \rho < 1)$ is the correlation between $X_1$ and $X_2$, i.e., $\rho = \rho_{12} = C_{12}/\sigma_1\sigma_2$. In this case the covariance matrix is

$$\mathbf{C} = \begin{pmatrix} \sigma_1^2 & \rho\sigma_1\sigma_2 \\ \rho\sigma_1\sigma_2 & \sigma_2^2 \end{pmatrix}, \tag{1.60}$$

and the lower Cholesky factorization of $\mathbf{C}$ is given by

$$\mathbf{L} = \begin{pmatrix} \sigma_1 & 0 \\ \rho\sigma_2 & \sigma_2\sqrt{1-\rho^2} \end{pmatrix}. \tag{1.61}$$

The correlation matrix is simply

$$\boldsymbol{\rho} = \begin{pmatrix} 1 & \rho \\ \rho & 1 \end{pmatrix}, \tag{1.62}$$

with Cholesky factorization $\boldsymbol{\rho} = \boldsymbol{\Lambda}\boldsymbol{\Lambda}^{\dagger}$,

$$\boldsymbol{\Lambda} = \begin{pmatrix} 1 & 0 \\ \rho & \sqrt{1-\rho^2} \end{pmatrix}. \tag{1.63}$$

The covariance matrix has inverse

$$\mathbf{C}^{-1} = \frac{1}{(1-\rho^2)} \begin{pmatrix} 1/\sigma_1^2 & -\rho/\sigma_1\sigma_2 \\ -\rho/\sigma_1\sigma_2 & 1/\sigma_2^2 \end{pmatrix}. \tag{1.64}$$

Conditional and marginal densities of the bivariate distribution are readily obtained by integrating over one of the variables in the foregoing joint density (see Problem 3).

For multivariate normal distributions one has the following general result, which we state without proof.

**Proposition.** *Consider the random vector* $\mathbf{X} \in \mathbb{R}^n$ *with partition* $\mathbf{X} = (\mathbf{X}_1, \mathbf{X}_2)$, $\mathbf{X}_1 \in \mathbb{R}^m$, $\mathbf{X}_2 \in \mathbb{R}^{n-m}$ *with* $1 \leq m < n$, $n \geq 2$. *Let* $\mathbf{X} \sim N_n(\boldsymbol{\mu}, \mathbf{C})$ *with mean* $\boldsymbol{\mu} = (\boldsymbol{\mu}_1, \boldsymbol{\mu}_2)$ *and* $n \times n$ *covariance*

$$\mathbf{C} = \begin{pmatrix} \mathbf{C}_{11} & \mathbf{C}_{12} \\ \mathbf{C}_{21} & \mathbf{C}_{22} \end{pmatrix}$$

*with nonzero determinant* $|\mathbf{C}_{22}| \neq 0$, *where* $\mathbf{C}_{11}$ *and* $\mathbf{C}_{22}$ *are* $m \times m$ *and* $(n-m) \times (n-m)$ *covariance matrices of* $\mathbf{X}_1$ *and* $\mathbf{X}_2$, *respectively, and* $\mathbf{C}_{12} = \mathbf{C}_{21}^{\dagger}$ *is the* $m \times (n-m)$ *cross-covariance matrix of the two subspace vectors. The conditional distribution of* $\mathbf{X}_1$, *given* $\mathbf{X}_2 = \mathbf{x}_2$, *is the m-dimensional normal density with mean* $\tilde{\boldsymbol{\mu}} = \boldsymbol{\mu}_1 + \mathbf{C}_{12}\mathbf{C}_{22}^{-1}(\mathbf{x}_2 - \boldsymbol{\mu}_2)$ *and covariance* $\tilde{\mathbf{C}} = \mathbf{C}_{11} - \mathbf{C}_{12}\mathbf{C}_{22}^{-1}\mathbf{C}_{21}$, *i.e.,* $\mathbf{x}_1 \sim N_m(\tilde{\boldsymbol{\mu}}, \tilde{\mathbf{C}})$ *conditional on* $\mathbf{X}_2 = \mathbf{x}_2$.

A relatively simple proof of this result follows by application of known identities for partitioned matrices. This result is useful in manipulating multidimensional integrals involving normal distributions.

In deriving analytical properties associated with expectations or conditional expectations of random variables, the concept of a *characteristic function* is useful. Given a pdf $p : \mathbb{R}^n \to \mathbb{R}$ for a continuous random vector $\mathbf{X} = (X_1, \ldots, X_n)$, the (joint) characteristic function is the function $\phi_{\mathbf{X}} : \mathbb{R}^n \to \mathbb{R}$ defined by

$$\phi_{\mathbf{X}}(\mathbf{u}) = E[e^{i\mathbf{u}\cdot\mathbf{X}}] = \int_{\mathbb{R}^n} e^{i\mathbf{u}\cdot\mathbf{x}} p(\mathbf{x}) d\mathbf{x}, \tag{1.65}$$

where $\mathbf{u} = (u_1, \ldots, u_n) \in \mathbb{R}^n$, $i \equiv \sqrt{-1}$. Since $\phi_{\mathbf{X}}$ is the Fourier transform of $p$, then from the theory of Fourier integral transforms we know that the characteristic function gives a complete characterization of the probabilitic laws of $\mathbf{X}$, equivalently as $p$ does. That is, any two random variables having the same characteristic function are identically distributed; i.e., the characteristic function uniquely determines the distribution. From the definition we observe that $\phi_{\mathbf{X}}$ is always a well-defined continuous function, given that $p$ is a bonafide distribution. Evaluating at the origin gives $\phi_{\mathbf{X}}(\mathbf{0}) = E[1] = 1$. The existence of derivatives $\partial^k \phi_{\mathbf{X}}(\mathbf{0})/\partial u_i^k$, $k \geq 1$ is dependent upon the existence of the respective *moments* of the random variables $X_i$. The $k$th moment of a single random variable $X \in \mathbb{R}$ is defined by

$$m_k = E[X^k] = \int_{-\infty}^{\infty} x^k p(x) dx, \tag{1.66}$$

while the $k$th centered moment is defined by

$$\mu^{(k)} = E[(X-\mu)^k] = \int_{-\infty}^{\infty} (x-\mu)^k p(x) dx, \tag{1.67}$$

$\mu = E[X]$, $k \geq 1$. [Note: for $X = X_i$ then $p \to p_i$ is the $i$th marginal pdf, $\mu \to \mu_i = E[X_i]$, $\mu^{(k)} \to \mu_i^{(k)} = E[(X_i - \mu_i)^k]$, etc.] From these integrals we thus see that the existence of the moments depends on the decay behavior of $p$ at the limits $x \to \pm\infty$. For instance, a distribution that exhibits asymptotic decay at least as fast as a decaying exponential has finite moments to all orders. Obvious examples of these include the distributions of normal, exponential, and uniform random variables. In contrast, distributions that decay as some polynomial to a negative power may, at most, only possess a number of finite moments. A classic case is the Student t distribution with integer $d$ degrees of freedom, which can be shown to possess only moments up to order $d$. This distribution is discussed in Chapter 4 with respect to modeling risk-factor return distributions.

The moments can be obtained from the derivatives of $\phi_{\mathbf{X}}$ at the origin. However, it is a little more convenient to work directly with the *moment-generating function* (mgf). The (joint) moment-generating function is given by

$$M_{\mathbf{X}}(\mathbf{u}) = E[e^{\mathbf{u}\cdot\mathbf{X}}] = \int_{\mathbb{R}^n} e^{\mathbf{u}\cdot\mathbf{x}} p(\mathbf{x}) d\mathbf{x}. \tag{1.68}$$

If the mgf exists (which is not always true), then it is related to the characteristic function: $M_{\mathbf{X}}(\mathbf{u}) = \phi_{\mathbf{X}}(-i\mathbf{u})$. It can be shown that if $E[|X|^r] < \infty$, then $M_X(u)$ (and $\phi_X(u)$) has continuous $r$th derivative at $u = 0$ with moments given by

$$m_k = E[X^k] = \frac{d^k M_X(0)}{du^k} = (-i)^k \frac{d^k \phi_X(0)}{du^k}, \qquad k = 1, \ldots, r. \tag{1.69}$$

Hence, a random variable $X$ has finite moments of all orders when $M_X(u)$ (or $\phi_X(u)$) is continuously differentiable to any order with $m_k = M_X^{(k)}(0) = (-i)^k \phi_X^{(k)}(0)$ , $k = 1, \ldots$ .

Given two independent random variables $X$ and $Y$, the characteristic function of the sum $X + Y$ simplifies into a product of functions: $\phi_{X+Y}(u) = E[e^{iu(X+Y)}] = E[e^{iuX}]E[e^{iuY}] = \phi_X(u)\phi_Y(u)$. Hence for $Z = \sum_{i=1}^n X_i$ we have $\phi_Z(u) = \Pi_{i=1}^n \phi_{X_i}(u)$ if all $X_i$ are independent. Characteristic functions or mgfs can be obtained in analytically closed form for various common distributions.

## Problems

**Problem 1.** Make use of equations (1.53) and (1.54) and the Schwarz inequality,

$$\left(\int_{\mathbb{R}^n} f(\mathbf{x})g(\mathbf{x})d\mathbf{x}\right)^2 \le \left(\int_{\mathbb{R}^n} (f(\mathbf{x}))^2 d\mathbf{x}\right)\left(\int_{\mathbb{R}^n} (g(\mathbf{x}))^2 d\mathbf{x}\right), \qquad \mathbf{x} \in \mathbb{R}^n, \tag{1.70}$$

to demonstrate the inequality $|C_{ij}| \le \sigma_i \sigma_j$, hence $|\rho_{ij}| \le 1$.

**Problem 2.** Consider two independent normal random variables $X$ and $Y$ with probability distributions

$$p_x(x) = \frac{1}{\sigma_x\sqrt{2\pi}} e^{-(x-\mu_x)^2/2\sigma_x^2} \quad \text{and} \quad p_y(y) = \frac{1}{\sigma_y\sqrt{2\pi}} e^{-(y-\mu_y)^2/2\sigma_y^2}, \tag{1.71}$$

respectively. Use convolution to show that $Z = X + Y$ is a normal random variable with probability distribution

$$p_z(z) = \frac{1}{\sigma_z\sqrt{2\pi}} e^{-(z-\mu_z)^2/2\sigma_z^2}, \tag{1.72}$$

where $\sigma_z^2 = \sigma_x^2 + \sigma_y^2$ and $\mu_z = \mu_x + \mu_y$.

**Problem 3.** Show that the joint density function for the bivariate normal has the form

$$p(x, y; \mu_1, \mu_2, \sigma_1, \sigma_2, \rho) = \frac{1}{2\pi\sigma_1\sigma_2\sqrt{1-\rho^2}} e^{-(y-\mu_2)^2/2\sigma_2^2}$$
$$\exp\left[-\frac{1}{2\sigma_1^2(1-\rho^2)}\left(x - \mu_1 - \rho\frac{\sigma_1}{\sigma_2}(y - \mu_2)\right)^2\right], \tag{1.73}$$

and thereby obtain the marginal and conditional distributions:

$$p_Y(Y) = \frac{1}{\sqrt{2\pi}\sigma_2} e^{-(Y-\mu_2)^2/2\sigma_2^2}, \tag{1.74}$$

$$p(x|Y) = \frac{1}{\sqrt{2\pi(1-\rho^2)}\sigma_1} \exp\left[-\frac{1}{2(1-\rho^2)\sigma_1^2}\left[x - \mu_1 - \rho\frac{\sigma_1}{\sigma_2}(Y - \mu_2)\right]^2\right]. \tag{1.75}$$

Verify that this same result follows as a special case of the foregoing proposition.

**Problem 4.** Find the moment-generating function for the following distributions:

(a) The uniform distribution on the interval $(a,b)$ with pdf: $p(x) = (b-a)^{-1}\mathbf{1}_{x\in(a,b)}$.
(b) The exponential distribution with parameter $\lambda > 0$ and pdf: $p(x) = \lambda e^{-\lambda x}\mathbf{1}_{x\geq 0}$.
(c) The gamma distribution with parameters $(n, \lambda)$, $n = 1, 2, \ldots, \lambda > 0$, and pdf: $p(x) = \frac{\lambda e^{-\lambda x}(\lambda x)^{n-1}}{(n-1)!}\mathbf{1}_{x\geq 0}$.

By differentiating the mgf, obtain the mean and variance of the random variable $X$ for each distribution (a)–(c).

**Problem 5.** Obtain the moment-generating function for:

(a) The multivariate normal with density given by equation (1.49).
(b) The chi-squared random variable with $n$ degrees of freedom: $Y = \sum_{i=1}^{n} Z_i^2$, where $Z_i \sim N(0, 1)$.

**Problem 6.** Rederive the result in problem 2 using an argument based solely on moment-generating functions.

**Problem 7.** Consider two independent exponential random variables $X_1$ and $X_2$ with respective parameters $\lambda_1$ and $\lambda_2$, $\lambda_1 \neq \lambda_2$. Find the pdf for $X_1 + X_2$ and the probability $P(X_1 < X_2)$. Hint: Use convolution and conditioning, respectively.

## 1.4 Brownian Motion, Martingales, and Stochastic Integrals

A particularly important example of a multivariate normal distribution is provided by a random path evaluated at a sequence of dates in the future. Consider a time interval $[0, t] = [t_0 = 0, t_1, \ldots, t_N = t]$, and subdivide it into $N \geq 1$ subintervals $[t_i, t_{i+1}]$ of length $\delta t_i = t_{i+1} - t_i$, $i = 0, \ldots, N-1$. The path points $(t, x_t)$ are defined for all $t = t_i$ by means of the recurrence relation

$$x_{t_{i+1}} = x_{t_i} + \mu(t_i)\delta t_i + \sigma(t_i)\delta W_{t_i}, \tag{1.76}$$

where the *increments* $\delta W_{t_i} = W_{t_i+1} - W_{t_i}$ are assumed uncorrelated (independent) normal random variables with probability density at $\delta W_{t_i} = \delta w_i$:

$$p_i(\delta w_i) = \frac{1}{\sqrt{2\pi\delta t_i}} e^{-(\delta w_i)^2/2\delta t_i}. \tag{1.77}$$

Since the increments are assumed independent, the joint pdf for all increments is

$$p(\delta w_0, \ldots, \delta w_{N-1}) = \prod_{i=0}^{N-1} p_i(\delta w_i). \tag{1.78}$$

This gives rise to two important unconditional expectations:

$$E\left[\delta W_{t_i}\delta W_{t_j}\right] = \delta_{ij}\delta t_i \qquad E\left[\delta W_{t_i}\right] = 0. \tag{1.79}$$

By usual convention we fix $W_0 = 0$. The joint pdf for the random variables $W_{t_1}, \ldots, W_{t_N}$ representing the probability density at the path points $W_{t_i} = w_i$ $(w_0 = 0)$ is then also a multivariate Gaussian function, which is obtained by simply setting $\delta w_i = w_{i+1} - w_i$ in equation (1.78). The set of real-valued random variables $(W_{t_i})_{i=0,\ldots,N}$ therefore represents the time-discretized *standard Brownian motion* (or *Wiener process*) at arbitrary discrete points in time. Iterating equation (1.76) gives

$$x_t = x_0 + \sum_{j=0}^{N-1} \left[\mu(t_j)\delta t_j + \sigma(t_j)\delta W_{t_j}\right], \tag{1.80}$$

where $x_{t_N} = x_t$ and $x_{t_0} = x_0$. The random variable $x_t$ is normal with mean

$$E_0[x_t] = x_0 + \sum_{i=0}^{N-1} \mu(t_i)\delta t_i \tag{1.81}$$

and variance

$$E_0[(x_t - E_0[x_t])^2] = E_0\left[\left(\sum_{i=0}^{N-1} \sigma(t_i)\delta W_{t_i}\right)^2\right] = \sum_{i=0}^{N-1} \sigma(t_i)^2 \delta t_i. \tag{1.82}$$

Note: We use $E_0[\,]$ to denote the expectation conditional only on the value of paths being fixed at initial time; i.e., $x_{t_0} = x_0 =$ fixed value. This is hence an unconditional expectation with respect to path values at any later time $t > 0$. Later, we will at times simply use the unconditional expectation $E[\,]$ to denote $E_0[\,]$. Sample paths of a process with zero mean and constant volatility are displayed in Figure 1.2.

Typical stochastic processes in finance are meaningful if time is discretized. The choice of the elementary unit of time is part of the modeling assumptions and depends on the applications at hand. In pricing theory, the natural elementary unit is often one day but can also be one week, one month as well as five minutes or one tick, depending on the objective. The mathematical theory, however, simplifies in the *continuous-time limit*, where the elementary time is infinitesimal with respect to the other time units in the problem, such as

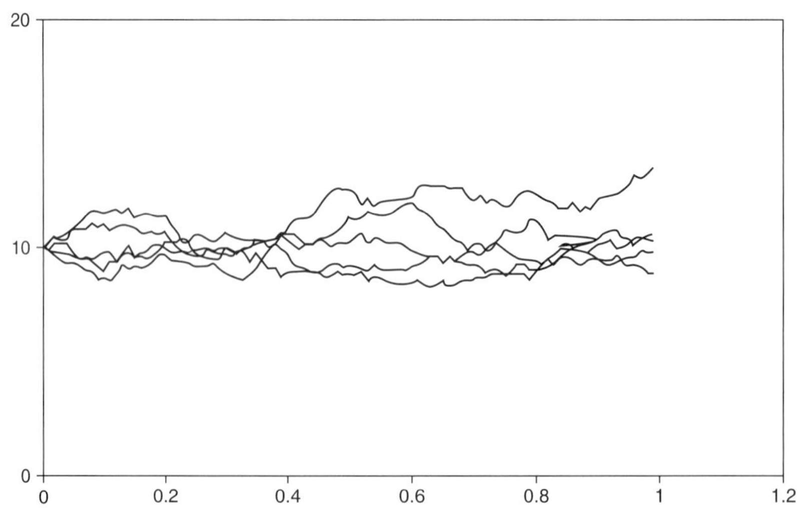


FIGURE 1.2 A simulation of five stochastic paths using equation (1.76), with $x_0 = 10$, constant $\mu(t) = 0.1$, $\sigma(t) = 0.2$, $N = 100$, and time steps $\delta t_i = 0.01$.

option maturities and cash flow periods. Mathematically, one can construct continuous-time processes by starting from a sequence of approximating processes defined for discrete-time values $i\delta t, i = 0, \ldots, N$, and then pass to the limit as $\delta t \to 0$. More precisely, one can define a continuous-time process in an interval $[t_0, t_N]$ by subdividing it into $N$ subintervals of equal length, defining a discrete time process $x_t^N \equiv x_{t_N}$ and then compute the limit

$$x_t = \lim_{N\to\infty} x_t^N \tag{1.83}$$

by assuming that the discrete-time process $x_t^N$ is constant over the partition subintervals. The elementary increments $\delta x_t = x_{t+\delta t} - x_t$ are random variables that obviously tend to zero as $\delta t \to 0$, but which are still meaningful in this case. The convention is to denote these increments as $dx$ in the limit $\delta t \to 0$ and to consider the straight $d$ as a reminder that, at the end of the calculations, one is ultimately interested in the limit as $\delta t \to 0$.[4]

The continuous-time limit is obtained by holding the terminal time $t = t_N$ fixed and letting $N \to \infty$, i.e.,

$$E_0[x_t] \equiv \lim_{N\to\infty} E_0[x_t^N] = x_0 + \int_0^t \mu(\tau)d\tau \equiv x_0 + \bar{\mu}(t)t, \tag{1.84}$$

and

$$E_0[(x_t - E_0[x_t])^2] \equiv \lim_{N\to\infty} E_0[(x_t^N - E_0[x_t^N])^2] = \int_0^t \sigma^2(\tau)d\tau \equiv \bar{\sigma}(t)^2 t, \tag{1.85}$$

where we introduced the time-averaged drift $\bar{\mu} = \bar{\mu}(t)$ and volatility $\bar{\sigma} = \bar{\sigma}(t)$ over the time period $[0, t], t \in \mathbb{R}_+$. Since $x_t$ is normally distributed, we finally arrive at the *transition probability density* for a stochastic path to attain value $x_t$ at time $t$, given an initially known value $x_0$ at time $t = 0$:

$$p(x_t, x_0; t) = \frac{1}{\bar{\sigma}\sqrt{2\pi t}} \exp\left( - \frac{(x_t - (x_0 + \bar{\mu}t))^2}{2\bar{\sigma}^2 t} \right). \tag{1.86}$$

This density, therefore, gives the distribution (conditional on a starting value $x_0$) for a process with continuous motion on the entire real line $x_t \in (-\infty, \infty)$ with constant drift and volatility. [Note: $x_0, x_t$ are real numbers (not random) in equation (1.86).]

A *Markov chain* is a discrete-time stochastic process such that for all times $t \in \mathcal{T}$ the increments $x_{t+\delta t} - x_t$ are random variables independent of $x_t$. A *Markov process* is the continuous-time limit of a Markov chain. The process just introduced provides an example of a Markov chain because the increments are independent.

The probability space for a general discrete-time stochastic process where calendar time can take on values $t_0 < t_1 < \cdots < t_N$ is the space of vectors $\mathbf{x} \in \mathbb{R}^N$ with an appropriate multivariate measure, such as $P(d\mathbf{x}) = p(x_1, \ldots x_N)d\mathbf{x}$, where $p$ is a probability density. By considering a process $x_t$ only up to an intermediate time $t_i$, $i < N$, we are essentially restricting the information set of possible events or probability space of paths. The family $(\mathcal{F}_t)_{t\geq 0}$ of all reduced (or filtered) probability spaces $\mathcal{F}_t$ up to time $t$, for all times $t \geq 0$, is called *filtration*. One can think of $\mathcal{F}_t$ as the set of all paths up to time $t$. A pay-off of a derivative contract

[4] These definitions are admittedly NOT entirely rigorous, but they are meant to allow the reader to quickly develop an intuition in case she doesn't have a formal probability education. In keeping with the purpose of this book, our objective is to have the reader learn how to master the essential techniques in stochastic calculus that are useful in finance without assuming that she first learn the formal mathematical theory.

occuring at time $t$ is a well-defined (measurable) random variable on all the spaces $\mathcal{F}_{t'}$ with $t' \geq t$ but not on the spaces with $t' < t$. Filtrations are essentially hierarchies of probability spaces (or information sets) through which more and more information is revealed to us as time progresses; i.e., $\mathcal{F}_{t'} \subset \mathcal{F}_t$ if $t' < t$ so that given a time partition $t_0 < t_1 < \cdots < t_N$, $\mathcal{F}_{t_0} \subset \mathcal{F}_{t_1} \subset \cdots \subset \mathcal{F}_{t_N}$. We say that a random variable or process is $\mathcal{F}_t$-measurable if its value is revealed at time $t$. Such a random variable or process is also said to be *nonanticipative* with respect to the filtration or *$\mathcal{F}_t$-adapted* (see later for a definition of nonanticipative functions, while a definition of an adapted process is also provided in Section 1.9 in the context of continuous-time asset pricing). *Conditional expectations with respect to a filtration $\mathcal{F}_t$ represent expectations conditioned on knowing all of the information about the process only up to time $t$.* It is customary to use the following shorthand notation for conditional probabilities:

$$E_t[\cdot] = E[\cdot\,|\mathcal{F}_t]. \tag{1.87}$$

**Definition 1.9. Martingale** *A real-valued $\mathcal{F}_t$-adapted continuous-time process $(x_t)_{t\geq 0}$ is said to be a P-martingale if the boundedness condition $E[|x_t|] < \infty$ holds for all $t \geq 0$ and*

$$x_t = E_t[x_T], \tag{1.88}$$

*for $0 \leq t < T < \infty$*

This definition implies that the conditional expectation for the value of a martingale process at a future time $T$, given all previous history up to the current time $t$ (i.e., adapted to a filtration $\mathcal{F}_t$), is its current time $t$ value. Our best prediction of future values of such a process is therefore just the presently observed value. [Note: Although we have used the same notation, i.e., $x_t$, this definition generally applies to arbitrary continuous-time processes that satisfy the required conditions; the pure Wiener process or standard Brownian motion is just a special case.] We remark that the expectation $E[\,] \equiv E^P[\,]$ and conditional expectation $E_t[\,] \equiv E_t^P[\,]$ are assumed here to be taken with respect to a given probability measure $P$. For ease of notation in what follows we drop the explicit use of the superscript $P$ unless the probability measure must be made explicit. If one changes filtration or the probability space associated with the process, then the same process may not be a martingale with respect to the new probability measure and filtration. However, the reverse also applies, in the sense that a process may be converted into a martingale by modifying the probability measure.

A more general property satisfied by a stochastic process $(x_t)_{t\geq 0}$ (regardless of whether the process is a martingale or not) is the so-called *tower property* for $s < t < T$:

$$E_s[E_t[x_T]] = E_s[x_T]. \tag{1.89}$$

This follows from the basic property of conditional expectations: The expectation of a future expectation must be equal to the present expectation or presently forecasted value. Another way to see this is that a recursive application of conditional expectations always gives the conditional expectation with respect to the smallest information set. In this case $\mathcal{F}_s \subset \mathcal{F}_t \subset \mathcal{F}_T$. A martingale process $f_t = f(x_t, t)$ can also be specified by considering a conditional expectation over some (payoff) function $\phi$ of an underlying process. In particular, consider an underlying process $x_t$ starting at time $t_0$ with some value $x_0$ and the conditional expectation

$$f_t = f(x_t, t) = E_t[\phi(x_T, T)], \tag{1.90}$$

for any $t_0 \le s \le t \le T$, then $f_t$ satisfies the martingale property. In fact

$$E_s[f(x_t, t)] = E_s\Big[E_t[\phi(x_T, T)]\Big] = E_s[\phi(x_T, T)] = f(x_s, s). \tag{1.91}$$

The process introduced in equation (1.76) is a martingale in case the drift function $\mu(t)$ is identically zero. In fact, in this case if $t_i < t_j$, we have

$$E_{t_i}[x_{t_j}] = E_{t_i}[\cdots E_{t_{j-1}}[E_{t_j}[x_{t_j}]]\cdots] \tag{1.92}$$

$$= E_{t_i}[\cdots E_{t_{j-1}}[x_{t_j}]\cdots] = x_{t_i}. \tag{1.93}$$

Bachelier was one of the pioneers of stochastic calculus, and he proposed to use a process similar to $x_t$ as defined by equation (1.76) in the continuous-time limit to model stock price processes.[5] A difficulty with the Bachelier model was that stock prices can attain negative values. The problem can be corrected by regarding $x_t$ to be the natural logarithm of stock prices; this conditional density turns out to be related to (although not equivalent to) the risk-neutral density used for pricing derivatives within the Black–Scholes formulation, as is seen in Section 1.6, where we take a close look at *geometric Brownian motion*. The density in equation (1.86) leads to Bachelier's formula for the expectation of the random variable $(x_t - K)_+$, with constant $K > 0$, where $(x)_+ \equiv x$ if $x > 0$, $(x)_+ \equiv 0$ if $x \le 0$ (see Problem 9). In passing to the continuous-time limit, we have, based on equation (1.86), arrived at an expression for the random variable $x_t$ in terms of the random variable $W_t$ for the standard Brownian motion (or Wiener process):

$$x_t = x_0 + \bar{\mu} t + \bar{\sigma} W_t. \tag{1.94}$$

The distribution for the zero-drift random variable $(W_t)_{t \ge 0}$, representing the real-valued standard Brownian motion (Wiener process) at time $t$ with $W_{t=0} \equiv W_0 = 0$, is given by

$$p_W(w, t) = \frac{1}{\sqrt{2\pi t}} e^{-w^2/2t} \tag{1.95}$$

at $W_t = w$. Note that this is also entirely consistent with the marginal density obtained by integrating out all intermediate variables $w_1, \ldots, w_{N-1}$ in the joint pdf of the discretized process $(W_{t_i})_{i=0,\ldots,N}$ with $w = w_N$, $t = t_N$.

According to the distributions given by equations (1.77) and (1.95), one concludes that standard Brownian motion (or the Wiener process) is a martingale process characterized by independent Gaussian (normal) increments with trajectories [i.e., path points $(t, x_t)$] that are continuous in time $t \ge 0$: $\delta W_t = W_{t+\delta t} - W_t \sim N(0, \delta t)$ (i.e., normally distributed with mean zero and variance $\delta t$) and $W_{t+\delta t} - W_t$ is independent of $W_s$ for $\delta t > 0$, $0 \le s \le t$, $0 \le t < \infty$. Moreover, specializing to the case of zero drift and $\bar{\sigma} = 1$ and putting $t_0 = s$, the corresponding

[5] The date March 29, 1900, should be considered as the birth date of mathematical finance. On that day, Louis Bachelier successfully defended at the Sorbonne his thesis *Théorie de la Spéculation*. As a work of exceptional merit, strongly supported by Henri Poincaré, Bachelier's supervisor, it was published in *Annales Scientifiques de l' Ecole Normale Supérieure*, one of the most influential French scientific journals. This model was a breakthrough that motivated much of the future work by Kolmogorov and others on the foundations of modern stochastic calculus. The stochastic process proposed by Bachelier was independently analyzed by Einstein (1905) and is referred to as *Brownian motion* in the physics literature. It is also referred to as the *Wiener–Bachelier process* in a book by Feller, *An Introduction to Probability Theory and Its Applications* [Fel71]. However, this terminology didn't affirm itself, and now the process is commonly called the *Wiener process*.

probability distribution given by equation (1.86) with shifted time $t \to (t-s)$ then gives the well-known property: $W_t - W_s \sim N(0, t-s)$, $W_t \sim N(0, t)$. In fact we have the homogeneity property for the increments: $W_{t+s} - W_s \sim W_t - W_0 = W_t \sim N(0, t)$. In particular, $E[W_t] = 0$ and $E[W_t^2] = t$. An additional property is $E[W_s W_t] = \min(s, t)$. This last identity obtains from the independence of the increments [i.e., equation (1.79)]. Indeed consider any $t_i < t_j$, $0 \le i < j \le N$, then:

$$\begin{aligned} E\big[W_{t_i} W_{t_j}\big] &= E\big[(W_{t_i} - W_0)((W_{t_j} - W_{t_i}) + (W_{t_i} - W_0))\big] \\ &= E\big[(W_{t_i} - W_0)^2\big] = E\big[W_{t_i}^2\big] = t_i. \end{aligned} \tag{1.96}$$

A similar argument with $t_j < t_i$ gives $t_j$, while for $t_i = t_j$ we obviously obtain $t_i$. All of these properties also follow by taking expectations with respect to the joint pdf for the Wiener paths.

An important aspect of martingales is whether or not their trajectories or paths are continuous in time. Consider any real-valued martingale $x_t$, then $\delta x_t = x_{t+\delta t} - x_t$ is a process corresponding to the change in a path over an arbitrary time difference $\delta t > 0$. From equation (1.88), $E_t\big[\delta x_t\big] = 0$, so, not surprisingly, the increments of a martingale path are unpredictable (irregular), even in the infinitesimal limit $\delta t \to 0$. However, the irregularity of paths can be either continuous or discontinuous. An example of a martingale with discontinuous paths is a *jump process*, where paths are generally right continuous at every point in time as a consequence of incorporating jump discontinuities in the process at a random yet countable number of points within a time period. We refer the interested reader to recent works on the growing subject of financial modeling with jump processes (see, for example, [CT04]). Here and throughout, we focus on continuous diffusion models for asset pricing; hence our discussion is centered on continuous martingales (i.e., martingales with continuous paths). Let $f(t) = x_t(\omega)$, $t \ge 0$, represent a particular realized path indexed by the scenario $\omega$, then continuity in the usual sense implies that the graph of $f(t)$ against time is continuous for all $t \ge 0$. Denoting the left and right limits at $t$ by $f(t-) = \lim_{s \to t^-} f(s)$ and $f(t+) = \lim_{s \to t^+} f(s)$, then $f(t) = f(t-) = f(t+)$. Every Brownian path or any path of a stochastic process generated by an underlying Brownian motion displays this property, as can be observed, for example, in Figure 1.2. [In contrast, a path of a jump diffusion process would display a similar continuity in piecewise time intervals but with the additional feature of vertical jump discontinuities at random points in time at which only right continuity holds. If $\bar{t}$ is a jump time, then $f(\bar{t}-)$, $f(\bar{t}+)$ both exist, yet $f(\bar{t}-) \ne f(\bar{t}+)$ with $f(\bar{t}) = f(\bar{t}+)$, where $f(\bar{t}) - f(\bar{t}-)$ is the size of the jump at time $\bar{t}$.]

*Stochastic continuity* refers to continiuty of sample paths of a process $(x_t)_{t \ge 0}$ in the probabilitistic sense as defined by

$$\lim_{s \to t} P\big(|x_s - x_t| > \epsilon\big) = 0, \quad s, t > 0 \tag{1.97}$$

for any $\epsilon > 0$. This is readily seen to hold for Brownian motion and for continuous martingales. The class of continuous-time martingales that are of interest are so-called *continuous square integrable martingales*, i.e., martingales with finite unconditional variance or finite second moment: $E[x_t^2] < \infty$ for $t \ge 0$. Such processes are closely related to Brownian motion and include Brownian motion itself. Further important properties of the paths of a continuous square integrable martingale (e.g., Brownian motion) then also follow. Consider again the time discretization $[0, t] = [t_0 = 0, t_1, \ldots, t_N = t]$ with subintervals $[t_i, t_{i+1}]$ and path points $(t_i, x_{t_i})$. The *variation* and *quadratic variation* of the path are, respectively, defined as:

$$V_1 = \lim_{N \to \infty} V_1^N \equiv \lim_{N \to \infty} \sum_{i=0}^{N-1} |\delta x_{t_i}| \tag{1.98}$$

and

$$V_2 = \lim_{N\to\infty} V_2^N \equiv \lim_{N\to\infty} \sum_{i=0}^{N-1} (\delta x_{t_i})^2, \tag{1.99}$$

$\delta x_{t_i} = x_{t_{i+1}} - x_{t_i}$. The properties of $V_1$ and $V_2$ provide two differing measures of how paths behave over time and give rise to important implications for stochastic calculus. Since the process is generally of nonzero variance, then $P(V_2^N > 0) = 1$ and $P(V_2 > 0) = 1$. In particular, if we let $\delta t_i = \delta t = t/N$ and consider the case of Brownian motion $x_t = W_t$, then by rewriting $V_2$ we have with probability 1:

$$V_2 = \lim_{N\to\infty} \left( \frac{1}{N} \sum_{i=0}^{N-1} (\delta x_{t_i})^2 \right) N = \lim_{N\to\infty} \left( \frac{1}{N} \sum_{i=0}^{N-1} (\delta W_{t_i})^2 \right) N = t. \tag{1.100}$$

Here we used the Strong law of large numbers and the fact that the $(\delta W_{t_i})^2$ are identically and independently distributed random variables with common mean of $\delta t$. Based on this important property of nonzero quadratic variation, Brownian paths, although continuous, are not differentiable. For finite $N$ the variation $V_1^N$ is finite. As the number $N$ of increments goes to infinity, $\delta t_i \to 0$ and, from property (1.97), we see that the size of the increments approaches zero. The question that arises then is whether $V_1$ exists or not. Except for the trivial case of a constant martingale, the result is that $V_1^N \to \infty$ as $N \to \infty$; i.e., the variation $V_1$ is in fact infinite. Without trying to provide any rigorous proof of this here, we simply state the usual heuristic and somewhat instructive argument for this fact based on the following observation:

$$V_2^N = \sum_{i=0}^{N-1} |\delta x_{t_i}|^2 \leq \left[ \max_{0\leq i\leq N} \{|\delta x_{t_i}|\} \right] \sum_{i=0}^{N-1} |\delta x_{t_i}| = \left[ \max_{0\leq i\leq N} \{|\delta x_{t_i}|\} \right] V_1^N. \tag{1.101}$$

Since the quadratic variation $V_2$ is greater than zero, taking the limit $N \to \infty$ on both sides of the inequality shows that the right-hand side must have a nonzero limit. Yet from equation (1.97) we have $\max\{|\delta x_{t_i}|\} \to 0$ as $N \to \infty$. Hence we must have that the right-hand side is a limit of an indeterminate form (of type $0 \cdot \infty$); that is, $V_1 = \lim_{N\to\infty} V_1^N = \infty$, which is what we wanted to show.

Once we are equipped with a standard Brownian motion and a filtered probability space, then the notion of stochastic integration arises by considering the concept of a nonanticipative function. Essentially, a (random) function $f_t$ is said to be nonanticipative w.r.t. a Brownian motion or process $W_t$ if its value at any time $t > 0$ is independent of future information. That is, $f_t$ is possibly only a function of the history of paths up to time $t$ and time $t$ itself: $f_t = f(\{(W_s)_{0\leq s\leq t}\}, t)$. The value of this function at time $t$ for a particular realization or scenario $\omega$ may be denoted by $f_t(\omega)$. Nonanticipative functions therefore include all deterministic (i.e., nonrandom) functions as a special case. Given a continuous nonanticipative function $f_t$ that satisfies the "nonexplosive" condition

$$E\left[ \int_0^t f_s^2 ds \right] < \infty, \tag{1.102}$$

the Itô (stochastic) integral is the random variable denoted by

$$I_t(f) = \int_0^t f_s dW_s < \infty \tag{1.103}$$

and is defined by the limit

$$I_t(f) = \lim_{N\to\infty} \sum_{i=0}^{N-1} f_{t_i} \delta W_{t_i} = \lim_{N\to\infty} \sum_{i=0}^{N-1} f_{t_i} [W_{t_{i+1}} - W_{t_i}]. \tag{1.104}$$

It can be shown that this limit exists for any choice of time partitioning of the interval $[0, t]$; e.g., we can choose $\delta t_i = \delta t = t/N$. Each term in the sum is given by a random number $f_{t_i}$ [but fixed over the next time increment $(t_i, t_{i+1})$] times a random Gaussian variable $\delta W_{t_i}$. Because of this, the Itô integral can be thought of as a random walk on increments with randomly varying amplitudes. Since $f_t$ is nonanticipative, then for each $i$th step we have the conditional expectation for each increment in the sum: $E_{t_i}[f_{t_i} \delta W_{t_i}] = f_{t_i} E_{t_i}[\delta W_{t_i}] = 0$. Given nonanticipative functions $f_t$ and $g_t$, the following formulas provide us with the first and second moments as well as the variance-covariance properties of Itô integrals:

$$(i)\ E[I_t(f)] = E\left[\int_0^t f_s dW_s\right] = 0, \tag{1.105}$$

$$(ii)\ E[(I_t(f))^2] = E\left[\left(\int_0^t f_s dW_s\right)^2\right] = E\left[\int_0^t f_s^2 ds\right], \tag{1.106}$$

$$(iii)\ E[I_t(f) I_t(g)] = E\left[\left(\int_0^t f_s dW_s\right)\left(\int_0^t g_s dW_s\right)\right] = E\left[\int_0^t f_s g_s ds\right]. \tag{1.107}$$

Based on the definition of $I_t(f)$ and the properties of Brownian increments, it is not difficult to obtain these relations. We leave this as an exercise for the reader. Of interest in finance are nonanticipative functions of the form $f_t = f(x_t, t)$, where $x_t$ is generally a continuous stochastic (price) process $(x_t)_{t\geq 0}$. The Itô integral is then of the form

$$I_t(f) = \int_0^t f(x_s, s) dW_s, \tag{1.108}$$

and, assuming that condition (1.102) holds, then properties (i)–(iii) also apply. Another notable property is that the Itô integral is a martingale, since $E_t[I_u(f)] = I_t(f)$, for $0 < t < u$.

The Itô integral leads us into important types of processes and the concept of a *stochastic differential equation* (SDE). In fact the general class of stochastic processes that take the form of sums of stochastic integrals are (not surprisingly) known as *Itô processes*. It is of interest to consider nonanticipative processes of the type $a_t = a(x_t, t)$ and $b_t = b(x_t, t)$, $t \geq 0$, where $(x_t)_{t\geq 0}$ is a random process. A stochastic process $(x_t)_{t\geq 0}$ is then an Itô process if there exist two nonanticipative processes $(a_t)_{t\geq 0}$ and $(b_t)_{t\geq 0}$ such that the conditions

$$P\left(\int_0^t |a_s| ds < \infty\right) = 1 \quad \text{and} \quad P\left(\int_0^t b_s^2 ds < \infty\right) = 1$$

are satisfied, and

$$x_t = x_0 + \int_0^t a(x_s, s) ds + \int_0^t b(x_s, s) dW_s, \tag{1.109}$$

for $t > 0$. These probability conditions are commonly imposed smoothness conditions on the drift and volatility functions. This stochastic integral equation is conveniently and formally abbreviated by simply writing it in SDE form:

$$dx_t = a(x_t, t) dt + b(x_t, t) dW_t. \tag{1.110}$$

We shall use SDE notation in most of our future discussions of Itô processes.

Itô integrals give rise to an important property, known as *Doob–Meyer decomposition*. In particular, it can be shown that if $(M_s)_{0\leq s\leq t}$ is a square integrable martingale process, then there exists a (nonanticipative) process $(f_s)_{0\leq s\leq t}$ that satisfies equation (1.102) such that

$$M_t = M_0 + \int_0^t f_s dW_s. \tag{1.111}$$

From this we observe that an Itô process $x_t$ as given by equation (1.109) is divisible into a sum of a martingale component and a (generally random) drift component.

## Problems

**Problem 1.** Show that the finite difference $\frac{x_{t_{i+1}} - x_{t_i}}{\delta t_i}$ of the Brownian motion in equation (1.76) is a normally distributed random variable with mean $\mu(t_i)$ and volatility $\sigma(t_i)/\sqrt{\delta t_i}$. Hint: Use equation (1.76) and take expectations while using equation (1.79).

**Problem 2.** Show that the random variable

$$\xi = \sum_{i=0}^{N-1} a(t_i)\delta x_{t_i}, \tag{1.112}$$

where $\delta x_{t_i} = x_{t_{i+1}} - x_{t_i}$, and $x_{t_i}$ defined by equation (1.76), is a normal random variable. Compute its mean and variance. Hint: Take appropriate expectations while using equation (1.79).

**Problem 3.** Suppose that the time intervals are given by $\delta t_i = t/N$, where $t$ is any finite time value and $N$ is an integer. Show that equations (1.84) and (1.85) follow in the continuous-time limit as $N \to \infty$ for fixed $t$.

**Problem 4.** Show that the random variable $\xi = \sum_{i=1}^{N} a(t_i)(\delta W_{t_i})^2$ has mean and variance given by

$$E[\xi] = \sum_{i=1}^{N} a(t_i)\delta t_i, \qquad E[(\xi - E[\xi])^2] = 2\sum_{i=1}^{N} a(t_i)^2(\delta t_i)^2 \tag{1.113}$$

Hint: Since $\delta W_{t_i} \sim N(0, \delta t_i)$ independently for each $i$, one can use the identity in Problem 2 of Section 1.6. That is, by considering $E[\exp(\alpha\delta W_{t_i})]$ for nonzero parameter $\alpha$ and applying a Taylor expansion of the exponential and matching terms in the power series in $\alpha^n$, one obtains $E[(\delta W_{t_i})^n]$ for any $n \geq 0$. For this problem you only need terms up to $n = 4$.

**Problem 5.** Show that the distribution $p(x, x_0; t)$ in equation (1.86) approaches the one-dimensional Dirac delta function $\delta(x - x_0)$ in the limit $t \to 0$.

**Problem 6.** (i) Obtain the joint marginal pdf of the random variables $W_s$ and $W_t$, $s \neq t$. Evaluate $E[(W_t - W_s)^2]$ for all $s, t \geq 0$. (ii) Compute $E_t[W_s^3]$ for $s > t$.

**Problem 7.** Let the processes $(x_t)_{t\geq 0}$ and $(y_t)_{t\geq 0}$ be given by $x_t = x_0 + \mu_x t + \sigma_x W_t$ and $y_t = y_0 + \mu_y t + \sigma_y W_t$, where $\mu_x$, $\mu_y$, $\sigma_x$, $\sigma_y$ are constants. Find:

(i) the means $E[x_t]$, $E[y_t]$;
(ii) the unconditional variances $\text{Var}(x_t)$, $\text{Var}(y_t)$;
(iii) the unconditional covariances $\text{Cov}(x_t, y_t)$ and $\text{Cov}(x_s, y_t)$ for all $s, t \geq 0$.

**Problem 8.** Obtain $E[X_t]$, $\text{Var}(X_t)$, and $\text{Cov}(X_s, X_t)$ for the processes

$$(a) X_t = X_0 e^{-\alpha t} + \sigma \int_0^t e^{-\alpha(t-s)} dW_s, \quad t \geq 0, \tag{1.114}$$

$$(b) X_t = \alpha(1 - t/T) + \beta(t/T) + (T - t)\int_0^t \frac{dW_s}{T - s}, \quad 0 \leq t \leq T, \tag{1.115}$$

where $\alpha$, $\beta$, $\sigma$ are constant parameters and time $T$ is fixed in (b). The process in (a) describes the so-called *Ornstein–Uhlenbeck process*, while (b) describes a *Brownian bridge*, whereby the process is Brownian in nature, yet it is also exactly pinned down at initial time and final time $T$, i.e., $X_0 = \alpha$, $X_T = \beta$. For (a) assume $X_0$ is a constant.

**Problem 9.** Assume that $x_t$ is described by a random process given by equation (1.94), or equivalently by the conditional density in equation (1.86). Show that the conditional expectation at time $t = 0$ defined by

$$C(t, K) = E_0[(x_t - K)_+], \tag{1.116}$$

where $(x)_+ = x$ if $x > 0$ and zero otherwise gives the formula

$$C(t, K) = (x_0 + \bar{\mu}t - K)N\left(\frac{x_0 + \bar{\mu}t - K}{\bar{\sigma}\sqrt{t}}\right) + \bar{\sigma}\sqrt{t}\varphi\left(\frac{x_0 + \bar{\mu}t - K}{\bar{\sigma}\sqrt{t}}\right), \tag{1.117}$$

where $N(\cdot)$ is the standard cumulative normal distribution function and

$$\varphi(x) = \frac{1}{\sqrt{2\pi}}e^{-x^2/2}. \tag{1.118}$$

By further restricting the drift, $\mu = 0$ gives Bachelier's formula. This corresponds (from the viewpoint of pricing theory) to the fair price of a standard call option struck at $K$, and maturing in time $t$, assuming a zero interest rate and *simple* Brownian motion for the underlying "stock" level $x_t$ at time $t$. Hint: One way to obtain equation (1.117) is by direct integration over all $x_t$ of the product of the density $p$ [of equation (1.86)] and the payoff function $(x_t - K)_+$. Use appropriate changes of integration variables and the property $1 - N(x) = N(-x)$ to arrive at the final expression.

## 1.5 Stochastic Differential Equations and Itô's Formula

For purposes of describing asset price processes it is of interest to consider SDEs for diffusion processes $x_t$ that are defined in terms of a lognormal drift function $\mu(x, t)$ and a lognormal volatility function $\sigma(x, t)$ and are written as follows:[6]

$$dx_t = \mu(x_t, t)x_t dt + \sigma(x_t, t)x_t dW_t. \tag{1.119}$$

Assuming the drift and volatility are smooth functions, the discretization process in the previous section extends to this case and produces a solution to equation (1.119) as the limit as $N \to \infty$ of the Markov chain $x_{t_0}, \ldots, x_{t_N}$ defined by means of the recurrence relations

$$x_{t_{i+1}} = x_{t_i} + \mu(x_{t_i}, t_i)x_{t_i}\delta t_i + \sigma(x_{t_i}, t_i)x_{t_i}\delta W_{t_i}. \tag{1.120}$$

[6] When the drift and volatility (or diffusion) terms in the SDE are written in the form given by equation (1.119) it is common to refer to $\mu$ and $\sigma$ as the lognormal drift and volatility, respectively. The reason for using this terminology stems from the fact that in the special case that $\mu$ and $\sigma$ are at most only functions of time $t$ (i.e., not dependent on $x_t$), the SDE leads to geometric Brownian motion, and, in particular, the conditional transition density is exactly given by a lognormal distribution, as discussed in the next section.

From this discrete form of equation (1.119) we observe that $x_{t+\delta t} - x_t = \delta x_t = \mu(x_t, t)x_t\delta t + \sigma(x_t, t)x_t\delta W_t$. Alternatively, the solution to equation (1.119) can be characterized as the process $x_t$ such that

$$\mu(x_t, t) = \lim_{\delta t \to 0} \frac{E_t[x_{t+\delta t} - x_t]}{x_t \delta t}, \qquad \sigma(x_t, t)^2 = \lim_{\delta t \to 0} \frac{E_t[(x_{t+\delta t} - x_t)^2]}{x_t^2 \delta t}. \tag{1.121}$$

These expectations follow from the properties $E_t[\delta W_t] = 0$ and $E_t[(\delta W_t)^2] = \delta t$. Notice that, although an SDE defines a stochastic process in a fairly constructive way, conditional distribution probabilities, such as the one for the Wiener process in equation (1.86), can be computed in analytically closed form only in some particular cases. Advanced methods for obtaining closed-form conditional (transition) probability densities for certain families of drift and volatility functions are discussed in Chapter 3, where the corresponding Kolmogorov (or Fokker–Planck) partial differential equation approach is presented in detail.

A method for constructing stochastic processes is by means of nonlinear transformations. The stochastic differential equation satisfied by a nonlinear transformation as a function of another diffusion process is given by Itô's lemma:

**Lemma 1.3. Itô's Lemma** *If the function $f_t = f(x_t, t)$ is smooth with continuous derivatives $\partial f/\partial t$, $\partial f/\partial x$, and $\partial^2 f/\partial x^2$ and $x_t$ satisfies the stochastic differential*

$$dx_t = a(x_t, t)dt + b(x_t, t)dW_t, \tag{1.122}$$

*where $a(x, t)$ and $b(x, t)$ are smooth functions of $x$ and $t$, then the stochastic differential of $f_t$ is given by*

$$\begin{aligned} df_t &= \left(\frac{\partial f}{\partial t} + a(x_t, t)\frac{\partial f}{\partial x} + \frac{b(x_t, t)^2}{2}\frac{\partial^2 f}{\partial x^2}\right)dt + b(x_t, t)\frac{\partial f}{\partial x}dW_t \\ &\equiv A(x_t, t)dt + B(x_t, t)dW_t. \end{aligned} \tag{1.123}$$

*In stochastic integral form:*

$$f_t = f_0 + \int_0^t A(x_s, s)ds + \int_0^t B(x_s, s)dW_s. \tag{1.124}$$

A nonrigorous, yet instructive, "proof" is as follows.[7]

**Proof.** Using a Taylor expansion we find

$$\begin{aligned} \delta f_t &= f(x_t + \delta x_t, t + \delta t) - f(x_t, t) \\ &= \frac{\partial f}{\partial t}(x_t, t)\delta t + \frac{\partial f}{\partial x}(x_t, t)\delta x_t + \frac{1}{2}\frac{\partial^2 f}{\partial x^2}(x_t, t)(\delta x_t)^2 + O((\delta t)^{\frac{3}{2}}), \end{aligned} \tag{1.125}$$

where the remainder has an expectation and variance converging to zero as fast as $(\delta t)^2$ in the limit $\delta t \to 0$. Inserting the finite differential form of equation (1.122) into equation (1.125) while replacing $(\delta W_t)^2 \to \delta t$ and retaining only terms up to $O(\delta t)$ gives

$$\begin{aligned} \delta f_t &= \left(\frac{\partial f}{\partial t}(x_t, t) + a(x_t, t)\frac{\partial f}{\partial x}(x_t, t) + \frac{b(x_t, t)^2}{2}\frac{\partial^2 f}{\partial x^2}(x_t, t)\right)\delta t \\ &\quad + b(x_t, t)\frac{\partial f}{\partial x}(x_t, t)\delta W_t + O((\delta t)^{\frac{3}{2}}). \end{aligned} \tag{1.126}$$

[7] For more formal rigorous treatments and proofs see, for example, [IW89, Øks00, JS87].

Taking the limit $N \to \infty$ ($\delta t \to 0$), the finite difference $\delta t$ is the infinitesimal differential $dt$, $\delta W_t$ is the stochastic differential $dW_t$, the remainder term drops out, and we finally obtain equation (1.123). Alternatively, with the use of equation (1.125) we can obtain the drift function of the $f_t$ process:

$$\begin{aligned} A(x_t, t) &= \lim_{\delta t \to 0} \frac{E_t[\delta f_t]}{\delta t} \\ &= \frac{\partial f}{\partial t} + \frac{\partial f}{\partial x} \lim_{\delta t \to 0} \frac{E_t[\delta x_t]}{\delta t} + \frac{1}{2}\frac{\partial^2 f}{\partial x^2} \lim_{\delta t \to 0} \frac{E_t[(\delta x_t)^2]}{\delta t} \\ &= \frac{\partial f}{\partial t}(x_t, t) + a(x_t, t)\frac{\partial f}{\partial x}(x_t, t) + \frac{b(x_t, t)^2}{2}\frac{\partial^2 f}{\partial x^2}(x_t, t); \end{aligned}$$

and the volatility function of the $f_t$ process:

$$\begin{aligned} B(x_t, t)^2 &= \lim_{\delta t \to 0} \frac{E_t[(\delta f_t)^2]}{\delta t} \\ &= \left(\frac{\partial f}{\partial x}\right)^2 \lim_{\delta t \to 0} \frac{E_t[(\delta x_t)^2]}{\delta t} = b(x_t, t)^2 \left(\frac{\partial f}{\partial x}(x_t, t)\right)^2. \end{aligned}$$

The drift and volatility functions therefore give equation (1.123), as required. Here we have made use of the expectations

$$a(x_t, t) = \lim_{\delta t \to 0} \frac{E_t[\delta x_t]}{\delta t}, \qquad b(x_t, t)^2 = \lim_{\delta t \to 0} \frac{E_t[(\delta x_t)^2]}{\delta t}$$

following from the finite differential form of equation (1.122). □

Note: Itô's formula is rather simple to remember if one just takes the Taylor expansion of the infinitesimal change $df$ up to second order in $dx$ and up to first order in the time increment $dt$ and then inserts the stochastic expression for $dx$ and replaces $(dx)^2$ by $b(x, t)^2 dt$.

As we will later see, in most pricing applications, $x_t$ represents some asset price process, and therefore it proves convenient to consider Itô's lemma applied to the SDE of equation (1.119); i.e., $a(x, t) = x\mu(x, t)$, $b(x, t) = x\sigma(x, t)$, written in terms involving the lognormal drift and volatility functions for the random variable $x$. Equation (1.123) then gives

$$df_t = \left(\frac{\partial f}{\partial t} + x\mu\frac{\partial f}{\partial x} + \frac{x^2\sigma^2}{2}\frac{\partial^2 f}{\partial x^2}\right)dt + x\sigma\frac{\partial f}{\partial x}dW_t \tag{1.127}$$

$$\equiv \mu_f f_t dt + \sigma_f f_t dW_t \tag{1.128}$$

From this form of the SDE we identify the corresponding lognormal drift $\mu_f = \mu_f(x, t)$ and volatility $\sigma_f = \sigma_f(x, t)$ for the process $f_t$.

The foregoing derivation of Itô's lemma for one underlying random variable can be extended to the general case of a function $f(x_1, \ldots, x_n, t)$ depending on $n$ random variables $x = (x_1, \ldots, x_n)$ and time $t$. [Note: To simplify notation, we shall avoid the use of subscript $t$ in the variables, i.e., $x_{1,t} = x_1$, etc.] We can readily derive Itô's formula by assuming that the $x_i$, $i = 1, \ldots, n$, satisfy the stochastic differential equations

$$dx_i = a_i dt + b_i \sum_{j=1}^{n} \Lambda_{ij} dW_t^j. \tag{1.129}$$

Here the coefficients $a_i = a_i(x_1, \ldots, x_n, t)$ and $b_i = b_i(x_1, \ldots, x_n, t)$ are any smooth functions of the arguments. Furthermore we assume that the Wiener processes $W_t^j$ are mutually independent, i.e.,

$$E[dW_t^i dW_t^j] = \delta_{ij}\ dt. \tag{1.130}$$

The constants $\rho_{ij} = \rho_{ji}$ (with $\rho_{ii} = 1$) are correlation matrix elements and are convenient for introducing correlations among the increments (e.g. see equation (1.176) of Section 1.6):

$$\begin{aligned} E[(dx_i)(dx_j)] &= b_i b_j \sum_{k=1}^{n}\sum_{l=1}^{n} \Lambda_{ik}\Lambda_{jl} E[dW_t^k dW_t^l] \\ &= b_i b_j \sum_{k=1}^{n} \Lambda_{ik}\Lambda_{jk} dt \equiv b_i b_j \rho_{ij} dt. \end{aligned} \tag{1.131}$$

When $i = j$ this gives $E[(dx_i)^2] = b_i^2 dt$. Taylor expanding $df$ up to second order in the $dx_i$ increments and to first order in $dt$ we have

$$df = \frac{\partial f}{\partial t} dt + \sum_{i=1}^{n} \frac{\partial f}{\partial x_i} dx_i + \frac{1}{2} \sum_{i,j=1}^{n} \frac{\partial^2 f}{\partial x_i \partial x_j}(dx_i)(dx_j) \tag{1.132}$$

Now replacing $(dx_i)(dx_j)$ by the right-hand side of equation (1.131) while substituting the above expression for $dx_i$ and collecting terms in $dt$ and the $dW_t^i$ gives the final expression:

$$\begin{aligned} df = &\left( \frac{\partial f}{\partial t} + \sum_{i=1}^{n} \left[ a_i \frac{\partial f}{\partial x_i} + \frac{b_i^2}{2} \frac{\partial^2 f}{\partial x_i^2} \right] + \sum_{i<j=1}^{n} b_i b_j \rho_{ij} \frac{\partial^2 f}{\partial x_i \partial x_j} \right) dt \\ &+ \sum_{j=1}^{n} \left( \sum_{i=1}^{n} \rho_{ij} b_i \frac{\partial f}{\partial x_i} \right) dW_t^j. \end{aligned} \tag{1.133}$$

This procedure can be straightforwardly applied or extended to stochastic differentials of various processes that are dependent on groups of underlying random variables.

As we shall see in the coming sections, where we cover derivatives pricing in continuous time, it is important to work out the stochastic differential of the quotient of two processes, namely; $f_t \equiv g_t / h_t$, where

$$\frac{dg_t}{g_t} = \mu_g dt + \sum_{i=1}^{n} \sigma_g^i dW_t^i, \qquad \frac{dh_t}{h_t} = \mu_h dt + \sum_{i=1}^{n} \sigma_h^i dW_t^i \tag{1.134}$$

are stochastic differential equations assumed satisfied by $g_t$ and $h_t$, respectively. Note that the drift and volatility functions[8] are generally considered functions of time and of the underlying processes, $\mu_g = \mu_g(g_t, h_t, t)$, $\mu_h = \mu_h(g_t, h_t, t)$, $\sigma_g^i = \sigma_g^i(g_t, h_t, t)$, $\sigma_h^i = \sigma_h^i(g_t, h_t, t)$. The function $\sigma_g^i$ is the volatility of the process $g_t$ with respect to the $i$th independent Wiener process (or $i$th risk factor).[9] The stochastic differential of the ratio $f_t = g_t / h_t$ can be obtained via the Taylor expansion of the differential $df$ up to first order in $dt$ and up to second order

[8] Here and throughout the rest of the book we shall sometimes take the liberty to refer to the lognormal drift and volatility functions simply as the drift and volatility so as to avoid excessive use of such terminology.

[9] In what follows we shall at times also refer to independent Brownian motions as risk factors.

in the $dg$ and $dh$ terms. Hence considering $f$ as function of $g$, $h$, and $t$ and taking appropriate partial derivatives gives

$$df = \frac{1}{h}dg - \frac{g}{h^2}dh - \frac{1}{h^2}(dg)(dh) + \frac{g}{h^3}(dh)^2. \tag{1.135}$$

Here $\frac{\partial f}{\partial t} = 0$, since there is no explicit time dependence. Moreover, since $\frac{\partial^2 f}{\partial g^2} = 0$, the $(dg)^2$ term is absent. This last SDE takes on a particularly simple form when we divide through by $f$:

$$\frac{df}{f} = \frac{dg}{g} - \frac{dh}{h} - \frac{dg}{g}\frac{dh}{h} + \left(\frac{dh}{h}\right)^2 = \left(\frac{dg}{g} - \frac{dh}{h}\right)\left(1 - \frac{dh}{h}\right) \tag{1.136}$$

Substituting equations (1.134), expanding out, and setting to zero any term containing $(dW_t^i)(dt)$ or $(dt)^2$ [i.e., terms of $O((dt)^{3/2})$ or higher] then gives

$$\frac{df}{f} = \left[\mu_g - \mu_h - \sum_{i=1}^{n} \sigma_h^i(\sigma_g^i - \sigma_h^i)\right]dt + \sum_{i=1}^{n}(\sigma_g^i - \sigma_h^i)dW_t^i. \tag{1.137}$$

Here we have also made use of the replacement $dW_t^i\, dW_t^j = \delta_{ij}\, dt$. This gives the stochastic differential of $f_t = g_t/h_t$. Note that this equation in compact form reads

$$\frac{df}{f} = \mu_f dt + \sum_{i=1}^{n} \sigma_f^i dW_t^i, \tag{1.138}$$

where the drift of $f$ is $\mu_f = \mu_g - \mu_h - \sum_{i=1}^{n} \sigma_h^i(\sigma_g^i - \sigma_h^i)$ and the volatility is given by $\sigma_f^i = \sigma_g^i - \sigma_h^i$. It is important to note that pricing formulas ultimately involve the absolute value or square of the volatilities, i.e., $\sigma_f^i = \left|\sigma_g^i - \sigma_h^i\right| = \sqrt{(\sigma_g^i)^2 + (\sigma_h^i)^2 - 2\sigma_g^i\sigma_h^i}$. This will become clear in the sections that follow. Namely, a rigorous justification of this arises from consideration of the partial differential equation (i.e., the forward or backward Kolmogorov equation) satisfied by the corresponding transition probability density function, which explicitly involves only terms in the square of the volatilities. Finally, note that for the case of only one risk factor, i.e., $n = 1$, we have equation (1.138) with $\mu_f = \mu_g - \mu_h - \sigma_h(\sigma_g - \sigma_h)$ and $\sigma_f = \sigma_g - \sigma_h$. For general $n$, using vector notation $(\boldsymbol{\sigma}_f = \boldsymbol{\sigma}_g - \boldsymbol{\sigma}_h,\ \mu_f = \mu_g - \mu_h - \boldsymbol{\sigma}_h \cdot (\boldsymbol{\sigma}_g - \boldsymbol{\sigma}_h))$ and equation (1.138) takes the form:

$$\frac{df}{f} = \mu_f dt + \boldsymbol{\sigma}_f \cdot d\mathbf{W}_t. \tag{1.139}$$

Recall that a martingale process, which we shall here simply denote by $f_t$, is a stochastic process for which $E_t^P[f_T] = f_t$, $t \le T$, under a given probability measure $P$. Recall that this is a driftless process, in the sense that its expected value, under $P$, is constant over all future times. We have already encountered a simple example of such a process, namely, the standard Brownian motion, or Wiener process $W_t$. Equation (1.90) provides a method of generating a martingale process. Based on Itô's Lemma we now have the following result.

**Theorem. (Feynman–Kac)** *If f(x,t) is the function given by the conditional expectation*

$$f(x, t) = E_t[\phi(x_T)], \tag{1.140}$$

*at time* $t \le T$, *with* $x_t = x$ *and underlying process obeying equation (1.122), then f(x,t) satisfies the partial differential equation*

$$\frac{\partial f(x,t)}{\partial t} + a(x,t)\frac{\partial f(x,t)}{\partial x} + \frac{b(x,t)^2}{2}\frac{\partial^2 f(x,t)}{\partial x^2} = 0, \tag{1.141}$$

*with terminal time condition* $f(x, T) = \phi(x)$.

**Proof.** The proof follows by considering the conditional expectation of equation (1.126) at time $t$, which leaves us with only the drift term in $\delta t$ (to order $\delta t$), since the Wiener term is Markovian. On the other hand,

$$E_t[\delta f_t] = E_t[f_{t+\delta t}] - f_t = 0. \tag{1.142}$$

The last equality is due to the martingale property of $f_t$. In the limit of infinitesimal time step we are left with the infinitesimal drift term, which vanishes identically only if equation (1.141) is satisfied. The terminal condition follows simply because $f(x, t=T) = E_T[\phi(x_T)] = \phi(x)$, with $x_T = x$ imposed when $t = T$. □

The Black–Scholes partial differential equation discussed in Section 1.13 is a special case of the Feynman–Kac result. The generalization of equation (1.141) to $n$ dimensions is also readily obtained by using Itô's lemma in $n$ dimensions.

## Problems

**Problem 1.** Consider the stochastic processes $g_t$ and $h_t$ defined earlier. Further assume that the volatilities of the two processes are identical with respect to all Brownian increments, i.e., $\sigma_g^i = \sigma_h^i$ for all $i$. Show that the process $f_t = g_t/h_t$ is deterministic with solution

$$f_T = f_t \exp\left(\int_t^T (\mu_g(g_s, s) - \mu_h(h_s, s))ds\right). \tag{1.143}$$

**Problem 2.** Consider two processes defined by $g_t = g_0 e^{\sigma_g W_t + \mu_g t}$ and $h_t = h_0 e^{\sigma_h W_t + \mu_h t}$, where $W_t$ is a standard Wiener process and $\mu_g$, $\mu_h$, $\sigma_g$, $\sigma_h$, $g_0$, and $h_0$ are constants. Use Itô's lemma to show that

$$\frac{dg_t}{g_t} = \left(\mu_g + \frac{\sigma_g^2}{2}\right)dt + \sigma_g dW_t, \qquad \frac{dh_t}{h_t} = \left(\mu_h + \frac{\sigma_h^2}{2}\right)dt + \sigma_h dW_t. \tag{1.144}$$

Then assume $df_t/f_t = \mu_f dt + \sigma_f dW_t$. Find these drift and volatility coefficients in terms of $\mu_g$, $\mu_h$, $\sigma_g$, and $\sigma_h$, for the cases $f_t = g_t/h_t$ and $f_t = g_t h_t$.

**Problem 3.** Obtain the stochastic differential equations satisfied by the Ornstein–Uhlenbeck and Brownian bridge processes in Problem 8 of Section 1.4.

# 1.6 Geometric Brownian Motion

Univariate geometric Brownian motion with time-dependent coefficients is characterized by the SDE of the form

$$dS_t = \mu(t)S_t\, dt + \sigma(t)S_t\, dW_t, \tag{1.145}$$

with initial condition $S_0$, where $\mu = \mu(t)$ and $\sigma = \sigma(t)$ are deterministic functions of time $t$. This equation can be solved by means of the change of variable

$$x_t = \log \frac{S_t}{S_0}. \tag{1.146}$$

The transformed equation is obtained using Itô's lemma,

$$dx_t = \left(\mu(t) - \frac{\sigma(t)^2}{2}\right) dt + \sigma(t) dW_t, \tag{1.147}$$

and is to be solved with initial condition $x_0 = 0$. Following the procedure in Section 1.4 we discretize this equation in the time interval $[0,T]$ using a partition in $N$ subintervals of length $\delta t = \frac{T}{N}$:

$$x_{t_{i+1}} = x_{t_i} + \left(\mu(t_i) - \frac{\sigma(t_i)^2}{2}\right) \delta t + \sigma(t_i)\ \delta W_{t_i}. \tag{1.148}$$

By iterating the recurrence relations up to time $T$, we find

$$x_T = \sum_{i=0}^{N-1} \left[\left(\mu(t_i) - \frac{\sigma(t_i)^2}{2}\right) \delta t + \sigma(t_i)\ \delta W_{t_i}\right]. \tag{1.149}$$

Hence $x_T$ is a normal random variable for all $N > 1$. In the limit as $N \to \infty$, the mean of $x_T$ is given by

$$E_0[x_T] = \lim_{N\to\infty} \sum_{i=0}^{N-1} \left(\mu(t_i) - \frac{\sigma(t_i)^2}{2}\right) \delta t = \int_0^T \left(\mu(t) - \frac{\sigma(t)^2}{2}\right) dt \tag{1.150}$$

and the variance is given by

$$E_0[x_T^2] - (E_0[x_T])^2 = \lim_{N\to\infty} \sum_{i=0}^{N-1} \sigma(t_i)^2\ \delta t = \int_0^T \sigma(t)^2\ dt. \tag{1.151}$$

Introducing the time-averaged drift and volatility

$$\bar{\mu}(T) \equiv \frac{1}{T} \int_0^T \mu(t) dt \tag{1.152}$$

and

$$\bar{\sigma}(T) \equiv \sqrt{\frac{1}{T} \int_0^T \sigma(t)^2\ dt}, \tag{1.153}$$

we conclude that $x_T = \log \frac{S_T}{S_0} \sim N\left(\left(\bar{\mu}(T) - \frac{\bar{\sigma}^2(T)}{2}\right) T, \bar{\sigma}^2(T)T\right)$. This result is also easily verified by directly applying properties (1.105) and (1.106) to the integrated form of equation (1.147).

The solution to stochastic differential equation (1.145) for all $t \geq 0$ is hence

$$S_t = S_0 \exp\left(\left(\bar{\mu}(t) - \frac{\bar{\sigma}^2(t)}{2}\right) t + \bar{\sigma}(t) W_t\right), \tag{1.154}$$

where $\bar{\mu}(t)$ and $\sigma(t)$ are given by equations (1.152) and (1.153), respectively. This solution (which is actually a strong solution) can also be verified by a direct application of Itô's lemma (see Problem 1). Note that this represents a solution, in the sense that the random variable denoted by $S_t$ and parameterized by time $t$ is expressed in terms of the underlying random variable, $W_t$, for the pure Wiener process.

This solution gives a closed-form expression for generating sample paths for geometric Brownian motion. Equation (1.154) provides a general expression for the case of time-dependent drift and volatility. It is very instructive at this point to compute expectations of functions of $S_t$. Let us consider the process in equation (1.145) and proceed now to compute the expectations $E_0[S_t]$ and $E_0[(S_t - K)_+]$, for some constant $K \geq 0$, where $(x)_+ \equiv \max(x, 0) = x$ if $x > 0$ and zero if $x \leq 0$. Using the solution in equation (1.154), the expectation of $S_t$ under the density of equation (1.95) (i.e., conditional on $S_{t=0} = S_0$, hence we write $E_0[\,]$) is

$$\begin{aligned} E_0[S_t] &= S_0 e^{(\bar{\mu}-\bar{\sigma}^2/2)t} E_0[e^{\bar{\sigma} W_t}] \\ &= S_0 e^{(\bar{\mu}-\bar{\sigma}^2/2)t} e^{\bar{\sigma}^2 t/2} = S_0 e^{\bar{\mu} t}. \end{aligned} \tag{1.155}$$

To compact notation we denote $\bar{\mu} \equiv \bar{\mu}(t)$, $\bar{\sigma} \equiv \bar{\sigma}(t)$. In the last step we have used an important identity derived in Problem 2 of this section. This result shows that the stock price is expected to grow exponentially at a rate of $\bar{\mu}$.

Using equation (1.154), the expectation $E_0[(S_t - K)_+]$ is given by

$$\begin{aligned} E_0[(S_t - K)_+] &= \int_{-\infty}^{\infty} p(y, t)\left(S_0 e^{(\bar{\mu}-\bar{\sigma}^2/2)t} e^{\bar{\sigma} y} - K\right)_+ dy \\ &= \frac{S_0 e^{(\bar{\mu}-\bar{\sigma}^2/2)t}}{\sqrt{2\pi t}} \int_{-\infty}^{\infty} e^{-y^2/2t} \left(e^{\bar{\sigma} y} - \frac{K}{S_0} e^{-(\bar{\mu}-\bar{\sigma}^2/2)t}\right)_+ dy \end{aligned} \tag{1.156}$$

The last step obtains from the identity $(ax - b)_+ = a(x - b/a)_+$, for $a > 0$. Changing integration variable $y = \sqrt{t}x$ while employing this identity again gives

$$E_0[(S_t - K)_+] = \frac{S_0 e^{(\bar{\mu}-\frac{\bar{\sigma}^2}{2})t}}{\sqrt{2\pi}} \int_{-\infty}^{\infty} e^{-x^2/2+\bar{\sigma}\sqrt{t}x} \left(1 - \frac{K}{S_0} e^{-[(\bar{\mu}-\frac{\bar{\sigma}^2}{2})t+\bar{\sigma}\sqrt{t}x]}\right)_+ dx \tag{1.157}$$

Since $e^{-\bar{\sigma}\sqrt{t}x}$ is a monotonically decreasing function of $x$, there is a value $x_K$ such that

$$\left(1 - \frac{K}{S_0} e^{-[(\bar{\mu}-\bar{\sigma}^2/2)t+\bar{\sigma}\sqrt{t}x]}\right)_+ = \begin{cases} 1 - \dfrac{K}{S_0} e^{-[(\bar{\mu}-\bar{\sigma}^2/2)t+\bar{\sigma}\sqrt{t}x]}, & x > x_K \\ 0, & x \leq x_K \end{cases} \tag{1.158}$$

where

$$x_K = -\frac{\log(S_0/K) + (\bar{\mu} - \bar{\sigma}^2/2)t}{\bar{\sigma}\sqrt{t}}. \tag{1.159}$$

Hence, the integral in equation (1.157) becomes a sum of two parts in the region $x \in (x_K, \infty)$:

$$E_0[(S_t - K)_+] = \frac{S_0 e^{(\bar{\mu}-\frac{\bar{\sigma}^2}{2})t}}{\sqrt{2\pi}} \int_{x_K}^{\infty} e^{-x^2/2+\bar{\sigma}\sqrt{t}x}\, dx - \frac{K}{\sqrt{2\pi}} \int_{x_K}^{\infty} e^{-x^2/2}\, dx. \tag{1.160}$$

Completing the square in the first integration gives

$$\begin{aligned} E_0[(S_t - K)_+] &= S_0 e^{\bar{\mu} t}(1 - N(x_K - \bar{\sigma}\sqrt{t})) - KN(-x_K) \\ &= S_0 e^{\bar{\mu} t} N(\bar{\sigma}\sqrt{t} - x_K) - KN(-x_K) \\ &= S_0 e^{\bar{\mu} t} N(d_+) - KN(d_-), \end{aligned} \tag{1.161}$$

where $N(\cdot)$ is the standard cumulative normal distribution function and

$$d_{\pm} = \frac{\log(S_0/K) + (\bar{\mu} \pm \bar{\sigma}^2/2)t}{\bar{\sigma}\sqrt{t}}. \tag{1.162}$$

Note that here we have used the property $N(-x) = 1 - N(x)$.

The Black–Scholes pricing formula for a plain European call option follows automatically. In particular, assuming a risk-neutral pricing measure, the drift is given by the instantaneous risk-free rate $\mu(t) = r(t)$. Hence, the price of a call at current time $(t = 0)$ with current stock level (or spot) $S_0$, strike $K$, and maturing in time $t$ is given by the discounted expectation

$$C_0(S_0, K, t) = e^{-\bar{r}t} E_0[(S_t - K)_+] = S_0 N(d_+) - e^{-\bar{r}t} K N(d_-), \tag{1.163}$$

where $\bar{r}$ is the time-averaged continuously compounded risk-free interest rate

$$\bar{r} = \bar{r}(t) \equiv \frac{1}{t}\int_0^t r(\tau)d\tau, \tag{1.164}$$

and $d_{\pm}$ is given by equation (1.162) with $\bar{\mu} = \bar{r}$. It is instructive to note the inherent difference between the Black–Scholes pricing formula in equation (1.163) and Bachelier's formula in equation (1.117). Bachelier's formula is a result of assuming a standard Brownian motion for the underlying stock price process [i.e., equation (1.94)]. In contrast, formulas of the Black–Scholes type are equivalent to the assumption of geometric Brownian motion for the underlying price process. Using equation (1.154) as defining a change of probability variables $W_t \to S_t$, the one-dimensional analogue of equation (1.48) together with equation (1.95) gives

$$p(S_t, S_0; t) = \frac{1}{S_t \bar{\sigma}\sqrt{2\pi t}} e^{-[\log(S_t/S_0) - (\bar{\mu} - \bar{\sigma}^2/2)t]^2/2\bar{\sigma}^2 t}. \tag{1.165}$$

This is the *lognormal distribution* function defined on positive stock price space $S_t \in (0, \infty)$. The log-returns $\log(S_t/S_0)$ are distributed normally with mean $(\bar{\mu} - \bar{\sigma}^2/2)t$ and variance $\bar{\sigma}^2 t$. Setting $\bar{\mu} = \bar{r}$ gives the *risk-neutral conditional probability density* for a stock attaining a value $S_t$ at time $t > 0$ given an initial value $S_0$ at time $t = 0$. Hence, the Black–Scholes pricing formula for European options can also be obtained by taking discounted expectations of payoff functions with respect to this risk-neutral density. In particular, a European-style claim having pay-off $\Lambda(S_T)$ as a function of the terminal stock level $S_T$, where $T > 0$ is a maturity time, has arbitrage-free price $f_0(S_0, T)$ at time $t = 0$ expressible as

$$f_0(S_0, T) = e^{-\bar{r}(T)T} E_0^Q[\Lambda(S_T)] = e^{-\bar{r}(T)T} \int_0^\infty p(S_T, S_0; T)\Lambda(S_T)dS_T. \tag{1.166}$$

Here the superscript $Q$ is used to denote an expectation with respect to the risk-neutral density given by equation (1.165) with drift $\bar{\mu} = \bar{r}(T)$. Note that within this probability measure, equation (1.155) shows that stock prices drift at the time-averaged risk-free rate $r(t)$ at time $t$. As will become apparent in the following sections, this must be the case in order to ensure arbitrage-free pricing.

For pricing applications, discussed in greater length in later sections of this chapter, it is useful to consider a slight extension of the foregoing closed-form solutions to geometric Brownian motion. Namely, we can extend equation (1.154) by a simple shift in time variables as follows:

$$S_T = S_t \exp\left(\left(\bar{\mu}(t, T) - \frac{\bar{\sigma}^2(t, T)}{2}\right)(T - t) + \bar{\sigma}(t, T)W_{T-t}\right), \tag{1.167}$$

with time-averaged drift and volatility over the period *[t,T]*

$$\bar{\mu}(t,T) \equiv \frac{1}{T-t}\int_t^T \mu(\tau)d\tau, \qquad \bar{\sigma}^2(t,T) \equiv \frac{1}{T-t}\int_t^T \sigma^2(\tau)d\tau. \tag{1.168}$$

Here $W_{T-t} = W_T - W_t$ is the Wiener normal random variable with mean zero and variance $T-t$; i.e., $W_{T-t} \sim \sqrt{T-t}x$, $x \sim N(0,1)$. For constant drift and volatility this solution simplifies in the obvious manner. The formula for the conditional expectation now extends to give

$$E_t[(S_T - K)_+] = e^{\bar{\mu}(T-t)} S_t N(d_+) - KN(d_-), \tag{1.169}$$

with

$$d_\pm = \frac{\log(S_t/K) + (\bar{\mu} \pm \bar{\sigma}^2/2)(T-t)}{\bar{\sigma}\sqrt{T-t}} \tag{1.170}$$

and $\bar{\mu} = \bar{\mu}(t,T)$, $\bar{\sigma} = \bar{\sigma}(t,T)$. A related expectation that is useful for pricing purposes is (see Problem 3)

$$E_t[(K - S_T)_+] = KN(-d_-) - e^{\bar{\mu}(T-t)} S_t N(-d_+). \tag{1.171}$$

Within the risk-neutral probability measure, $\bar{\mu} = \bar{r}$. Hence discounting this expectation by $e^{-\bar{r}(T-t)}$ gives the analogue of equation (1.163) for the Black–Scholes price of a put option at calendar time $t$, spot $S_t$, and maturing at time $T$ with strike $K$:

$$P_t(S_t, K, T) = e^{-\bar{r}(T-t)} KN(-d_-) - S_t N(-d_+), \tag{1.172}$$

where $d_\pm$ is given by equation (1.170) with $\bar{\mu} = \bar{r} \equiv r(t,T)$.

In closing this section, we consider the more general multidimensional case of geometric Brownian motion. Multivariate geometric Brownian motions describe $n$-dimensional state spaces of vector valued processes $S_t^1, \ldots, S_t^n$ and can be described with two different but equivalent sets of notations. Let's consider $n$ uncorrelated standard Wiener processes

$$W_t^1, \ldots, W_t^n, \text{with } E_t[dW_t^i dW_t^j] = \delta_{ij}\, dt. \tag{1.173}$$

A simple way to introduce correlations among the price processes is to allow for correlated Wiener processes by defining a new set of $n$ processes $W_t^{S^i}$ as

$$dW_t^{S^i} = \sum_{j=1}^{n} \Lambda_{ij}\, dW_t^j, \tag{1.174}$$

or, in matrix-vector notation,

$$d\mathbf{W}_t^S = \boldsymbol{\Lambda} \cdot d\mathbf{W}_t. \tag{1.175}$$

Using equation (1.174) we have

$$E_t\left[dW_t^{S^i} dW_t^{S^j}\right] = \sum_{k,l=1}^{n} \Lambda_{ik}\Lambda_{jl}\, \delta_{kl}\, dt = \sum_{k=1}^{n} \Lambda_{ik}\Lambda_{jk}\, dt \equiv \rho_{ij}\, dt, \tag{1.176}$$

where the last relation defines a correlation matrix $\boldsymbol{\rho}$, with elements $\rho_{ij}$, and lower Cholesky decomposition given by

$$\boldsymbol{\rho} = \boldsymbol{\Lambda}\boldsymbol{\Lambda}^\dagger. \tag{1.177}$$

Throughout this section, superscript † denotes matrix transpose.

Stochastic differential equations for the stock price processes can be written as follows:

$$\frac{dS_t^i}{S_t^i} = \mu_i\, dt + \sigma_i\, dW_t^{S^i} \tag{1.178}$$

$$= \mu_i\, dt + \sigma_i \sum_{j=1}^{n} \Lambda_{ij}\, dW_t^j \equiv \mu_i\, dt + \sum_{j=1}^{n} L_{ij}\, dW_t^j \tag{1.179}$$

where the last expression defines the matrix $\mathbf{L}$, $L_{ij} = \sigma_i \Lambda_{ij}$. Note that the lognormal drifts $\mu_i$ and volatilities $\sigma_i$ can generally depend on time, although to simplify notation we have chosen not to denote this explicitly. The last relation in equation (1.179) defines a lower Cholesky factorization of the covariance matrix

$$\mathbf{C} = \mathbf{L}\mathbf{L}^{\dagger} = \mathbf{\Sigma}\mathbf{\Lambda}\mathbf{\Lambda}^{\dagger}\mathbf{\Sigma} = \mathbf{\Sigma}\boldsymbol{\rho}\mathbf{\Sigma}. \tag{1.180}$$

Here $\mathbf{\Sigma}$ is the diagonal matrix of lognormal volatilities with *(ij)*-elements given by $\delta_{ij}\ \sigma_i$, $\mathbf{L} = \mathbf{\Sigma}\mathbf{\Lambda}$ and $\mathbf{\Sigma} = \mathbf{\Sigma}^{\dagger}$. In vector notation we can write equations (1.179) in a compact form as

$$\frac{dS_t^i}{S_t^i} = \mu_i\, dt + \boldsymbol{\sigma}_i \cdot d\mathbf{W}_t, \tag{1.181}$$

where $\boldsymbol{\sigma}_i = (\sigma_{i1}, \ldots, \sigma_{in})$ is the volatility vector for the *i*th stock, whose *j*th component $\sigma_{ij} = L_{ij}$ gives the lognormal volatility with respect to the *j*th risk factor.

Equation (1.61) in Section 1.2 gives $\mathbf{L}$ for the case $n = 2$. In particular, in the case of two stocks we can introduce a correlation $\rho$, where equations (1.179) now take the specific form

$$\frac{dS_t^1}{S_t^1} = \mu_1\, dt + \sigma_1\, dW_t^1, \tag{1.182}$$

$$\frac{dS_t^2}{S_t^2} = \mu_2\, dt + \rho\sigma_2\, dW_t^1 + \sqrt{1-\rho^2}\sigma_2\, dW_t^2, \tag{1.183}$$

with infinitesimal variances and covariances

$$E_t\left[\left(\frac{dS_t^1}{S_t^1}\right)^2\right] = \sigma_1^2\, dt,\ E_t\left[\left(\frac{dS_t^2}{S_t^2}\right)^2\right] = \sigma_2^2\, dt,\ E_t\left[\frac{dS_t^1}{S_t^1}\frac{dS_t^2}{S_t^2}\right] = \rho\sigma_1\sigma_2\, dt. \tag{1.184}$$

For this case the volatility vectors are given by $\boldsymbol{\sigma}_1 = (\sigma_1, 0)$ and $\boldsymbol{\sigma}_2 = \left(\rho\sigma_2, \sigma_2\sqrt{1-\rho^2}\right)$ for stock prices $S_t^1$ and $S_t^2$, respectively.

More generally, equations (1.179) [or (1.181)] describe geometric Brownian motion for an arbitrary number of $n$ stocks with infinitesimal correlations and variances:

$$E_t\left[\frac{dS_t^i}{S_t^i}\frac{dS_t^j}{S_t^j}\right] = C_{ij}\, dt, \qquad E_t\left[\left(\frac{dS_t^i}{S_t^i}\right)^2\right] = \sigma_i^2\, dt. \tag{1.185}$$

The vectors $\boldsymbol{\sigma}_i$ are seen to be given by the *i*th rows of matrix $\mathbf{L}$, i.e., the matrix of the lower Cholesky factorization of the covariance matrix.

A solution to the system of stochastic differential equations (1.179) [or (1.181)] is readily obtained by employing a simple change-of-variable approach (see Problem 4). In particular,

$$S_T^i = S_t^i \exp\left(\left(\mu_i - \frac{\sigma_i^2}{2}\right)(T-t) + \sigma_i \sum_{j=1}^{n} \Lambda_{ij} W_{T-t}^j\right); \qquad i = 1, \ldots, n, \tag{1.186}$$

where we denote $W^j_{T-t} = W^j_T - W^j_t$, for each $j$th independent Wiener normal random variable with mean zero and variance $T-t$; i.e., $W^j_{T-t} = \sqrt{T-t}x_j$, $x_j \sim N(0,1)$ independently for all $j = 1, \ldots, n$. From this result one readily obtains the multivariate lognormal distribution function $p(\mathbf{S}_T, \mathbf{S}_t; T-t)$, i.e., the analogue of equation (1.165) [see equation (1.198) in Problem 5]. The pricing of European-style options whose pay-offs depend on a group of $n$ stocks, i.e., European basket options, can then proceed by computing expectations of such pay-offs over this density, where the drifts are set by risk neutrality. That is, let's assume a money-market account $B_t = e^{rt}$ with constant risk-free rate $r$, then within the risk-neutral measure the stock prices must all drift at the same rate, giving $\mu_i = r$.[10] Let $V_t$ denote the option price at time $t$ for a European-style contract with payoff function at maturity time $T$ given by $V_T = \Pi(\mathbf{S}_T)$, $\mathbf{S}_T = (S^1_T, \ldots, S^n_T)$. The arbitrage-free price is then given by the expectation

$$\begin{aligned} V_t &= e^{-r(T-t)} E_t^{Q(B)}[\Pi(\mathbf{S}_T)] \\ &= e^{-r(T-t)} \int_{\mathbb{R}^n_+} p(\mathbf{S}_T, \mathbf{S}_t; T-t)\Pi(\mathbf{S}_T) d\mathbf{S_T} \\ &= \frac{e^{-r(T-t)}}{(2\pi)^{n/2}} \int_{\mathbb{R}^n} e^{-\frac{1}{2}|\mathbf{x}|^2} \Pi(\mathbf{S}_T(\mathbf{x})) d\mathbf{x}, \end{aligned} \tag{1.187}$$

where $\mathbf{S}_T(\mathbf{x})$ has components $S^i_T(\mathbf{x})$ given by equation (1.186), $\mathbf{x} = (x_1, \ldots, x_n)$. The price hence involves an $n$-dimensional integral over a multivariate normal times some payoff function. Exact analytical expressions for basket options are generally difficult to obtain, depending on the type of payoff function as well as the number of dimensions $n$. Numerical integration methods can be used in general. Monte Carlo simulation methods are very useful for this purpose. The reader interested in gaining insight into the numerical implementation of standard Monte Carlo methods for pricing such options is referred to Project 8 on Monte Carlo pricing of basket options in Part II of this book.

Exact analytical pricing formulas for certain types of elementary basket options, however, can be obtained, as demonstrated in the following worked-out example.

**Example. Chooser basket options on two stocks.**

Consider a basket of two stocks with prices $S^1_t$ (for stock 1) and $S^2_t$ (for stock 2) modeled as before with constants $\mu_1$, $\mu_2$, $\rho$, $\sigma_1$, $\sigma_2$. Specifically, the risk-neutral geometric Brownian motions of the two stocks are given by

$$S^1_T = S^1_T(x_1, x_2) = S^1_0 e^{(r - \frac{\sigma_1^2}{2})T + \sigma_1\sqrt{T}x_1}, \tag{1.188}$$

$$S^2_T = S^2_T(x_1, x_2) = S^2_0 e^{(r - \frac{\sigma_2^2}{2})T + \sigma_2\sqrt{T}(\rho x_1 + \sqrt{1-\rho^2}x_2)}, \tag{1.189}$$

where $S^1_0$, $S^2_0$ are initially known stock prices at current time $t = 0$. The earlier pricing formula gives

$$V_0 = \frac{e^{-rT}}{2\pi} \int_{-\infty}^{\infty} \int_{-\infty}^{\infty} e^{-\frac{1}{2}(x_1^2 + x_2^2)} \Pi(S^1_T(x_1, x_2), S^2_T(x_1, x_2)) dx_1\, dx_2 \tag{1.190}$$

[10] This drift restriction is further clarified later in the chapter where we discuss the asset-pricing theorem in continuous time.

for the general payoff function. A *simple chooser* option is a European contract defined by the payoff $\max(S_T^1, S_T^2)$. This pay-off has a simple relation to other elementary pay-offs; i.e., $\max(S_T^1, S_T^2) = (S_T^2 - S_T^1)_+ + S_T^1 = (S_T^1 - S_T^2)_+ + S_T^2$. The current price $V_0$ of the simple chooser is hence given by $V_0 = C_0 + S_0^1$, where $C_0$ denotes the price of the contract with payoff $(S_T^2 - S_T^1)_+$. This follows since an expectation of a sum is the sum of expectations and from the fact that the stock prices drift at rate $r$; i.e., $e^{-rT} E_0^{Q(B)}[S_T^i] = S_0^i$. The problem remains to find the price $C_0$ given by the integral

$$C_0 = \frac{e^{-rT}}{2\pi} \int_{-\infty}^{\infty} \int_{-\infty}^{\infty} e^{-\frac{1}{2}(x_1^2 + x_2^2)} \left(S_T^2(x_1, x_2) - S_T^1(x_1, x_2)\right)_+ \, dx_1 \, dx_2. \tag{1.191}$$

The integrand is nonzero on the domain $\{(x_1, x_2) \in \mathbb{R}^2; S_T^2(x_1, x_2) > S_T^1(x_1, x_2)\}$. From equations (1.188) and (1.189) we find the domain is $\{(x_1, x_2) \in \mathbb{R}^2; x_1 < ax_2 + b\}$, where

$$a \equiv \frac{\sigma_2\sqrt{1-\rho^2}}{(\sigma_1 - \rho\sigma_2)}, \qquad b \equiv \frac{\log(S_0^2/S_0^1) + \frac{1}{2}(\sigma_1^2 - \sigma_2^2)T}{(\sigma_1 - \rho\sigma_2)\sqrt{T}}.$$

Here we assume $\sigma_1 - \rho\sigma_2 > 0$ and leave it to the reader to verify that a similar derivation of the same price given next also follows for the case $\sigma_1 - \rho\sigma_2 \leq 0$. Using this integration domain and inserting expressions (1.188) and (1.189) into the last integral gives

$$C_0 = \frac{S_0^2 e^{-\frac{1}{2}\sigma_2^2 T}}{2\pi} \int_{-\infty}^{\infty} e^{-\frac{1}{2}x_2^2 + \sqrt{1-\rho^2}\sigma_2\sqrt{T}x_2} \left[\int_{-\infty}^{ax_2+b} e^{-\frac{1}{2}x_1^2 + \rho\sigma_2\sqrt{T}x_1} dx_1\right] dx_2$$
$$- \frac{S_0^1 e^{-\frac{1}{2}\sigma_1^2 T}}{2\pi} \int_{-\infty}^{\infty} e^{-\frac{1}{2}x_2^2} \left[\int_{-\infty}^{ax_2+b} e^{-\frac{1}{2}x_1^2 + \sigma_1\sqrt{T}x_1} dx_1\right] dx_2$$

By completing the square in the exponents, the integrals in $x_1$ give cumulative normal functions $N(\cdot)$. In particular,

$$C_0 = \frac{S_0^2 e^{-\frac{1}{2}(1-\rho^2)\sigma_2^2 T}}{\sqrt{2\pi}} \int_{-\infty}^{\infty} e^{-\frac{1}{2}x_2^2 + \sqrt{1-\rho^2}\sigma_2\sqrt{T}x_2} N(ax_2 + b - \rho\sigma_2\sqrt{T}) dx_2$$
$$- \frac{S_0^1}{\sqrt{2\pi}} \int_{-\infty}^{\infty} e^{-\frac{1}{2}x_2^2} N(ax_2 + b - \sigma_1\sqrt{T}) dx_2.$$

At this point we make use of the integral identity (see Problem 6),

$$\frac{1}{\sqrt{2\pi}} \int_{-\infty}^{\infty} e^{-\frac{1}{2}x^2 + Cx} N(Ax + B) dx = e^{\frac{1}{2}C^2} N\left(\frac{AC + B}{\sqrt{1 + A^2}}\right), \tag{1.192}$$

for any constants $A$, $B$, and $C$, giving

$$C_0 = S_0^2 N\left(\frac{(a\sqrt{1-\rho^2} - \rho)\sigma_2\sqrt{T} + b}{\sqrt{1+a^2}}\right) - S_0^1 N\left(\frac{b - \sigma_1\sqrt{T}}{\sqrt{1+a^2}}\right).$$

After a bit of algebra, using $a$ and $b$ just defined, we finally obtain the exact expression for the price in terms of the initial stock prices and the effective volatility $\nu$ as

$$C_0 = S_0^2 N(d_+) - S_0^1 N(d_-), \tag{1.193}$$

with

$$d_\pm = \frac{\log(S_0^2/S_0^1) \pm \frac{1}{2}\nu^2 T}{\nu\sqrt{T}}, \tag{1.194}$$

$\nu^2 = \sigma_1^2 + \sigma_2^2 - 2\rho\sigma_1\sigma_2.$

Changes of numeraire methods for obtaining exact analytical solutions in the form of Black–Scholes–type formulas for basket options on two stocks, as well as other options involving two correlated underlying random variables, are discussed later in this chapter.

## Problems

**Problem 1.** Use Itô's lemma to verify that equation (1.154) provides a solution to equation (1.145).

**Problem 2.** Consider an exponential function of a normal random variable $X$, $e^{aX}$ for any parameter $a$, where $X \in (-\infty, \infty)$ has probability density at $X = x$ given by

$$p(x,t) = \frac{1}{\sqrt{2\pi t}} e^{-x^2/2t}, \qquad (t > 0).$$

Show that

$$E[e^{aX}] = \exp\left(a^2 t/2\right).$$

Hint: make use of the integral identity

$$\int_{-\infty}^{\infty} e^{-ax^2+bx}\, dx = \sqrt{\frac{\pi}{a}}\, e^{b^2/4a},$$

where $a > 0$ and $b$ are constants.

**Problem 3.** Derive the expectation in equation (1.171) by making use of the identity $(a-b)_+ = (b-a)_+ + a - b$.

**Problem 4.** Consider the general correlated $n$-dimensional geometric Brownian process discussed in this section. Use Itô's lemma to show that the processes $Y_t^i \equiv \log S_t^i$ obey

$$dY_t^i = (\mu_i - \sigma_i^2/2)dt + \sigma_i \sum_{j=1}^{n} \lambda_{ij} dW_t^j. \tag{1.195}$$

Assuming all volatilities are nonzero, the correlation matrix is positive definite. Hence, $\boldsymbol{\lambda}$ has an inverse $\boldsymbol{\lambda}^{-1}$. Define new random variables $X_t^j \equiv \sum_{i=1}^{n} \sigma_i^{-1} \lambda_{ji}^{-1} Y_t^i$ and show that

$$dX_t^j = \tilde{\mu}_j\, dt + dW_t^j, \tag{1.196}$$

with $\tilde{\mu}_j \equiv \sum_{i=1}^{n} \sigma_i^{-1} \lambda_{ji}^{-1} (\mu_i - \frac{1}{2}\sigma_i^2)$, has solution

$$X_T^j = X_t^j + \tilde{\mu}_j (T-t) + W_T^j - W_t^j, \qquad j = 1, \ldots, n. \tag{1.197}$$

Invert this solution back into the old random variables, hence obtaining equation (1.186).

**Problem 5.** Treat $W_{T-t}^j$ and $\log(S_T^i / S_t^i)$ as two sets of $n$ independent variables in equation (1.186) and thereby compute the Jacobian of the transformation among the variables. Then invert equation (1.186) and use the identity in equation (1.48) with the distribution function for the $n$ independent uncorrelated Wiener processes to show that the analytical formula for the transition probability density for geometric Brownian motion is given by

$$p(\mathbf{S}_T, \mathbf{S}_t; T-t) = (2\pi(T-t))^{-\frac{n}{2}} |\mathbf{C}|^{-\frac{1}{2}} \exp\left(-\tfrac{1}{2}\mathbf{z}\cdot\mathbf{C}^{-1}\cdot\mathbf{z}\right), \tag{1.198}$$

where the $n$-dimensional vector $\mathbf{z}$ has components

$$z_i \equiv \frac{\log(S_T^i/S_t^i) - (\mu_i - \frac{1}{2}\sigma_i^2)(T-t)}{\sqrt{T-t}}. \tag{1.199}$$

**Problem 6.** Using the definition of the cumulative normal function, write

$$\frac{1}{\sqrt{2\pi}} \int_{-\infty}^{\infty} e^{-\frac{1}{2}x^2+Cx} N(Ax+B)dx = \frac{1}{2\pi} \int_{-\infty}^{\infty} e^{-\frac{1}{2}x^2+Cx} \left[ \int_{-\infty}^{Ax+B} e^{-\frac{1}{2}y^2} dy \right] dx. \tag{1.200}$$

Introduce a change of variables $(\eta, \xi) \equiv (y - Ax, y + Ax)$ and integrate while completing squares to obtain equation (1.192).

## 1.7 Forwards and European Calls and Puts

Consider a situation with a stock price that at current time $t = 0$ has price $S_0$ while at time $T > 0$ in the future is described by a certain random variable $S_T$. Suppose that there is also a *zero-coupon bond* maturing at time $T$, i.e., a riskless claim to one unit of account at time $T$. Let

$$Z_t(T) = e^{-r(T-t)} \tag{1.201}$$

be its price at time $t$. Here $r$ is the yield up to time $T$. Unlike the rate introduced in equation (1.5), in this case $r$ is defined with the continuously compounded rule; we refer again to Chapter 2 for a more systematic discussion of fixed-income terminology.

Let's consider a situation where $S_t$ is contained in the half-line of positive real numbers $\mathbb{R}_+$. Let $P$ be the real-world measure with density $p(S)$; $P$ is inferred through statistical estimations based on historical data. Pricing measures, instead, are evaluated as the result of a calibration procedure starting from option prices. Also, as discussed in detail later in this chapter, pricing measures depend on the choice of a *numeraire asset*. In our framework, a numeraire asset is given by an asset price process, $g_t$, that is strictly positive at initial time $t = 0$ and any other future time $t$, $t \leq T$. The corresponding pricing measure is denoted by $Q(g)$, specifying the fact that the asset price $g_t$ is the chosen numeraire. A possible choice of numeraire is given by the bond $g_t = Z_t(T)$; this choice corresponds to the pricing measure denoted by $Q(Z(T))$, which is called the *forward measure*. Note that since $r$ is constant, this also coincides with the risk-neutral measure. Technically speaking the name for the risk-neutral measure corresponds to using the continuously compounded money-market account $B_t = e^{rt}$ (i.e., the continuously compounded value of one unit of account deposited at time $t = 0$ earning interest rate $r$) as numeraire.[11] For constant interest rate, the two measures are then easily shown to be equivalent since $Z_t(T) = B_t/B_T$. This point is further clarified in Chapter 2. Other choices of numeraire asset are also possible; for example, $g_t = S_t$ corresponds to using the stock price as numeraire. As mentioned earlier and also described in detail later in the chapter, expectations taken based on the information available up to current time $t$ with respect to the pricing measure $Q(g)$, with $g_t$ as numeraire asset price, are denoted by $E_t^{Q(g)}[\,]$. In this section, note that (without loss in generality) we are simply setting $t = 0$ as current time and allowing $T$ to be any future time.

[11] Note that we previously used the symbol $B_t$ to denote the bond price. However, here we instead use $B_t$ to denote the value of the money-market account.

By applying risk-neutral valuation to the zero-coupon bond, we find that

$$Z_0(T) = e^{-rT} = \alpha E_0^{Q(Z(T))}[Z_T(T)] = \alpha E_0^{Q(Z(T))}[1] = \alpha, \tag{1.202}$$

where $Z_T(T) = 1$. Hence, the discount factor $\alpha$ can be interpreted as the initial price of the zero-coupon bond. Although we have not yet formally introduced continuous-time financial models at this point in the chapter, the arguments presented in this section are generally valid if we assume dynamic trading is allowed in continuous time.

Risky assets are modeled by a function $\phi : \mathbb{R}_+ \to \mathbb{R}$ of the stock price at time $T$. Let $(A_t)_{0\le t\le T}$ be a price process such that $A_T = \phi(S_T)$; such an asset is called a *European-style option on the stock S with maturity T and payoff function* $\phi(S_T)$. Applying the asset-pricing theorem, the arbitrage-free price $A_0$ at time $t = 0$ of this option can be written as a discounted expectation under a pricing measure *Q(Z(T))*,

$$A_0 = e^{-rT} E_0^{Q(Z(T))}[\phi(S_T)]. \tag{1.203}$$

An alternative and instructive way of writing this equation is

$$\frac{A_0}{Z_0(T)} = E_0^{Q(Z(T))}\left[\frac{A_T}{Z_T(T)}\right]. \tag{1.204}$$

Although the numeraire asset in equation (1.204) is the riskless bond $Z_t(T)$, the pricing formula can be extended to the case of a generic numeraire asset $g$. Let's denote *Q(g)* as the probability measure, with $g_t$ as numeraire asset price at time $t$, and defined so that

$$\frac{A_0}{g_0} = E_0^{Q(g)}\left[\frac{A_T}{g_T}\right] \tag{1.205}$$

for all random variables $A_T = \phi(S_T)$ and for all $T > 0$. Assuming the price is unique, equating the price $A_0$ in equation (1.204) with that in this last equation gives a relationship for the equivalence of the two pricing (or probability) measures:

$$g_0 E_0^{Q(g)}\left[\frac{\phi(S_T)}{g_T}\right] = Z_0(T) E_0^{Q(Z(T))}\left[\frac{\phi(S_T)}{Z_T(T)}\right]. \tag{1.206}$$

A variety of numeraire assets can be chosen for derivative pricing. Depending on the pay-off, one choice over another may be more convenient for evaluating the expectation and hence obtaining the derivative price, as seen in detail in the examples of pricing derivations in Section 1.12.

A *forward contract* on an underlying stock $S$ stipulated at initial time $t = 0$ and with maturity time $t = T$ is a European-style claim with payoff $S_T - F_0$ at time $T$. Here $F_0$ is the *forward price* at time $t = 0$. Forward contracts are entered at the equilibrium forward price $F_0$, for which their present value is zero. A simple arbitrage argument gives a (*model-independent*) forward price $F_0$ as

$$F_0 = Z_0(T)^{-1} S_0. \tag{1.207}$$

Indeed, to replicate the pay-off of a forward contract one can buy the underlying stock at price $S_0$ and carry it to maturity while funding the purchase with a loan to be returned also at maturity. The nominal of the loan to be paid back at time $T$ is then $Z_0(T)^{-1} S_0$ (e.g., this equals $e^{rT} S_0$ if we assume a constant interest rate).

Since the forward contract is initially worthless, the valuation formula yields

$$0 = E_0^{Q(Z(T))}[S_T - F_0]. \tag{1.208}$$

Since $F_0$ is constant, we have that

$$E_0^{Q(Z(T))}[S_T] = F_0 = Z_0(T)^{-1} S_0 = e^{rT} S_0. \tag{1.209}$$

The interpretation of this formula is that, under the pricing measure *Q(Z(T))*, the expected return on a stock is the risk-free yield $r$ over the maturity $T$. The argument just outlined is model independent and can be shown to extend to all assets with no intermediate cash flows, thus no carry costs, before maturity time $T$. The expected return on any asset under the pricing measure $Q(Z(T))$ is the risk-free rate, no matter how volatile they are. Also notice that the expected return with respect to the real-world measure is quite different.

The popular geometric Brownian motion model, also called the *Black–Scholes model*, gives a lognormal risk-neutral probability density for the stock price process. As derived in Section 1.6, the stock price at time $T$ is a lognormal random variable,

$$S_T = S_0 \exp\left(\left(r - \frac{\sigma^2}{2}\right)T + \sigma\sqrt{T}x\right), \tag{1.210}$$

where $x \sim N(0, 1)$ and $\sigma > 0$ is the model volatility parameter. As we have seen, the risk-neutral distribution for $S_T$ is defined in such a way as to satisfy the growth condition in equation (1.209)

$$E_0^{Q(Z(T))}[S_T] = \frac{1}{\sqrt{2\pi}} \int_{-\infty}^{\infty} S_0 \exp\left(\left(r - \frac{\sigma^2}{2}\right)T + \sigma\sqrt{T}x\right) e^{-\frac{x^2}{2}} dx = S_0 e^{rT}. \tag{1.211}$$

Two important examples of European-style securities are the *call option struck at K and of maturity T* with price process $C_t$ and payoff function

$$C_T \equiv (S_T - K)_+ \tag{1.212}$$

and the *put option struck at K and of maturity T* with price process $P_t$ and payoff function

$$P_T \equiv (K - S_T)_+. \tag{1.213}$$

**Theorem 1.3. (Put-Call parity).** *If $C_0(S_0, K, T)$ and $P_0(S_0, K, T)$ denote the prices at time $t = 0$ of a plain European call and a plain European put, respectively, both maturing at a later time T and both struck at K, then we have the put-call parity relationship, namely,*

$$C_0(S_0, K, T) - P_0(S_0, K, T) = S_0 - KZ_0(T). \tag{1.214}$$

The proof of the put-call parity relationship descends from the fact that a portfolio with a long position in a call struck at $K$ and maturing at $T$ and a short position in a put struck at $K$ and maturing at $T$ has the same pay-off as a forward contract stipulated at the forward price $K$. (See Section 1.8.)

In contrast to the put-call parity relationship in equation (1.214), the evaluation of the price of a call or put option requires making an assumption on the measure *Q(Z(T))* and the stock price process. Under the Black–Scholes model, where the stock at time $T$ is given by equation (1.210), the expectation $E_0^{Q(Z(T))}[(S_T - K)_+]$ can be reduced to a simple integral. As shown in a detailed derivation in Section 1.6,

$$E_0^{Q(Z(T))}[(S_T - K)_+] = S_0 e^{rT} N(d_+) - KN(d_-), \tag{1.215}$$

where $N(\cdot)$ is the standard cumulative normal distribution function,

$$d_\pm = \frac{\log(S_0/K) + (r \pm \sigma^2/2)T}{\sigma\sqrt{T}}, \tag{1.216}$$

and the pricing formula for a plain European call option (with constant interest rate) in the Black–Scholes model is

$$\begin{aligned} C_{BS}(S_0, K, T, \sigma, r) &= e^{-rT} E_0^{Q(Z(T))}[(S_T - K)_+] \\ &= S_0 N(d_+) - Ke^{-rT} N(d_-). \end{aligned} \tag{1.217}$$

European put options are priced analytically in similar fashion by computing the expectation $e^{-rT} E_0^{Q(Z(T))}[(K - S_T)_+]$, as seen in the derivation of equation (1.172) of Section 1.6. From this formula, or by applying the put-call parity relation (1.214) using equation (1.217), we have the equivalent formulas for the put option price:

$$\begin{aligned} P_{BS}(S_0, K, T, \sigma, r) &= e^{-rT} E_0^{Q(Z(T))}[(K - S_T)_+] \\ &= S_0 N(d_+) - Ke^{-rT} N(d_-) - S_0 + Ke^{-rT} \\ &= Ke^{-rT} N(-d_-) - S_0 N(-d_+). \end{aligned} \tag{1.218}$$

A direct calculation shows that the functions $C_{BS}$ and $P_{BS}$ satisfy the *Black–Scholes partial differential equation* (BSPDE). Analytical and numerical methods for solving this equation are discussed at length throughout later sections and chapters of this book. The numerical projects in Part II provide implementation details for finite-difference lattice approaches to option pricing. A derivation of the BSPDE based on a dynamic replication strategy is provided in Section 1.9 (and a general derivation is given in Section 1.13), but here we simply quote it for the purposes of the present discussion. In terms of the partial derivatives with respect to the time to maturity $T$ and current stock price $S_0$ (with $r$ and $\sigma$ constants) this equation can be rewritten in the form

$$\frac{\partial V}{\partial T} = \frac{\sigma^2 S_0^2}{2} \frac{\partial^2 V}{\partial S_0^2} + rS_0 \frac{\partial V}{\partial S_0} - rV, \tag{1.219}$$

where the option value $V = V(S_0, T)$. The original Black–Scholes equation is really a backward-time equation involving $\partial V/\partial t$ in calendar time $t$, where the price $V$ is expressed in terms of $t$ and equals the pay-off at maturity (or expiry) $t = T$. That is, if we were to express the option value explicitly in terms of such a function of calendar time $t$, then, for example, for the case of a call struck at $K$, $C(S, t = T) = (S - K)_+$. Note that in the present context, however, since we are expressing the option value with respect to the time to maturity, denoted here by the variable $T$, the option price equals the pay-off when $T = 0$ (i.e., at zero time to expiry): $C_{BS}(S, K, T = 0) = (S - K)_+$ and $P_{BS}(S, K, T = 0) = (K - S)_+$, as is easily verified via equations (1.217) and (1.218) in the limit $T \to 0$. Since the Black–Scholes equation is time homogeneous for time-independent interest rate and volatility, option prices are generally functions of $T - t$ (where $t$ and $T \geq t$ represent actual calendar times), so $\partial/\partial t = -\partial/\partial T$

in the original Black–Scholes equation. By replacing $T - t \to T$ (without loss in generality this corresponds to setting current time $t = 0$), we further simplify all expressions, wherein $T$ now represents the time to maturity. The form in equation (1.219) is convenient for the following discussion.

Whether the pricing measure $Q(Z(T))$ is unique or not depends on the choice of hedging instruments. The asset-pricing theorem (in the single-period setting as stated earlier and in the continuous-time case discussed later in this chapter) only implies that — assuming absence of arbitrage — there exists such a measure and that this measure prices all pay-offs. Indeterminacies in $Q(Z(T))$ arise in case there is no perfect replication strategy for the given pay-off, which can be priced independently. The Black–Scholes model provides the most basic pricing model that captures option prices through the single volatility parameter $\sigma$. Since in finance there is no fundamental theory ruling asset price processes, all models are inaccurate to some degree. The Black–Scholes model is perhaps the most inaccurate among all those used, but also the most basic because of its simplicity. Inaccuracies in the Black–Scholes model are captured by the *implied volatility surface*, defined as the function $\sigma_{BS}(K, T)$ such that

$$C_{BS}(S_0, K, T, \sigma_{BS}(K, T), r) = C_0(K, T), \tag{1.220}$$

where $C_0(K, T)$ is the *observed market price* of the call option struck at $K$ and maturing at time $T$. This describes a surface $\sigma_I = \sigma_{BS}(K, T)$ in which the implied volatility $\sigma_I$ is graphed as a function of two variables $K$, $T$, i.e., across a range of strikes $K$ and time to maturity values $T$. For any fixed pair of values $(K,T)$ (and assumed fixed $S_0$, $r$), the function $C_{BS}$ is monotonically increasing in $\sigma$ [see equation (1.222)], hence the preceding equation can be uniquely inverted to give a value for the so-called *Black–Scholes implied volatility* $\sigma_I$ for any observed market price of a call. If the Black–Scholes (i.e., lognormal) model were accurate, the implied volatility surface would be flat and constant, for one single volatility parameter would price all options. Empirical evidence shows that implied volatility surfaces are instead curved (not flat!).

A practical and widely used approach to risk management involving the Black–Scholes pricing formulas is based on the calculation of portfolio sensitivities. Sensitivities of option prices in the Black–Scholes model with respect to changes in the underlying parameters $r, T, S, \sigma$ are of importance to hedging and computing risk for nonlinear portfolios. Within the Black–Scholes formulation, these sensitivities are easily obtained analytically by taking the respective partial derivatives of the European-style option price $V$ for a given pay-off. The list of sensitivities (also known as the *Greeks*) are defined as follows, where we specialize to provide the exact expressions for the case of a plain-vanilla call under the Black–Scholes model:

- The *delta*, denoted by $\Delta$, is defined as the derivative

$$\Delta = \frac{\partial V}{\partial S_0} = \frac{\partial C_{BS}}{\partial S_0} = N(d_+). \tag{1.221}$$

- The *vega*, denoted by $\Lambda$, is defined as the derivative

$$\Lambda = \frac{\partial V}{\partial \sigma} = \frac{\partial C_{BS}}{\partial \sigma} = S_0 \sqrt{T} \frac{e^{-d_+^2/2}}{\sqrt{2\pi}}. \tag{1.222}$$

- The *gamma*, denoted by $\Gamma$, is defined as the second derivative

$$\Gamma = \frac{\partial^2 V}{\partial S_0^2} = \frac{\partial^2 C_{BS}}{\partial S_0^2} = \frac{e^{-d_+^2/2}}{\sigma S_0 \sqrt{2\pi T}}, \tag{1.223}$$

- The *rho*, denoted by $\rho$, is defined as the derivative

$$\rho = \frac{\partial V}{\partial r} = \frac{\partial C_{BS}}{\partial r} = K T e^{-rT} N(d_-), \tag{1.224}$$

- The *theta*, denoted by $\Theta$, is defined as the derivative[12]

$$\Theta = \frac{\partial V}{\partial T} = \frac{\partial C_{BS}}{\partial T} = (\sigma^2 S_0^2/2)\Gamma + r(S_0 \Delta - C_{BS}). \tag{1.225}$$

The numerical project called "The Black–Scholes Model" in Part II provides the interested reader with an in-depth implementation of such formulas for calls as well as for puts and so-called butterfly spread options. The corresponding spreadsheet is then useful for numerically graphing and analyzing the dependence of the various option prices and their sensitivities as functions of either $r$, $\sigma$, $S_0$, $K$, or $T$.

Given the sensitivities, one can approximate the change in price $\delta C$ of a call option due to small changes $T \to T + \delta T$, $S_0 \to S_0 + \delta S_0$, $\sigma \to \sigma + \delta\sigma$, $r \to r + \delta r$ by means of a truncated Taylor expansion,

$$\delta C \cong \Delta(\delta S_0) + \Lambda(\delta\sigma(K,T)) + \frac{1}{2}\Gamma(\delta S_0)^2 + \rho(\delta r) + \Theta(\delta T). \tag{1.226}$$

Here, $\delta S_0$, $\delta r$, $\delta\sigma(K,T)$, and $\delta T$ are small changes in the stock price, the interest rate, the implied Black–Scholes volatility $\sigma = \sigma(K,T)$, and the time to maturity $T$ of the option at hand. In the Black–Scholes model, $\sigma(K,T)$ does not depend on the two arguments and these parameters are constant, so the only source of randomness is the price of the underlying. However, in practice one observes that implied volatilities and interest rates also change over time and affect option values.

As we discuss in more detail in Chapter 4, the risk of option positions is hedged on a portfolio basis and risk-reducing trades are placed in such a way as to decrease portfolio sensitivities to the underlyings. In particular:

- The delta can be reduced by taking a position in the stock or, more commonly, in a forward or futures contract on the stock.
- The vega and gamma can be reduced by taking a position in another option.
- The rho can be reduced by taking a position in a zero-coupon bond of maturity $T$.

## Problems

**Problem 1.** Derive the formulas in equations (1.221)–(1.225).

**Problem 2.** Obtain formulas analoguous to equations (1.221)–(1.225) for the corresponding put option with value $P_{BS}$.

[12] In other literature this is sometimes defined as $-\partial V/\partial T$.

**Problem 3.** Consider a portfolio with positions $\theta_i$ in $N$ securities, each with price $f_i$, $i = 1, \ldots, N$, respectively. Assume the security prices are functions of the same spot $S_0$ at current time $t_0$ and that each price function $f_i = f_i(S_0, T_i - t_0)$ satisfies the time-homogeneous BSPDE with constant interest rate and volatility. The contract maturity dates $T_i$ are allowed to be distinct. Find the relation between the $\Theta$, $\Delta$, and $\Gamma$ of the portfolio.

## 1.8 Static Hedging and Replication of Exotic Pay-Offs

Options other than the calls and puts considered in the previous section are often called *exotic*. In this section, we consider the replication of arbitrary pay-offs via portfolios made up of standard instruments (i.e., consisting of calls, puts, underlying stock, and cash). In finance, such replicating portfolios are useful for the static hedging of European-style options.

A *butterfly spread* option maturing in time $T$ is a portfolio of three calls with current value

$$B_0(S_0, K, T, \epsilon) = \frac{1}{\epsilon^2}(C_0(S_0, K-\epsilon, T) + C_0(S_0, K+\epsilon, T) - 2C_0(S_0, K, T)), \tag{1.227}$$

for some $\epsilon > 0$, where $C_0(S_0, K, T)$ represents the (model-independent) price of a European call with current stock price $S_0$, strike $K$, and time to maturity $T$. We observe that (apart from the normalization constant) this option consists of a long position in a call struck at $K+\epsilon$, a long position in a call struck at $K-\epsilon$, and two short positions in a call struck at $K$, with all calls maturing at the same time. At expiry $T \to 0$ we simply have the payoff function for the butterfly spread:

$$\begin{aligned}
\delta_\epsilon(S_T - K) &= \frac{1}{\epsilon^2}(C_T(S_T, K-\epsilon) + C_T(S_T, K+\epsilon) - 2C_T(S_T, K)) \\
&= \frac{1}{\epsilon^2}\begin{cases} (S_T - (K-\epsilon))_+, & S_T \le K \\ ((K+\epsilon) - S_T)_+, & S_T > K. \end{cases}
\end{aligned} \tag{1.228}$$

Here we have used $C_T(S_T, K) \equiv (S_T - K)_+$ for the pay-off of a call. The normalization factor hence ensures that the area under the graph of the pay-off (as function of $S_T$) is unity, for all choices of $\epsilon$ (see Figure 1.3). In the limit $\epsilon \to 0$, the function $\delta_\epsilon(S_T - K)$ converges to the Dirac delta function $\delta(S_T - K)$ (see Problem 1).

From the one-dimensional version of equation (1.27), we have

$$\lim_{\epsilon \to 0} \int_0^\infty \delta_\epsilon(S_T - K) f(K) dK = \int_0^\infty \delta(S_T - K) f(K) dK = f(S_T), \tag{1.229}$$

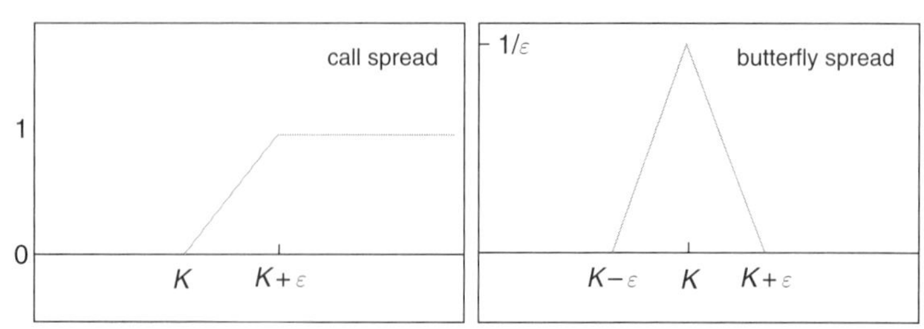


FIGURE 1.3 Payoff functions for a call spread and a corresponding unit butterfly spread struck at $K$, where $2\epsilon$ is the width of the butterfly spread.

for any $S_T > 0$ and any continuous function $f$. From the linearity property of expectations and risk-neutral pricing we must have

$$B_0(S_0, K, T, \epsilon) = e^{-rT} E_0^Q[\delta_\epsilon(S_T - K)]. \tag{1.230}$$

In particular, we find that in the limit $\epsilon \to 0$,

$$\begin{aligned}\lim_{\epsilon\to 0} B_0(S_0, K, T, \epsilon) &= \lim_{\epsilon\to 0} e^{-rT} E_0^Q\big[\delta_\epsilon(S_T - K)\big] \\ &= e^{-rT} \lim_{\epsilon\to 0} \int_0^\infty p(S_0, 0; S_T, T)\delta_\epsilon(S_T - K)dS_T \\ &= e^{-rT} \int_0^\infty p(S_0, 0; S_T, T)\delta(S_T - K)dS_T \\ &= e^{-rT} p(S_0, 0; K, T),\end{aligned} \tag{1.231}$$

where $p(S_0, 0; K, T)$ is the risk-neutral probability density that the stock price $S_T$ equals $K$ at time $t = T$, conditional to its equaling $S_0$ at initial time $t = 0$. This result basically tells us that the price of an infinitely narrow butterfly spread is the price of a so-called Arrow–Debreu security, i.e., the value of a security that pays one unit of account if the stock price (i.e., the state) $S_T = K$ is attained at maturity. One concludes that knowledge of the prices of European calls at all strikes is equivalent to the knowledge of the risk-neutral transition probability density $p(S_0, 0; S_T, T)$ for all $S_T$. Notice, though, that this does not uniquely identify the price process under the risk-neutral measure because all possible transition probabilities $p(S_t, t; K, T)$ for any $t > 0$ are not uniquely determined.[13] By recognizing that equation (1.227) is in fact a representation of the finite difference for the second derivative, we obtain from the last equation

$$\frac{\partial^2 C_0(S_0, K, T)}{\partial K^2} = e^{-rT} p(S_0, 0; K, T). \tag{1.232}$$

We will arrive at this equation again in Section 1.13 when we discuss the Black–Scholes partial differential equation and its dual equation.

Other common portfolios of trades include the following.

- *Covered calls* consist of a long position in the underlying and a short position in a call, typically struck above the spot at the contract inception. This position is meant to trade potential returns above the strike at future time for the option price. The pay-off at the option maturity is

$$S_T - (S_T - K)_+. \tag{1.233}$$

- *Bull spreads* are option spread positions consisting of one long call struck at $K_1$ and one short call struck at $K_2$ with payoff function

$$(S_T - K_1)_+ - (S_T - K_2)_+, \tag{1.234}$$

  $K_1 < K_2$. This portfolio is designed to profit from a rally in the price of the underlying security.

[13] There are in general a variety of models involving jumps, stochastic or state-dependent volatility, or a combination of all that result in the same prices for European options but yield different valuations for path-dependent pay-offs.

- *Bear spreads* are option spread positions in one short put struck at $K_1$ and one long put struck at $K_2$ with payoff function

$$-(K_1 - S_T)_+ + (K_2 - S_T)_+, \tag{1.235}$$

$K_1 < K_2$. This portfolio profits from a decline in price of the underlying security.
- *Digitals* obtain in the limit that $(K_2 - K_1) \to 0$ in a spread option with positions scaled by the strike spread $(K_2 - K_1)^{-1}$. A digital is also called a *binary*. For instance, the pay-off of a bull digital (or digital call) is a unit step function obtained when such a limit is taken in a bull spread with $(K_2 - K_1)^{-1}$ long positions in a call struck at $K_1$ and $(K_2 - K_1)^{-1}$ short positions in a call struck at $K_2$, with $K_1 < K_2$:

$$\theta(S_T - K) = \begin{cases} 1 & \text{if } S_T \geq K \\ 0 & \text{otherwise} \end{cases}. \tag{1.236}$$

The bear digital (or digital put) obtains similarly by considering the limiting case of the bear spread, and the pay-off is $\theta(K - S_T) = 1 - \theta(S_T - K)$, giving 1 if $S_T < K$ and zero otherwise.
- *Wingspreads (also called Condors)* consist of two long and two short positions in calls. These are similar to butterfly spreads, except the body of the payoff function has a flat maximum instead of a vertex; in formulas, the payoff function is

$$(S_T - K_1)_+ - (S_T - K_2)_+ - (S_T - K_3)_+ + (S_T - K_4)_+, \tag{1.237}$$

with $K_1 < K_2 < K_3 < K_4$ and $K_2 - K_1 = K_4 - K_3$.
- *Straddles* involve the simultaneous purchase or sale of an equivalent number of calls and puts on the same underlying with the same strike and same expiration. The straddle buyer speculates that the realized volatility up to the option's maturity will be large and cause large deviations for the price of the underlying asset. The pay-off is

$$(S_T - K)_+ + (K - S_T)_+. \tag{1.238}$$

- *Strangles* are similar to straddles, except the call is struck at a different level than the put; i.e.,

$$(S_T - K_1)_+ + (K_2 - S_T)_+, \tag{1.239}$$

with $K_1 > K_2$ or $K_1 < K_2$. The case $K_1 < K_2$ is an in-the-money strangle, and $K_1 > K_2$ is an out-of-the-money strangle, since the minimum payoff values attained are $K_2 - K_1$ and zero, respectively.
- *Calendar spreads* are spread options where the expiration dates are different and the strike prices are the same, for example:

$$(S_{T_1} - K)_+ - (S_{T_2} - K)_+, \tag{1.240}$$

with $T_1 \neq T_2$. This option strategy is added here for completeness, although it differs from all of the foregoing because the portfolio involves options of varying expiry dates.

Consider the problem of replicating a generic payoff function $\phi(S)$, $0 < S < \infty$, assumed throughout to be twice differentiable. By virtue of equation (1.229), one can achieve replication by means of positions in infinitely narrow butterfly spreads of all possible strikes.

A perhaps more instructive replication strategy involves positions in the underlying stock, a zero-coupon bond and European call options, of all possible strikes and fixed expiration time $T$. Assuming $\phi(0)$ $\phi'(0)$ exist, the formula is

$$\phi(S) = \phi(0) + \phi'(0)S + \int_0^\infty n(K)C_T(S, K)dK. \tag{1.241}$$

$n(K)dK$ represents the size of the position in the call of strike $K$. The function $n(S)$ is related to the payoff function and can be evaluated by differentiating equation (1.241) twice:

$$\phi''(S) = \int_0^\infty n(K)\delta(S-K)dK = n(S). \tag{1.242}$$

Here we make use of the identity

$$\frac{\partial^2}{\partial S^2}(S-K)_+ = \frac{\partial^2}{\partial K^2}(S-K)_+ = \delta(S-K). \tag{1.243}$$

As shown in Problem 3 of this section, equation (1.241) can be derived via an integration-by-parts procedure. The conclusion we can draw is that if calls of all strikes are available, the arbitrage-free price $f_0 = f_0(S_0, T)$ at time $t = 0$ of a contingent European claim with payoff $\phi(S_T)$ at maturity $t = T$ is

$$f_0 = \phi(0)Z_0(T) + \phi'(0)S_0 + \int_0^\infty \phi''(K)C_0(S_0, K, T)dK. \tag{1.244}$$

Besides the basic assumption that asset prices satisfy equation (1.205), it is crucial to point out that the foregoing replication formulas follow without any assumption on the model of the underlying stock motion; i.e., the replication equations are also true by assuming a stochastic process of a more general form that includes the lognormal model as a special case. Moreover, these equations can be extended to apply to a payoff $\phi(S)$ defined on a region $S \in [S_0, S_1]$, where $S_0$, $S_1$ may be taken as either finite or infinite. Specifically, let us consider the space $[S_0, S_1]$, then, using the delta function integration property[14] and assuming $\phi(S_0)$, $\phi'(S_0)$ exist, one can derive

$$\phi(S) = \phi(S_0) + \phi'(S_0)(S - S_0) + \int_{S_0}^{S_1} \phi''(K)(S-K)_+ \, dK. \tag{1.245}$$

The discretized form of this formula reads

$$\phi(S) \approx \phi(S_0) + \phi'(S_0)(S - S_0) + \sum_{i=1}^{N}(\Delta K_i)\phi''(K_i)(S-K_i)_+, \tag{1.246}$$

where $K_i$ are chosen as $S_0 < K_1 < K_2 < \cdots < K_N < S_1$. Let us assume that the strikes are chosen as equally spaced, $\Delta K_i = K_i - K_{i-1} = \Delta K$. Hence, the replication consists of a cash position of size $\phi(S_0) - \phi'(S_0)S_0$, a stock position of size $\phi'(S_0)$, and $N$ call positions of size $(\Delta K_i)\phi''(K_i)$ in calls struck at $K_i$. In most practical cases, this formula actually offers a more accurate discrete representation than the analogous form obtained from discretizing the integral in equation (1.241). This is especially the case when considering a pay-off whose nonzero values are localized to a region $[S_0, S_1]$ for finite $S_1$ or to a region $[S_0, \infty)$, with $S_0 > 0$.

[14] Here one uses the general property $\int_{S-\beta}^{S+\eta} \delta(S-K)\phi(K)dK = \phi(S)$ for any real constants $\beta, \eta > 0$.

This is the situation for pay-offs of the general form $\Lambda(S, X)\mathbf{1}_{\mathcal{A}}$, for some function $\Lambda(S, X)$ with strike $X > 0$. Here $\mathbf{1}_{\mathcal{A}}$ is the indicator function having nonzero value only if condition $\mathcal{A}$ is satisfied. If $\mathcal{A}$ is chosen as the condition $S > X$, then $\mathbf{1}_{S>X} = \theta(S - X)$. The plain European call pay-off obtains with the obvious choice $\Lambda(S, X) = S - X$. It should also be noted that an alternate replication formula involving puts at various strikes (instead of calls) is readily obtained in a manner similar as before or by a simple application of put-call parity (see Problem 6), giving

$$\phi(S) = \phi(S_1) + \phi'(S_1)(S - S_1) + \int_{S_0}^{S_1} \phi''(K)(K - S)_+ \, dK, \tag{1.247}$$

assuming that $\phi(S_1)$, $\phi'(S_1)$ exist.

Note that these formulas assume that the payoff function is well behaved at either the lower endpoint or the upper endpoint. A formula that is valid irrespective of whether the payoff function is singular at either endpoint can be obtained by subdividing the interval $[S_0, S_1]$ into two regions: a lower region $[S_0, \bar{S}]$ and an upper region $[\bar{S}, S_1]$ for any $\bar{S}$ with $S_0 < \bar{S} < S_1$. In the lower region we use puts, while calls are used for the upper region. In particular, via a straightforward integration-by-parts procedure one can derive (see Problem 7)

$$\phi(S) = \phi(\bar{S}) + \phi'(\bar{S})(S - \bar{S}) + \int_{S_0}^{\bar{S}} \phi''(K)(K - S)_+ \, dK + \int_{\bar{S}}^{S_1} \phi''(K)(S - K)_+ \, dK. \tag{1.248}$$

One is then at liberty to choose $\bar{S}$, which acts as a kind of separation boundary for whether calls or puts are used. Note that in the limit $\bar{S} \to S_0$ the formula reduces to that in equation (1.245), with only calls being used, while the opposing limit $\bar{S} \to S_1$ gives equation (1.247), with only puts used for replication. A similar approximate discretization scheme as discussed earlier may be used for these integrals, giving rise to a replication in terms of a finite number of calls and puts at appropriate strikes. This last formula may hence prove advantageous in practice when liquidity issues are present. In particular, this replication can be exploited to better balance the use of available market contracts that are either in-the-money or out-of-the-money puts or calls.

We now give some examples of applications of the foregoing replication theory.

**Example 1. Exponential Pay-Off.**

As a first example, let

$$\phi(S) = (e^{S-X} - 1)_+ = [e^{S-X} - 1]\theta(S - X) = \begin{cases} e^{S-X} - 1, & S \geq X \\ 0, & S < X. \end{cases} \tag{1.249}$$

One can readily verify that this payoff function can be exactly replicated using the right-hand side of either equation (1.241) or equation (1.245) with $S_1 = \infty$. Using $\phi(X) = 0$, $\phi'(K) = \phi''(K) = e^{K-X}$ (for $K > X$), and adopting the replication formula in equation (1.246) with $S_0 = X$ and any $S_1 > X$ gives

$$\phi(S) \approx S - X + \sum_{i=1}^{N} w_i (S - K_i)_+, \tag{1.250}$$

with call positions (i.e., weights) $w_i = (\Delta K)e^{K_i - X}$ and strikes $K_i = X + i\,\Delta K$. Note that one may also use slightly different subdivisions, all of which converge to the same result in the

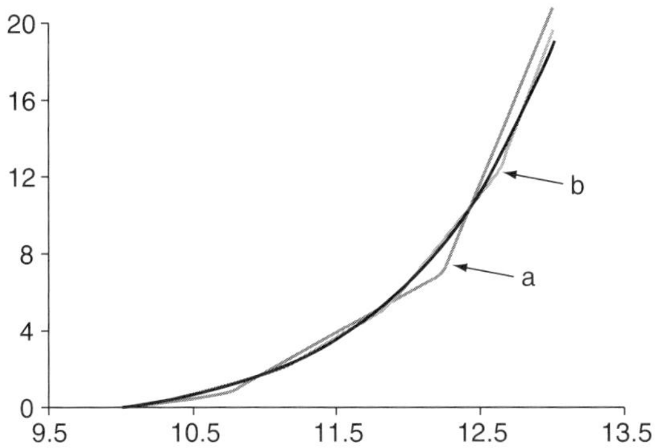


FIGURE 1.4 Rapid convergence of the static replication of the exponential pay-off defined in equation (1.249) (in the region $[X, X+L]$ with $X=10$, $L=3$) using equation (1.250) with a sum of (a) two calls with $K_1 = 10.75$, $K_2 = 12.25$ versus (b) four calls with $K_1 = 10.375$, $K_2 = 11.125$, $K_3 = 11.875$, $K_4 = 12.625$.

limit of infinitesimal spacing $\Delta K \to 0$. Figure 1.4 partly shows the result of this replication strategy in practice. Nearly exact replication is already achieved with only eight strikes.

**Example 2. Sinusoidal Pay-Off.**

Consider the sinusoidal pay-off

$$\phi(S) = \sin\left(\frac{\pi(S-X)}{L}\right)\mathbf{1}_{X \le S \le X+L}, \qquad X, L > 0. \tag{1.251}$$

The choice of strikes $K_i = X + iL/N$, $i = 1, \ldots, N$, with $S_0 = X$ and $S_1 = X + L$, within equation (1.246) gives

$$\phi(S) \approx \frac{\pi}{L}(S-X) + \sum_{i=1}^{N} w_i (S - K_i)_+, \tag{1.252}$$

where $w_i = -(\pi^2/NL)\sin(i\pi/N)$. Figure 1.5 shows the convergence using this replication strategy.

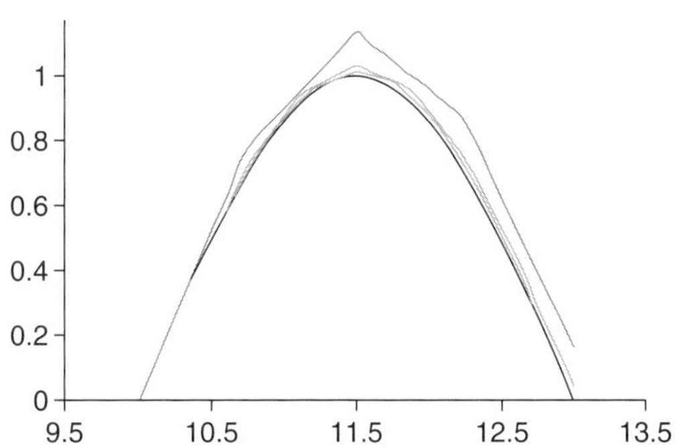


FIGURE 1.5 A comparison of three replication curves and the exact sine pay-off defined in equation (1.251) (in the region $[X, X+L]$ with $X=10$, $L=3$) with $N=4$, $N=8$, and $N=12$ short calls, a long position in the stock, and a short cash position using equation (1.252). With $N=12$ the replication is already very accurate.

**Example 3. Finite Number of Market Strikes.**

In realistic applications there typically is only a select number of strikes available in the market, so the trader has no control over the values of $K_i$ to be used in the replication strategy. In this situation the set of calls (puts) with strikes $K_i$, $i = 1, \ldots, N$, is already given (i.e., preassigned) for some fixed $N$, and the spacing between strikes is not necessarily uniform. A solution to this problem is to consider a slight variation to equation (1.246) and write the finite expansion

$$\phi(S) \approx w_{-1} + w_0 S + \sum_{i=1}^{N} w_i (S - K_i)_+. \tag{1.253}$$

The coefficient $w_{-1}$ gives the cash position, while the weight $w_0$ gives the stock position, and the weights $w_i$ give the positions in the calls struck at values $K_i$. The goal is to find the positions $w_i$ providing the best fit, in the linear least squares sense, as follows. By subdividing the stock price space $[S_0, S_1]$ into $M$ interval slices $S^{(j)}$, with $S^{(j)} < S^{(j+1)}$, $j = 1, \ldots, M$, the $N+2$ positions $w_i$ can be determined by matching the approximate payoff function on the right-hand side of equation (1.253) to the value of the exact payoff function $\phi(S^{(j)})$ at these $M$ stock points. This leads to a linear system of $M$ equations in the $N+2$ unknown weights $w_i$:

$$\phi(S^{(j)}) = w_{-1} + w_0 S^{(j)} + \sum_{i=1}^{N} w_i (S^{(j)} - K_i)_+, \qquad j = 1, \ldots, M. \tag{1.254}$$

One can always make the choice $M \geq N+2$ so that there are at least as many equations as unknown weights. A solution to this system can be found within the linear least squares sense, giving the $w_i$. This technique is fairly robust and also offers a rapidly convergent replication. The reader interested in gaining further experience with the actual numerical implementation of this procedure as applied to logarithmic pay-offs is referred to the numerical project in Part II of this book dealing specifically with the replication of the static component of variance swap contracts.

## Problems

**Problem 1.** A particular representation of the Dirac delta function $\delta(x)$ is given by the limit $\epsilon \to 0$ of the sequence of functions $f_\epsilon(x) = (1/\epsilon^2)(\epsilon - |x|)_+$. Using this fact, demonstrate that the butterfly spread pay-off defined in equation (1.228) gives the Dirac delta function $\delta(S_T - K)$ in the limit $\epsilon \to 0$.

**Problem 2.** Consider the bull spread portfolio with maximum pay-off normalized to unity:

$$\frac{C_T(S, K+\epsilon) - C_T(S, K)}{\epsilon}, \tag{1.255}$$

$C_T(S, K) = (S - K)_+$. Compute the limit $\epsilon \to 0$ and thereby obtain the pay-off of a bull digital.

**Problem 3.** Show that under suitable assumptions on the function $\phi$ [i.e., $\phi(0)$ and $\phi'(0)$ exist] we have

$$\int_0^\infty \phi''(K)(S - K)_+ \, dK = \phi(S) - \phi'(0)S - \phi(0), \tag{1.256}$$

hence verifying equation (1.241). For this purpose use integration by parts twice, together with the property in equation (1.243) as well as the identity

$$\frac{\partial}{\partial S}(S - K)_+ = \theta(S - K), \tag{1.257}$$

where $\theta(x)$ is the Heaviside unit step function having value 1, or 0 for $x \geq 0$, or $x < 0$, respectively. Note that the derivative of this function gives the Dirac delta function.

**Problem 4.** Demonstrate explicitly that the pay-offs of Examples 1 and 2 of this section satisfy equation (1.245) with $S_0 = X$, $S_1 = X + L$, $L > 0$.

**Problem 5.** Assume that calls of all strikes are available for trade and have a known price. Express the present value of the log payoff $\phi(S_T) = \log \frac{S_T + a}{S_0}$, with constant $a > 0$, in terms of call option prices of all strikes $K > 0$. Find a similar expression in terms of put option prices.

**Problem 6.** Apply equation (1.241) to a call payoff $\phi(S) = (S - X)_+$, with constant $X$, to obtain the put-call parity relation

$$(S - X)_+ = S - X + (X - S)_+, \tag{1.258}$$

for all $S > 0$. In deriving this result, the property in equation (1.243) is useful. Now make use of the right-hand side of this put-call parity formula into equation (1.245) and integrate by parts to arrive at equation (1.247).

**Problem 7.** Consider the interval $S \in [S_0, S_1]$. Integrate by parts twice while using the general properties stated earlier for the functions $\theta(x)$, $(x)_+$, and the delta function $\delta(x)$ to arrive at the identities

$$\int_{S_0}^{\bar{S}} \phi''(K)(K - S)_+ dK = \phi(S)\mathbf{1}_{S_0 < S < \bar{S}} - \phi(\bar{S})\theta(\bar{S} - S) + \phi'(\bar{S})(\bar{S} - S)_+ \tag{1.259}$$

and

$$\int_{\bar{S}}^{S_1} \phi''(K)(S - K)_+ dK = \phi(S)\mathbf{1}_{\bar{S} \leq S < S_1} - \phi(\bar{S})\theta(S - \bar{S}) - \phi'(\bar{S})(S - \bar{S})_+ \tag{1.260}$$

where $\mathbf{1}_{\mathcal{D}}$ is the indicator function having unit value for the domain $\mathcal{D}$ and zero otherwise. Add these two expressions to finally obtain equation (1.248).

**Problem 8.** Using risk-neutral valuation, i.e., equation (1.166), derive the Black–Scholes pricing formula for the price of a European digital call and that of a digital put struck at $K$ with time to maturity $T$. For simplicity assume geometric Brownian motion with constant interest rate and volatility. Interpret the meaning of the digital option prices in terms of the price of a standard call. Hint: The derivation of the European digital call boils down to computing the risk-neutral probability $P(S_T \geq K)$, where the algebraic steps are similar to what is used to derive a standard call price.

**Problem 9.** Derive the Greeks $\Delta$, $\Gamma$, and vega for a European digital call.

## 1.9 Continuous-Time Financial Models

In this section, we introduce the basic concepts in continuous-time finance. Derivative claims are structured as contracts written on underlying assets that can be used as hedging instruments. An elegant mathematical structure underlying these financial concepts is reviewed in this section.

In perfect-markets models, a basic asset price process is given by a *money-market account* on which we can deposit and out of which we can borrow without limits. The value at time $t$ of one dollar deposited in a money-market account at initial time $t = 0$ with continuously compounded interest up to time $t$, is denoted by $B_t$.

**Definition 1.10. Money-Market Account.** *Assuming continuous compounding, a money-market account is an asset price process $B_t$ that is monotonically increasing in time, has zero volatility, and follows an equation of the form*

$$dB_t = r_t B_t \, dt, \tag{1.261}$$

*where $r_t$ is a stochastic process that is positive at all times.*[15] *By integrating equation (1.261) we find the stochastic integral representation*

$$B_t = e^{\int_0^t r_s ds}. \tag{1.262}$$

The *instantaneous rate* (or *short rate*) $r_t$ is assumed positive at all times. This is a way to implicitly account for an important restriction: If interest rates were negative, an arbitrage strategy would be to borrow money at negative interest and hold the cash in a safety deposit instead of in an interest-bearing account. Assuming that security costs to store money in a safety deposit are negligible, the existence of such a strategy constrains interest rates to stay positive.

**Definition 1.11. Financial Model: Continuous Time.** *A continuous-time financial model $\mathcal{M} = (\mathcal{F}_t, A_t^1, \ldots, A_t^n)$ is given by a filtration $\mathcal{F}_t$ and n price processes as basic hedging instruments:*

$$(A_t^1, \ldots, A_t^n), \quad t \in \mathbb{R}_+. \tag{1.263}$$

*The value $A_0^i$ can be used to model the current (or spot) price of the ith asset if current time is set as $t = 0$ and the random variable $A_t^i$ models the price of the ith asset at any time $t > 0$.*

**Definition 1.12. Diffusion Pricing Model.** *In a diffusion model the price processes of all hedging instruments (or securities) obey stochastic differential equations of the form*

$$\frac{dA_t^i}{A_t^i} = \mu_t^{A^i} dt + \sum_{\alpha=1}^{M} \sigma_{\alpha,t}^{A^i} \, dW_t^\alpha. \tag{1.264}$$

*Here, the $dW_t^\alpha$, $\alpha = 1, \ldots, M$, are independent Brownian motions (or Wiener processes) with $E[dW_t^\alpha] = 0$ and $E[dW_t^\alpha \, dW_t^\beta] = \delta_{\alpha\beta} \, dt$. The functions $\sigma_{\alpha,t}^{A^i}$ are so-called lognormal volatilities of the ith asset price process $(A_t^i)_{t \geq 0}$ with respect to the αth Brownian motion (i.e., with respect to the αth risk factor), and the functions $\mu_t^{A^i}$ are lognormal drifts of the ith asset price process. These are generally functions of the asset values $A_t^1, \ldots, A_t^n$ and time t.*

Note: We can assume further that one of the assets, e.g., $A_t^1$, is the money-market account, which is the only asset characterized by having zero volatility; in this case $\sigma_{\alpha,t}^{A^1} = 0$ for all $\alpha = 1, \ldots, M$.

**Definition 1.13. Adapted Process.** *A stochastic process $\xi_t$ is adapted to the filtration $\mathcal{F}_t$ if $\xi_t$ is a random variable in the probability space generated by $\mathcal{F}_t$. In other words, the value of $\xi_t$ depends only on the values taken by the paths $(A_s^1, \ldots, A_s^n)$ for $0 \leq s \leq t$, as they were realized up to time t, i.e., $\xi_t$ is $\mathcal{F}_t$-measurable.*

[15] Technically, $B_t$ is of zero quadratic variation because the differential contains no term with $dW_t$; however, $r_t$ can generally be stochastic.

**Definition 1.14. Stopping Time.** *A stopping time* $\tau \in (0, T]$, *for any finite time* $T$, *is an* $\mathcal{F}_t$*-measurable positive random variable such that the time event* $\{t = \tau\}$, *with probability* $P(\tau < \infty) = 1$, *corresponds to a decision to stop and is determined entirely by the information set* $\mathcal{F}_t$ *up to time* $t = \tau$. *That is, given the filtration* $\mathcal{F}_t$ *we know whether or not* $\tau \le t$.

Note that for asset-pricing purposes the information set $\mathcal{F}_t$ basically derives from the set of all asset price paths $(A_t^1, \ldots, A_t^n)$, $0 \le t \le \tau$. This rather technical definition and abstract concept of a stopping time is best illustrated with examples. For instance, let $x_t$ be some real-valued diffusion process (e.g., a Wiener process) and let $[a, b] \subset \mathbb{R}$ be a given fixed finite interval. Assume initially $x_0 \notin [a, b]$ at time $t = 0$ and allow the process to evolve in time $t > 0$ up to time $T$. The random variable defined by

$$\tau = \begin{cases} \min\{t;\ \text{such that } x_t \in [a, b]\}, & \text{if } 0 < t < T \\ T, & \text{otherwise} \end{cases} \tag{1.265}$$

is then a stopping time and corresponds to the first entry time $t < T$ of the process $x_t$ into the interval *[a,b]*. Some basic useful properties of stopping times follow readily, such as additivity: If $\tau_1$ and $\tau_2$ are two stopping times in a given time interval, then $\tau = \tau_1 + \tau_2$ is also a stopping time and, moreover, $\min(\tau_1, \tau_2)$ and $\max(\tau_1, \tau_2)$ are also stopping times. In the pricing of European-style options the expiration time is an example of a stopping time that is actually known at contract inception. In contrast, for American-style options the expiration period (or lifetime of the contract) is still finite, yet there is the added freedom of early exercise. As we shall see in Section 1.14, the early-exercise time is actually an example of an optimal stopping time that is (dynamically) determined by the level of the asset or stock price at the time of early exercise. Other examples of stopping times and derivative instruments are given by *barrier contracts*, for which the pay-off depends on whether or not a certain price process crosses a given barrier in the future. Suppose $H$ is a fixed number, and define $\tau$ as the time $t = \tau$ at which $A_t = H$ for the first time, subject to the initial condition $A_0$. Then $\tau$ is a stopping time. Cash flows for barrier options can occur at the time the barrier is crossed or at maturity. A counterexample to a stopping time is the time $\tau'$, defined as the last time before a given maturity date $T$ for which $A_{\tau'} = H$. $\tau'$ is not a stopping time because knowledge about when $\tau'$ occurs requires information on the full path $x_t$ for all $t \in [0, T]$ and in particular for times after $\tau'$ itself.

**Definition 1.15. Derivative instrument.**[16] *A derivative instrument, or contingent claim, is a contractual agreement between two parties who agree to exchange a cash flow stream in the future, where the cash flow amounts are adapted processes and the timings are stopping times in the given financial model. A discrete cash flow stream is modeled by a sequence of pairs* $(\tau_j, c_j)$, $j = 1, \ldots, m$, *where the* $\tau_j$ *are stopping times and the* $c_j$ *are cash flow amounts depending on the price processes* $(A_t^1, \ldots, A_t^n)$ *up to time* $\tau_j$. *Continuous cash flow streams are modeled by more general adapted processes* $\gamma_t$ *such that* $d\gamma_t$ *is the cash flow occurring in the time interval* $[t, t + dt)$. *In the particular case of a discrete cash flow stream* $(\tau_j, c_j)$, $\tau_j = \tau_1, \ldots, \tau_m < t$, *the continuous-time representation* $c_t$ *is given by*

$$\int_0^t d\gamma_t = \sum_{j=1}^m c_j. \tag{1.266}$$

[16] It should be clearly understood that we are throughout assuming all claims or assets are nondefaultable; e.g., the money-market account is assumed nondefaultable. The definition must be modified in the case of defaultable (credit) derivatives, where pricing depends on time of default and recovery, quantities not directly observable from market-traded instruments.

An example of a continuous cash flow stream is given by exchange-traded futures and options contracts. These contracts have the same final pay-off as forward and ordinary option contracts. However, to reduce credit risk to a minimum, exchanges ask investors to hold a margin account and mark-to-market gains and losses on a daily basis based on realized prices or to unwind the position. This results in a daily stream of cash flows that can be modeled as continuous.

**Definition 1.16. Self-Financing Trading Strategy.** *A self-financing trading strategy in the hedging instruments* $A_t^1, \ldots, A_t^n$ *is a zero cash flow–replicating strategy for all time* $t \in [0, T]$. *That is, this strategy consists of a portfolio of positions* $\xi_t^i$ *in the assets* $A_t^i$, *with value* $V_t = \sum_{i=1}^n \xi_t^i A_t^i$, *where the* $\xi_t^i$, $i = 1, \ldots, n$, *are adapted processes such that at all times* $t \in [0, T]$ *we have*

$$\sum_{i=1}^{n} (A_t^i + dA_t^i) d\xi_t^i = 0. \tag{1.267}$$

The meaning of the self-financing condition is that the cash flow $d\gamma_t$ resulting at time $t + dt$ are reinvested in the underlying assets by adjusting the positions $\xi_{t+dt}^i$ by purchasing or selling the corresponding hedging instruments at the prices $A_t^i + dA_t^i$ at an infinitesimally later time $t + dt$ (i.e., positions are readjusted only after the prices have changed during time $dt$). In this sense the positions are adapted, i.e., nonanticipative with respect to the stochastic changes in the asset prices. The infinitesimal change in the portfolio value $V_t$ of a self-financing strategy is only due to changes in the prices of the underlying instruments since there are no allowed additional cash inflows or outflows after initial time; hence,[17]

$$dV_t = \sum_{i=1}^{n} \xi_t^i dA_t^i. \tag{1.268}$$

In integral form this is written as

$$V_t = V_0 + \sum_{i=1}^{n} \int_0^t \xi_s^i dA_s^i. \tag{1.269}$$

Using Itô's lemma, the change in portfolio value, $dV_t = V_{t+dt} - V_t$, must also satisfy

$$dV_t = \sum_{i=1}^{n} \left[ \xi_t^i dA_t^i + A_t^i d\xi_t^i + (d\xi_t^i)(dA_t^i) \right]. \tag{1.270}$$

Equating these two expressions then gives the self-financing condition rewritten in the form contained in equation (1.267).

**Definition 1.17. Self-Financing Replicating Strategy.** *A self-financing replicating strategy (or perfect hedge) in the hedging instruments* $A_t^1, \ldots, A_t^n$ *that replicates a given cash flow stream* $d\gamma_t$, *where* $\gamma_t$ *is a given contingent claim at time t in some time interval* $t \in [0, T]$, *is defined as a family of adapted processes* $\xi_t^i$, $i = 1, \ldots, n$, *such that at all times* $t \in [0, T]$ *we have*

$$\gamma_t = \gamma_0 + \sum_{i=1}^{n} \int_0^t \xi_s^i \, dA_s^i, \tag{1.271}$$

[17]Note: We assume throughout that the assets do not pay dividends, although in the case of dividends the appropriate formulas extend in a simple manner.

*or, equivalently in differential form,*

$$d\gamma_t = \sum_{i=1}^{n} \xi_t^i \, dA_t^i. \tag{1.272}$$

In the case of a European-style option with payoff $\phi(S_T)$ at time $T$, where $S_t$ is the underlying stock price process, a self-financing replication strategy in the stock and the money-market account, with value $\xi_t^1 B_t + \xi_t^2 S_t$ at time $t$, would satisfy

$$B_t \, d\xi_t^1 + (S_t + dS_t)d\xi_t^2 = 0 \tag{1.273}$$

for all times $t \in [0, T)$. [Note that the term $dB_t = r_t B_t \, dt$ vanishes since it gives rise to a term of $O((dt)d\xi_t^1)$, i.e., of order greater than $dt$.] At time $T$, the position is unwound so that the payout $\phi(S_T)$ [i.e., $\gamma_T = \phi(S_T)$ in this case] is generated; i.e., the portfolio has terminal value

$$\xi_T^1 B_T + \xi_T^2 S_T = \phi(S_T). \tag{1.274}$$

In the case of a barrier or American option, where the payout occurs at a stopping time $0 \le \tau \le T$, the equation (1.273) is valid until time $\tau$, at which point we have

$$B_\tau \xi_\tau^1 + S_\tau \xi_\tau^2 = \phi(S_\tau). \tag{1.275}$$

One of the main problems in pricing theory is whether or not the cash flow streams associated with a contingent claim can be replicated by means of a self-financing trading strategy. If a self-financing trading strategy exists and reproduces all the cash flows of a given contingent claim, then the present value of the cash flow stream can (uniquely in case of no arbitrage) be identified as the cost of setting up the self-financing trading strategy. The question of whether such a self-financing strategy exists relates to attainability and market completeness.

The practical implementation of trading strategies is limited by the existence of transaction costs, by liquidity effects, which pose restrictions on the amounts of a given instrument that can be traded at the posted price, and by the delays with which information reaches market participants. To a first approximation, these effects can be taken into account implicitly by assuming that there are no imperfections. A key role is played by the condition of *absence of arbitrage*, which is stated next and which implies that all portfolios with the same payoff structure have the same price. Asking for absence of arbitrage is a way of accounting for finite market liquidity since, in fact, if an asset had two different prices, trades to exploit the opportunity would cause the prices to realign.

**Definition 1.18. Arbitrage: Continuous Time.** *The self-financing trading strategy* $(\xi_t^1, \ldots, \xi_t^n)$, $0 \le t \le T$, *in the hedging assets* $(A_t^1, \ldots, A_t^n)$ *is an arbitrage strategy if either of the following two conditions holds.*
**A1.** *The portfolio value process*

$$V_t = \sum_{i=1}^{n} \xi_t^i A_t^i \tag{1.276}$$

*is such that* $V_0 < 0$ *and with probability* $P(V_T \ge 0) = 1$.
**A2.** *The value process* $V_t$ *is such that* $V_0 = 0$ *and* $P(V_T > 0) > 0$ *with* $P(V_t \ge 0) = 1$ *for all* $t \in [0, T]$.

In plain language, condition A2 says that an arbitrage opportunity is a self-financed strategy that can generate a profit at zero cost and with no possibility of a loss at any time during the strategy.

Typically, when solving the replication problem for a cash flow stream, the current price of the stream is not known, a priori. Knowledge of the cash flow stream, however, is sufficient because if a trading strategy replicates the cash flows, in virtue of the hypothesis of absence of arbitrage, the value of this strategy at all times yields the price or value process $V_t$. Next we consider a couple of examples of replication (or hedging) strategies. One is static in time; the other is dynamic.

**Example 1. Perpetual Double Barrier Option.**

Suppose there are no carry costs such as interest rates or dividends for holding a position in the stock. Consider a perpetual option with two barriers: a lower barrier at stock value $L$ and an upper barrier at $H$, with $L < H$. If the stock price touches the lower barrier before it touches the upper barrier, the holder receives $R_L$ dollars and the contract terminates. Otherwise, whenever the upper barrier is hit first, the holder receives $R_H$ dollars and the contract terminates. The problem is to find the price and a hedging strategy for this contract.

To solve this problem, let $\tau_L$ be the stopping time for hitting the lower barrier and $\tau_H$ be the stopping time for hitting the higher barrier. The stopping time $\tau$ at which the option expires is the minimum of these times,

$$\tau = \min(\tau_L, \tau_H). \tag{1.277}$$

If one considers a replicating portfolio $f_t = aS_t + b$ at any time $t$, then the barrier levels give rise to two equations:

$$aH + b = R_H, \qquad aL + b = R_L, \tag{1.278}$$

corresponding to the portfolio value (i.e., payout) for hitting either barrier. The value $f_\tau$ of the perpetual double barrier contract evaluated at the stopping time $t = \tau$ is

$$f_\tau = aS_\tau + b. \tag{1.279}$$

Solving the system in equation (1.278) for the portfolio weights $a$ and $b$, we find that

$$a = \frac{R_H - R_L}{H - L}, \qquad b = R_H - aH. \tag{1.280}$$

Absence of arbitrage therefore implies that the price process followed by $f_t$ is given by the value of the portfolio $aS_t + b$ that replicates the cash flows.

**Example 2. Dynamic Hedging in the Black–Scholes Model**

Consider the Black–Scholes model with a stock price following geometric Brownian motion,

$$\frac{dS_t}{S_t} = \mu\, dt + \sigma\, dW_t. \tag{1.281}$$

In this model, the price at time $t$ of a call struck at $K$ and maturity at calendar time $T > t$ is given by the function $C_{BS}(S_t, K, T - t, \sigma, r)$ in equation (1.217). Let's assume that in this economy interest rates are constant and equal to $r$.

One can show that the pay-off of the call can be replicated by means of a self-financing trading strategy that costs $C_t = C_{BS}(S_t, K, T - t, \sigma, r)$ to set up at calendar time $t$. This strategy involves two adapted processes $a_t$ and $b_t$ for the hedge ratios that give the positions at calendar time $t$ in two assets: the stock of price $S_t$ and a zero-coupon bond maturing at time $T$ of price $Z_t(T) = e^{-r(T-t)}$. Namely,

$$C_t = a_t S_t + b_t Z_t(T). \tag{1.282}$$

To show this, we need to find the two processes $a_t$ and $b_t$. Let us note that self-financing condition (1.267) in this case reads

$$(S_t + dS_t)da_t + Z_t(T)db_t = 0. \tag{1.283}$$

By the differential of equation (1.282) and using the self-financing condition we find

$$dC_t = a_t\, dS_t + rb_t Z_t(T)dt. \tag{1.284}$$

On the other hand, applying Itô's lemma (in one dimension) to the price process $C_t$ (considered as function of $t$ and $S_t$) we find

$$dC_t = \left(\frac{\partial C_{BS}}{\partial t} + \frac{\sigma^2 S^2}{2}\frac{\partial^2 C_{BS}}{\partial S^2}\right)dt + \frac{\partial C_{BS}}{\partial S} dS_t,$$

where $S = S_t$. By equating coefficients in $dt$ and $dS_t$ with the previous equation we find

$$a_t = \frac{\partial C_{BS}}{\partial S} \tag{1.285}$$

and

$$rb_t Z_t(T) = \frac{\partial C_{BS}}{\partial t} + \frac{\sigma^2 S^2}{2}\frac{\partial^2 C_{BS}}{\partial S^2}. \tag{1.286}$$

Solving for $b_t$ from replication equation (1.282) gives

$$b_t = Z_t(T)^{-1}(C_t - a_t S_t). \tag{1.287}$$

Substituting $b_t$ as given by equation (1.287), as well as $a_t$ from equation (1.285) into equation (1.286), we arrive at the Black–Scholes partial differential equation in current time $t$ and spot price $S = S_t$:

$$\frac{\partial C_{BS}}{\partial t} + rS\frac{\partial C_{BS}}{\partial S} + \frac{\sigma^2 S^2}{2}\frac{\partial^2 C_{BS}}{\partial S^2} - rC_{BS} = 0 \tag{1.288}$$

This is precisely the equation satisfied by the function $C_{BS}(S_t, K, T-t, \sigma, r)$ given by equation (1.217) with $T \to T-t$.

Notice that the parameter $\mu$ in the equation for the stock price process (1.281) appears in neither the Black–Scholes formula, the Black–Scholes equation, nor the hedge ratios $a_t$ and $b_t$. Section 1.10 provides a more general explanation of this very notable simplification.

## 1.10 Dynamic Hedging and Derivative Asset Pricing in Continuous Time

In this section, we present the main theorem for pricing derivative assets within the continuous-time framework.

**Theorem 1.4. Fundamental Theorem of Asset Pricing (Continuous-Time Case). Part I.** *Consider a diffusion continuous-time financial model $\mathcal{M} = (\mathcal{F}_t, A_t^1, \ldots, A_t^n)$, where the hedging instruments are assumed to satisfy a diffusion equation of the form (1.264), i.e.,*

$$\frac{dA_t^i}{A_t^i} = \mu_t^{A^i} dt + \sum_{\alpha=1}^{M} \sigma_{\alpha,t}^{A^i} dW_t^{\alpha}, \qquad i = 1, \ldots, n, \tag{1.289}$$

*where $dW_t^{\alpha}$ are understood to be standard Brownian increments with respect to a specified probability measure. Also, suppose there exists a money-market account $B_t$ with*

$$dB_t = r_t B_t \, dt. \tag{1.290}$$

*Finally, suppose there are no arbitrage opportunities. Then:*

*(i) Under all equivalent probability measures, there exists a family of adapted processes $q_{\alpha,t}$, $\alpha = 1, \ldots, M$ (one for each risk factor), such that, for any asset price process $A_t$ obeying an equation similar to equation (1.289) with drift $\mu_t^A$ and volatilities $\sigma_{\alpha,t}^A$, the drift term is linked to the corresponding volatilities by the equation*

$$\mu_t^A = r_t + \sum_{\alpha=1}^{M} q_{\alpha,t} \sigma_{\alpha,t}^A, \tag{1.291}$$

*where $q_{\alpha,t}$ are independent of the asset A in question.*

In finance parlance, the adapted processes $q_{\alpha,t}$ are known as the *price of risk* for the $\alpha$th risk factor (or $\alpha$th Brownian motion). Note that this result applies to any asset obeying a diffusion process: In particular, the drifts $\mu_t^{A^i}$ and volatilities $\sigma_{\alpha,t}^{A^i}$ of the base asset prices $A_t^i$ are themselves also linked by an equation similar to equation (1.291), with $q_{\alpha,t}$ independent of the prices $A_t^i$.

**Definition 1.19. Numeraire Asset.** *Any asset $g_t$ whose price process is positive, in the sense that $g_t > 0$ for all t, is chosen as the numeraire for pricing. That is, $g_t$ is an asset price relative to which the value of all other assets $A_t$ are expressed using the ratio $\frac{A_t}{g_t}$.*

**Theorem 1.5. Fundamental Theorem of Asset Pricing (Continuous-Time Case). Part II.** *Under the hypotheses in Part I of the theorem, we have the following: (ii) If $g_t$ is a numeraire asset, then there exists a probability measure $Q(g)$ for which the price $A_t$ at time t of any attainable instrument without cash flows up to a stopping time $\tau > t$ is given by the martingale condition*

$$\frac{A_t}{g_t} = E_t^{Q(g)}\left[\frac{A_\tau}{g_\tau}\right]. \tag{1.292}$$

*Under the measure $Q(g)$ the prices of risk in equation (1.291) for the $\alpha$th factors are given by the volatilities of $g_t$ for the corresponding $\alpha$th factors:*

$$q_{\alpha,t}^g = \sigma_{\alpha,t}^g. \tag{1.293}$$

Note that we are throughout assuming that the contingent claim or derivative instrument to be priced is *attainable*, meaning that one can find a self-financing replicating strategy that exactly replicates the cash flows of the claim. If one also assumes that the financial model satisfies market completeness, then every contingent claim or cash flow stream is assumed attainable.

**Definition 1.20. Pricing Measure: Continuous Time.** *Given a numeraire asset price process $g_t$, the pricing measure associated with $g$ is the martingale measure $Q(g)$ for which pricing formula (1.292) holds for any asset price process $A_t$.*

**Definition 1.21. Risk-Neutral Measure.** *Assuming continuous compounding, the risk-neutral measure $Q(B)$ is the martingale measure with the money-market account as numeraire asset $g_t = B_t = e^{\int_0^t r_s ds}$.*

**Theorem 1.6. Fundamental Theorem of Asset Pricing (Continuous-Time Case). Part III.** *Under the hypotheses in Part I of the theorem, we have the following: (iii) Under the risk-neutral measure $Q(B)$ all the components of the price-of-risk vector, $q^g_{\alpha,t}$, $\alpha = 1, \ldots, M$, vanish, and the drift $\mu^A_t$ of any asset price $A_t$ at time $t$ is equal to the riskless rate $r_t$. The price process for any attainable instrument without cash flows up to any stopping time $\tau > t$ is given by the expectation at time $t$:*

$$A_t = E^{Q(B)}_t \left[ e^{-\int_t^\tau r_u du} A_\tau \right]. \tag{1.294}$$

*(iv) Any attainable price process $A_t$ can be replicated by means of a self-financing trading strategy with portfolio value $V_t = \zeta^{(0)}_t B_t + \sum_{i=1}^n \zeta^{(i)}_t A^i_t$ in the base assets $A^i_t$ and in the money-market account $B_t$:*

$$dA_t = dV_t = \zeta^{(0)}_t r_t B_t \, dt + \sum_{i=1}^n \zeta^{(i)}_t \, dA^i_t, \tag{1.295}$$

*where the positions $\zeta^{(i)}_t$ satisfy the self-financing condition*

$$B_t \, d\zeta^{(0)}_t + \sum_{i=1}^n (A^i_t + dA^i_t) d\zeta^{(i)}_t = 0. \tag{1.296}$$

**Proof.**
**(i)**. Assume no arbitrage and consider a self-financing trading strategy, with components $\zeta^{(1)}_t, \ldots, \zeta^{(n)}_t$ as adapted positions in the family of base assets $A^1_t, \ldots, A^n_t$. Then

$$\sum_{i=1}^n (A^i_t + dA^i_t) d\zeta^{(i)}_t = 0 \tag{1.297}$$

holds. This strategy has portfolio value at time $t$ given by

$$\Pi_t = \sum_{i=1}^n \zeta^{(i)}_t A^i_t. \tag{1.298}$$

This strategy is instantaneously riskless if the stochastic component is zero, i.e., $d\Pi_t = r_t \Pi_t \, dt$. Given our assumptions, a riskless strategy exists and can be explicitly constructed as follows. Using the self-financing condition in equation (1.297) and Itô's lemma for the stochastic differential $d\Pi_t$ we obtain the infinitesimal change in portfolio value in time $[t, t+dt)$:

$$d\Pi_t = \sum_{i=1}^n \left[ (A^i_t + dA^i_t) d\zeta^{(i)}_t + \zeta^{(i)}_t dA^i_t \right] = \sum_{i=1}^n \zeta^{(i)}_t dA^i_t. \tag{1.299}$$

Due to the assumption of no arbitrage, the rate of return on this portfolio over the period $[t, t+dt)$ must equal the riskless rate of return on the money-market account, i.e., $d\Pi_t = r_t \Pi_t \, dt$.[18] Substituting equation (1.289) into the foregoing stochastic differential and setting the coefficients in all the stochastic terms $dW_t^\alpha$ to zero gives

$$\sum_{i=1}^{n} \sigma_{\alpha,t}^{A^i} \zeta_t^{(i)} A_t^i = 0, \tag{1.300}$$

for all $\alpha = 1, \ldots, M$. Here the functions $\sigma_{\alpha,t}^{A^i}$ are volatilities in the $\alpha$th factor for each asset $A^i$. This equation states that the $\mathbb{R}^n$-dimensional vector of components $\zeta_t^{(i)} A_t^i$ is orthogonal to the subspace of $M$ vectors (labeled by $\alpha = 1, \ldots, M$) in $\mathbb{R}^n$ having components $\sigma_{\alpha,t}^{A^i}$, $i = 1, \ldots, n$.

Absence of arbitrage also implies that the portfolio earns a risk-free rate, $d\Pi_t = r_t \Pi_t dt$; hence, setting the drift coefficient in the stochastic differential $d\Pi_t$ to $r_t \Pi_t$ while using equation (1.298) gives this additional condition:

$$\sum_{i=1}^{n} (\mu_t^{A^i} - r_t) \zeta_t^{(i)} A_t^i = 0. \tag{1.301}$$

Here, the quantities $\mu_t^{A^i}$ are drifts for each $i$th asset. Hence equation (1.300) must imply equation (1.301) for all arbitrage-free strategies satisfying the self-financing condition. Equation (1.301) states that the $\mathbb{R}^n$-dimensional vector of components $\zeta_t^{(i)} A_t^i$ must be orthogonal to the $\mathbb{R}^n$-dimensional vector with components $(\mu_t^{A^i} - r_t)$. This means that if the vector with components $\zeta_t^{(i)} A_t^i$ is orthogonal to the $M$ vectors of components $\sigma_{\alpha,t}^{A^i}$, then it is also orthogonal to the vector of components $(\mu_{\alpha,t}^{A^i} - r_t)$. From linear algebra we know that this is possible if and only if the vector of components $(\mu_t^{A^i} - r_t)$ is a linear combination of the $M$ vectors of components $\sigma_{\alpha,t}^{A^i}$ (i.e., is contained in the linear subspace spanned by the $M$ vectors). Hence for any given time $t$, we have

$$\mu_t^{A^i} = r_t + \sum_{\alpha=1}^{M} q_{\alpha,t} \sigma_{\alpha,t}^{A^i}, \tag{1.302}$$

with coefficients $q_{\alpha,t}$ *independent of the asset* $A^i$, for all $i = 1, \ldots, n$. Since this is true for all self-financing strategies and choices of base assets, this implies that the same relation must follow for any asset $A_t$; namely, equation (1.291) obtains.

**(ii)** Let $g$ be a numeraire asset. The measure $Q(g)$ is specified by the condition in equation (1.292). At this point we make use of a previously derived result contained in equation (1.138). Applying that formula now to the quotient $A_t/g_t$, where $A_t$ satisfies an equation of the form (1.264) (with $A^i$ replaced by $A$) and the numeraire asset $g_t$ satisfies a similar equation,

$$\frac{dg_t}{g_t} = \mu_t^g \, dt + \sum_{\alpha=1}^{M} \sigma_{\alpha,t}^g \, dW_t^\alpha, \tag{1.303}$$

[18] A simple argument shows that if the portfolio return is greater than $r_t$, then an arbitrage strategy exists by borrowing money at the lower rate $r_t$ at time $t$ and investing in the portfolio until time $t + dt$. On the other hand, if the portfolio return is less than $r_t$, then an arbitrage strategy also exists by short-selling the portfolio at time $t$ and investing the earnings in the money-market account. Both strategies yield a zero-cost profit.

immediately gives the drift component:

$$E_t\left[d\frac{A_t}{g_t}\right] = \frac{A_t}{g_t}\left[\mu_t^A - \mu_t^g - \sum_{\alpha=1}^{M} \sigma_{\alpha,t}^g(\sigma_{\alpha,t}^A - \sigma_{\alpha,t}^g)\right]dt \tag{1.304}$$

$$= \frac{A_t}{g_t}\sum_{\alpha=1}^{M}(q_{\alpha,t} - \sigma_{\alpha,t}^g)(\sigma_{\alpha,t}^A - \sigma_{\alpha,t}^g)dt. \tag{1.305}$$

In the last equation we have used equation (1.291) for both $g_t$ and $A_t$. In order for the ratio $A_t/g_t$ to be a martingale process for all (arbitrary) choices of the asset $A_t$, this expectation must be zero. This is the case if and only if the process for the price of risk $q_t^\alpha$ is related to the numeraire asset $g_t$, $q_{\alpha,t} = q_{\alpha,t}^g$, as follows:

$$q_{\alpha,t}^g = \sigma_{\alpha,t}^g, \qquad \alpha = 1, \ldots, M. \tag{1.306}$$

That is, the prices of risk $q_t^\alpha$ are equal to the volatilities of the numeraire asset for each respective risk factor.

**(iii)** This is a particular case of (ii) and follows when money-market account $B_t$ is chosen as numeraire asset. Since $dB_t = r_t B_t\, dt$, the prices of risk in this case are all zero, i.e., $q_{\alpha,t}^B = 0$, and therefore $\mu_t^A = r_t$ for all asset price processes $A_t$. In particular, we have that

$$A_t = E_t^{Q(B)}\left[A_T\frac{B_t}{B_T}\right] = E_t^{Q(B)}\left[A_T e^{-\int_t^T r_s ds}\right], \tag{1.307}$$

giving the result. Here we have used the fact that $B_t$ at time $t$ is a known (i.e., nonstochastic) quantity that can be taken inside the expectation.

**(iv)**. Consider the trading strategy with positions $\zeta_t^{(i)}$ in the base assets $A_t^i$. A long position in this trading strategy and a short position in the generic asset $A_t$ is a riskless combination that accrues at the risk-free rate. By adjusting the position in the money-market account $\zeta_0^{(0)}$ so that the trading strategy has the same value of asset $A_0$ at initial time $t = 0$, the resulting trading strategy will track the price process $A_t$ for all times. This trading strategy is also self-financing. In fact

$$dA_t = d\left(\zeta_t^{(0)}B_t + \sum_{i=1}^{n}\zeta_t^{(i)}A_t^i\right)$$

$$= \zeta_t^{(0)}r_t B_t\, dt + B_t\, d\zeta_t^{(0)} + \sum_{i=1}^{n}[(A_t^i + dA_t^i)d\zeta_t^{(i)} + \zeta_t^{(i)}dA_t^i]. \tag{1.308}$$

Hence equation (1.295) obtains from equation (1.296). □

In summary, we observe that the asset pricing theorem is connected to the evaluation of conditional expectations of martingales (i.e., relative asset price processes) within a filtered probability space and under a choice of an equivalent probability measure (also called an *equivalent martingale measure*). A measure is specified by the chosen numeraire asset $g$ obeying a stochastic price process of its own, given by equation (1.303). Given a numeraire $g$, the relative asset price process $A_t/g_t$, for a generic asset price $A_t$, is a martingale under the corresponding measure $Q(g)$. Equivalent martingale measures then arise by considering different choices of numeraire assets. In particular, consider another numeraire asset, denoted by $\tilde{g}$, with price process $\tilde{g}_t$, and suppose that measure $Q(\tilde{g})$ is equivalent to $Q(g)$, then prices computed under any two equivalent measures must be equal:

$$A_t = g_t E_t^{Q(g)}\left[\frac{A_T}{g_T}\right] = \tilde{g}_t E_t^{Q(\tilde{g})}\left[\frac{A_T}{\tilde{g}_T}\right]. \tag{1.309}$$

Rearranging terms gives

$$E_t^{Q(g)}\left[\frac{A_T}{g_T}\right] = \frac{\tilde{g}_t}{g_t} E_t^{Q(\tilde{g})}\left[\frac{A_T}{\tilde{g}_T}\right]$$
$$= E_t^{Q(\tilde{g})}\left[\frac{g_T/\tilde{g}_T}{g_t/\tilde{g}_t}\frac{A_T}{g_T}\right]. \tag{1.310}$$

Note that this holds true for an arbitrary random variable $X_T = A_T/g_T$. We hence obtain the general property under two equivalent measures:

$$E_t^{Q(g)}[X_T] = \rho_t^{-1} E_t^{Q(\tilde{g})}[X_T \rho_T], \tag{1.311}$$

where $\rho_t = g_t/\tilde{g}_t \equiv \left(\frac{dQ(g)}{dQ(\tilde{g})}\right)_t$, $t \in [0, T]$, is a Radon–Nikodym derivative of $Q(g)$ with respect to $Q(\tilde{g})$ (with both measures being restricted to the filtration $\mathcal{F}_t$). For $t = T$ we write $\left(\frac{dQ(g)}{dQ(\tilde{g})}\right)_T = \frac{dQ(g)}{dQ(\tilde{g})}$. Choosing $X_t = 1$ in the foregoing equation shows that $\rho_t$ is also a martingale with respect to $Q(\tilde{g})$.

Let's now fix our choice for one of the numeraires; i.e., let $\tilde{g}_t = B_t$ be the value process of the money-market account so that $Q(\tilde{g}) = Q(B)$ is the risk-neutral measure. Taking the stochastic differential of the quotient process $\rho_t = g_t/B_t$ gives

$$\frac{d\rho_t}{\rho_t} = (\mu_t^g - r_t)dt + \sum_{\alpha=1}^{M} \sigma_{\alpha,t}^g \, dW_t^\alpha. \tag{1.312}$$

Under the risk-neutral measure with $dW_t^\alpha$ as Brownian increments under *Q(B)*, this process must be driftless so that we have $\mu_t^g = r_t$. In particular, this martingale takes the form of an *exponential martingale*,

$$\rho_t = \frac{g_t}{B_t} = \exp\left(-\frac{1}{2}\int_0^t ||\boldsymbol{\sigma}_s^g||^2 ds + \int_0^t \boldsymbol{\sigma}_s^g \cdot d\mathbf{W}_s\right), \tag{1.313}$$

where $||\boldsymbol{\sigma}_s^g||^2 = \boldsymbol{\sigma}_s^g \cdot \boldsymbol{\sigma}_s^g = \sum_{\alpha=1}^M (\sigma_{\alpha,s}^g)^2$ and $\boldsymbol{\sigma}_s^g \cdot d\mathbf{W}_s = \sum_{\alpha=1}^M \sigma_{\alpha,s}^g \, dW_s^\alpha$. At this point we can implement the Girsanov theorem for exponential martingales, which tells us that the $\mathbb{R}^M$-valued vector increment defined by

$$d\mathbf{W}_t^g = -\boldsymbol{\sigma}_t^g \, dt + d\mathbf{W}_t \tag{1.314}$$

is a standard Brownian vector increment under the measure *Q(g)*. In the risk-neutral measure the base assets must all drift at the same risk-free rate,

$$\frac{dA_t^i}{A_t^i} = r_t \, dt + \sum_{\alpha=1}^{M} \sigma_{\alpha,t}^{A^i} \, dW_t^\alpha, \qquad i = 1, \ldots n. \tag{1.315}$$

Substituting for $d\mathbf{W}_t$ using equation (1.314) into this equation and compacting to vector notation gives

$$\frac{dA_t^i}{A_t^i} = (r_t + \boldsymbol{\sigma}_t^g \cdot \boldsymbol{\sigma}_t^{A^i})dt + \boldsymbol{\sigma}_t^{A^i} \cdot d\mathbf{W}_t^g, \qquad i = 1, \ldots, n. \tag{1.316}$$

This last equation is therefore entirely consistent with the formulation presented earlier in terms of the prices of risk. In particular, equation (1.316) is precisely equation (1.289), wherein the Brownian increments are understood to be w.r.t. $Q(g)$, with $g_t$ as an arbitrary choice of numeraire asset-price process. From equation (1.316) we again see that the vector of the prices of risk is $\mathbf{q}_t = \boldsymbol{\sigma}_t^g$. In financial terms, each component of $\mathbf{q}_t$ essentially represents the excess return on the risk-free rate (per unit of risk or volatility for the component risk factor) required by investors in a fair market.

**Example 1. Perpetual Double Barrier Option — Risk-Neutral Measures.**

Reconsider the case of the perpetual double barrier option with zero interest rates discussed previously. The pricing formula for $f_t$ is independent of the real-world stock price drift, although this drift does in fact affect the real-world probability of hitting one barrier before the other. Since interest rates vanish, no discounting is required, and the price process $f_t$ has the following representation under the risk-neutral measure $Q = Q(B)$:

$$f_t = E_t^Q[f_\tau]. \tag{1.317}$$

In this case, the price process $f_t$ is a martingale under the risk-neutral measure because interest rates are zero for all time and the value of the money-market account is constant, i.e., unity. Hence the martingale property gives

$$f_t = R_H \text{ Prob}^Q[S_\tau = H|S_t] + R_L \text{ Prob}^Q[S_\tau = L|S_t], \tag{1.318}$$

where the probabilities are conditional on the current stock price's value $S_t$. These probabilities of hitting either barrier must also sum to unity,

$$\text{Prob}^Q[S_\tau = H|S_t] + \text{Prob}^Q[S_\tau = L|S_t] = 1. \tag{1.319}$$

Note that $f_t = aS_t + b$, where $a$ and $b$ are given by equations (1.280). Hence, the probability of hitting either barrier under the risk-neutral measure can be found by solving equations (1.318) and (1.319). Notice that these probabilities do not depend on the drift of the stock price under the real-world measure.

## Problems

**Problem 1.** Find explicit expressions for the preceding risk-neutral probabilities $P_L = \text{Prob}^Q[S_\tau = L|S_t]$ and $P_H = \text{Prob}^Q[S_\tau = H|S_t]$. Find the limiting expressions for the case that $H >> L$ (i.e., $H \to \infty$ for fixed $L$). What is the price of the perpetual double barrier for this case?

# 11 Hedging with Forwards and Futures

Let $A_t$ be an asset price process for the asset $A$. A *forward contract*, with value $V_t$ at time $t$, on the underlying asset $A$ (e.g., a stock) is a contingent claim with maturity $T$ and pay-off at time $T$ equal to

$$V_T = A_T - F, \tag{1.320}$$

where $F$ is a fixed amount. According to the fundamental theorem of asset pricing (FTAP), the price of this contract at time $t < T$ prior to maturity is equal to $A_t - FZ_t(T)$, where $Z_t(T)$ is the value at calendar time $t$ of a zero-coupon (discount) bond maturing at time $T$. This can be seen in several ways. The first is the following. The payout $A_T$ can be replicated by holding a position in the asset $A$ at all times, while the cash payment $F$ at time $T$ is equivalent to holding a zero-coupon bond of nominal $F$ and maturing at time $T$. Alternatively, to assess the current price $V_t$ of the forward contract using FTAP of Section 1.10, we can evaluate the following expectation at time $t$ of the pay-off under the forward measure with $g_t = Z_t(T)$ as numeraire, giving

$$V_t = Z_t(T) E_t^{Q(Z(T))}[A_T - F] = A_t - F Z_t(T). \tag{1.321}$$

Here we used the facts that at maturity $Z_T(T) = 1$ and that $E_t^{Q(Z(T))}[A_T] = A_t / Z_t(T)$, $E_t^{Q(Z(T))}[F] = F$. The *equilibrium forward price* (at time $t$), denoted by $F_t(A, T)$, is the so-called *forward price* such that the value $V_t$ of the forward contract at time $t$ is zero. Setting $V_t = 0$ in equation (1.321), we find

$$F_t(A, T) = \frac{A_t}{Z_t(T)}. \tag{1.322}$$

Let's assume stochastic interest rates, i.e., a diffusion process for the zero-coupon bond [satisfying equation (1.349) of Problem 1], as well as diffusion processes for the asset $A_t$ [satisfying equation (1.348) of Problem 1] and the equilibrium forward price satisfying

$$\frac{dF_t(A, T)}{F_t(A, T)} = \mu_t^{F(A,T)} dt + \sigma_t^{F(A,T)} dW_t. \tag{1.323}$$

Then a relatively straightforward calculation using Itô's lemma yields the following form for the lognormal volatility of the forward price (see Problem 1 of this section):

$$\sigma_t^{F(A,T)} = \sigma_t^A - \sigma_t^{Z(T)}, \tag{1.324}$$

and its drift

$$\mu_t^{F(A,T)} = \mu_t^A - \mu_t^{Z(T)} - \sigma_t^{Z(T)}(\sigma_t^A - \sigma_t^{Z(T)}), \tag{1.325}$$

where $\sigma_t^{Z(T)}$ is the lognormal volatility of the zero-coupon bond price and $\sigma_t^A$ that of the asset. We note that the foregoing drift and volatility functions are generally functions of the underlying asset price $A_t$, calendar time $t$, and maturity $T$. Moreover, these relationships hold for any choice of numeraire asset $g_t$. As part of Problem 1 of this section, the reader is also asked to derive more explicit expressions for the drifts and volatilities of the forward price under various choices of numeraire.

**Definition 1.22. Futures Contract.** *Futures contracts are characterized by an underlying asset of price process $A_t$ and a maturity $T$. Let us partition the lifetime interval [0,T] in $N$ subintervals of length $\delta t = \frac{T}{N}$. Let $t_i = i \cdot \delta t$ be the endpoints of the intervals. The futures contract with reset period $\delta t$ is characterized by a futures price $F_{t_i}^*(A, T)$ for all $i = 0, \ldots, N$, and at all times $t_i$ the following cash flow occurs at time $t_{i+1}$:*

$$c_{t_{i+1}} = F_{t_{i+1}}^*(A, T) - F_{t_i}^*(A, T). \tag{1.326}$$

*Furthermore, the futures price at time $t_N = T$ equals the asset price $F_T^*(A, T) = A_T$, while at previous times the futures price is set in such a way that the present value of the futures contract is zero.*

Recall that under the risk-neutral measure *Q(B)*, the price of risk is zero (i.e., the numeraire $g_t$ is the money-market account $B_t$ with zero volatility with respect to all risk factors — $\sigma^g_{\alpha,t} = 0$). Hence, according to equation (1.291) of the asset pricing theorem, all asset prices $A_t$ drift at the riskless rate $\mu^A_t = r_t$ under *Q(B)*:

$$\frac{dA_t}{A_t} = r_t\, dt + \sum_{\alpha=1}^{M} \sigma^A_{\alpha,t}\, dW^\alpha_t, \tag{1.327}$$

where we have assumed *M* risk factors or, in the case of one risk factor, we simply have

$$\frac{dA_t}{A_t} = r_t\, dt + \sigma^A_t\, dW_t. \tag{1.328}$$

**Proposition.** *In the limit as $\delta t \to 0$, futures prices behave as (zero-drift) martingales under the risk-neutral measure.*

**Proof.** By definition, the futures price is such that the present value of a futures contract is zero at all reset times *t*, and the cash flows at the subsequent times $t + \delta t$ are given by the random variable $\delta F^*_t(A, T) = F^*_{t+\delta t}(A, T) - F^*_t(A, T)$, so the following condition holds under the risk-neutral measure:

$$E^{Q(B)}_t \left[ \frac{\delta F^*_t(A, T)}{B_{t+\delta t}} \right] = 0, \tag{1.329}$$

where we discount at times $t + \delta t$. Taking the limit $\delta t \to 0$, gives $B^{-1}_t E^{Q(B)}_t [dF^*_t(A, T)] = 0$. Since $B_t \neq 0$, the stochastic differential $dF^*_t(A, T)$ has zero-drift terms for all *t*; i.e., $F^*_t(A, T)$ is a martingale under the measure *Q(B)*, with $E^{Q(B)}_t [dF^*_t(A, T)] = 0$. □

The price spread between futures and forwards is given by

$$F^*_t(A, T) - F_t(A, T) = E^{Q(B)}_t [A_T] - \frac{A_t}{Z_t(T)}, \tag{1.330}$$

with $F^*_T(A, T) = F_T(A, T)$ (i.e., at maturity the two prices are the same). In Chapter 2 we shall derive a formula for this spread based on a simple diffusion model for the asset and discount bond. The topic of stochastic interest rates and bond pricing will be covered in Chapter 2. However, we note here that when interest rates are deterministic (nonstochastic), where $r_t$ is a known ordinary function of *t*, then the discount bond price is simply given by a time integral: $Z_t(T) = \exp(-\int_t^T r_s\, ds) = B_t / B_T$. When interest rates are stochastic (i.e., nondeterministic), as is more generally the case, then we can use equation (1.294) of the asset-pricing theorem, for the case $Z_t(T)$ as asset, to express the discount bond price as an expectation of the payoff $Z_T(T) = 1$ (i.e., the payout of exactly one dollar for certain at maturity) under the measure with the money-market account as numeraire:

$$Z_t(T) = E^{Q(B)}_t [e^{-\int_t^T r_s ds} Z_T(T)] = E^{Q(B)}_t [e^{-\int_t^T r_s ds}]. \tag{1.331}$$

[This expectation is not a simple integral (as arises in the pricing of European options) and can in fact generally be expressed as a multidimensional path integral. See, for example, the project on interest rate trees in Part II.] In the case that the interest rate process is a deterministic function of time or, more generally, when the underlying asset price process $A_t$ is statistically independent of the interest rate process (where both processes may be nondeterminsitic), then forward and future prices coincide and the spread vanishes. In fact, in this case

$$E_t^{Q(B)}[A_T] = \frac{E_t^{Q(B)}[e^{-\int_t^T r_s ds}]}{Z_t(T)} E_t^{Q(B)}[A_T] = \frac{E_t^{Q(B)}[e^{-\int_t^T r_s ds} A_T]}{Z_t(T)} = \frac{A_t}{Z_t(T)}, \tag{1.332}$$

where we have used equations (1.331) and (1.294).

**Definition 1.23. European-Style Futures Options.** *European-style futures options are contracts with a payoff function $\phi(A_T)$ at maturity T. They are similar to the regular earlier European-style option contracts, except those are written on the underlying and traded over the counter with upfront payment, while futures options are traded using a margin account mechanism similar to that of futures contracts. Namely, futures options are traded in terms of a futures option price $A_t^*$ that equals $\phi(A_T)$ at maturity $t = T$, while the associated cash flow stream to the holder's margin account is given by*

$$c_t = A_t^* - A_{t-dt}^*. \tag{1.333}$$

Notice that, similar to an ordinary futures contract, futures option prices $A_t^*$ follow martingale processes under the risk-neutral measure.

**Example 1. European Futures Options.**

The futures option price $V_t^*$ for a European-style option with payoff function $\phi(A_T)$ is thus given by the martingale condition

$$V_t^* = E_t^{Q(B)}[\phi(A_T)]. \tag{1.334}$$

The analogue of the Black–Scholes (i.e., lognormal) model can be written as follows under the risk-neutral measure

$$dA_t^* = \sigma A_t^* \, dW_t, \tag{1.335}$$

where the drift is zero because of the martingale property. We remark here that, in case interest rates are stochastic, the implied Black–Scholes volatility on the futures option does not necessarily coincide with the implied Black–Scholes volatility for plain vanilla equivalents.

Let $A_t^* = F_t^*(A, T)$ be the futures price on the asset. At maturity we have $A_T = F_T^*(A, T) = F_T(A, T)$. The pricing formula for a futures call option struck at the futures price $K$ is given by

$$C_t^*(K, T) = E_t^{Q(B)}[(A_T - K)_+] = F_t^*(A, T)N(d_+) - KN(d_-), \tag{1.336}$$

where

$$d_\pm = \frac{\log(F_t^*(A, T)/K) \pm (\sigma^2/2)(T-t)}{\sigma\sqrt{T-t}} \tag{1.337}$$

and we have used the standard expectation formula in equation (1.169) for the case of zero drift, and where the underlying variable $S_t$ in that formula is now replaced by $F_t^*(A, T)$. Notice that this formula carries no explicit dependence on interest rates.

**Example 2. Variance Swaps.**

An example of a dynamic trading strategy involving futures contracts and the static hedging strategies discussed in Section 1.8 is provided by variance swaps. Variance swaps are defined as contracts yielding the pay-off at maturity time $T$:

$$\mathcal{N}\left[\frac{1}{T}\int_0^T \sigma_t^2\, dt - \Sigma^2\right], \tag{1.338}$$

where $\mathcal{N}$ is a fixed notional amount in dollars per annualized variance. Assuming that technical upfront fees are negligible, variance swaps are priced by specifying the variance $\Sigma^2$, which, as we show, is computed in such a way that the value of the variance swap contract is zero at contract inception ($t = 0$); i.e., since this is structured as a forward contract, it must have zero initial cost. Computing the expectation of the pay-off at initial time $t = 0$ and setting this to zero therefore gives the fair value of this variance in terms of a stochastic integral:

$$\Sigma^2 = \frac{1}{T} E_0^{Q(B)}\left[\int_0^T \sigma_t^2\, dt\right]. \tag{1.339}$$

We shall compute this expectation by recasting the integrand as follows. Assuming a diffusion process for futures prices and assuming that European call and put options of all strikes and maturity $T$ are available, such a contract can be replicated exactly.[19]

More precisely, assume that futures prices $F_t^* \equiv F_t^*(A, T)$ on a contract maturing at time $T$ with underlying asset price $A_t$ (e.g., a stock price) at time $t$ obeys the following zero-drift process under the risk-neutral measure $Q(B)$:

$$\frac{dF_t^*}{F_t^*} = \sigma_t\, dW_t, \tag{1.340}$$

where the volatility $\sigma_t$ is a random process that can generally depend on time as well as on other stochastic variables.

Then consider the dynamic trading strategy, whereby at time $t$ one holds $\frac{1}{F_t^*}$ futures contracts. If one starts implementing the strategy at time $t = 0$ and accumulates all the gains and losses from the futures position into a money-market account, then the worth $\Pi_T$ accumulated at time $T$ is

$$\Pi_T = \int_0^T \frac{dF_t^*}{F_t^*} = \int_0^T \sigma_t\, dW_t. \tag{1.341}$$

Due to Itô's lemma we have

$$d\log F_t^* = \frac{dF_t^*}{F_t^*} - \frac{1}{2}\left(\frac{dF_t^*}{F_t^*}\right)^2 = \frac{dF_t^*}{F_t^*} - \frac{\sigma_t^2}{2}dt, \tag{1.342}$$

and integrating from time $t = 0$ to $T$ we find

$$\log F_T^* - \log F_0^* = \int_0^T \sigma_t\, dW_t - \frac{1}{2}\int_0^T \sigma_t^2\, dt = \Pi_T - \frac{1}{2}\int_0^T \sigma_t^2\, dt, \tag{1.343}$$

[19] We point out that in actuality the price of a variance swap is largely model independent. That is, it is possible to replicate the cash flows as long as the trader can set up a static hedge and trade futures on the underlying.

where equation (1.340) has been used. Rearranging this equation gives the integrand in equation (1.339) as

$$\frac{1}{T}\int_0^T \sigma_t^2\, dt = \frac{2}{T}\Pi_T - \frac{2}{T}\log\frac{F_T^*}{F_0^*}. \tag{1.344}$$

This last expression demonstrates the precise nature of the replication. This contains (i) a static part given by the logarithmic payoff function and (ii) a dynamic part given by the stochastic time integral $\Pi_T$. Substituting this last expression into equation (1.339) and using the fact that $\Pi_t$ is a martingale,[20] i.e., $E_0^{Q(B)}[\Pi_T]=0$, we obtain

$$\Sigma^2 = -\frac{2}{T}E_0^{Q(B)}\left[\log\frac{F_T^*}{F_0^*}\right]. \tag{1.345}$$

Replicating the logarithmic payoff function in terms of standard call and/or put pay-offs of various strikes using the replication schemes described in Section 1.8 then gives a formula for $\Sigma^2$ in terms of futures calls and/or puts. In particular, by applying replication equation (1.248) on the domain $F_T^* \in (0,\infty)$ and taking expectations, equation (1.345) takes the form (see Problem 2)

$$\Sigma^2 = \frac{2}{T}\left[1 - \frac{F_0^*}{\bar{F}} - \log\frac{\bar{F}}{F_0^*} + \int_0^{\bar{F}} P_0^*(K,T)\frac{dK}{K^2} + \int_{\bar{F}}^{\infty} C_0^*(K,T)\frac{dK}{K^2}\right], \tag{1.346}$$

with any choice of nonzero parameter $\bar{F}\in(0,\infty)$, and where $C_0^*(K,T)$ and $P_0^*(K,T)$ represent the current $t=0$ prices of a futures call and put option, respectively, at strike $K$ and maturity $T$. Note that this formula holds irrespective of what particular assumed form for the volatility $\sigma_t$. In the cases of analytically solvable diffusion models, such as some classes of state-dependent models studied in Chapter 3, the call and put options can be expressed in closed analytical form. Of course, if $\sigma_t = \sigma(t)$, i.e., a deterministic function of only time, then the futures price obeys a geometric Brownian motion, and in this case, according to our previous analysis, we have simple analytical expressions of the Black–Scholes type, with $C_t^*(K,T)$ given by equation (1.336), and

$$P_t^*(K,T) = E_t^{Q(B)}[(K - F_T^*)_+] = KN(-d_-) - F_t^*(A,T)N(-d_+), \tag{1.347}$$

with $d_\pm$ given by equation (1.337), wherein $\sigma \to \bar{\sigma} \equiv \sqrt{(T-t)^{-1}\int_t^T \sigma^2(s)ds}$. For a numerical implementation of the efficient replication of logarithmic pay-offs for variance swaps in cases where only a select number of market call contracts is assumed available, the reader is encouraged to complete the project on variance swaps in Part II.

## Problems

**Problem 1.** Derive the equations for the drift and volatility of the forward price as discussed in this section. For the domestic asset assume the process

$$\frac{dA_t}{A_t} = \mu_t^A dt + \sigma_t^A\, dW_t. \tag{1.348}$$

[20] Here we recall the property for the first moment $E_0[\int_0^t f_s\, dW_s]=0$, which is valid under a suitable measure and conditions on the adapted process $f_t$.

Let $Z_t(T)$ be the price process of a domestic discount bond of maturity $T$. For any fixed maturity $T > t$, the discount bond price process is assumed to obey a stochastic differential equation of the form

$$\frac{dZ_t(T)}{Z_t(T)} = \mu^Z dt + \sigma^Z dW_t, \tag{1.349}$$

where shorthand notation is used ($\mu^Z \equiv \mu_t^{Z(T)}$, $\sigma^Z \equiv \sigma_t^{Z(T)}$) to denote the lognormal drift and volatility functions of the discount bond. Find the drift of the forward price process $F_t(A, T)$, defined by equation (1.322), within the following three different choices of numeraire asset $g_t$: (i) the money-market account: $g_t = B_t = e^{\int_0^t r_s ds}$, where $r_t$ is the domestic short rate at time $t$, (ii) the discount bond: $g_t = Z_t(T)$, and (iii) the asset : $g_t = A_t$. Hint: Make use of the formula for the stochastic differential of a quotient of two processes that was derived in a previous section.

**Problem 2.** Use equation (1.248) with payoff function $\phi(F) = -\log \frac{F}{F_0^*}$, $F \equiv F_T^*$, $\bar{S} = \bar{F}$, $S_0 = 0$, $S_1 = \infty$, with $0 < \bar{F} < \infty$, to show

$$\phi(F) = 1 - \frac{F}{\bar{F}} - \log \frac{\bar{F}}{F_0^*} + \int_0^{\bar{F}} (K - F)_+ \frac{dK}{K^2} + \int_{\bar{F}}^{\infty} (F - K)_+ \frac{dK}{K^2}. \tag{1.350}$$

Now, arrive at the formula in equation (1.346) by taking the expectation of this pay-off at $t = 0$ under the measure *Q(B)* while making use of the fact that an expectation can be taken inside any integral over $K$ and the fact that $E_t^{Q(B)}[F_T^*] = F_t^*$, i.e., that $F_t^*$ is a martingale within this measure.

## 12 Pricing Formulas of the Black–Scholes Type

In this section we apply the fundamental theorem of asset pricing of Section 1.10 to derive a few exact pricing formulas. The worked-out examples are meant to demonstrate the use of different numeraire assets for option pricing.

**Example 1. Plain European Call Option.**

As a first example, let's revisit the problem of pricing the plain European call. Consider the Black–Scholes model (i.e., geometric Brownian motion) for a stock of constant volatility $\sigma$ and in an economy with a constant interest rate $r$. Under the risk-neutral measure with money-market account $g_t = B_t = e^{rt}$ as numeraire, the expected return on the stock is just the risk-free rate $r$; hence,

$$dS_t = rS_t\, dt + \sigma S_t\, dW_t. \tag{1.351}$$

The stock price process is given in terms of a standard normal random variable [i.e., equation (1.154)]: $S_T = S_t e^{\left(r - \frac{\sigma^2}{2}\right)(T-t) + \sigma\sqrt{T-t}x}$, $x \sim N(0, 1)$. Using equation (1.292), the arbitrage-free price at time $t$ of a European call option struck at $K > 0$ with maturity $T > t$ is hence the discounted expectation under the risk-neutral measure *Q(B)*:

$$\begin{aligned} C_t(S_t, K, T) &= e^{-r(T-t)} E_t^{Q(B)}[(S_T - K)_+] \\ &= \frac{e^{-r(T-t)}}{\sqrt{2\pi}} \int_{-\infty}^{\infty} e^{-\frac{x^2}{2}} \left( S_t e^{\left(r - \frac{\sigma^2}{2}\right)(T-t) + \sigma\sqrt{T-t}x} - K \right)_+ dx \\ &= S_t N(d_+) - Ke^{-r(T-t)} N(d_-), \end{aligned} \tag{1.352}$$

where

$$d_\pm = \frac{\log(S_t/K) + \left(r \pm \frac{1}{2}\sigma^2\right)(T-t)}{\sigma\sqrt{T-t}}. \tag{1.353}$$

Note that the details of this integral expectation were presented in Section 1.6.

This Black–Scholes pricing formula plays a particularly important role because it is the prototype for a large number of pricing formulas. As we shall see in a number of examples in this and the following chapters, analytically solvable pricing problems for European-style options often lead to pricing formulas of a similar structure. In the case that the underlying asset pays *continuous dividends*, the foregoing pricing formula for a European call (and the corresponding put) must be slightly modified. A similar derivation procedure also applies, as shown at the end of this section.

If the drift and the volatility are deterministic functions of time, $r = r(t)$ and $\sigma = \sigma(t)$, the Black–Scholes formula extends thanks to the formula in equation (1.167) of Section 1.6. Using again the money-market account $g_t = B_t = \exp(\int_0^t r(s)ds)$ as numeraire asset and setting

$$\bar{r}(t, T) = \frac{1}{(T-t)} \int_t^T r(u)du$$

gives $B_t/B_T = e^{-\bar{r}(t,T)(T-t)}$, and we find

$$\begin{aligned} C_t(S_t, K, T) &= e^{-\bar{r}(t,T)(T-t)} E_t^{Q(B)}\left[(S_T - K)_+\right] \\ &= \frac{e^{-\bar{r}(t,T)(T-t)}}{\sqrt{2\pi}} \int_{-\infty}^{\infty} e^{-\frac{x^2}{2}} \left( S_t e^{\left(\bar{r}(t,T) - \frac{\bar{\sigma}(t,T)^2}{2}\right)(T-t) + \bar{\sigma}(t,T)\sqrt{T-t}x} - K \right)_+ dx \\ &= S_t N(d_+) - K e^{-\bar{r}(t,T)(T-t)} N(d_-), \end{aligned} \tag{1.354}$$

where

$$\bar{\sigma}(t, T)^2 = \frac{1}{(T-t)} \int_t^T \sigma(t)^2 \, dt, \tag{1.355}$$

and

$$d_\pm = \frac{\log(S_t/K) + \left(\bar{r}(t, T) \pm \frac{1}{2}\bar{\sigma}(t, T)^2\right)(T-t)}{\bar{\sigma}(t, T)\sqrt{T-t}}. \tag{1.356}$$

Note that, in agreement with the results obtained in Section 1.6, the Black–Scholes pricing formula now involves the time-averaged interest rate and volatility over the maturity time $T - t$.

**Example 2. A Currency Option.**

Let

$$dX_t = \mu_X X_t \, dt + \sigma_X X_t \, dW_t \tag{1.357}$$

be a model for the foreign exchange rate $X_t$ at time $t$, assuming that the lognormal volatility $\sigma_X$ of the exchange rate and drift $\mu_X$ are constants. Suppose that the domestic risk-free interest rate $r^d$ and the foreign interest rate $r^f$ are both constant, and let $B_t^d = e^{r^d t}$ and $B_t^f = e^{r^f t}$ be the worth of the two money-market accounts, respectively. The drift $\mu_X$ can be computed as follows. First we note that the foreign currency money-market account, after conversion into domestic currency, is a domestic asset and therefore must obey a price process of the form

$$d(X_t B_t^f) = (r^d + \sigma_g \sigma_{XB^f})(X_t B_t^f)dt + \sigma_{XB^f}(X_t B_t^f)dW_t, \tag{1.358}$$

where $\sigma_g$ and $\sigma_{XB^f}$ are lognormal volatilities of the numeraire $g_t$ and $X_t B_t^f$, respectively. We shall choose $g_t = B_t^d$ (i.e., the domestic risk-neutral measure) giving $\sigma_g = 0$. By direct application of Itô's lemma for the product of two processes we also have the stochastic differential

$$d(X_t B_t^f) = X_t\, dB_t^f + B_t^f\, dX_t + (dX_t)(dB_t^f) = X_t\, dB_t^f + B_t^f\, dX_t, \tag{1.359}$$

where the third term in the middle expression is of order $dt\, dW_t$ and hence set to zero. This follows since both domestic and foreign money-market accounts satisfy a deterministic differential equation, in particular,

$$dB_t^f = r^f B_t^f dt. \tag{1.360}$$

Plugging this and equation (1.357) into equation (1.359) gives

$$d(X_t B_t^f) = (r^f + \mu_X)(X_t B_t^f)dt + \sigma_X (X_t B_t^f)dW_t. \tag{1.361}$$

Hence, comparing equations (1.358) and (1.361) gives $\mu_X = r^d - r^f$. The foreign exchange rate therefore follows a geometric Brownian motion with this constant drift and constant volatility $\sigma_X$. The pricing formula for a foreign exchange call option struck at exchange rate $K$ is then

$$\begin{aligned} C_t(X_t, K, T) &= e^{-r^d(T-t)} E_t^{Q(B^d)}\left[(X_T - K)_+\right] \\ &= \frac{e^{-r^d(T-t)}}{\sqrt{2\pi}} \int_{-\infty}^{\infty} e^{-\frac{x^2}{2}} \left( X_t e^{((r^d - r^f) - \frac{\sigma_X^2}{2})(T-t) + \sigma_X \sqrt{T-t}x} - K \right)_+ dx \\ &= e^{-r^d(T-t)}\left[e^{(r^d - r^f)(T-t)} X_t N(d_+) - K N(d_-)\right], \\ &= e^{-r^f(T-t)} X_t N(d_+) - K e^{-r^d(T-t)} N(d_-), \end{aligned} \tag{1.362}$$

where

$$d_\pm = \frac{\log(X_t/K) + (r^d - r^f \pm \frac{1}{2}\sigma_X^2)(T-t)}{\sigma_X \sqrt{T-t}}. \tag{1.363}$$

**Example 3. A Quanto Option.**

Consider the case of a quanto option, in which we have a stock denominated in a foreign currency with geometric Brownian process

$$dS_t^f = \mu S_t^f\, dt + \sigma_S S_t^f\, dW_t^S, \tag{1.364}$$

and the foreign exchange process is also a geometric Brownian motion, with

$$dX_t = (r^d - r^f)X_t\, dt + \sigma_X X_t\, dW_t^X, \tag{1.365}$$

under the risk-neutral measure with numeraire $g_t = B_t^d$. Note that the drift rate $\mu_X = r^d - r^f$ was derived in the previous example. The constants $\sigma_S$ and $\sigma_X$ are the lognormal volatilities of the stock and foreign exchange rate, respectively. These Brownian increments are not independent; however, the foregoing equations can also be written equivalently in terms of two independent Brownian increments $dW_t^1$, $dW_t^2$, where

$$dW_t^X = \rho\, dW_t^1 + \sqrt{1-\rho^2}\, dW_t^2, \qquad dW_t^S = dW_t^1.$$

Here $\rho$ is a correlation between the stock price and the foreign exchange rate at time $t$, with

$$dW^S\, dW^X = \rho\, dt. \tag{1.366}$$

In vector notation, $d\mathbf{W}_t = (dW_t^1, dW_t^2)$ and

$$\frac{dX_t}{X_t} = (r^d - r^f)dt + \boldsymbol{\sigma}_X \cdot d\mathbf{W}_t, \tag{1.367}$$

$$\frac{dS_t^f}{S_t^f} = \mu dt + \boldsymbol{\sigma}_S \cdot d\mathbf{W}_t, \tag{1.368}$$

where $\boldsymbol{\sigma}_X = (\rho\sigma_X, \sigma_X\sqrt{1-\rho^2})$, $\boldsymbol{\sigma}_S = (\sigma_S, 0)$. Suppose one wants to price a call option on the stock $S_t^f$ struck at $K$ and then to convert this into domestic currency at a preassigned fixed rate $\bar{X}$. Since $g_t = B_t^d$, the prices of all domestic assets (as well as the prices of foreign assets denominated in domestic currency) drift at the domestic risk-free rate. Hence the return on the price process $X_t S_t^f$ must be $r^d$. This also follows because the price of risk $q^g = q^{B^d} = \sigma_{B^d} = 0$. By direct application of Itô's lemma we also have

$$\frac{d(X_t S_t^f)}{X_t S_t^f} = \frac{dS_t^f}{S_t^f} + \frac{dX_t}{X_t} + \frac{dS_t^f}{S_t^f}\frac{dX_t}{X_t}. \tag{1.369}$$

Plugging the preceding expressions into this equation gives

$$\begin{aligned}\frac{d(X_t S_t^f)}{X_t S_t^f} &= (\mu + r^d - r^f + \boldsymbol{\sigma}_X \cdot \boldsymbol{\sigma}_S)dt + (\boldsymbol{\sigma}_X + \boldsymbol{\sigma}_S)\cdot d\mathbf{W}_t \\ &= (\mu + r^d - r^f + \rho\sigma_X\sigma_S)dt + \sigma_S dW_t^S + \sigma_X dW_t^X \end{aligned} \tag{1.370}$$

Since the drift must equal $r^d$,

$$\mu = r^f - \rho\sigma_S\sigma_X \tag{1.371}$$

is the constant drift of $S_t^f$ in equation (1.364). The arbitrage-free price of a quanto call option struck at foreign price $K$ is then given by

$$\begin{aligned} C_t(S_t^f, K, T) &= \bar{X}e^{-r^d(T-t)}E_t^{Q(B^d)}\left[(S_T^f - K)_+\right] \\ &= \bar{X}e^{-r^d(T-t)}\left[e^{\mu(T-t)}S_t^f N(d_+) - KN(d_-)\right] \\ &= \bar{X}\left[e^{-(r^d - r^f + \rho\sigma_S\sigma_X)(T-t)}S_t^f N(d_+) - e^{-r^d(T-t)}KN(d_-)\right], \end{aligned} \tag{1.372}$$

where

$$d_\pm = \frac{\log(S_t^f/K) + \left(r^f - \rho\sigma_S\sigma_X \pm \frac{1}{2}\sigma_S^2\right)(T-t)}{\sigma_S\sqrt{T-t}}. \tag{1.373}$$

**Example 4. Elf-X Option (Equity-Linked Foreign Exchange Option).**

Assume equation (1.364), as in the previous example, and now write

$$dX_t = \mu_X X_t\, dt + \sigma_X X_t\, dW_t^X \tag{1.374}$$

for the foreign exchange process, with $\mu_X$ dependent on the choice of numeraire. Consider the case where the pay-off is

$$C_T = (X_T - K)_+ S_T^f. \tag{1.375}$$

The foreign asset price $S_t^f$ cannot be used as a domestic numeraire asset, but the converted process $g_t = X_t S_t^f$ can. Indeed this is a positive price process denominated in domestic currency. Under the measure with $g_t$ as numeraire we first need to compute the drift $\mu_X$ explicitly. This is done by considering the process $X_t B_t^f$, which must drift at the domestic risk-free rate plus a price-of-risk component

$$\frac{d(X_t B_t^f)}{X_t B_t^f} = (r^d + \boldsymbol{\sigma}_{XS^f} \cdot \boldsymbol{\sigma}_{XB^f})dt + \boldsymbol{\sigma}_{XB^f} \cdot d\mathbf{W}_t, \tag{1.376}$$

where $\boldsymbol{\sigma}_{XS^f}$ and $\boldsymbol{\sigma}_{XB^f}$ are volatility vectors of the price processes $X_t S_t^f$ and $X_t B_t^f$, respectively. These are expressible in the basis of either $(dW_t^1, dW_t^2)$ or $(dW_t^S, dW_t^X)$, as described in the previous example. [Note also that the Brownian increments, written still as $d\mathbf{W}_t$ in the SDE are actually w.r.t. the measure $Q(XS^f)$.] From equation (1.370) we have $\boldsymbol{\sigma}_{XS^f} = \boldsymbol{\sigma}_X + \boldsymbol{\sigma}_S$. From a direct application of Itô's lemma we also have

$$\frac{d(X_t B_t^f)}{X_t B_t^f} = (r^f + \mu_X)dt + \boldsymbol{\sigma}_X \cdot d\mathbf{W}_t. \tag{1.377}$$

By equating drifts and the volatility vectors in these two expressions we find $\boldsymbol{\sigma}_{XB^f} = \boldsymbol{\sigma}_X$ and

$$r^d + (\boldsymbol{\sigma}_S + \boldsymbol{\sigma}_X) \cdot \boldsymbol{\sigma}_X = r^f + \mu_X. \tag{1.378}$$

Hence,

$$\mu_X = r^d - r^f + \boldsymbol{\sigma}_S \cdot \boldsymbol{\sigma}_X + ||\boldsymbol{\sigma}_X||^2.$$

The drift $\mu_{X^{-1}}$ and volatility of the inverse exchange rate $X_t^{-1}$ under the same measure are computed using Itô's lemma [i.e., apply equation (1.138) with numerator = 1 and denominator = $X_t$]:

$$\frac{dX_t^{-1}}{X_t^{-1}} = (-\mu_X + \sigma_X^2)dt - \sigma_X\, dW_t^X.$$

Hence,

$$\mu_{X^{-1}} = -\mu_X + \sigma_X^2 = r^f - r^d - \boldsymbol{\sigma}_S \cdot \boldsymbol{\sigma}_X = r^f - r^d - \rho\sigma_X\sigma_S,$$

where the square of the volatility is the same as that of $X_t$, namely $\sigma_X^2$. Using the measure $Q(XS^f)$, we therefore have the arbitrage-free price:

$$\begin{aligned} C_t(S_t^f, X_t, K, T) &= (X_t S_t^f) E_t^{Q(XS^f)}\left[\frac{S_T^f (X_T - K)_+}{X_T S_T^f}\right] \\ &= K X_t S_t^f E_t^{Q(XS^f)}[(K^{-1} - X_T^{-1})_+] \\ &= K X_t S_t^f \left[K^{-1} N(-d_-) - e^{\mu_{X^{-1}}(T-t)} X_t^{-1} N(-d_+)\right] \\ &= S_t^f \left[X_t N(-d_-) - e^{-(r^d - r^f + \rho\sigma_X\sigma_S)(T-t)} K N(-d_+)\right], \end{aligned} \tag{1.379}$$

where

$$d_{\pm} = \frac{\log(K/X_t) + (r^f - r^d - \rho\sigma_X\sigma_S \pm \frac{1}{2}\sigma_X^2)(T-t)}{\sigma_X\sqrt{T-t}}. \tag{1.380}$$

Let us now consider Black–Scholes pricing formulas as well as symmetry relations for European calls and puts under an economy whereby the underlying asset pays continuous dividends. This will be useful for the discussion on American options in Section 1.14. In particular, let us assume that the asset price $S_t$ follows geometric Brownian motion, as in Example 1, but with an additional drift term due to a constant dividend yield $q$:

$$dS_t = (r-q)S_t\,dt + \sigma S_t\,dW_t. \tag{1.381}$$

Note that from equation (1.165) we readily have the risk-neutral lognormal transition density for this asset price process,

$$p(S_T, S_t; \tau) = \frac{1}{\sigma S_T\sqrt{2\pi\tau}} e^{-[\log(S_T/S_t)-(r-q-\frac{\sigma^2}{2})\tau]^2/2\sigma^2\tau}, \tag{1.382}$$

$\tau = T-t$. We follow Example 1 and choose $B_t = e^{rt}$ as numeraire. Then, using equation (1.169) with drift $(r-q)$ as given by equation (1.381), the price of a European call struck at $K$ with underlying asset paying continuous dividend $q$ is

$$\begin{aligned} C_t(S_t, K, T) &= e^{-r(T-t)}E_t^{Q(B)}\left[(S_T-K)_+\right] \\ &= e^{-r(T-t)}\left[e^{(r-q)(T-t)}S_tN(d_+) - KN(d_-)\right] \\ &= e^{-q(T-t)}S_tN(d_+) - Ke^{-r(T-t)}N(d_-), \end{aligned} \tag{1.383}$$

with

$$d_{\pm} = \frac{\log(S_t/K) + \left(r - q \pm \frac{1}{2}\sigma^2\right)(T-t)}{\sigma\sqrt{T-t}}. \tag{1.384}$$

The corresponding European put price is easily derived in similar fashion, giving

$$P_t(S_t, K, T) = Ke^{-r(T-t)}N(-d_-) - S_te^{-q(T-t)}N(-d_+). \tag{1.385}$$

The previous put-call parity relation for plain European calls and puts, i.e., equation (1.214), is now modified to read

$$C_t(S_t, K, T) - P_t(S_t, K, T) = e^{-q(T-t)}S_t - Ke^{-r(T-t)} \tag{1.386}$$

for generally nonzero $q$.

This put-call parity is a rather general property that obtains whenever relative asset prices are martingales. Within the geometric Brownian motion model, we can further establish *another* special symmetry property that relates a call price to its corresponding put price. In particular, explicitly denoting the dependence on the interest rate $r$ and dividend yield $q$, we have

$$C_t(S, K, T; r, q) = P_t(K, S, T; q, r). \tag{1.387}$$

This relation states that the Black–Scholes pricing formula for a call, with spot $S_t = S$, strike $K$, interest rate $r$, and dividend $q$, is the same as the Black–Scholes pricing formula for a put where one inputs the strike as $S$, spot as $S_t = K$, interest rate as $q$, and dividend as $r$. That is, by interchanging $r$ and $q$ and interchanging $S$ and $K$, the call and put pricing formulas give the same price. For this reason we can also refer to identity (1.387) as a *put-call reversal symmetry*. This result can be established by relating expectations under different numeraires as follows. Consider the modified asset price process defined by $\tilde{S}_t \equiv e^{qt} S_t$, then Itô's lemma gives

$$d\tilde{S}_t = r\tilde{S}_t \, dt + \sigma \tilde{S}_t \, dW_t \tag{1.388}$$

within the risk-neutral measure. By alternatively choosing $\tilde{g}_t = \tilde{S}_t$ as numeraire, equation (1.292) gives the arbitrage-free price of the call as

$$C_t(S, K, T; r, q) = SKe^{-q(T-t)} E_t^{Q(\tilde{g})}\left[(K^{-1} - X_T)_+\right] \tag{1.389}$$

where we have used the spot value $S_t = S$ and defined the process $X_t \equiv S_t^{-1}$. From equation (1.388), we see that the lognormal volatility of $\tilde{g}_t$ (or the price of risk) is $\sigma$; therefore, under the new measure, $Q(\tilde{g})$, equation (1.381) becomes

$$dS_t = (r - q + \sigma^2) S_t \, dt + \sigma S_t \, d\tilde{W}_t, \tag{1.390}$$

where $d\tilde{W}_t$ denotes the Brownian increment under measure $Q(\tilde{g})$. Using this equation and applying Itô's lemma to $X_t = S_t^{-1}$ gives

$$dX_t = (q - r) X_t \, dt - \sigma X_t \, d\tilde{W}_t \tag{1.391}$$

Under $Q(\tilde{g})$, the transition density $\tilde{p}$ for the process $X_t$ is hence given by equation (1.382) with $r$ and $q$ interchanged and the replacement $S_t \to X_t$, $S_T \to X_T$:

$$\tilde{p}(X_T, X_t; \tau) = \frac{1}{\sigma X_T \sqrt{2\pi\tau}} e^{-[\log(X_T/X_t) - (q - r - \frac{\sigma^2}{2})\tau]^2 / 2\sigma^2 \tau}. \tag{1.392}$$

Under $Q(\tilde{g})$, the drift of the lognormal diffusion $X_t$ is $q - r$. Using equations (1.171) and (1.392) with spot $X_t = 1/S_t = 1/S$ at current time $t$, the expectation in equation (1.389) is evaluated to give

$$C_t(S, K, T; r, q) = SKP_t(1/S, 1/K, T; q, r). \tag{1.393}$$

This establishes the identity, which is actually equivalent to equation (1.387), as can be verified using equation (1.385). Finally, note that equation (1.387) is also verified by directly manipulating equation (1.385) or (1.383).

A class of slightly more sophisticated options that can also be valued analytically within the Black–Scholes model are European-style *compound options*. Such contracts are options on an option. Examples are a *call-on-a-call* and a *call-on-a-put*. Such compound options are hence characterized by two expiration dates, $T_1$ and $T_2$, and two strike values. Let us specifically consider a call-on-a-call option. This contract gives the holder the right (not the obligation) to purchase an underlying call option for a fixed strike price $K_1$ at calendar time $T_1$. The underlying call is a call option on an asset or stock with strike $K_2$ and expiring at a later calendar time $T_2 > T_1$ — we denote its value by $C_{T_1}(S_{T_1}, K_2, T_2)$, where $S_{T_1}$ denotes the stock price at $T_1$. Hence at time $T_1$ this underlying call will be purchased (i.e., the compound call-on-a-call will be exercised at time $T_1$) only if $C_{T_1}(S_{T_1}, K_2, T_2) > K_1$. Let $t$ denote current calendar time, $t < T_1 < T_2$, then the pay-off of the call-on-a-call at $T_1$ is $\left(C_{T_1}(S_{T_1}, K_2, T_2) - K_1\right)_+$.

Since $C_{T_1}$ is a monotonically increasing function of $S_{T_1}$, this pay-off is nonzero only for values of $S_{T_1}$ above a (critical) value $S_1^*$ defined as the unique solution to the (nonlinear) equation $C_{T_1}(S_1^*, K_2, T_2) = K_1$. Hence $\left(C_{T_1}(S_{T_1}, K_2, T_2) - K_1\right)_+ = C_{T_1}(S_{T_1}, K_2, T_2) - K_1$ for $S_{T_1} > S_1^*$ and zero otherwise.

Denoting the value of the call-on-a-call option by $V^{cc}(S, t)$, where $S_t = S$ is the spot at time $t$, and assuming a constant interest rate with $g_t = e^{rt}$ as numeraire asset price, we generally have

$$V^{cc}(S, t) = e^{-r(T_1 - t)} E_t^{Q(B)}\left[\left(C_{T_1}(S_{T_1}, K_2, T_2) - K_1\right)_+\right]. \tag{1.394}$$

Specializing to the case where the stock price process obeys equation (1.381) within the risk-neutral measure *Q(B)*, this expectation is readily evaluated in terms of the standard univariate cumulative normal and bivariate cumulative normal functions. Inserting the price of the call from equation (1.383) gives

$$V^{cc}(S, t) = e^{-r\tau_1} \int_{S_1^*}^{\infty} \left[e^{-q(T_2 - T_1)} S_1 N(d_+) - K_2 e^{-r(T_2 - T_1)} N(d_-) - K_1\right] p(S_1, S; \tau_1) dS_1, \tag{1.395}$$

$\tau_1 = T_1 - t$, where $p$ is the transition density function defined in equation (1.382) and

$$d_\pm = \frac{\log(S_1/K_2) + \left(r - q \pm \frac{1}{2}\sigma^2\right)(T_2 - T_1)}{\sigma\sqrt{T_2 - T_1}}. \tag{1.396}$$

Equation (1.395) is a sum of three integrals. The third integral term involves the risk-neutral probability that the stock price is above $S_1^*$ after a time $\tau_1$ and having initiated at $S$. This integral is reduced to a standard cumulative normal function by changing the integration variable to $x = \log S_1$:

$$\int_{S_1^*}^{\infty} p(S_1, S; \tau_1) dS_1 = N(a_-), \tag{1.397}$$

where we define

$$a_\pm = \frac{\log(S/S_1^*) + \left(r - q \pm \frac{1}{2}\sigma^2\right)\tau_1}{\sigma\sqrt{\tau_1}}. \tag{1.398}$$

The second integral term in equation (1.395) can be rewritten using

$$N(d_-) = \int_{K_2}^{\infty} p(S_2, S_1; T_2 - T_1) dS_2, \tag{1.399}$$

giving

$$\int_{S_1^*}^{\infty} N(d_-) p(S_1, S; \tau_1) dS_1 = \int_{S_1^*}^{\infty} \int_{K_2}^{\infty} p(S_2, S_1; T_2 - T_1) p(S_1, S; \tau_1) dS_2\, dS_1. \tag{1.400}$$

This double integral can be recast in terms of a standard bivariate cumulative normal function

$$N_2(a, b; \rho) = \frac{1}{2\pi\sqrt{1 - \rho^2}} \int_{-\infty}^{a} \int_{-\infty}^{b} \exp\left[-\frac{x^2 + y^2 - 2\rho x y}{2(1 - \rho^2)}\right] dy\, dx, \tag{1.401}$$

where $\rho$ is a correlation coefficient. For this purpose it proves useful to define

$$\tau_2 = T_2 - t \qquad \text{and} \qquad \rho = \sqrt{\tau_1/\tau_2}, \tag{1.402}$$

hence $T_2 - T_1 = \tau_2 - \tau_1$. Introducing the change of variables

$$-x = \frac{\log(S_1/S) - \left(r - q - \frac{1}{2}\sigma^2\right)\tau_1}{\sigma\sqrt{\tau_1}}, \quad -y = \frac{\log(S_2/S) - \left(r - q - \frac{1}{2}\sigma^2\right)\tau_2}{\sigma\sqrt{\tau_2}}$$

Equation (1.400) then becomes (after some algebraic manipulation)

$$\begin{aligned}\int_{S_1^*}^{\infty} N(d_-)p(S_1, S; \tau_1)dS_1 &= \frac{1}{2\pi\sqrt{1-\rho^2}} \int_{-\infty}^{a_-} \int_{-\infty}^{b_-} \exp\left[ -\frac{x^2}{2} - \frac{(y-\rho x)^2}{2(1-\rho^2)} \right] dy\, dx, \\ &= N_2(a_-, b_-; \rho),\end{aligned} \tag{1.403}$$

where

$$b_{\pm} = \frac{\log(S/K_2) + \left(r - q \pm \frac{1}{2}\sigma^2\right)\tau_2}{\sigma\sqrt{\tau_2}}. \tag{1.404}$$

We leave it to the reader to verify that the first integral term in equation (1.395) can be reduced, using similar manipulations as earlier, to give

$$\int_{S_1^*}^{\infty} S_1 N(d_+)p(S_1, S; \tau_1)dS_1 = Se^{(r-q)\tau_1} N_2(a_+, b_+; \rho). \tag{1.405}$$

Combining the three integrals in equation (1.395) finally gives

$$V^{cc}(S, t) = Se^{-q\tau_2} N_2(a_+, b_+; \rho) - K_2 e^{-r\tau_2} N_2(a_-, b_-; \rho) - K_1 e^{-r\tau_1} N(a_-). \tag{1.406}$$

Derivations of similar pricing formulas for related types of compound options are left to the interested reader (see Problem 10).

## Problems

Within the problems involving a single underlying asset or stock, assume we are in a Black–Scholes world where the asset price process obeys geometric Brownian motion of the form

$$dA_t = (r + q \cdot \sigma_A)A_t\, dt + \sigma_A A_t\, dW_t, \tag{1.407}$$

where $dW_t$ is the assumed Brownian increment under the given measure, the interest rate $r$ (in the appropriate economy) and volatility $\sigma_A$ are constants, and $q$ is a market price of risk.

**Problem 1.** Find the price of a call option on foreign stock struck in foreign currency, i.e., of the contract with payoff

$$C_T = X_T(S_T^f - K)_+. \tag{1.408}$$

**Problem 2.** Find the price of a call option on foreign stock struck in domestic currency with payoff

$$C_T = (X_T S_T^f - K)_+, \tag{1.409}$$

where $X_t$ is the exchange rate at time $t$.

**Problem 3.** Consider again the example of the quanto option in Example 3. Compute the coefficient $\alpha$ in such a way that the price process

$$g_t \equiv \bar{X} e^{\alpha t} S_t^f \tag{1.410}$$

is a domestic asset price process. Further, price the quanto option in Example 3 using $g_t$ as a numeraire asset. Describe the replication strategy for the numeraire asset $g_t$.

**Problem 4.** Derive the price of an Elf-X option from the point of view of the foreign investor taking as payoff

$$C_T = (S_T^f - K S_T^f Y_T)_+, \tag{1.411}$$

where $Y_T = 1/X_T$.

**Problem 5.** A forward starting call on a stock $S$ is structured as follows. The holder will receive at a preassigned future time $T_1$ a call struck at $K = \alpha S_{T_1}$ and maturing at time $T_2 > T_1$. Here, $\alpha$ is a positive preassigned constant and $S_{T_1}$ is the stock price realized at time $T_1$.

Find (i) the present time $t = t_0 \leq T_1$ price of the forward starting call prior to maturity $T_1$ and (ii) a static hedging strategy that applies up to time $T_1$. Using the result in (i), show that the price of the contract simplifies to that of a standard call struck at $K = \alpha S_0$ with time to maturity $T_2 - t_0$ in the limiting case that $T_1 \to t_0$ (with $t_0$, $T_2$ held fixed). On the other hand, show that in the limit $T_2 \to T_1$ (with $t_0$, $T_1$ held fixed) the contract price is simply given by $S_0(1-\alpha)_+$. This last result is consistent with the price of a standard call with maturity $t = T_1$ and strike $K = \alpha S_{T_1}$.

**Problem 6.** Consider two stocks $S^1$ and $S^2$ described by correlated geometric Brownian motion with constant volatilities $\sigma_1$ and $\sigma_2$ and with correlation $\rho$. As seen in Section 1.6, a simple chooser option yields the pay-off as the maximum of the two stock levels,

$$\max(S_T^1, S_T^2), \tag{1.412}$$

at the maturity date $T$. Find the price of this instrument at time $t < T$. Find the relationship between the price of this chooser option and that of the chooser with payoff $(S_T^2 - S_T^1)_+$.

*One Solution*: To solve for either option price, pick the price of stock 1 as numeraire, $g_t = S_t^1$. So, for instance, to price the latter option, show that the price $C_t$ is given by an expectation

$$C_t = S_t^1 E_t^p[(f_T - 1)_+], \tag{1.413}$$

where the random variable $f_t = S_t^2 / S_t^1$ obeys

$$\frac{df_t}{f_t} = (\rho\sigma_2 - \sigma_1) dW_t^1 + \sigma_2 \sqrt{1-\rho^2}\, dW_t^2. \tag{1.414}$$

From this, show that we have

$$\log \frac{f_T}{f_t} = -\frac{\nu^2}{2}(T-t) + (\rho\sigma_2 - \sigma_1)(W_T^1 - W_t^1) + \sigma_2\sqrt{1-\rho^2}(W_T^2 - W_t^2), \tag{1.415}$$

where $W_t^i$ are independent Wiener processes at time $t$ and

$$\nu \equiv \sqrt{\sigma_1^2 + \sigma_2^2 - 2\rho\sigma_1\sigma_2} = |\boldsymbol{\sigma}_f|.$$

Since $\log(f_T/f_t)$ is normally distributed, find its mean and variance and thereby obtain the lognormal drift and volatility of $f_t$, i.e., the lognormal density $p = p(f_T, f_t; T-t)$, giving the price

$$C_t = S_t^2 N(d_+) - S_t^1 N(d_-), \tag{1.416}$$

where

$$d_\pm \equiv \frac{\log(S_t^2/S_t^1) \pm \frac{1}{2}\nu^2(T-t)}{\nu\sqrt{T-t}}. \tag{1.417}$$

**Problem 7.** Derive the standard call option-pricing formula of Example 1 of this section, but this time use the stock price as numeraire, i.e., $g_t = S_t$. In particular, show that with this choice of numeraire,

$$\frac{d(1/S_t)}{(1/S_t)} = -r\,dt - \sigma\,dW_t, \tag{1.418}$$

where $dW_t$ stands for the Brownian increment under the measure $Q(S)$ with $S_t$ as numeraire. Then show that this leads to

$$C_t(S_t, K, T) = KS_t E_t^{Q(S)}\left[(1/K - 1/S_T)_+\right]. \tag{1.419}$$

Note: This is related to the price of an European put contract where the random variable is now the inverse of the stock price struck at the inverse of the strike, i.e., $1/K$, and with drift $= -r$. Compute this expectation to obtain the final expression.

**Problem 8.** Consider a foreign money-market account $B_t^f = e^{\int_0^t r_s^f ds}$ (with interest rate in foreign currency given by $r_t^f$ at time $t$), a domestic asset with price $A_t^d$, and a foreign asset with price $A_t^f$. Let $X_t$ be the exchange rate process in converting foreign currency into domestic. Suppose we choose $g_t = A_t^d$ as our numeraire asset. Compute the drift of the following processes: $X_t$, $B_t^f$, and $A_t^f$, within the $Q(g)$ measure.

**Problem 9.** Consider a domestic asset with price $A_t^d$ and a foreign asset with price $A_t^f$. Let the constant $\kappa$ be the conversion factor

$$\kappa = \frac{A_0^d}{A_0^f}. \tag{1.420}$$

[Note that this is given in terms of the asset prices at some current time $t = 0$.]

(i) Find a pricing formula for the contract at current time $t = 0$ with payoff function

$$\max(A_T^d, \kappa A_T^f) \tag{1.421}$$

at maturity $t = T$. Assume that all relevant lognormal volatilities and correlations are constant.

(ii) How can one hedge this contract? Is it necessary to trade the foreign currency dynamically?

**Problem 10.** Derive pricing formulas analogous to equation (1.406) for (i) a call-on-a-put, (ii) a put-on-a-put, and (iii) a put-on-a-call.

## 1.13 Partial Differential Equations for Pricing Functions and Kernels

Consider the continuous-time model with *state-dependent volatility*

$$\frac{dS_t}{S_t} = (r(t) + q\sigma(S_t, t))dt + \sigma(S_t, t)dW_t, \tag{1.422}$$

where $q$ is the price of risk (also equal to the volatility of the numeraire asset). Here, $r(t)$ is a deterministic, time-dependent short rate consistent with the term structure of interest rates. The state-dependent volatility $\sigma(S, t)$ is sometimes called the *local volatility*.

The asset price process $A_t$ of an European-style option contingent on the asset $S$ in the model described by equation (1.422) is given by a *pricing function* $A(S,t)$ through a formula of the form

$$A_t = A(S_t, t). \tag{1.423}$$

The existence of a pricing function is an expression of the fact that the current price of an European option depends only on current calendar time and on the current (i.e., spot) price $S = S_t$ for the underlying asset (assuming all other contract parameters are held fixed as the maturity time, etc.).

**Theorem. (Black–Scholes Equation)** *The pricing function $A(S, t)$ of a European claim contingent on the asset $S$ in equation (1.422) satisfies the Black–Scholes equation*

$$\frac{\partial A}{\partial t} + \frac{\sigma^2 S^2}{2}\frac{\partial^2 A}{\partial S^2} + rS\frac{\partial A}{\partial S} - rA = 0, \tag{1.424}$$

where $r = r(t)$, $\sigma = \sigma(S, t)$.

This is a backward time parabolic partial differential equation related closely to the backward Kolmogorov equation, as we shall see later.

**Proof.** Choosing as numeraire asset the money-market account $B_t = e^{\int_0^t r(s)ds}$, the price of risk $q = 0$ and the risk-neutral pricing formula yields

$$E^{Q(B)}[dA_t] = r(t)A_t\, dt \tag{1.425}$$

Equation (1.424) follows by applying Itô's lemma to the calculation of $dA_t = dA(S_t, t)$. Namely,

$$\frac{\partial A}{\partial t} + rS\frac{\partial A}{\partial S} + \frac{\sigma^2 S^2}{2}\frac{\partial A}{\partial S^2} = rA, \tag{1.426}$$

$r = r(t)$, $\sigma = \sigma(S, t)$. Lastly, note that this follows simply from the Feynman–Kac theorem. □

A second important partial differential equation concerns the probability density function $P(S, t)$ under the risk-neutral measure for the stock price values $S$ at time $t$, given an initial Dirac delta function distribution at time $t = t_0$:

$$P(S, t = t_0) = \delta(S - S_0). \tag{1.427}$$

More explicitly, this function is given by $P(S,t) \equiv p(S,t;S_0,t_0)$; i.e., this is the risk-neutral transition probability density for the price of the underlying asset to begin at value $S_0$ at initial time $t_0$ and end with value $S_t = S$ at time $t$. The function $p(S,t;S_0,t_0)$ is also commonly referred to as a *pricing kernel*. We have already seen a specific example of this as the lognormal transition density for geometric Brownian motion. In general, the resulting equation, called the *Fokker–Planck* (or *forward Kolmogorov*) equation, is contained in the following statement.

**Theorem 1.7. (Fokker–Planck Equation)** *The probability density function P(S, t) under the risk-neutral measure for the stock price values S at time t satisfying initial condition (1.427) obeys the following equation:*

$$\frac{\partial P}{\partial t} = \frac{1}{2}\frac{\partial^2}{\partial S^2}(\sigma^2 S^2 P) - r\frac{\partial}{\partial S}(SP), \tag{1.428}$$

where $r = r(t)$, $\sigma = \sigma(S,t)$.

**Proof.** This result can be derived as a consequence of the Black–Scholes equation. Consider a generic asset with pricing function $A(S,t)$ defined in the interval $t \in (t_0, T)$, we then have from risk-neutral valuation that at any time $t$,

$$A(S_0,t_0) = e^{-\int_{t_0}^{t} r(s)ds}\int_0^\infty P(S,t)A(S,t)dS. \tag{1.429}$$

Note here that we assume that the range of solution is $S \in (0,\infty)$, although the derivation can be extended to cases with different ranges. Taking the partial derivative with respect to calendar time $t$ on both sides of this equation, we find

$$\int_0^\infty \left[-rPA + A\frac{\partial P}{\partial t} + P\left(rA - rS\frac{\partial A}{\partial S} - \frac{S^2\sigma^2}{2}\frac{\partial^2 A}{\partial S^2}\right)\right]dS = 0,$$

where $r = r(t)$, $\sigma = \sigma(S,t)$, and the Black–Scholes equation (1.424) has been used for $\frac{\partial A}{\partial t}$. Integrating the last two terms by parts we obtain:

$$-\int_0^\infty PS\frac{\partial A}{\partial S}dS = -(PSA)\Big|_0^\infty + \int_0^\infty A\frac{\partial}{\partial S}(SP)dS = \int_0^\infty A\frac{\partial}{\partial S}(SP)dS,$$

and

$$\begin{aligned}
-\int_0^\infty \frac{\sigma^2 S^2}{2}P\frac{\partial^2 A}{\partial S^2}dS &= -\frac{1}{2}\frac{\partial}{\partial S}(\sigma^2 S^2 P)\frac{\partial A}{\partial S}\Big|_0^\infty + \int_0^\infty \frac{\partial A}{\partial S}\frac{\partial}{\partial S}\left(\frac{\sigma^2 S^2 P}{2}\right)dS \\
&= -\frac{1}{2}\left(\frac{\partial A}{\partial S} - A\right)\frac{\partial}{\partial S}(\sigma^2 S^2 P)\Big|_0^\infty - \frac{1}{2}\int_0^\infty A\frac{\partial^2}{\partial S^2}(\sigma^2 S^2 P)dS \\
&= -\frac{1}{2}\int_0^\infty A\frac{\partial^2}{\partial S^2}(\sigma^2 S^2 P)dS.
\end{aligned}$$

In the last equation we have integrated by parts twice. Notice that the nonintegral terms all vanish, due to the boundary conditions on the probability density function $P$, namely, that the function $P$ and the first and second partial derivatives with respect to $S$, are assumed to be rapidly decaying functions of $S$ as $S \to 0$ and $S \to \infty$. Collecting terms gives that for *any* derivative pricing function $A(S,t)$,

$$\int_0^\infty A(S,t)\left[\frac{\partial P}{\partial t} + r\frac{\partial}{\partial S}(SP) - \frac{1}{2}\frac{\partial^2}{\partial S^2}(\sigma^2 S^2 P)\right]dS = 0.$$

This can only occur if the integrand term in brackets is identically zero; hence equation (1.428) is fulfilled. □

The corresponding backward Kolmogorov equation for the density is given by the so-called *Lagrange adjoint* of equation (1.428). By combining equations (1.429) [with $P(S,t) \equiv P \equiv p(S,t;S_0,t_0)$] and (1.424), we readily see that $e^{-\int_{t_0}^{t} r(s)ds} P$ must satisfy the same equation as $A(S_0, t_0)$ for all initial times $t_0 < t$. Simplifying the equation in terms of $P$ only, we find the backward Kolmogorov equation:

$$\frac{\partial P}{\partial t_0} + \frac{1}{2}\sigma^2(S_0, t_0)S_0^2 \frac{\partial^2 P}{\partial S_0^2} + r(t_0)S_0 \frac{\partial P}{\partial S_0} = 0. \tag{1.430}$$

This is a backward-time parabolic partial differential equation of the form of the Black–Scholes equation [i.e., replacing *(S,t)* by $(S_0, t_0)$ in equation (1.424)]. The only term missing is the compounding term $r(t_0)A$. However, as just mentioned, the function $e^{-\int_{t_0}^{t} r(s)ds} P$ does exactly satisfy the Black–Scholes equation. This is, not surprisingly, consistent with our discussion in Section 1.8, where we showed [see equation (1.231)] that the discounted transition density gives the current price of a European butterfly option with inifinitely narrow spread (i.e., the price of an Arrow–Debreu security).

A partial differential equation satisfied by the pricing function of European-style call options *C(S, t; K, T)* regarded explicitly as functions of the strike and maturity time arguments *(K,T)* [instead of functions of the arguments *(S, t)*, which are held fixed] can now be derived as follows.

**Theorem 1.8. (Dual Black–Scholes Equation)** *The pricing function for a European call option C(S, t; K, T) satisfies the following equation:*

$$\frac{\partial C}{\partial T} = -r(T)K\frac{\partial C}{\partial K} + \frac{1}{2}K^2\sigma^2(K,T)\frac{\partial^2 C}{\partial K^2}. \tag{1.431}$$

**Proof.** European-style call prices admit the following representation in terms of the risk-neutral transition probability density [i.e., the density for the risk-neutral measure *Q(B)*]:

$$C(K,T) = Z_0(T)E_0^{Q(B)}[(S-K)_+] = Z_0(T)\int_0^\infty P(S,T)(S-K)_+ \, dS, \tag{1.432}$$

where $Z_0(T) = e^{-\int_0^T r(s)ds}$. Without loss of generality we simply set current time $t = 0$. Using the property $\partial(S-K)_+/\partial K = -\theta(S-K)$, where $\theta(x)$ is the Heaviside step function with value 1 for $x \geq 0$ and value 0 for $x < 0$, the first and second derivatives of equation (1.432) with respect to the strike $K$ give

$$\frac{\partial C}{\partial K} = -Z_0(T)\int_K^\infty P(S,T)dS, \tag{1.433}$$

and

$$\frac{\partial^2 C}{\partial K^2} = Z_0(T)P(K,T). \tag{1.434}$$

The derivative with respect to maturity is given by

$$\begin{aligned} \frac{\partial C}{\partial T} &= -rZ_0(T)\int_0^\infty (S-K)_+ P \, dS + Z_0(T)\int_0^\infty \frac{\partial P}{\partial T}(S-K)_+ \, dS \\ &= -rC + Z_0(T)\int_0^\infty \left[-r\frac{\partial}{\partial S}(SP) + \frac{1}{2}\frac{\partial^2}{\partial S^2}(\sigma^2 S^2 P)\right](S-K)_+ \, dS, \end{aligned}$$

where $r = r(T)$, $\sigma = \sigma(S, T)$. Note that we have used equation (1.428) with $t = T$. The integral containing the first derivative with respect to $S$ can be evaluated by parts as follows:

$$\begin{aligned}\int_0^\infty (S-K)_+ \frac{\partial}{\partial S}(SP)dS &= -\int_K^\infty SP\ dS \\ &= -\int_0^\infty (S-K)_+ P\ dS - K\int_K^\infty P\ dS \\ &= \left[-C + K\frac{\partial C}{\partial K}\right] Z_0(T)^{-1},\end{aligned}$$

where we used the identity $S = (S-K)_+ + K$, for $S \in [K, \infty)$, and equations (1.432) and (1.433). The integral containing the second derivative can again be evaluated by parts:

$$\begin{aligned}\int_0^\infty (S-K)_+ \frac{\partial^2}{\partial S^2}(\sigma^2 S^2 P)dS &= -\int_K^\infty \frac{\partial}{\partial S}(\sigma^2 S^2 P)dS \\ &= \sigma^2(K,T)K^2 P(K,T) \\ &= Z_0(T)^{-1}\sigma^2(K,T)K^2\frac{\partial^2 C}{\partial K^2}.\end{aligned}$$

Collecting the intermediate results obtained so far, we arrive at the following dual Black–Scholes equation:

$$\begin{aligned}\frac{\partial C}{\partial T} &= -rC + rC - rK\frac{\partial C}{\partial K} + \frac{1}{2}K^2\sigma^2(K,T)\frac{\partial^2 C}{\partial K^2} \\ &= -rK\frac{\partial C}{\partial K} + \frac{1}{2}K^2\sigma^2(K,T)\frac{\partial^2 C}{\partial K^2}.\end{aligned}$$

□

A consequence of this result is the following, which may be used in practice to calibrate a local volatility surface $\sigma_l = \sigma(K, T)$ via market European call option prices across a range of maturities and strikes.

**Theorem 1.9. (Derman–Kani)** *If a local volatility function exists, then it is unique and it can be expressed in analytical closed form as follows in terms of call option prices:*

$$\sigma^2(K,T) = \frac{2}{K^2}\frac{\frac{\partial C}{\partial T} + rK\frac{\partial C}{\partial K}}{\frac{\partial^2 C}{\partial K^2}}. \tag{1.435}$$

This PDE pricing formalism extends readily into arbitrary dimensions. A general connection between a system of SDEs and the corresponding forward (backward) Kolmogorov PDEs that govern the transition probability density is as follows. Consider a diffusion model with $n$ correlated random processes $\mathbf{x}_t = (x_t^1, \ldots, x_t^n) \in \mathbb{R}^n$ satisfying the system of SDEs:

$$\frac{dx_t^i}{x_t^i} = \mu_i(\mathbf{x}_t, t)dt + \sum_{\alpha=1}^{M} \sigma_{i,\alpha}(\mathbf{x}_t, t)dW_t^\alpha; \qquad i = 1, \ldots, n, \tag{1.436}$$

with $M \geq 1$ independent Brownian motions, $dW_t^\alpha\, dW_t^\beta = \delta_{\alpha,\beta}\, dt$, and where the drifts and volatilites are generally functions of time $t$ and $\mathbf{x}_t$. Let us define the differential operator $\mathcal{L}$ by

$$\mathcal{L}_{\mathbf{x},t} f \equiv \sum_{i=1}^{n} x_i\mu_i(\mathbf{x},t)\frac{\partial f}{\partial x_i} + \frac{1}{2}\sum_{i,j=1}^{n} x_i x_j \nu_{i,j}(\mathbf{x},t)\frac{\partial^2 f}{\partial x_i \partial x_j}, \tag{1.437}$$

with Lagrange adjoint operator $\mathcal{L}^\dagger$ given by

$$\mathcal{L}^\dagger_{\mathbf{x},t} f \equiv -\sum_{i=1}^{n} \frac{\partial[x_i\mu_i(\mathbf{x},t)f]}{\partial x_i} + \frac{1}{2}\sum_{i,j=1}^{n} \frac{\partial^2[x_i x_j \nu_{i,j}(\mathbf{x},t)f]}{\partial x_i\, \partial x_j}, \tag{1.438}$$

where the functions $\nu_{i,j}$, $i, j = 1, \ldots, n$, are defined by

$$\nu_{i,j}(\mathbf{x},t) = \sum_{\alpha=1}^{M} \sigma_{i,\alpha}(\mathbf{x},t)\sigma_{j,\alpha}(\mathbf{x},t). \tag{1.439}$$

These operators act on any sufficiently differentiable function $f = f(\mathbf{x}, t)$. The transition probability density $p = p(\mathbf{x}, t; \mathbf{x}_0, t_0)$ associated with the foregoing diffusion process then satisfies the forward (Fokker–Planck) Kolmogorov PDE,

$$\frac{\partial p}{\partial t} = \mathcal{L}^\dagger_{\mathbf{x},t} p \tag{1.440}$$

as well as the corresponding backward PDE,

$$\frac{\partial p}{\partial t_0} + \mathcal{L}_{\mathbf{x}_0,t_0} p = 0, \tag{1.441}$$

for all $t_0 < t$, with initial (or final) time condition

$$p(\mathbf{x}, t = t_0; \mathbf{x}_0, t_0) = p(\mathbf{x}, t; \mathbf{x}_0, t_0 = t) = \delta(\mathbf{x} - \mathbf{x}_0).$$

Assuming that a diffusion path starting at some point $\mathbf{x}_0$ at time $t_0$ and ending at a point $\mathbf{x}$ at time $t$ must be at all possible points $\bar{\mathbf{x}}$ at any intermediate time $\bar{t}$, $t_0 \le \bar{t} \le t$, then a consistency requirement in the theory is the so-called *Chapman–Kolmogorov integral equation*:

$$p(\mathbf{x}, t; \mathbf{x}_0, t_0) = \int_{\mathbb{R}^n} p(\mathbf{x}, t; \bar{\mathbf{x}}, \bar{t}) p(\bar{\mathbf{x}}, \bar{t}; \mathbf{x}_0, t_0) d\bar{\mathbf{x}}. \tag{1.442}$$

Prices of European-style contingent claims can then be computed by taking integrals over an appropriate pricing kernel as follows. Suppose we are within a certain measure $Q(g)$ where underlying assets depend on random variables $x^i_t$ that have appropriate drift and volatilites in accordance with equation (1.436). Assuming the existence of a martingale measure where the numeraire is, for example, of the form $e^{\int_0^t \lambda(\mathbf{x}_s,s)ds}$ (i.e., with $\lambda$ as a discounting function), then according to the asset pricing theorem of the previous section, the price of a contingent claim $A(\mathbf{x}, t)$ with payoff $\phi(\mathbf{x})$ is given by the expectation

$$A(\mathbf{x}, t) = E^Q_t\left[e^{-\int_t^T \lambda(\mathbf{x}_s,s)ds}\phi(\mathbf{x})\right]. \tag{1.443}$$

Then due to the Feynman–Kac formula (in $n$ dimensions) we have the corresponding Black–Scholes PDE:

$$\frac{\partial A(\mathbf{x},t)}{\partial t} + \mathcal{L}_{\mathbf{x},t} A(\mathbf{x},t) - \lambda(\mathbf{x},t)A(\mathbf{x},t) = 0, \tag{1.444}$$

$t < T$, with terminal condition $A(\mathbf{x}, T) = \phi(\mathbf{x})$, as required. From this analysis we see that the price of the contingent claim satisfying this Black–Scholes type of PDE can in fact be expressed as an integral over the set of diffusion paths. With the particular choice

$\lambda(\mathbf{x}_t, t) = r(t)$ (the risk-free rate), then, the density $p$ is the risk-neutral density expressed in the $\mathbf{x}$-space variables. The claim's price is then simply given by an integral in $\mathbb{R}^n$:

$$A(\mathbf{x}, t) = e^{-\int_t^T r(s)ds} \int p(\mathbf{x}_T, T; \mathbf{x}, t)\phi(\mathbf{x}_T)d\mathbf{x}_T. \tag{1.445}$$

This is a multidimensional extension of equation (1.429). Note also that here, variables $\mathbf{x}$ do not necessarily represent prices. In general, asset prices are functions of $\mathbf{x}$ and time $t$. A nice feature of such integral equations, among others, is the fact that they provide a solution whereby the kernel $p$ and hence the expected values can be propagated forward in the time variable $T$, starting from $T = t$, where the delta function condition is employed.

### Problems

**Problem 1.** Consider the one-dimensional lognormal density $p(S, S_0; t - t_0)$ given by equation (1.165). Show that it satisfies forward and backward equations of the form (1.428) and (1.430) as well as the Chapman–Kolmogorov equation,

$$\int_0^\infty p(S, \bar{S}; t - \bar{t})p(\bar{S}, S_0; \bar{t} - t_0)d\bar{S} = p(S, S_0; t - t_0), \tag{1.446}$$

$t_0 \leq \bar{t} \leq t$.

**Problem 2.** Consider the $n$-dimensional lognormal density given by equation (1.198). Verify that this density satisfies the appropriate Kolmogorov equations.

## 14 American Options

In this section we briefly present the theory for pricing American, or *early-exercise*, options. The distinction between an American-style option and its European counterpart is that the holder of the American option has the additional freedom or right to exercise the option at *any* date from contract inception until expiration. This additional *time optionality* generally gives rise to an additional worth, appropriately also referred to as the *early-exercise premium*. We mainly focus our discussion on calls and puts, although the theory is also useful for treating other types of pay-offs. Throughout this section, we shall assume that we are within a Black–Scholes world with only one underlying asset. Although the formal theory readily extends into the multiasset case, the practical implementation and analysis issues are nontrivial and not within the scope of our present discussion. The development of numerical methods for pricing multiasset American options remains a topic of active research (see, for example, [BD96, BG97b, BG97a, BKT01, Gla04]).

### 14.1 *Arbitrage-Free Pricing and Optimal Stopping Time Formulation*

To begin our discussion, we consider the case where the underlying asset (or stock) price process $(S_t)_{t\geq 0}$ follows the geometric Brownian motion model as given by equation (1.381) in the risk-neutral measure, where $r$ is the risk-free interest rate and $q$ is a continuous dividend yield. We therefore assume that $r \geq 0$, $q \geq 0$, $\sigma$ are constants (i.e., state and time independent), although the formalism (i.e., the governing equations) readily extends to the case of state-dependent drift and volatility functions. Let $t_0$ be the present time (i.e., contract inception). An American call (or put) option struck at $K$ with expiration at time $T$ is a claim to a payoff $(S_t - K)_+$ (or $(K - S_t)_+$) that the holder can exercise at any intermediate time

$t$ prior to maturity, i.e., $t_0 \le t \le T$. The time at which the option is exercised is a stopping time. Recall the simpler situation in which the stopping time is initially known (i.e., as in the case of a European-style claim), then from the theorem of asset pricing the arbitrage-free price of a claim with a given pay-off occurring at time $t$ is simply given by the discounted expectation via equation (1.294). In particular, the value at present time $t_0$ of a cash flow $(S_t - K)_+$ delivered at a later time $t$ is given by

$$e^{-r(t-t_0)} E_0\big[(S_t - K)_+\big],$$

where $E_0[\cdot] = E^Q[\cdot|\mathcal{F}_{t_0}] = E^Q[\cdot|S_{t_0} = S_0]$ is used as a simplified notation to denote the expectation at time $t_0$ within the risk-neutral measure $Q(B)$, with $B_t = e^{rt}$ as numeraire, conditional on $S_{t=t_0} = S_0$. This expectation gives us the fair value of the cash flow as long as the delivery time $t$ is a given stopping time, which may either be deterministic or random. For the case in which the stopping time is given by the maturity, e.g., $t = T$, the foregoing expectation obviously corresponds to the price of an European call [as given by equation (1.383), with $t$, $S_t$ replaced by $t_0$, $S_0$].

For American contracts the holder has the freedom to exercise at any time within the continuous set of values $\mathcal{T} = \{t : t_0 \le t \le T\}$, giving rise to an *optimal stopping time* (i.e., *early-exercise time*) at which the holder should exercise the option for maximal gain. In particular, we shall see that an *early-exercise boundary* arises on the $(t, S_t)$-plane (i.e., time-spot plane) that separates the domain $[t_0, T] \times \mathbb{R}_+$ into two subdomains. These consist of a so-called *continuation* domain, for which the option is not yet exercised, and a *stopping* domain, whereby the option is exercised early. Hence, a main distinction from the European case is that the exercise time is *not known prematurely* and must be optimally determined as part of the solution to the pricing problem. As observed later, the basic financial reasoning for the emergence of an early-exercise boundary is that the holder can either claim a profit from the underlying dividend income by opting to purchase the asset (e.g., for the case of a call) or profit from the interest that arises from selling the underlying asset and investing the proceeds in a money-market account (e.g., for a put).

More generally, let us consider a nonnegative payoff function $\phi(S)$, $S \in \mathbb{R}_+$. The values of the European and corresponding American claim to such a pay-off are given, respectively, by

$$V_E(S_0, T - t_0) = E_0\big[e^{-r(T-t_0)}\phi(S_T)\big] \tag{1.447}$$

and

$$V(S_0, T - t_0) = \sup_{t\in\mathcal{T}} E_0\big[e^{-r(t-t_0)}\phi(S_t)\big]. \tag{1.448}$$

Throughout this section we use $V_E$ to distinguish the European price from its American counterpart. In equation (1.448) the supremum is taken over all possible stopping times in the set $\mathcal{T}$. Note that both pricing functions are functions of the current time to maturity $T - t_0$, as is generally true when the drift and volatility terms have no explicit time dependence. We remark that although various theoretical frameworks exist for the determination of optimal stopping times, exact analytical formulas for such quantities as well as for American option values in terms of known transcendental functions have not been found to date. This is the case for the geometric Brownian motion model and, of course, for the more complex state-dependent models. In Section 1.14.4 we develop an integral-equations approach for computing the early-exercise boundary and the American option value, whereas in this section we provide a discrete-time backward induction formulation, which is useful for approximating the continuous-time quantities.

Formally, the optimal stopping time, denoted by $t^*$, is given by the infimum over the set $\mathcal{T}$ such that the value of the American option is equal to its *intrinsic value* (or *face value*) as given by the pay-off at the observed asset price:

$$t^* = \inf\{t \in \mathcal{T}, V(S_t, T-t) = \phi(S_t)\}. \tag{1.449}$$

The *stopping domain*, corresponding to spot and time values for which it is optimal to exercise prematurely, consists of the set of points

$$\mathcal{D} = \{(t, S) : t \in \mathcal{T}, V(S, T-t) = \phi(S)\}, \tag{1.450}$$

while the *continuation domain*, corresponding to spot and time values for which the option is not exercised prematurely, is the set of points

$$\mathcal{C} = \{(t, S) : t \in \mathcal{T}, V(S, T-t) > \phi(S)\}. \tag{1.451}$$

Assuming there exists an optimal stopping time $t^*$, then from asset-pricing theory this time is given implicitly by

$$E_0\left[e^{-r(t^*-t_0)}\phi(S_{t^*})\right] = V(S_0, T-t_0). \tag{1.452}$$

This is a result that is not practical as it stands since the equation involves the American option value on the right-hand side, which is itself not yet known and dependent upon the stopping domain. This is a common feature among optimal stopping problems for Markov processes in continuous time, because they are essentially *free-boundary* value problems as shown shortly.

The structure of the stopping domains may be quite complicated for certain classes of payoff functions and diffusion models. However, for standard piecewise call/put types of pay-offs considered here, the domains turn out to be simply connected. In particular, the boundary of $\mathcal{D}$ is an early-exercise boundary curve given by

$$\partial\mathcal{D} = \{(\tau, S) : 0 \le \tau \le T - t_0, S = S^*(\tau)\}, \tag{1.453}$$

with $S^*(\tau)$ given by a smooth curve

$$S^*(\tau) = \min\{S > 0 : V(S, \tau) = (S-K)_+\} \tag{1.454}$$

for a call and

$$S^*(\tau) = \max\{S > 0 : V(S, \tau) = (K-S)_+\} \tag{1.455}$$

for a put struck at $K$. Here the function $V(S, \tau)$ represents the value of the American call $C(S, K, \tau)$ or put $P(S, K, \tau)$, respectively, where $S$ is the value of the underlying spot. From equation (1.451) it is obvious that the continuation domain is the set of all points $(\tau, S)$ such that $V(S, \tau)$ is greater than the respective payoff function at $S$. As we will see, the subscript $+$ signs are actually redundant in equations (1.454) and (1.455). Note that here we have simply expressed the boundary and the option price in terms of the time-to-maturity variable $\tau = T - t \in [0, T - t_0]$ rather than the calendar time $t \in [t_0, T]$. This is convenient for what follows since the diffusion models are assumed to be time homogeneous. The optimal-exercise decision for the holder therefore depends on the observed spot (or stock price level) and the time to maturity (or calendar time) of the observation. In this sense, Amercian options can be characterized as having a kind of path dependence.

Before any further analysis, we make note of one very basic and important property of the early-exercise premium (or value): *The European option value $V_E$ satisfies the condition (i) $V_E(S, \tau) \geq \phi(S)$ for all $(S, \tau)$ if and only if the corresponding American option value V satisfies (ii) $V(S, \tau) = V_E(S, \tau)$ for all $(S, \tau)$.* That is, if the corresponding European price is always above its intrinsic value during the contract lifetime, then it is never optimal to exercise the American option at any time earlier than expiry; i.e., there is no early-exercise premium and $V = V_E$. To show this, note that equation (1.448) implies $V(S, \tau) \geq V_E(S, \tau)$. Hence condition (i) gives $V(S, \tau) \geq \phi(S)$, so the American option is always above the intrinsic value, implying that the holder would not exercise earlier for a lower value. The optimal exercise (stopping) time is therefore at expiry $T$; hence (i) implies (ii). To prove the converse, observe that since the American option value must satisfy $V(S, \tau) \geq \phi(S)$ for all $(S, \tau)$, condition (ii) implies (i). This result is essentially a statement of the fact that an early-exercise boundary (and premium) arises only if the corresponding European option value falls below the intrinsic (payoff function) value. Because of this we have the following rather well-known result.

**Proposition.**
*(i) An Amercian call has a nonzero early-exercise premium if and only if $q > 0$.*
*(ii) An Amercian put has a nonzero early-exercise premium if and only if $r > 0$.*

This result will be seen to follow explicitly from the early-exercise boundary properties and the formulas for the early-exercise premiums developed in the following subsections. However, a simple and instructive proof goes as follows.

**Proof.** The put-call parity relation for European calls and puts gives

$$C_E(S, K, \tau) - P_E(S, K, \tau) = e^{-q\tau}S - e^{-r\tau}K. \tag{1.456}$$

Rewriting this we have

$$C_E(S, K, \tau) = S - K + P_E(S, K, \tau) + [(e^{-q\tau} - 1)S - (e^{-r\tau} - 1)K]. \tag{1.457}$$

Since $P_E(S, K, \tau) > 0$, then for $q = 0$ either of these expressions gives $C_E(S, K, \tau) > S - e^{-r\tau}K \geq S - K$. Hence $C_E$ is always above its intrinsic value, and from the previous property we conclude that the European call value is equal to the American call value, $C_E(S, K, \tau) = C(S, K, \tau)$, so the early-exercise premium is zero. For the case $q > 0$, we use equation (1.457) and note that since the European put is a decreasing function of $S$, there exist large enough values of $S > K$ such that $P_E(S, K, \tau) + [(e^{-q\tau} - 1)S - (e^{-r\tau} - 1)K] < 0$, i.e., $C_E(S, K, \tau) < S - K$ for some $S > K$. From the previous result we therefore have $C(S, K, \tau) \neq C_E(S, K, \tau)$ and hence conclude that the early-exercise premium is nonzero for $q > 0$. This proves (i), while statement (ii) is proved in a similar fashion by reversing the roles of $S,q$ with $K,r$ and is left as an exercise. □

An obvious consequence of this proposition is that: (i) for an American call on a non-dividend-paying stock the exercise boundary is trivial (i.e., it is never optimal to exercise early), and (ii) for an American put on a nondividend-paying stock the exercise boundary is nontrivial (i.e., there is an optimal early-exercise time) if the interest rate is positive. In what follows (and also from the framework of Section 1.14.4) we will be able to further assess such properties.

### Pricing by Recurrence: Dynamic Programming Approach

We now consider specifically the recursive formulation for pricing American options. This involves an iteration method that goes backward in calendar time (or forward in time to maturity). Formally, the American option price is given by equation (1.448). In order to actually implement this formula in a practical manner, we subdivide the time interval $[t_0, T] = [t_0, t_1, \ldots, t_N = T]$ into $N \geq 1$ subintervals $[t_i, t_{i+1}]$, $\delta t_i = t_{i+1} - t_i > 0$, $i = 0, \ldots, N-1$. For notational purposes it is useful to introduce the price function $V_t(S)$. For the case of time-homogeneous diffusions we have

$$V_t(S) \equiv V(S, T-t) = V(S, \tau), \tag{1.458}$$

with $\tau = T - t$ being the time remaining to maturity. We therefore assume that exercise can only occur at a fixed set of (intermediate stopping) times given by $\{t_i : i = 0, \ldots, N\}$. Equation (1.448) can then be approximated by

$$V_0(S_0) = \sup_{t \in \{t_i : i=0,\ldots,N\}} E_0\big[e^{-r(t-t_0)}\phi(S_t)\big], \tag{1.459}$$

$V_0(S_0) \equiv V_{t_0}(S_0) = V(S_0, T - t_0)$. For small $\delta t_i$ values we expect equation (1.459) to be a good approximation to equation (1.448). From the theory of optimal stopping rules, one can show that in the limit $\delta t_i \to 0$ ($N \to \infty$) this approximation approaches the exact American option value in equation (1.448), which allows for continuous-time exercise. We remark that equation (1.459) actually gives the exact price of a *Bermudan* option with payoff function $\phi$. Bermudans are bonafide contracts that essentially lie in between European and American contracts and are in reality structured specifically with only a fixed set of allowable exercise dates. Moreover, in any realistic trading strategy it is interesting to note that the actual information on asset price levels can only be accessible to the trader at intermittent times (i.e., at best one obtains "tick-by-tick"data). Hence, for the holder of an American option the exercise decision times, although approaching the continuum limit, essentially occur at discretely spaced points in time.

By discretizing time, the underlying asset price process with values $S_{t_i} \in \mathbb{R}_+$, $i = 0, \ldots, N$, is then a Markov chain. Iterating backward in calendar time starting from maturity, equation (1.459) is readily shown to imply that the option price at any intermediate time satisfies the recurrence relation

$$V_{t_i}(S) = \max\big\{\phi(S), E_{t_i}\big[e^{-r\delta t_i} V_{t_{i+1}}(S_{t_{i+1}}) | S_{t_i} = S\big]\big\}, \tag{1.460}$$

$i = N-1, \ldots, 0$, where $V_T(S) = \phi(S)$. This result states that the option price at each date $t_i$ is given by the maximum of the pay-off (or the immediate-exercise value) and the discounted expected value of continuing without early exercise at time $t_i$. Note that at each $i$th step the expectation is conditional on $S_{t_i} = S$. [Remark: Equation (1.460) can also be rewritten as a forward recurrence relation in terms of a discretized time to maturity variable $\tau_i = T - t_i$ using equation (1.458)]. This formulation can be applied to asset prices that obey diffusion processes with generally state- and time-dependent drift and volatility functions. Here and in the following subsections, however, we are assuming time-homogeneous solutions; i.e., the drift and volatility functions of the asset price process are only allowed to be explicitly state dependent. Assuming a generally state-dependent Markov diffusion process $(S_t)_{t \geq 0}$, $S_t \in \mathbb{R}_+$ with assumed risk-neutral transition probability density function $p(S', S; \tau)$, the earlier expectation then gives

$$V_{t_i}(S) = \max\big\{\phi(S), \tilde{V}_{t_i}(S)\big\}, \tag{1.461}$$

where

$$\tilde{V}_{t_i}(S) = e^{-r\delta t_i}\int_0^\infty p(S', S; \delta t_i)V_{t_{i+1}}(S')dS' \tag{1.462}$$

represents the *continuation value* of the option at time $t_i$. For the particular process, of equation (1.381), $p$ is specifically the lognormal density function given by equation (1.382). In this iteration approach, the American (or Bermudan) option prices are obtained without necessarily computing the early-exercise boundary. However, this can also be obtained simultaneously. From equation (1.461) we see that equations (1.449), (1.450), and (1.451) give the stopping rule

$$t^* = \min\{t_i; i = 0, \ldots, N: \phi(S_{t_i}) = \tilde{V}_{t_i}(S_{t_i})\}, \tag{1.463}$$

the early-exercise (stopping) domain as the union of line segments

$$\mathcal{D} = \cup_{i=0,\ldots,N}\{(t_i, S) : \phi(S) \geq \tilde{V}_{t_i}(S)\}, \tag{1.464}$$

and the continuation domain

$$\mathcal{C} = \cup_{i=0,\ldots,N}\{(t_i, S) : \phi(S) < \tilde{V}_{t_i}(S)\}. \tag{1.465}$$

### Relation to Lattice (Tree) Methods

The dynamic programming approach provides a basis for implementing a number of different numerical methods for computing option prices using either Monte Carlo simulations, quadrature rules of integration, lattice methods, or a combination of such methods. In particular, the dynamic programming formulation can be directly related to the simplest of the lattice models — the binomial and trinomial lattices. For a detailed exposition on the implementations of lattice methods for pricing American options (as well as their European counterparts) the reader is urged to take a close look at the relevant numerical projects in Part II. The intricate details as well as the relevant equations and algorithms are explicitly described in those projects — the reader is also given the opportunity to numerically program the option-pricing applications. Here we shall simply give a very brief and generic discussion, meant only to emphasize the basic connection between the dynamic programming formulation and the lattice pricing models without having to repeat the underlying details.

Lattice methods can be viewed as either: (i) approximate solutions to recurrence relation (1.460) (or alternatively as approximate solutions to the equivalent option-pricing PDE by way of finite differences) or (ii) option-pricing models in their own right. Lattice models can accommodate time-inhomogeneous processes, as is the case for time-dependent drift and/or volatility functions. However, let's assume time-homogeneous models, where the underlying asset or stock price process is essentially modeled as a Markov chain on a discrete set of possible states. Generally, one assumes that the stock price can only move on a set of *nodes*, each denoted by a pair of integers *(i,j)* corresponding to a stock price value $S_j^i$. The lattice is a mesh or grid made up of all such nodes, where the integer $j$ is an index for the spatial position of the stock price on the lattice at time $t_i$, $i = 0, \ldots, N$. Lattice models allow for the implementation of time steps of fixed or variable size, but for the sake of simplicity let's assume a fixed time step of size $\delta t = (T - t_0)/N$. In fact, most implementations are based on equal-size time steps. Then conditional on $S_{t_i} = S_j^i$, the probability of a movement of the stock price within a single time step $\delta t$ from a node $(i, j)$ into a successor node $(i+1, j')$, with value $S_{t_{i+1}} = S_{j'}^{i+1}$, is given by the transition probability value $P(S_{t_{i+1}} = S_{j'}^{i+1} | S_{t_i} = S_j^i) \equiv p_{j\to j'} > 0$.

Although not critical to the present discussion, we note that for the binomial model there are only two successor nodes with $j' = j, j+1$, whereas the trinomial model has three successor nodes with $j' = j-1, j, j+1$, and so on.

The positive quantities $p_{j\to j'}$ are risk-adjusted probabilities and must obviously obey probability conservation,

$$\sum_{j'} p_{j\to j'} = 1, \qquad \text{for all } j, \tag{1.466}$$

where the sum is over all successor nodes in the model. Assuming the risk-neutral measure with money market as numeraire, the expected rate of return of the stock must equal the risk-free rate; i.e., $E_t[S_{t+\delta t}] = S_t e^{r\delta t}$. This is the risk-neutrality or no-arbitrage condition. For the lattice model it takes the form

$$\sum_{j'} p_{j\to j'} S_{j'}^{i+1} = e^{\mu\delta t} S_j^i, \tag{1.467}$$

for all $(i, j)$ nodes, where $\mu = r$ or $\mu = r - q$ for nondividend- or dividend-paying stock. In order to capture the variance in the asset price returns, the lattice model is also built to take into account the asset price volatilty. For instance, one can relate the variation either of stock prices or of the log-returns that are computed separately using the diffusion model and the lattice model. If the variation or second moment of the log-returns are considered, then we have $E_t[(\delta \log S_t)^2] = (\sigma(S_t))^2\,\delta t$ within order $\delta t$, where $\sigma(S_t)$ is the *local* volatility function for the general case of a state-dependent diffusion model of the form $\delta S_t = \mu(S_t)S_t\,\delta t + \sigma(S_t)S_t\,\delta W_t$. Applying this same expectation at each node within the lattice model and equating the two expectations gives

$$(\sigma_j^i)^2 \delta t = \sum_{j'} p_{j\to j'} \log^2(S_{j'}^{i+1}/S_j^i), \tag{1.468}$$

where $\sigma_j^i = \sigma(S_j^i)$ forms a set of volatility parameters. This is just one possible way of introducing lattice volatility parameters into the model. Equations (1.466), (1.467), and (1.468) are therefore collective constraints on the lattice geometry and the nodal transition probabilities. These form an integral part of the construction of the lattice model and its parameters — this is part of the model *calibration* procedure. Further steps in the calibration can also be undertaken by fitting the lattice parameters so that certain computed option prices exactly match the corresponding market prices. In most applications the number of adjustable lattice parameters is greatly reduced. In particular, for geometric Brownian motion there is only one volatility parameter, i.e., $\sigma_j^i \to \sigma$. Moreover, most lattice models are simplified by assuming that the nodal transitions are independent of the starting node, as is the case for constant local volatilities, i.e., $p_{j\to j'} \to p_{j'}$. For specific details on the contruction of lattices and on implementing various calibration schemes for American and European option pricing within the binomial and trinomial models, we again refer the reader to the relevant projects in Part II.

Once the lattice geometry and transition probabilities are determined, i.e., the lattice is calibrated, the option prices at each node in the lattice, $V_j^i = V_{t_i}(S_j^i)$, can be determined by recurrence:

$$V_j^i = \max\left\{\phi(S_j^i),\, e^{-r\delta t}\sum_{j'} p_{j\to j'} V_{j'}^{i+1}\right\}. \tag{1.469}$$

The current option price $V_0^0 = V_0(S_0)$ at spot $S_0^0 \equiv S_0$ is obtained by simply iterating over $N$ time steps, starting from the known payoff $V_j^N = \phi(S_j^N)$ at the terminal node values $S_j^N$.

Equation (1.469) also divides up the lattice into two groups of nodes: (i) a stopping domain as the set $\{(i, j) : V_j^i = \phi(S_j^i)\}$ and (ii) a continuation domain as the set $\{(i, j) : V_j^i > \phi(S_j^i)\}$. This second set gives the times $t_i$ and spot values $S_j^i$ for which the option should not be exercised early. According to equation (1.463), the optimal stopping time is

$$t^* = \min\{t_i = i\delta t : V_j^i = \phi(S_j^i)\}. \tag{1.470}$$

The early-exercise boundary is then also readily obtained. For instance, for a call this is the set of points $(i\delta t, S_*^i)$, $i = 0, \ldots, N$, where $S_*^i = \max\{S_j^i : V_j^i > S_j^i - K\}$; for a put, $S_*^i = \min\{S_j^i : V_j^i > K - S_j^i\}$. This offers a simple approach for approximating the early-exercise boundary curve in the continuous diffusion model corresponding to the limit $\delta t \to 0$. However, the resulting curve will not be smooth, even for relatively small time steps. More accurate calculations are afforded by applying more advanced techniques, such as the integral-equation approach discussed in Section 1.14.4. For the case of a trinomial lattice, equation (1.469) is related to the *explicit finite-difference* scheme for solving the Black–Scholes PDE. Alternative PDE solvers are based on *implicit finite-difference schemes*. Implicit schemes require the solution of a linear system of equations (or matrix inversion) for each time step in the propagation, yet they may offer more flexibility in the allowable range of lattice parameters for achieving accuracy and numerical stability. We refer the reader to the "Crank–Nicolson Option Pricer" project in Part II, which discusses a special type of implementation of the Crank–Nicolson implicit scheme for calibration and option pricing on a mesh.

### The Smooth Pasting Condition and PDE Approach

Although the free-boundary curve is not analytically computable as a function of time, one can generally establish the *smooth pasting condition*. This property guarantees that the price function for an American option has a continuous derivative at the exercise boundary and that the derivative is equal to the derivative of the payoff function at the exercise boundary. The following proposition summarizes this result.

**Proposition.** *Let $\mathcal{D}_\tau$, with time to expiry $\tau = T - t > 0$, be the early-exercise domain for which $V_t(S) \equiv V(S, \tau) = \phi(S)$ when $S \in \mathcal{D}_\tau$, where $\phi$ is any differentiable payoff function. Then the American option price function $V$ satisfies the smooth pasting condition at the boundary denoted by $S^*(\tau) \equiv S_t^*$:*

$$\left.\frac{\partial V(S, \tau)}{\partial S}\right|_{S=S^*(\tau)} = \phi'(S^*(\tau)), \tag{1.471}$$

*and the zero-time-decay condition obtains on the early-exercise domain,*

$$\frac{\partial V(S, \tau)}{\partial \tau} = 0, \qquad \textit{for } S \in \mathcal{D}_\tau. \tag{1.472}$$

Remark: The condition in equation (1.471) is also obviously valid for $S \in \mathcal{D}_\tau$ (excluding the boundary) since $V(S, \tau) = \phi(S)$ on that domain. What is important to emphasize here is that the derivative is continuous at the boundary of the stopping and continuation domains. These properties are valid under general proper Itô diffusion models. For a call (or put), then, equation (1.471) simply gives $\frac{\partial V(S^*(\tau),\tau)}{\partial S} = 1$ (or $-1$). This is illustrated in Figure 1.6. Although this proposition can be formally proven from the PDE approach, we shall instead demonstrate how it arises based on a dynamic hedging strategy argument, which turns out to be financially more insightful. First we note that the graph of the American option value

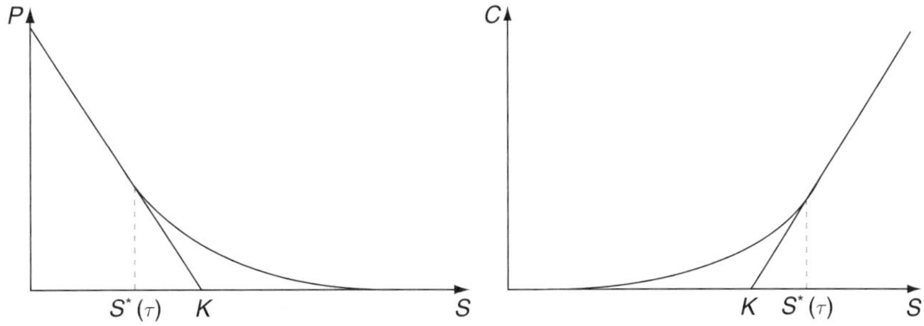


FIGURE 1.6 The pricing functions for an American put (left) and an American call (right) with continuous dividend yield satisfy the smooth pasting condition with slope equal to $-1$ and 1, respectively, at the optimal exercise boundary $S^*(\tau)$ for given time to expiry $\tau > 0$.

is never below that of the payoff function. Moreover, for given calendar time $t$ (or time to maturity $\tau$), the slope of the graph of $V_t(S) = V(S, \tau)$ at the exercise boundary point $S = S^*(\tau) \equiv S_t^*$ must be less (greater) than or equal to that of the payoff function if the latter is an increasing (decreasing) function at the boundary. That is: (i) $\frac{\partial V_t(S)}{\partial S}|_{S=S_t^{*(-)}} \leq \phi'(S_t^*)$ for the case $\phi'(S_t^*) \geq 0$ or (ii) $\frac{\partial V_t(S)}{\partial S}|_{S=S_t^{*(+)}} \geq \phi'(S_t^*)$ for the case $\phi'(S_t^*) \leq 0$. Here we use $S_t^{*(\pm)}$ to denote the limiting values from the right (+) or left ($-$) of $S_t^*$. Our objective is to show that these inequalities in the slopes are actually strict equalities. We now show this for case (i) as the argument follows in identical fashion for case (ii). In particular, let us assume that the asset or stock price at calendar time $t$ is at the boundary; i.e., let $S_t = S_t^*$. After an infinitesimally small time lapse $\delta t$, the stock price can move either up into the exercise domain $\mathcal{D}_\tau$ or down into the (no-exercise) domain of continuation. If the stock price moves upward, then its change is $\delta S_t = S_{t+\delta t} - S_t^* > 0$, so $S_{t+\delta t} > S_t^*$ and it remains in the exercise domain. In this case, $V_{t+\delta t}(S_{t+\delta t}) = \phi(S_{t+\delta t})$ and the option value changes by an amount $\delta V_t = \phi(S_{t+\delta t}) - \phi(S_t^*) = \phi'(S_t^*)\delta S_t$, to leading order in $\delta t$. So to achieve a delta hedge for an upward tick over time $\delta t$, the option writer has to buy $\Delta_t = \phi'(S_t^*)$ shares of the stock. The writer's delta-hedge portfolio at time $t$ consists of one short position in the option and $\Delta_t$ shares in the stock. Hence for an upward tick the hedge portfolio has value $\pi_t = -V_t(S_t^*) + \Delta_t S_t^* = -V_t(S_t^*) + \phi'(S_t^*)S_t^*$, and the change in portfolio value is $\delta\pi_t = -\delta V_t + \phi'(S_t^*)\delta S_t = 0$, to leading order in $\delta t$. On the other hand, if at time $t$ the stock ticks down, then $\delta S_t < 0$, $S_{t+\delta t} < S_t^*$; hence the stock price falls into the domain of continuation. Now assume the SDE in equation (1.381) holds. [Note: The same argument also readily follows if we assume a more general Itô diffusion with state- and time-dependent drift and volatility.] To leading order, then,

$$\delta S_t = \sigma S_t^* \delta W_t = -\sigma S_t^* \sqrt{\delta t}|z|, \tag{1.473}$$

where $z \sim N(0, 1)$, since $\delta W_t < 0$ for a downward tick. Now, $\delta V_t = \frac{\partial V_t(S)}{\partial S}|_{S=S_t^{*(-)}} \delta S_t$ and, using the foregoing expression, the hedge portfolio changes by

$$\begin{aligned}\delta\pi_t &= -\delta V_t + \phi'(S_t^*)\delta S_t \\ &= \left[\frac{\partial V_t(S)}{\partial S}\bigg|_{S=S_t^{*(-)}} - \phi'(S_t^*)\right]\sigma S_t^* \sqrt{\delta t}|z|.\end{aligned} \tag{1.474}$$

Taking expectations and using $E[|z|] = \sqrt{2/\pi}$ gives the expected change in the hedge portfolio:

$$E[\delta\pi_t] = \sqrt{\frac{2}{\pi}}\left[\left.\frac{\partial V_t(S)}{\partial S}\right|_{S=S_t^{*(-)}} - \phi'(S_t^*)\right]\sigma S_t^* \sqrt{\delta t}. \tag{1.475}$$

We hence conclude that the writer cannot exactly set up a delta hedge portfolio and in particular is expected to suffer a loss every time the underlying stock is in the vicinity of the boundary *unless* $\frac{\partial V_t(S_t^{*(-)})}{\partial S} = \phi'(S_t^*)$. Since $\frac{\partial V_t(S_t^{*(+)})}{\partial S} = \phi'(S_t^*)$, the function $\frac{\partial V_t(S)}{\partial S} \equiv \frac{\partial V(S,\tau)}{\partial S}$ is continuous at the boundary and we have established equation (1.471).

The zero-time-decay condition is shown by simply considering the total change in the American option value *along* the boundary $S = S^*(\tau)$ as the calendar time (or time to maturity) changes and the boundary point moves accordingly. Along the boundary we have $V(S^*(\tau), \tau) = \phi(S^*(\tau))$, and differentiating both sides of this relation w.r.t. $\tau$ gives (Note: The analysis in terms of $t$ is the same):

$$\frac{\partial V(S^*(\tau), \tau)}{\partial S}\frac{dS^*(\tau)}{d\tau} + \frac{\partial V(S^*(\tau), \tau)}{\partial \tau} = \phi'(S^*(\tau))\frac{dS^*(\tau)}{d\tau}. \tag{1.476}$$

Hence, using equation (1.471) gives $\frac{\partial V(S^*(\tau),\tau)}{\partial \tau} = 0$, and since the option is given by the time-independent payoff function everywhere else on the stopping domain, we have equation (1.472).

Delta hedging and continuous-time replication arguments apply to American options in the same way they apply to European options. Within the (no-exercise) continuation domain we therefore expect and require that the option price function satisfy the Black–Scholes PDE. The connection between the optimal stopping time formulation and the PDE approach can be shown as follows. Consider recurrence relation (1.460) with time step $\delta t > 0$ for any calendar time $t < T$,

$$V_t(S) = \max\left\{\phi(S), e^{-r\delta t}E_t\left[V_{t+\delta t}(S_{t+\delta t})|S_t = S\right]\right\}. \tag{1.477}$$

Assuming $V_t(S)$ is sufficiently smooth with continuous derivatives then to leading order $O(\delta t)$, we can Taylor-expand $V_{t+\delta t}(S_{t+\delta t})$ while using Itô's lemma. For a generally state- and time-dependent process obeying $\delta S_t = \mu(S_t, t)\delta t + \sigma(S_t, t)\delta W_t$, we have

$$\begin{aligned} V_t(S) &= \max\left\{\phi(S), (1 - r\delta t)E_t\left[V_t(S_t) + \left(\frac{\partial V_t(S_t)}{\partial t} + \mu(S_t, t)\frac{\partial V_t(S_t)}{\partial S_t}\right.\right.\right. \\ &\quad \left.\left.\left. + \frac{1}{2}\sigma^2(S_t, t)\frac{\partial^2 V_t(S_t)}{\partial S_t^2}\right)\delta t + \sigma(S_t, t)\frac{\partial V_t(S_t)}{\partial S_t}\delta W_t\middle| S_t = S\right]\right\} + O((\delta t)^2) \\ &= \max\left\{\phi(S), V_t(S) + \left[\frac{\partial V_t(S)}{\partial t} + \mathcal{L}_{BS}V_t(S)\right]\delta t\right\} + O((\delta t)^2). \end{aligned} \tag{1.478}$$

The second equation obtains by evaluating the conditional expectation (which sets $S_t = S$ and eliminates the $\delta W_t$ term) and then collecting terms up to $O(\delta t)$. This expression has been written more compactly using the Black–Scholes differential operator (for general drift and volatility functions) defined by

$$\mathcal{L}_{BS}V \equiv \frac{1}{2}\sigma^2(S, t)\frac{\partial^2 V}{\partial S^2} + \mu(S, t)\frac{\partial V}{\partial S} - rV \equiv (\mathcal{L}_{S,t} - r)V. \tag{1.479}$$

For values of $S$ in the continuation domain, the inequality $V_t(S) > \phi(S)$ is satisfied, and hence, from equation (1.478) we must have the Black–Scholes PDE:

$$\frac{\partial V_t(S)}{\partial t} + \mathcal{L}_{BS} V_t(S) = 0, \qquad \text{for all } S \notin \mathcal{D}_\tau. \tag{1.480}$$

By specializing to the geometric Brownian motion model, then, $\mu(S, t) = (r - q)S$, $\sigma(S, t) = \sigma S$ and the Black–Scholes PDE is

$$\frac{\partial V}{\partial \tau} = \frac{\sigma^2 S^2}{2} \frac{\partial^2 V}{\partial S^2} + (r - q)S \frac{\partial V}{\partial S} - rV \equiv \mathcal{L}_{BS} V, \qquad \text{for all } S \notin \mathcal{D}_\tau. \tag{1.481}$$

Thanks to the time-homogeneous property of the solution in this case, we have a PDE in terms of the time-to-maturity variable, $V = V(S, \tau)$, which will be convenient in subsequent discussions.

## 1.14.2 Perpetual American Options

An option with infinite time to maturity is called a *perpetual option*. Here we consider perpetual American calls and puts. These options are instructive since simple analytic solutions exist. Moreover, since the exercise boundary $S^*(\tau)$ is a monotonic function of time to maturity $\tau$ (i.e., increasing for a dividend-paying American call and decreasing for an American put), the perpetual option price provides us with the asymptotic limit $\lim_{\tau \to \infty} S^*(\tau) \equiv S^*$ of the exercise boundary for times infinitely far from maturity. We again consider an asset price process $S_t$ following geometric Brownian motion with constant interest rate $r$ and continuous dividend yield at constant rate $q$. Since a perpetual option has infinite time to maturity, its value does not depend on the passage of time; i.e., the price function is independent of time. Hence the time derivative of the price function is zero and the Black–Scholes partial differential equation (1.481) for the price of a perpetual option reduces to a time-independent ordinary differential equation (ODE).

We first consider the case of a perpetual put struck at $K$. The price function denoted by $P(S)$ must satisfy the ODE

$$\frac{1}{2}\sigma^2 S^2 \frac{d^2 P}{dS^2} + (r - q)S \frac{dP}{dS} - rP = 0 \tag{1.482}$$

for values away from the exercise boundary, $S^* < S < \infty$. The optimal exercise price $S^*$ is therefore the asset price at which the perpetual American put should be exercised. Since the value of the perpetual put must be equal to the intrinsic value at all values of $S \leq S^*$ and $S^* < K$, (see Figure 1.6) the boundary conditions on $P(S)$ are

$$\lim_{S \to \infty} P(S) = 0, \qquad P(S^*) = K - S^*. \tag{1.483}$$

$S^*$ is yet unknown but uniquely determined once $P(S)$ is obtained in terms of $S^*$ as described just next. Equation (1.482) is an ODE of the Cauchy–Euler (equidimensional) type and therefore has the general solution

$$P(S) = a_+ S^{\gamma_+} + a_- S^{\gamma_-}, \tag{1.484}$$

where $a_\pm$ are arbitrary constants and $\gamma_\pm$ are roots of the auxiliary quadratic equation

$$\tfrac{\sigma^2}{2}\gamma^2 + (r - q - \tfrac{\sigma^2}{2})\gamma - r = 0. \tag{1.485}$$

Solving for the roots gives

$$\gamma_{\pm} = \frac{-(r-q-\frac{\sigma^2}{2}) \pm \sqrt{(r-q-\frac{\sigma^2}{2})^2 + 2\sigma^2 r}}{\sigma^2}. \tag{1.486}$$

Assuming positive interest rate $r$, then $\gamma_-$ and $\gamma_+$ are negative and positive roots, respectively. To satisfy the first condition at infinity in equation (1.483) we must have $a_+ = 0$. By satisfying the second boundary condition in equation (1.483), $a_- = (K - S^*)/(S^*)^{\gamma_-}$, we obtain the price function in the form

$$P(S) = (K - S^*)\left(\frac{S}{S^*}\right)^{\gamma_-}, \qquad S \geq S^*. \tag{1.487}$$

The exercise boundary value $S^*$ can now be determined as the optimal value that maximizes the price $P(S)$ for all possible choices of $S^*$. The derivative w.r.t. the parameter $S^*$ of this price function gives

$$\frac{\partial P}{\partial S^*} = -\left(\frac{S}{S^*}\right)^{\gamma_-}\left(1 + \frac{K - S^*}{S^*}\gamma_-\right). \tag{1.488}$$

Setting this derivative to zero yields the extremum

$$S^* = \frac{K\gamma_-}{\gamma_- - 1}. \tag{1.489}$$

Computing the second derivative at this extremum gives $\frac{\partial^2 P}{\partial S^{*2}} = \frac{K\gamma_-}{(S^*)^2}(\frac{S}{S^*})^{\gamma_-} < 0$. Hence $S^*$ in equation (1.489) is a maximum, and inserting its value into equation (1.487) gives the price of the perpetual American put in the equivalent forms

$$\begin{aligned} P(S) &= \frac{K}{1-\gamma_-}\left(\frac{\gamma_- - 1}{\gamma_-}\right)^{\gamma_-}\left(\frac{S}{K}\right)^{\gamma_-} \\ &= -\frac{S^*}{\gamma_-}\left(\frac{S}{S^*}\right)^{\gamma_-}, \end{aligned} \tag{1.490}$$

for $S \geq S^*$. This solution is easily shown to satisfy the required smooth pasting condition

$$\left.\frac{dP}{dS}\right|_{S=S^*} = -1. \tag{1.491}$$

Next we consider the perpetual American call struck at $K$. As in the case of the put, the price function now denoted by $C(S)$ also satisfies equation (1.482), but for values $0 < S < S^*$. The optimal value $S^*$ is therefore the asset price at which the call should be exercised. The value $C(S)$ must be given by the intrinsic value of the call pay-off for values on the boundary $S \geq S^*$, where $S^* > K$; hence the boundary conditions are

$$\lim_{S\to 0} C(S) = 0, \quad C(S^*) = S^* - K. \tag{1.492}$$

The general solution is again given by equations (1.484) and (1.486). However, by satisfying the boundary conditions in equation (1.492) we now instead have $a_- = 0$ and $a_+ = (S^* - K)/(S^*)^{\gamma_+}$, giving

$$C(S) = (S^* - K)\left(\frac{S}{S^*}\right)^{\gamma_+}, \qquad 0 < S < S^*. \tag{1.493}$$

Using the same procedure as for the put, the optimal exercise boundary is determined by finding the maximum of $C(S)$ w.r.t. $S^*$, giving

$$S^* = \frac{K\gamma_+}{\gamma_+ - 1}. \tag{1.494}$$

Using $S^*$ from equation (1.494) in equation (1.493) gives the price of the perpetual American call, written equivalently in terms of $K$ or $S^*$:

$$\begin{aligned} C(S) &= \frac{K}{\gamma_+ - 1}\left(\frac{\gamma_+ - 1}{\gamma_+}\right)^{\gamma_+}\left(\frac{S}{K}\right)^{\gamma_+} \\ &= \frac{S^*}{\gamma_+}\left(\frac{S}{S^*}\right)^{\gamma_+}. \end{aligned} \tag{1.495}$$

This satisfies the required smooth pasting condition

$$\left.\frac{dC}{dS}\right|_{S=S^*} = 1. \tag{1.496}$$

It is instructive to examine what happens to the exercise boundary in the two separate limiting cases: (i) zero interest rate $r = 0$ and (ii) zero dividend yield $q = 0$. In case (i) we have from equation (1.486) that $\gamma_- = 0$ (assuming $q \geq -\sigma^2/2$, which is the case if $q \geq 0$). From equation (1.489) we see that $S^* = 0$; hence, for zero interest rate the perpetual put is never exercised early. This is consistent with the property of an American put for r = 0 and for any finite time to maturity, as shown in the next section. From a financial standpoint, there is no time value gained from an early pay-off with zero interest. For case (ii): Equation (1.486) gives $\gamma_+ = 1$ (assuming $r \geq -\sigma^2/2$, which is the case for $r \geq 0$). Moreover, $\gamma_+ \to 1^+$ as $q \to 0^+$ and from equation (1.494) we have $S^* \to \infty$. Hence in the limit of zero dividend yield the perpetual call is never exercised early, irrespective of the interest rate. This feature is also consistent with the plain American call of finite maturity, as shown in the next section.

## 1.14.3 Properties of the Early-Exercise Boundary

The perpetual American option formulas of the previous section already allowed us to determine the precise behavior of the optimal exercise boundary in the asymptotic limit of infinite time to expiry, i.e., as $\tau \to \infty$. To further complete the analysis of the boundary we now consider the opposite limit, of infinitesimally small positive time to maturity $\tau \to 0^+$. In particular, let us consider the case of the Amercian call struck at $K$ with continuous dividend yield $q$ and price function denoted by $C(S, K, \tau)$ at spot $S$. Since $C(S, K, \tau)$ is an increasing function of $\tau$, for $\tau > 0$, the graph of the American call price (plotted as a function of $S$) with greater time to maturity $\tau_2$ must lie above the graph of the price function for the corresponding call with time to maturity $\tau_1 < \tau_2$. Furthermore, the smooth pasting condition guarantees that the price functions join the intrinsic line at levels $S^*(\tau_1) - K$ and $S^*(\tau_2) - K$, respectively, giving $S^*(\tau_1) < S^*(\tau_2)$. Hence, we conclude that $S^*(\tau)$ is a continuously increasing function of positive $\tau$. To put this in financial terms, an American call with greater time to maturity should be exercised deeper in the money to account for the loss of time value on the strike $K$. Due to the fact that one would never prematurely exercise at a spot value below the strike level (i.e., exercising for a nonpositive pay-off), the early-exercise boundary for an Amercian call must, in addition, satisfy the property $S^*(\tau) > K$ for all $\tau > 0$.

To determine the boundary in the limit $\tau \to 0^+$, note that the option value approaches the intrinsic value; i.e., at expiry it is exactly given by the payoff function $C(S, K, \tau = 0) = S - K$ for values on the exercise boundary. Inserting this function into the right-hand side of equation (1.481) and taking derivatives gives

$$\frac{\partial C(S, K, 0^+)}{\partial \tau} = rK - qS \tag{1.497}$$

for $S > K$. Since the condition $\partial C(S, K, 0^+)/\partial \tau > 0$ ensures that the option is still alive (i.e., not yet exercised), the spot value $S$ at which $\partial C(S, K, 0^+)/\partial \tau$ becomes negative and hence for which the call is exercised at an instant just before expiry is given by $S = \frac{r}{q}K$. This is the case, however, if the value $\frac{r}{q}K$ is in the interval $S > K$, that is, if $r > q > 0$. In this case, just prior to expiry the call is not yet exercised if the spot is in the region $K < S < \frac{r}{q}K$ but would be exercised if $S \geq \frac{r}{q}K$. Hence, $S^*(0^+) = \frac{r}{q}K$ for $r > q > 0$. In the other case, $r \leq q$, so $\frac{r}{q}K \leq K$. Yet $S > K$, so $S^*(0^+) = K$ for $r \leq q$. Note that the condition $S^*(0^+) > K$ is not possible in this case because this leads to a suboptimal early exercise, since the loss in dividends would have greater value than the interest earned over the infinitesimal time interval until expiry. Combining these arguments we arrive at the general limiting condition for the exercise boundary of an American call just prior to expiry:

$$\lim_{\tau \to 0^+} S^*(\tau) = \max(K, \frac{r}{q}K). \tag{1.498}$$

From this property we see that $S^*(0^+) \to \infty$ as $q \to 0$. Hence, for zero dividend yield the American call is never exercised early, which is consistent with the fact that the plain (nondividend) American call has exactly the same worth as the plain European call.

Similar arguments can also be employed in the case of the Amercian put struck at $K$ with continuous dividend yield $q$. At expiry the put has value $P(S, K, \tau = 0) = K - S$ for values on the exercise boundary. We leave it as an exercise for the reader to show that the exercise boundary of an American put just prior to expiry is given by

$$\lim_{\tau \to 0^+} S^*(\tau) = \min(K, \frac{r}{q}K). \tag{1.499}$$

For $r = 0$ we therefore have $S^*(0^+) = 0$, irrespective of the value of $q$. Since $S^*(\tau)$ is a decreasing function of $\tau$, we conclude that the early-exercise boundary is always at zero, meaning that the American put with zero interest rate is never exercised before maturity. This is consistent with the conclusion we arrived at earlier, where we considered the perpetual American put. For $q \leq r$ we observe that the early-exercise boundary just before expiry is at the strike, $S^*(0^+) = K$. A special case of this is the vanilla American put, i.e., when $r > 0$ and $q = 0$. Figure 1.7 gives an illustration of typical early-exercise boundaries for a call and put. Given a time to maturity of $T$ at contract inception, we see that the American call with nonzero dividend is not yet exercised (i.e., is still alive) on the domain of points $(S, \tau)$ below the exercise curve: $S \in [0, S^*(\tau))$ and $\tau \in (0, T]$. In contrast, the American put is kept alive above the exercise curve: $S \in (S^*(\tau), \infty)$ and $\tau \in (0, T]$.

## 1.14.4 The Partial Differential Equation and Integral Equation Formulation

The problem of pricing an American option can be formulated as an initial-value partial differential equation (PDE) with a time-dependent free boundary. The early-exercise boundary is an unknown function of time, which must also be determined as part of the solution. In particular, let $V(S, \tau)$ represent the pricing function of an American option with spot $S$

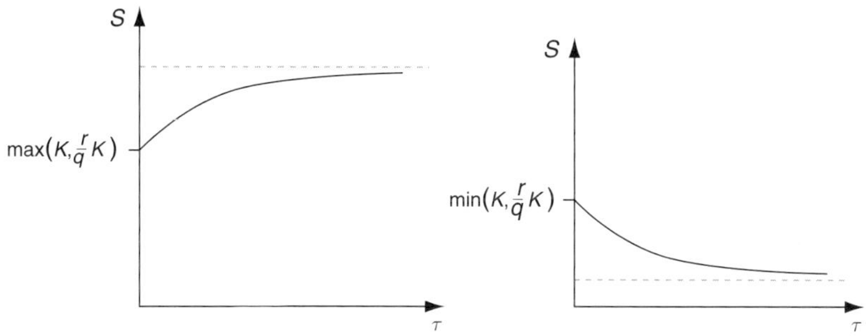


FIGURE 1.7 Early-exercise smooth boundary curves $S = S^*(\tau)$ for the American call (left), with $q > 0$, and put (right), with values depicted just before expiry $\tau \to 0^+$. In the limit of infinite time to expiry, the curves approach the horizontal asymptotes at $S = S^*$, where $S^*$ is given by equation (1.494) or equation (1.489) for the call or put, respectively.

and time to maturity $\tau$, $0 \leq \tau \leq T$, and having payoff or intrinsic function $V(S, 0) = \phi(S)$. Here we assume the pay-off is time independent, although the formulation also extends to the case of a known time-dependent payoff function. For given $\tau$, the solution domain is divisible into a union of two regions: (1) a continuation region $(S, \tau) \in \mathcal{D}'_\tau \times [0, T]$, for which the option is still alive or not exercised, and (2) a stopping region $(S, \tau) \in \mathcal{D}_\tau \times [0, T]$, where $\mathcal{D}_\tau$ is the complement of $\mathcal{D}'_\tau$ within $\mathbb{R}_+$, for which the American option is already exercised. The domains depend on $\tau$. As seen in the previous section, in the case of the American call, $\phi(S) = S - K$ on $\mathcal{D}_\tau = [S^*(\tau), \infty)$ (and $\mathcal{D}'_\tau = (0, S^*(\tau))$, while for the put, $\phi(S) = K - S$ on $\mathcal{D}_\tau = (0, S^*(\tau)]$ (and $\mathcal{D}'_\tau = (S^*(\tau), \infty)$. Assuming the underlying asset follows equation (1.381), equation (1.481) holds for $S \in \mathcal{D}'_\tau$. In contrast, the homogeneous Black–Scholes PDE does not hold on the domain of the early-exercise boundary, where the American option is given by the time-independent payoff function $V(S, \tau) = \phi(S)$. Since $\frac{\partial \phi(S)}{\partial \tau} = 0$, the solution on $\mathcal{D}_\tau$ satisfies $\frac{\partial V}{\partial \tau} = 0$. Combining regions and assuming the pay-off is twice differentiable gives a *nonhomogeneous* Black–Scholes PDE:

$$\frac{\partial V(S, \tau)}{\partial \tau} = \mathcal{L}_{BS} V(S, \tau) + f(S, \tau), \tag{1.500}$$

with (source) function

$$f(S, \tau) = \begin{cases} 0, & S \in \mathcal{D}'_\tau \\ -\mathcal{L}_{BS}\phi(S), & S \in \mathcal{D}_\tau, \end{cases} \tag{1.501}$$

where $\mathcal{L}_{BS}$ is the Black–Scholes differential operator. For geometric Brownian motion, $\mathcal{L}_{BS}$ is defined by equation (1.481). Given the function $f(S, \tau)$, whose time dependence is determined in terms of the free boundary, the solution to equation (1.500), subject to the initial condition $V(S, \tau = 0) = \phi(S)$ and boundary conditions $V(S = 0, \tau) = \phi(0)$, $V(S = \infty, \tau) = \phi(\infty)$, can be obtained in terms of the solution to the corresponding homogeneous Black–Scholes PDE. Recall from previous discussions that the transition probability density function $p(S', S; \tau)$ solves the forward Kolmogorov PDE in the $S'$ variable and the backward PDE in the spot variable $S$ with zero boundary conditions at $S = 0, \infty$ for all $\tau > 0$. As already mentioned, for process (1.381) $p$ is just the lognormal density given by equation (1.382). We also know that $e^{-r\tau} p$ solves the homogeneous Black–Scholes PDE. Combining these facts and applying

Laplace transforms, one arrives at the well-known Duhamel's solution to equation (1.500) in the form

$$V(S,\tau) = e^{-r\tau}\int_0^\infty p(S',S;\tau)\phi(S')dS' + \int_0^\tau e^{-r\tau'}\left[\int_0^\infty p(S',S;\tau')f(S',\tau-\tau')dS'\right]d\tau' \equiv V_E(S,\tau) + V^e(S,\tau). \tag{1.502}$$

One can readily verify that this solves equation (1.500), even for the more general case of state-dependent models (see Problem 1). An important aspect of this result is that the American option value $V(S,\tau)$ is expressible as a sum of two components. The first term is simply the European option value $V_E$, as given by the discounted risk-neutral expectation of the pay-off. Hence the second term, denoted by $V^e(S,\tau)$, must represent the early-exercise premium, which gives the holder the additional liberty of early exercise.

Assuming geometric Brownian motion for the underlying asset, equations (1.500) and (1.501) for the American call and put specialize to

$$\frac{\partial C}{\partial \tau} - \frac{\sigma^2 S^2}{2}\frac{\partial^2 C}{\partial S^2} - (r-q)S\frac{\partial C}{\partial S} + rC = \begin{cases} 0, & S < S^*(\tau) \\ qS - rK, & S \geq S^*(\tau) \end{cases} \tag{1.503}$$

and

$$\frac{\partial P}{\partial \tau} - \frac{\sigma^2 S^2}{2}\frac{\partial^2 P}{\partial S^2} - (r-q)S\frac{\partial P}{\partial S} + rP = \begin{cases} rK - qS, & S \leq S^*(\tau) \\ 0, & S > S^*(\tau) \end{cases}, \tag{1.504}$$

respectively. Here we used $\mathcal{L}_{BS}(S-K) = rK - qS$, and $S^*(\tau)$ denotes the early-exercise boundary for the respective call and put with strike $K$. The right-hand sides of these nonhomogeneous PDEs are nonzero only within the respective stopping regions. Using equation (1.502), the solutions to equations (1.503) and (1.504) for the American call and put price are given by

$$C(S,K,\tau) = C_E(S,K,\tau) + C^e(S,K,\tau) \tag{1.505}$$

and

$$P(S,K,\tau) = P_E(S,K,\tau) + P^e(S,K,\tau), \tag{1.506}$$

where the respective early-exercise premiums take on the integral forms

$$C^e(S,K,\tau) = \int_0^\tau e^{-r\tau'}\left[\int_{S^*(\tau-\tau')}^\infty p(S',S;\tau')(qS'-rK)dS'\right]d\tau' \tag{1.507}$$

and

$$P^e(S,K,\tau) = \int_0^\tau e^{-r\tau'}\left[\int_0^{S^*(\tau-\tau')} p(S',S;\tau')(rK-qS')dS'\right]d\tau'. \tag{1.508}$$

These premiums can also be recast as

$$C^e(S,K,\tau) = \int_0^\tau e^{-r\tau'} E_0\left[(qS_{\tau'} - rK)\mathbf{1}_{\{S_{\tau'} \geq S^*(\tau-\tau')\}}\right]d\tau' \tag{1.509}$$

and

$$P^e(S, K, \tau) = \int_0^{\tau} e^{-r\tau'} E_0\big[(rK - qS_{\tau'})\mathbf{1}_{\{S_{\tau'} \leq S^*(\tau-\tau')\}}\big]d\tau', \tag{1.510}$$

where $E_0$ denotes the current-time expectation, conditional on asset paths starting at $S_0 = S$ under the risk-neutral measure with density $p(S_{\tau'}, S; \tau')$. The time integral is over all intermediate times to maturity, and the indicator functions ensure that all asset paths fall within the early-exercise region. The properties of the early-exercise boundaries established in the previous section guarantee that the early-exercise premiums are nonnegative. For a dividend-paying call, equation (1.498), together with the indicator function condition, leads to $S_{\tau'} \geq \max(\frac{r}{q}K, K) \geq \frac{r}{q}K$; hence $qS_{\tau'} - rK \geq 0$ and $C^e$ is positive. A similar analysis follows for the put premium. The exercise premiums hence involve a continuous stream of discounted expected cash flows, beginning from contract inception until maturity. This lends itself to an interesting financial interpretation, as follows. Consider the case of the American put (a similar argument applies to the dividend-paying call) and an infinitesimal intermediate time interval $[\tau', \tau' + d\tau']$. Then from the holder's perspective the option should be optimally exercised if the asset price, given by $S_{\tau'}$ at time $\tau'$, attains the stopping region (i.e., reaches the early-exercise boundary with $S_{\tau'} \leq S^*(\tau - \tau')$ and $\tau - \tau'$ as the remaining time to maturity). Assuming that the holder is instead forced to keep the American put alive until expiry, the holder would have to be fairly compensated for the loss due to the delay in exercising during the time interval $d\tau'$. The value of this compensation is the difference between the interest on $K$ dollars and the dividend earned on the asset value $S_{\tau'}$, continuously compounded over time $d\tau'$. This cash flow is an amount $(rK - qS_{\tau'})d\tau'$, and corresponds to the early-exercise gain if the holder in fact had the privilege to optimally exercise. Allowing for all possible asset price scenarios from $S$ to $S_{\tau'}$ that attain the boundary gives rise to the expectation integral under the risk-neutral density for all intermediate times $0 \leq \tau' \leq \tau$. Summing up all of these infinitesimal cash flows and discounting their values to present time by an amount $e^{-r\tau'}$ gives the time integral, as in equation (1.508) or (1.510). We conclude that the early-exercise premium has an equivalent and alternative interpretation as a delay-exercise compensation.

The foregoing integral representations for the American call and put price can also be applied to cases where the volatility of the asset price process $S_t$ is considered generally state dependent. In order to implement the integral formulas, we need to be able to compute the transition density function $p$, either analytically or numerically. Moreover, the integrals can only be computed after having determined the early-exercise boundary $S^*(\tau')$ for $0 \leq \tau' \leq \tau$. For the geometric Brownian motion model (with constants $r$,$q$,$\sigma$), $p$ is given by the lognormal density, and the foregoing double integrals readily simplify to single time integrals in terms of standard cumulative normal functions. In particular, one readily derives explicit integral representations for the price of the American call and put (see Problem 2):

$$\begin{aligned} C(S, K, \tau) &= Se^{-q\tau}N(d_+) - Ke^{-r\tau}N(d_-) \\ &\quad + \int_0^{\tau} \big[qSe^{-q(\tau-\tau')}N(d_+^*(\tau')) - rKe^{-r(\tau-\tau')}N(d_-^*(\tau'))\big]d\tau', \end{aligned} \tag{1.511}$$

$$\begin{aligned} P(S, K, \tau) &= Ke^{-r\tau}N(-d_-) - Se^{-q\tau}N(-d_+) \\ &\quad + \int_0^{\tau} \big[rKe^{-r(\tau-\tau')}N(-d_-^*(\tau')) - qSe^{-q(\tau-\tau')}N(-d_+^*(\tau'))\big]d\tau', \end{aligned} \tag{1.512}$$

where

$$d_{\pm} = \frac{\log \frac{S}{K} + \left(r - q \pm \frac{1}{2}\sigma^2\right)\tau}{\sigma\sqrt{\tau}}, \tag{1.513}$$

$$d^*_{\pm}(\tau') = \frac{\log \frac{S}{S^*(\tau')} + \left(r - q \pm \frac{1}{2}\sigma^2\right)(\tau - \tau')}{\sigma\sqrt{\tau - \tau'}}. \tag{1.514}$$

These integral representations are valid for $S \in (0, \infty)$, $\tau \geq 0$. By setting $S = S^*(\tau)$ and applying the respective boundary conditions, $C(S^*(\tau), K, \tau) = S^*(\tau) - K$ for the call and $P(S^*(\tau), K, \tau) = K - S^*(\tau)$ for the put, equations (1.511) and (1.512) give rise to integral equations for the early-exercise boundary. For the call,

$$\begin{aligned} S^*(\tau) - K &= Se^{-q\tau}N(\tilde{d}_+) - Ke^{-r\tau}N(\tilde{d}_-) \\ &\quad + \int_0^{\tau} \left[qSe^{-q(\tau-\tau')}N(\tilde{d}^*_+(\tau')) - rKe^{-r(\tau-\tau')}N(\tilde{d}^*_-(\tau'))\right]d\tau'. \end{aligned} \tag{1.515}$$

and separately for the put,

$$\begin{aligned} K - S^*(\tau) &= Ke^{-r\tau}N(-\tilde{d}_-) - Se^{-q\tau}N(-\tilde{d}_+) \\ &\quad + \int_0^{\tau} \left[rKe^{-r(\tau-\tau')}N(-\tilde{d}^*_-(\tau')) - qSe^{-q(\tau-\tau')}N(-\tilde{d}^*_+(\tau'))\right]d\tau', \end{aligned} \tag{1.516}$$

where

$$\tilde{d}_{\pm} = \frac{\log \frac{S^*(\tau)}{K} + \left(r - q \pm \frac{1}{2}\sigma^2\right)\tau}{\sigma\sqrt{\tau}}, \tag{1.517}$$

$$\tilde{d}^*_{\pm}(\tau') = \frac{\log \frac{S^*(\tau)}{S^*(\tau')} + \left(r - q \pm \frac{1}{2}\sigma^2\right)(\tau - \tau')}{\sigma\sqrt{\tau - \tau'}}. \tag{1.518}$$

Note that equations (1.515) and (1.516) involve a variable upper integration limit and the integrands are nonlinear functions of $S^*(\tau)$, $S^*(\tau')$, $\tau$ and $\tau'$. From the theory of integral equations, equations (1.515) and (1.516) are known as *nonlinear Volterra integral equations*. Note that the solution $S^*(\tau)$, at time to maturity $\tau$, is dependent on the solution $S^*(\tau')$ from zero time to maturity $\tau' = 0$ up to $\tau' = \tau$. Although equations (1.515) and (1.516) are not analytically tractable, simple and efficient algorithms can be employed to solve for $S^*(\tau)$ numerically. For detailed descriptions on various numerical algorithms for solving these types of integral equations, see, for example, [DM88]. A typical procedure divides the solution domain into a regular mesh: $\tau_0 = 0$, $\tau_i = ih$, $i = 1, \ldots, n$, with $n$ steps spaced as $h = \tau/n$. By approximating the time integral via a quadrature rule (e.g., the trapezoidal rule), one obtains a system of algebraic equations in the values $S^*(\tau_i)$, which can be iteratively solved starting from the known value $S^*(\tau_0) = S^*(\tau = 0^+)$ at zero time to maturity. Alternatively, popular Runge–Kutta methods usually used for solving initial-value nonlinear ODEs can be also adapted to these integral equations. Once the early-exercise boundary is determined, the integral in equation (1.511) or (1.512) for the respective call or put can be computed. In particular, a quadrature rule that makes use of the computed points $S^*(\tau_i)$ can be implemented. Accurate approximations to the early-exercise boundary are obtained by choosing the number $n$ of points to be sufficiently large.

## Problems

**Problem 1.** Consider the state-dependent model $dS_t = \mu(S_t)dt + \sigma(S_t)dW_t$. Assuming $f(S, \tau)$ is differentiable w.r.t. $\tau$, show that equation (1.502) satisfies equation (1.500) for the appropriate operator $\mathcal{L}_{BS}$. Hint: Since $V_E$ satisfies the homogeneous Black–Scholes PDE, from superposition one need only show that $V^e$ satisfies equation (1.500). Use the property of interchanging order of differentiation and integration, integration by parts, and the fact that $e^{-r\tau}p$ satisfies the homogeneous Black–Scholes PDE with initial condition $p(S', S; 0) = \delta(S' - S)$. Provide an extension to equation (1.502), if possible, for the more general case of explicitly time-dependent drift and volatility.

**Problem 2.** (a) By employing similar manipulations as were used to obtain the standard Black–Scholes formulas in Section 1.6, derive equations (1.511) and (1.512) from equations (1.507) and (1.508). (b) Show that the pricing formulas for the American call and put in equations (1.511) and (1.512) satisfy the required boundary conditions at $S = 0$ and $S = \infty$.

**Problem 3.** Find an analytical formula for the price as well as the early-exercise boundaries of a perpetual American butterfly option with payoff function $\delta_\epsilon(S - K)$ given by equation (1.228) of Section 1.8. Assume $K - \epsilon > 0$ and that the underlying asset price obeys geometric Brownian motion with constant interest rate $r$ and continuous dividend yield $q$.

**Problem 4.** Using equations (1.511) and (1.512), derive integral representations for the delta, gamma, and vega sensitivities of the American call and put.

**Problem 5.** Let $V(S, \tau)$ and $V_E(S, \tau)$ denote the American and European option values, respectively, with spot $S$, time to maturity $\tau$, and payoff function $\phi(S)$. Assume a constant interest rate $r$ and continuous dividend yield $q$ under the geometric Brownian motion model for the process $S_t$. Prove the equivalence of these two statements:

(i) $V(S, \tau) > V_E(S, \tau)$ for all $S > 0$, $\tau > 0$.
(ii) $\phi(S) > e^{-r\tau}\phi(e^{(r-q)\tau}S)$ for some point $(S, \tau)$. Explain why American options on futures have a nonzero early-exercise premium.

**Problem 6.** Consider a Bermudan put option with strike $K$ at maturity $T$ with only a single intermediate early-exercise date $T_1 \in [0, T]$. Assume the underlying stock price obeys equation (1.381) within the risk-neutral measure, and let $P(S_t, K, T - t)$ denote the option value at calendar time $t$ with spot $S_t$. Find an analytically closed-form expression for the present-time $t = 0$ price $P(S_0, K, T)$. Hint: This problem is very closely related to the valuation of a compound option discussed at the end of Section 1.12. In particular, proceed as follows. From backward recurrence show that

$$P(S_0, K, T) = e^{-rT_1} E_0\big[P(S_{T_1}, K, T - T_1)\big], \tag{1.519}$$

with

$$P(S_{T_1}, K, T - T_1) = \begin{cases} P_E(S_{T_1}, K, T - T_1), & S_{T_1} > S^*_{T_1} \\ K - S_{T_1}, & S_{T_1} \le S^*_{T_1}, \end{cases} \tag{1.520}$$

where $P_E$ is the European put price function, $E_0[\,]$ is the risk-neutral expectation at time 0, and the critical value $S^*_{T_1}$ for the early-exercise boundary at calendar time $T_1$ solves

$$P_E(S^*_{T_1}, K, T - T_1) = K - S^*_{T_1}.$$

Compute this expectation as a sum of two integrals, one over the domain $S_{T_1} > S^*_{T_1}$ and the other over $0 < S_{T_1} \leq S^*_{T_1}$ while using equations (1.382) and (1.385) to finally arrive at the expression for $P(S_0, K, T)$ in terms of univariate and bivariate cumulative normal functions. Show whether $S^*_{T_1}$ is a strictly increasing or decreasing function of the volatility $\sigma$, and explain your answer. What is this functional dependency for the case of a Bermudan call? Explain.

CHAPTER • 2

# Fixed-Income Instruments

## 2.1 Bonds, Futures, Forwards, and Swaps

### 2.1.1 Bonds

A *bond* is paper issued by a corporate or sovereign entity promising a cash flow stream at future dates. In this chapter, we make the important assumption that credit risk is negligible, meaning that the probability that bond issuers default on their promise of making payments is zero.

Mathematically, a bond is modeled as a cash flow stream with a present value. The *cash flow map* of a bond is given by a sequence of pairs $(\mathbf{c}, \mathbf{T}) = (c_i, T_i), i = 1, \ldots, n$, where $T_1 < \cdots < T_n$ are *future* cash flow dates in increasing order and $c_1, \ldots, c_n$ are the corresponding cash flow amounts. A cash flow stream $(\mathbf{c}, \mathbf{T})$ has a present value at calendar time $t$ denoted by $PV_t(\mathbf{c}, \mathbf{T})$. *Pure discount bonds*, or *zero-coupon bonds*, are securities with one single cash flow of fixed amount, i.e., the nominal amount $N$ at maturity $T$; see Figure 2.1. The *continuously compounded yield* $y_t(T)$ for the period $[t, T]$ is often used to express the value $Z_t(T)$ at time $t$ of a zero-coupon bond maturing at time $T$ and is defined as follows:

$$Z_t(T) = \exp\big(-y_t(T)(T-t)\big). \tag{2.1}$$

Note that $Z_T(T) = 1$. Simple-compounding rules are often used. The *simply compounded yield* $y_t^\alpha$ with period $\alpha \le T - t$ is defined as follows:

$$Z_t(T) = \left(1 + \alpha y_t^{(\alpha)}(T)\right)^{-\frac{(T-t)}{\alpha}}. \tag{2.2}$$

For example, letting $\alpha = (T-t)/n, n \ge 1$, gives n simple compounding periods in $[t, T]$. Notice that in an economy where one postulates that the cost of holding a cash position is negligible — which is the case if one neglects security costs — one obtains the inequality

$$Z_t(T_1) \ge Z_t(T_2), \tag{2.3}$$

for all maturities $T_1 \le T_2$ and any fixed present time $t \le T_1$.

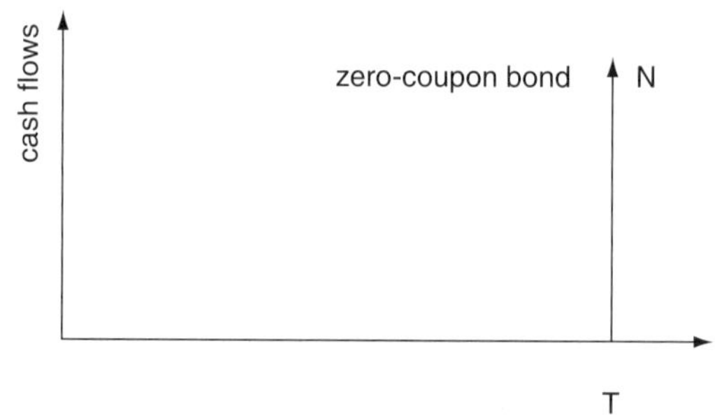


FIGURE 2.1 Zero-coupon bond with one cash flow at maturity $T$.

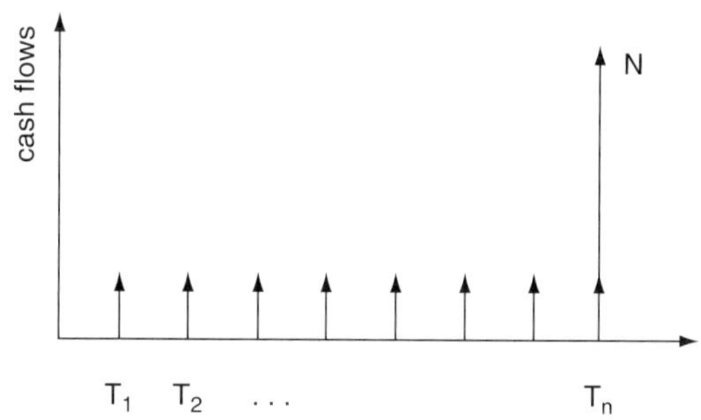


FIGURE 2.2 Cash flow stream for an $n$-coupon bond.

A cash flow stream $(\mathbf{c}, \mathbf{T})$ of multiple $n$-coupon payments can be replicated by means of a portfolio of zero-coupon bonds. Figure 2.2 depicts such a cash flow stream with equal payments until maturity, at which time a nominal payment in the amount of $N$ is made. Assuming that zero-coupon bonds of all maturities are traded, the present value of the given cash flow stream is given by the sum of discounted cash flows:

$$PV_t(\mathbf{c}, \mathbf{T}) = \sum_{i=1}^{n} c_i e^{-y_t(T_i)(T_i - t)} \overset{(\text{or})}{=} \sum_{i=1}^{n} c_i \left(1 + \alpha y_t^{(\alpha)}(T_i)\right)^{-\frac{(T_i - t)}{\alpha}}, \tag{2.4}$$

where the first sum in the equation assumes continuous compounding and the second assumes simple compounding. One defines *yields* of a coupon bond with cash flow map $(\mathbf{c}, \mathbf{T})$ to be the quantities $y_t(\mathbf{c}, \mathbf{T})$ [or $y_t^{(\alpha)}(\mathbf{c}, \mathbf{T})$ for simple compounding] such that

$$PV_t(\mathbf{c}, \mathbf{T}) = \sum_{i=1}^{n} c_i e^{-y_t(\mathbf{c}, \mathbf{T})} \overset{(\text{or})}{=} \sum_{i=1}^{n} c_i \left(1 + y_t^{(\alpha)}(\mathbf{c}, \mathbf{T})\right)^{-\frac{(T_i - t)}{\alpha}}, \tag{2.5}$$

where, again, the first sum in the equation assumes continuous compounding and the second assumes simple compounding.

Besides coupon bonds, some instruments with uncertain cash flows can also be priced in terms of the zero-coupon bonds. An example is a *bond-forward* contract. This is a forward contract on a zero-coupon bond of given maturity $T_2$, with a future settlement date $T_1$. Two parties $A$ and $B$ agree, at present time $t$, that a prescribed interest rate will apply within some interval $[T_1, T_2]$ in the future, with $t < T_1 < T_2$. A bond-forward of nominal $N$ is equivalent

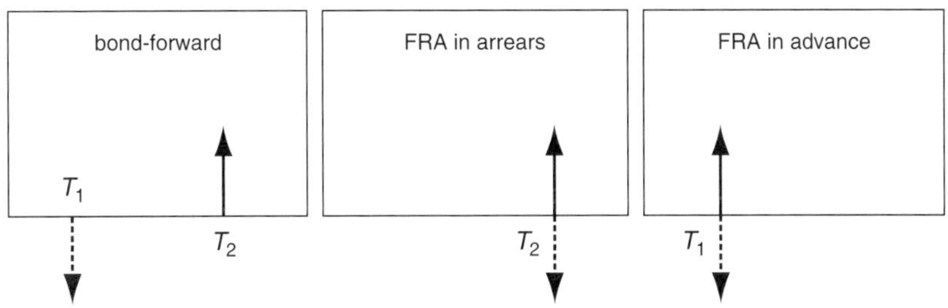


FIGURE 2.3 A comparison of equivalent present-value cash flows for an FRA with payments in arrears and in advance. The three figures correspond to the three possibilities of designing the cash flows: either both occurring at $T_1$, or both at $T_2$, or one at $T_1$ and one at $T_2$.

to the combination of two cash flows, as depicted in Figure 2.3. Party $A$ pays an amount $N$ at time $T_1$, and after a time $\tau$ she receives an amount

$$N(1+\alpha f_t^{(\alpha)}(T_1, T_2))^{\frac{\tau}{\alpha}} \stackrel{\text{(or)}}{=} N \exp\left(\tau f_t(T_1, T_2)\right) \tag{2.6}$$

at time $T_2$. Here, $\tau = (T_2 - T_1)$ is the *tenor* and $f_t^{(\alpha)}(T_1, T_2)$ is the *forward rate* computed with a simple-compounding rule of period $\alpha \leq \tau$, while $f_t(T_1, T_2)$ uses continuous compounding as further explained below. Notice that in the limit when the forward maturity is at current time, i.e., when $T_1 = t$, forward rates coincide with yields, i.e.,

$$f_t^{(\alpha)}(t, T_2) = y_t^{(\alpha)}(T_2), \tag{2.7}$$

and $y_t(T_2) = f_t(t, T_2)$ if continuous-compounding is assumed instead. The most convenient compounding convention for forward rates is the one with an intermediate compounding period equal to the tenor, i.e., $\alpha = \tau$. The *equilibrium value* of the forward rate is the rate for which the present value of the bond-forward contract is zero. Assuming continuous compounding, the present value of the two cash flows is

$$PV_t = -NZ_t(T_1) + Ne^{\tau f_t(T_1, T_2)} Z_t(T_2), \tag{2.8}$$

whereas for simple compounding

$$PV_t = N(Z_t(T_2) - Z_t(T_1)) + N\tau f_t^{(\tau)}(T_1, T_2) Z_t(T_2). \tag{2.9}$$

The equilibrium rate corresponds to the value for which $PV_t = 0$, hence giving

$$f_t(T_1, T_2) = \frac{1}{\tau} \log\left(\frac{Z_t(T_1)}{Z_t(T_2)}\right). \tag{2.10}$$

This coincides with the *continuously compounded forward rate* for the interval $[T_1, T_2]$ as viewed at present time $t$. In contrast, for simple compounding the equilibrium rate (or forward rate) denoted by $f_t^{(\tau)}(T_1, T_2)$ satisfies

$$1 + \tau f_t^{(\tau)}(T_1, T_2) = \frac{Z_t(T_1)}{Z_t(T_2)}. \tag{2.11}$$

Note that the forward rate is also related to the *forward price* for a unit zero-coupon bond maturing at time $T_2$ with settlement at time $T_1$. Forward rates and forward prices are further discussed in later sections.

### 2.1.2 Forward Rate Agreements

A *forward rate agreement* (FRA) is an instrument with the same risk profile, cash flow map, and present value of a bond-forward, but with only one actual cash flow. Such FRAs are struck at the equilibrium forward rate at the time of issue and come in two flavors, since payments can be either in *advance* or in *arrears*. In an FRA with payments in arrears, struck at the equilibrium rate $f_t(T_1, T_2)$, there is only one cash flow (with positive and negative components) at time $T_2$. Using equation (2.8), or (2.9), and inflating the cash flow at time $T_1$ into a cash flow at time $T_2$ gives only one cash flow at time $T_2$, of amount

$$N\tau\left[f_t^{(\tau)}(T_1, T_2) - y_{T_1}^{(\tau)}(T_2)\right] \tag{2.12}$$

for simple compounding or

$$N\left[e^{f_t(T_1,T_2)\tau} - e^{y_{T_1}(T_2)\tau}\right] \tag{2.13}$$

for continuous compounding. In contrast, in a similar FRA with payments in advance, the cash flow occurs only at time $T_1$. Discounting the cash flow at time $T_2$ back to time $T_1$ gives the following payoff amount for an FRA with payments in advance:

$$N\left(e^{[f_t(T_1,T_2)-y_{T_1}(T_2)]\tau} - 1\right) \tag{2.14}$$

for continuous compounding or

$$N\left(\frac{1+\tau f_t^{(\tau)}(T_1, T_2)}{1+\tau y_{T_1}^{(\tau)}(T_2)} - 1\right) \tag{2.15}$$

for simple compounding. The cash flows for these FRAs are depicted in Figure 2.3.

#### Problems

**Problem 1.** Prove that the condition (2.3) implies that all forward rates are nonnegative.

**Problem 2.** Conversely, prove that if all forward rates are positive, then the discount function is monotonically decreasing, i.e., that condition (2.3) holds.

### 2.1.3 Floating Rate Notes

A *floating rate note* (FRN) is an instrument with a series of settlement dates $T_j = T_0 + j\tau$, $j = 0, \ldots, n$, at which cash flows occur. In contrast to a bond, the size of a cash flow $c(T_j)$ (i.e., the coupon payment) at the generic date $T_j$ depends on the interest rate prevailing at time $T_j$ or earlier. In the simplest, so-called *plain-vanilla* structures, cash flow amounts are defined in a manner that the FRN can be associated to a cash flow map and priced directly off the yield curve, i.e., with no volatility risk. There are two variations of FRNs. Either the coupon payments are settled in arrears, i.e., paid out at time $T_j$ based on the rate for the period that just ended, $(T_j - \tau, T_j]$, or they are settled in advance with payments at time $T_{j-1}$. A plain-vanilla FRN with payments in arrears has cash flows given by

$$c(T_j) = \tau N y_{T_j-\tau}^{(\tau)}(T_j) + N\delta_{jn}. \tag{2.16}$$

Here $N$ is the notional amount of the FRN and $\delta_{jn}$ equals 1 in case $j = n$ and 0 otherwise; hence, the second term in equation (2.16) represents the notional repayment, which takes place only at the time of maturity $T_n$. For an FRN with payments in advance, the cash flows for times $T_j < T_n$ are obtained by discounting at the rate $y^{(\tau)}_{T_j-\tau}(T_j)$; hence,

$$c(T_j) = \frac{N\tau y^{(\tau)}_{T_j-\tau}(T_j)}{1+\tau y^{(\tau)}_{T_j-\tau}(T_j)} \tag{2.17}$$

if $j < n$ and $c(T_n) = N$ at maturity. Note that here we are assuming simple compounding with fixed period $\tau$. The present value at time $t \le T_0$ is the same in either case. In particular, with payments in arrears we have

$$\begin{aligned}\mathbf{FRN}_t &= \sum_{j=0}^{n} c(T_j)Z_t(T_j)\\ &= c(T_0)Z_t(T_0) + N\tau\sum_{j=1}^{n} y^{(\tau)}_{T_j-\tau}(T_j)Z_t(T_j) + NZ_t(T_n).\end{aligned} \tag{2.18}$$

This expression simplifies by using the relation

$$(1+\tau y^{(\tau)}_{T_j-\tau}(T_j))Z_t(T_j) = Z_t(T_{j-1}) \tag{2.19}$$

in the above sum, which collapses to give

$$\mathbf{FRN}_t = NZ_t(T_0) + c(T_0)Z_t(T_0). \tag{2.20}$$

In financial terms, this follows from the fact that if one has the notional amount available at time $T_0$ and invests it in a series of term deposits of tenor $\tau$ until maturity, one generates all the cash flows corresponding to the coupon payments starting from the initial and the principal repayment. This is depicted in Figure 2.4.

## 1.4 Plain-Vanilla Swaps

A *payer's interest rate swap* can be regarded as a combination of a short position in a floating rate (the floating leg) and a long position in a bond (the fixed leg) with the same nominal or

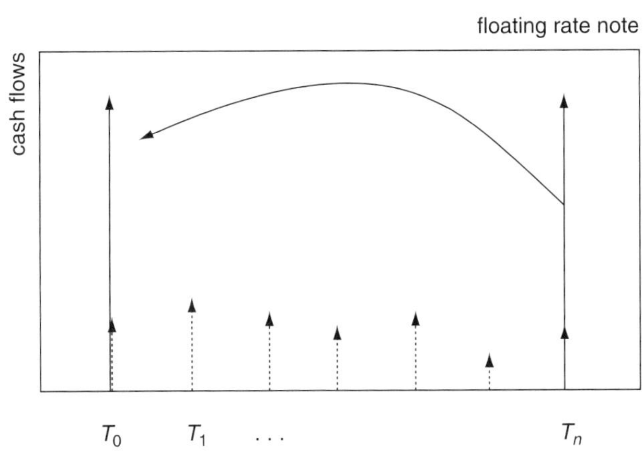


FIGURE 2.4 Equivalent cash flows for an FRN.

principal amount $N$ and paying coupons at a preassigned fixed rate $r^s$. A *receiver's interest rate swap* can be regarded as a short payer's swap. Cash flow dates are at times $T_j = T_0 + j\tau$, $j = 0, \ldots, n$, with period $\tau$. Clearly, swaps can be priced directly from the yield curve, and their replication does not involve any volatility risk. Swaps come in two variations, with the floating rate (typically a six-month LIBOR) agreed to be the rate prevailing either at the beginning or at the end of each period $(T_{j-1}, T_j]$. Assuming a principal repayment of $N$ at time $T_n$, the present value at time $t$ of the fixed leg is

$$\mathbf{PV}_t^{\text{fixed}} = c^{\text{fixed}}(T_0)Z_t(T_0) + Nr^s \sum_{j=1}^{n} \tau Z_t(T_j) + NZ_t(T_n) \tag{2.21}$$

and that for the floating leg is

$$\mathbf{PV}_t^{\text{float}} = c^{\text{float}}(T_0)Z_t(T_0) + N \sum_{j=1}^{n} \tau y_{T_j-\tau}^{(\tau)}(T_j)Z_t(T_j) + NZ_t(T_n), \tag{2.22}$$

with simple compounding at the floating rate assumed. From arbitrage arguments it also follows that the yields in this equation are given by the forward rates $f_t^{(\tau)}(T_{j-1}, T_j)$.

The swap rate $r_t^s$ is said to be at *equilibrium* at time $t$ if the present value to the receiver or payer of the swap at time $t$ is zero, i.e., if $\mathbf{PV}_t^{\text{fixed}} = \mathbf{PV}_t^{\text{float}}$. More precisely, using algebra similar to what was used in the preceding section, on FRNs [i.e., using equation (2.19)], the equilibrium swap rate of a swap with payments in arrears can be shown to satisfy the following equation:

$$N(Z_t(T_0) - Z_t(T_n)) + (c^{\text{float}}(T_0) - c^{\text{fixed}}(T_0))Z_t(T_0) = Nr_t^s \sum_{j=1}^{n} \tau Z_t(T_j). \tag{2.23}$$

Assuming equal initial coupons $c^{\text{float}}(T_0) = c^{\text{fixed}}(T_0)$, we have

$$r_t^s = \frac{Z_t(T_0) - Z_t(T_n)}{\sum_{j=1}^{n} \tau Z_t(T_j)}. \tag{2.24}$$

It is important to note that this result is independent of any assumed short rate model. Also, from the cash flow structure one can observe that interest rate swaps may be decomposed in terms of FRAs. Figure 2.5 shows the basic cash flow map of a receiver's swap with variable positive cash flows and the corresponding negative fixed amounts.

### 2.1.5 Constructing the Discount Curve

In this section, we describe the most liquid classes of interest-sensitive assets. These instruments can be priced directly from the discount curve and owe their popularity to the relative ease of replication, which results in liquid, efficient markets. Conversely, prices of such assets are used to reverse information on the discount curve. The discount curve is found by an interpolation algorithm, subject to the requirement that the present values $P_i$ of a series of cash flow maps $\mathbf{c}_i$, $i = 1, \ldots, n$, is reproduced, that is, subject to

$$P_i = \sum_j c_{ij}(T_{ij})Z_0(T_{ij}), \tag{2.25}$$

where $T_{ij}$ is the time when the $j$th cash flow of the $i$th cash flow map occurs and $c_{ij}$ is the corresponding amount.

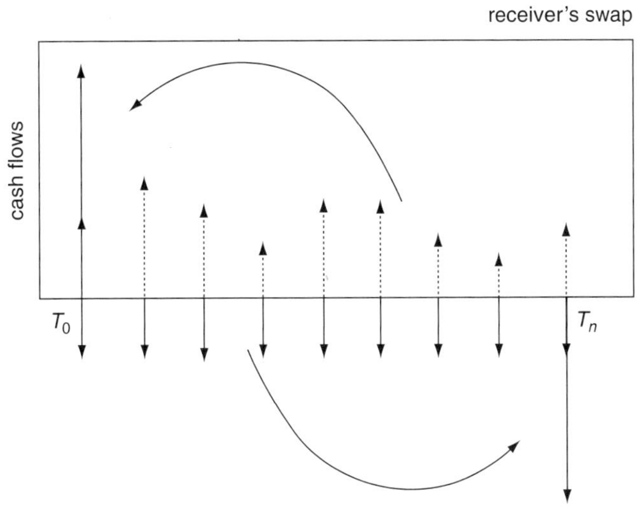


FIGURE 2.5 Fixed-leg and floating-leg cash flows for a receiver's swap.

A variety of analytical methods can be used to imply the discount curve. The following is a possible strategy that works quite well for the LIBOR curve. The method consists of two steps. In the first step one finds a best fit in a special parameterized family of meaningful discount functions. A possibility is to use the CIR discount function $Z_0^{\text{CIR}}(T)$, introduced in the following sections, but other choices would work as well. As a second step, one can represent the discount curve as

$$Z_0(T) = Z_0^{\text{CIR}}(T) + \delta Z_0(T) \tag{2.26}$$

and find the correction, $\delta Z_0(T) = Z_0(T) - Z_0^{\text{CIR}}(T)$, in such a way that the present values of the cash flow map in equation (2.25) are exactly reproduced, forward rates are positive, and the function $\delta Z_0(T)$ is as smooth as possible.

Cubic splines can be used to represent the function $\delta Z_0(T)$. A cubic spline is parameterized by the function values and the second derivatives on a time grid $T_1, \ldots, T_n$. The value of $\delta Z_0(T)$ for time $T \in (T_\alpha, T_{\alpha+1})$ falling in between the grid points can be interpolated as follows, using a cubic polynomial:

$$\delta Z_0(T) = a_\alpha(T - T_\alpha)^3 + b_\alpha(T - T_\alpha)^2 + c_\alpha(T - T_\alpha) + d_\alpha. \tag{2.27}$$

The constants $a_\alpha, b_\alpha, c_\alpha, d_\alpha$ solve the equations

$$d_\alpha = \delta Z_0(T_\alpha), \qquad 2b_\alpha = \delta Z_0''(T_\alpha), \tag{2.28}$$

$$a_\alpha(T_{\alpha+1} - T_\alpha)^3 + b_\alpha(T_{\alpha+1} - T_\alpha)^2 + c_\alpha(T_{\alpha+1} - T_\alpha) + d_\alpha = \delta Z_0(T_{\alpha+1}), \tag{2.29}$$

$$6a_\alpha(T_{\alpha+1} - T_\alpha) + 2b_\alpha = \delta Z_0''(T_{\alpha+1}). \tag{2.30}$$

This set of equations, in the given coefficients for each $\alpha$ grid point, involves function evaluations at both times $T_\alpha$ and $T_{\alpha+1}$, some of which correspond to points outside the discount curve. Hence, the equations constitute an underdetermined linear system. A good way to select a satisfactory solution is to further require that the weighted sum of squares

$$\sum_{\alpha=1}^{n} \left(Z_0(T_\alpha)^2 + \lambda Z_0''(T_\alpha)^2\right) \tag{2.31}$$

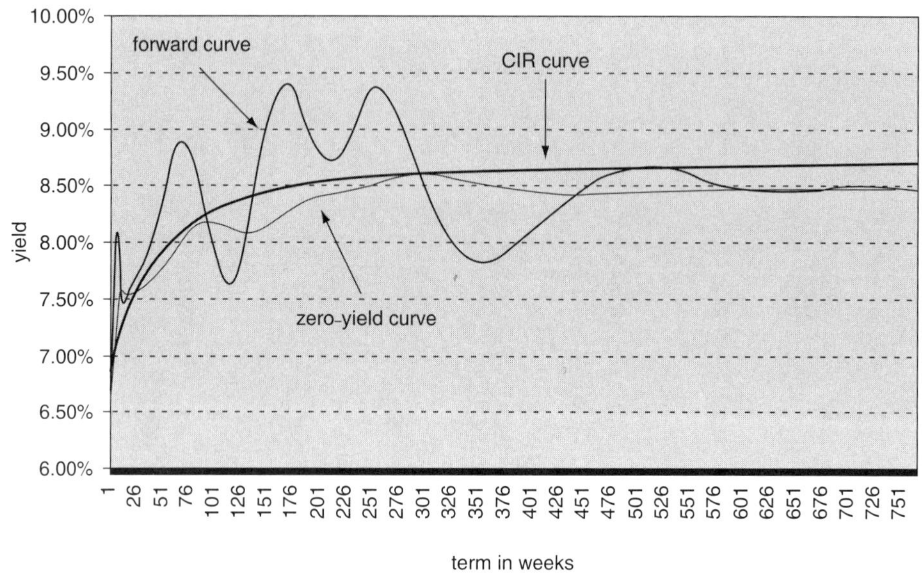


FIGURE 2.6 An actual-yield curve versus the yield curve obtained using a CIR discount function. The actual forward rates curve is also drawn for comparison.

be minimal. The parameter $\lambda$ adjusts the so-called *tension* of the yield curve. The limit $\lambda \to 0$ corresponds to an infinitely tense curve, in which the discount factors are linearly interpolated between the vertices. In the limit $\lambda \to \infty$, sharp turns in the curve are highly penalized. The spreadsheet (related to the "Interest Rate Trees: Calibration and Pricing" project of Part II) can be worked out by the reader interested in implementing the details of this fitting scheme, as depicted in Figure 2.6.

## 2.2 Pricing Measures and Black–Scholes Formulas

In Section 1.12 we derived pricing formulas of the Black–Scholes type assuming interest rates are deterministic functions of time. In this section, we lift this restriction and find Black–Scholes type of models that are solvable, giving explicit pricing formulas for stock options with stochastic interest rates and a number of interest rate derivatives. Pricing models for interest rate derivatives are based mostly on the postulate that interest rates and the discount function follow a diffusion process, thus *ruling out jumps*. In a general diffusion model, the price process for discount bonds $Z_t(T)$ of the various maturity dates $T$ obeys a stochastic differential equation of the following form:

$$dZ_t(T) = (r_t + q_t\sigma_t^{Z(T)})Z_t(T)dt + Z_t(T)\sigma_t^{Z(T)}dW_t. \tag{2.32}$$

Here, $q_t$ is a price of risk component dependent on the chosen numeraire, while $\sigma_t^{Z(T)}$ is the zero-coupon bond (lognormal) volatility.

Recall that the pricing formula in the asset pricing theorem (covered in Chapter 1) provides a way to express prices in terms of discounted expectations of future pay-offs with respect to a pricing measure:

$$A_t = g_t E_t^{Q(g)}\left[\frac{A_T}{g_T}\right]. \tag{2.33}$$

In this formula, "discounting" is achieved through a numeraire asset $g$, whose volatility is the price of risk for the pricing measure denoted by $Q(g)$. The actual asset price $A_t$ is independent of $g$; changing the numeraire is equivalent to changing coordinates in path space. Recall that all domestic assets drift at the instantaneous domestic risk-free rate plus a price of risk component given by the dot product $\boldsymbol{\sigma}_g \cdot \boldsymbol{\sigma}_A$, where $\boldsymbol{\sigma}_g$ and $\boldsymbol{\sigma}_A$ are lognormal volatility vectors of the chosen numeraire $g_t$ and the asset price $A_t$. As the following example demonstrates, it is useful to select the appropriate numeraire asset in order to derive pricing formulas in analytically closed form. The choices of numeraire asset we use in this section are:

- *Risk-neutral measure*, corresponding to selecting $g_t = B_t \equiv e^{\int_0^t r_s ds}$, the money-market or savings account
- *Forward measure* with maturity $T$, (also called the T-forward measure) corresponding to selecting $g_t = Z_t(T)$, the zero-coupon bond price with maturity date $T$
- *Bond-forward measure* with cash flow map $(\mathbf{c}, \mathbf{T})$, corresponding to selecting the bond's present value:

$$g_t = \sum_{i=1}^{n} c_i Z_t(T_i). \tag{2.34}$$

To achieve solvability, it is also necessary to identify an appropriate stochastic process whose expectation at maturity time one proposes to compute. As the following examples show, sometimes the obvious choice of the process is not the most convenient for the calculations. Furthermore, one needs to postulate a stochastic differential equation for the selected process whereby the drift is simple to compute (possibly zero) and the volatility is a deterministic function of time under the chosen measure. In the following sections we argue that there is a large class of models — known as Gaussian models — that naturally lead to deterministic volatilities in several important cases.

## .2.1 Stock Options with Stochastic Interest Rates

Consider a call option on the stock with price $S_t$ at time $t$, strike $K$, and maturity $T$. Let $F_t(S, T) = S_t/Z_t(T)$ be the forward price for the stock, with delivery at time $T$. Since $S_T = F_T(S, T)$, the pay-off for the call option can be written as follows:

$$\mathbf{C}_T = (F_T(S_T, T) - K)_+, \tag{2.35}$$

$(x)_+ \equiv \max(x, 0)$. In the forward measure $Q(g)$ with numeraire $g_t = Z_t(T)$, the forward price $F_t(S, T)$ is a martingale. Hence, we suppose that the process for $F_t(S, T)$ is given by

$$\frac{dF_t(S, T)}{F_t(S, T)} = \sigma(t) dW_t, \tag{2.36}$$

where the volatility $\sigma(t)$ of the forward price is a deterministic function of time. Recall from Section 1.6 that the transition probability distribution for such a process is lognormal:

$$p(F_t, t; F_T, T) = \frac{1}{\bar{\sigma} F_T \sqrt{2\pi(T-t)}} e^{-[\log(F_t/F_T) - \bar{\sigma}^2(T-t)/2]^2 / 2\bar{\sigma}^2(T-t)}, \tag{2.37}$$

where $F_t \equiv F_t(S, T)$ and $\bar{\sigma}$ involves the time-averaged square of the lognormal volatility,

$$\bar{\sigma}^2 = \frac{1}{T-t} \int_t^T \sigma(u)^2 \, du. \tag{2.38}$$

Putting equation (2.37) with equation (2.33) and using the fact that $Z_T(T) = 1$, the pricing formula for the value $\mathbf{C}_t$ of the call option at time $t$ is then given by

$$\frac{\mathbf{C}_t}{Z_t(T)} = E_t^{Q(Z_t(T))}[(F_T(S,T)-K)_+] = \int_0^\infty p(F_t, t; F_T, T)(F_T - K)_+ \, dF_T$$
$$= F_t(S,T)N(d_+) - KN(d_-), \tag{2.39}$$

where

$$d_\pm = \frac{\log(F_t(S,T)/K) \pm \frac{1}{2}\bar{\sigma}^2(T-t)}{\bar{\sigma}\sqrt{T-t}} \tag{2.40}$$

and $N(\cdot)$ is the cumulative standard normal distribution function.

### 2.2.2 Swaptions

Consider a payer swaption (or call swaption) struck at rate $r_K$ and of maturity $T$. The underlying is the fixed leg with pay-off as present value of all future cash flows if the swap rate $r_T^s > r_K$:

$$\mathbf{PSO}_T = \tau(r_T^s - r_K)_+ \sum_{j=1}^{n} Z_T(T_j), \tag{2.41}$$

where $\tau = T_{j+1} - T_j$ is the tenor. As a numeraire, select the present value of a stream of unit cash flows occurring at the coupon dates, $T_1 = T + \tau, \ldots, T_n = T + n\tau$, of the fixed leg:

$$g_t = \sum_{j=1}^{n} Z_t(T_j), \quad t < T. \tag{2.42}$$

Recalling the expression in equation (2.24) we see that the swap rate $r_t^s$ is a ratio of two assets, with denominator corresponding to the numeraire $g_t$. In this case one can easily show from the formula in equation (1.137) that $r_t^s$ is a martingale (i.e., has zero drift $\mu_t^{r^s} = 0$) with respect to the pricing measure $Q(g_t)$. Assuming that the lognormal volatility of the swap rate is a deterministic function of time, we set $\sigma_t^{(r^s)} = \sigma(t)$. The transition probability distribution function for the swap rate is then a lognormal function $p(r_t^s, t; r_T^s, T)$, similar to equation (2.37). Using steps similar to those in the previous section, one obtains the following Black–Scholes pricing formula for the swaption price $\mathbf{PSO}_t$ at time $t$:

$$\frac{\mathbf{PSO}_t}{\sum_{j=1}^{n} Z_t(T_j)} = \tau E_t^{Q(g)}[(r_T^s - r_K)_+]$$
$$= \tau[r_t^s N(d_+) - r_K N(d_-)], \tag{2.43}$$

where

$$d_\pm = \frac{\log(r_t^s/r_K) \pm \frac{1}{2}\bar{\sigma}^2(T-t)}{\bar{\sigma}\sqrt{T-t}}, \tag{2.44}$$

$N(\cdot)$ is the cumulative standard normal distribution function, and $\bar{\sigma}$ is defined as in equation (2.38), with time average taken over the squared lognormal volatility of the swap rate.

## 2.2.3 Caplets

Consider a caplet struck at fixed interest rate $r_K$, maturing at time $T$, on a floating rate $y_T^{(\tau)}(T+\tau)$ of tenor $\tau$ applied to the period $[T, T+\tau]$ in the future. The floating rate is typically the three- or six-month LIBOR. The pay-off of this caplet is given by a capped-rate differential compounded in time $\tau$ multiplied by the discount function over that period:

$$\mathbf{Cpl}_T = (y_T^{(\tau)}(T+\tau) - r_K)_+ \tau Z_T(T+\tau), \tag{2.45}$$

where the simply compounded yield is given by

$$y_T^{(\tau)}(T+\tau) = \tau^{-1}\big(Z_T(T+\tau)^{-1} - 1\big) = f_T^{(\tau)}(T, T+\tau). \tag{2.46}$$

Hence in terms of forward rates we have

$$\mathbf{Cpl}_T = \tau(f_T^{(\tau)}(T, T+\tau) - r_K)_+ Z_T(T+\tau). \tag{2.47}$$

In the measure $Q(g)$ with numeraire asset

$$g_t = Z_t(T+\tau), \tag{2.48}$$

the simply compounded forward rate

$$f_t^{(\tau)}(T, T+\tau) = \frac{1}{\tau}\left(\frac{Z_t(T)}{Z_t(T+\tau)} - 1\right) \tag{2.49}$$

is readily seen to be a martingale. Note that this follows because the forward rate is (besides the constant term $\tau^{-1}$) a ratio of two assets $Z_t(T)$ and $Z_t(T+\tau)$, where the denominator is $g_t$. As in the previous examples, the transition probability distribution $p(f_t, t; f_T, T)$ for the forward rate $f_t \equiv f_t^{(\tau)}(T, T+\tau)$ can be assumed lognormal and of the form in equation (2.37), with lognormal volatility $\sigma_t^f = \sigma(t)$ of the forward rate taken as a deterministic function of time. Hence, the pricing formula at time $t < T$ for the caplet with value $\mathbf{Cpl}_t$ is

$$\begin{aligned}
\mathbf{Cpl}_t &= \tau Z_t(T+\tau) E_t^{Q(g)}\big[(f_T - r_K)_+\big] \\
&= Z_t(T+\tau)[\tau f_t^{(\tau)}(T, T+\tau) N(d_+) - \tau r_K N(d_-)] \\
&= [Z_t(T) - Z_t(T+\tau)] N(d_+) - \tau r_K Z_t(T+\tau) N(d_-),
\end{aligned} \tag{2.50}$$

where

$$d_\pm = \frac{\log(f_t^{(\tau)}(T, T+\tau)/r_K) \pm \frac{1}{2}\bar{\sigma}^2(T-t)}{\bar{\sigma}\sqrt{T-t}}, \tag{2.51}$$

$N(\cdot)$ is the cumulative standard normal distribution function, and $\bar{\sigma}$ is defined as in equation (2.38), with time average taken over the squared lognormal volatility of the forward rate.

### 2.2.4 Options on Bonds

Consider a European call option struck at exercise $K$, of maturity date $T$, written on a coupon-bearing bond. The option pay-off can be written

$$\mathbf{BO}_T = (P_T - K)_+, \tag{2.52}$$

where $P_t$ is the present value of the bond,

$$P_t = \sum_{j=1}^{n} c_j Z_t(T_j), \tag{2.53}$$

with cash flows $c_n, \ldots, c_1$ at times $T_n > T_{n-1} > \cdots > T_1 > T$. Note that the sum in this present value involves only cash flows at future times past the maturity of the option. As numeraire asset, we choose $g_t = Z_t(T)$, and we assume a lognormal volatility for the forward price of the bond: $F_t \equiv F_t(P, T) = P_t / Z_t(T)$. Note that with this choice of numeraire the forward price is a zero-drift lognormal process, where we assume the lognormal volatility as a deterministic function of time, $\sigma_t^F = \sigma(t)$. Noting also that $P_T = F_T(P, T) = F_T$, the resulting pricing formula for the call option on the bond is obtained using steps similar to those in the previous examples:

$$\begin{aligned}\mathbf{BO}_t &= Z_t(T) E_t^{Q(Z_t(T))}\left[(F_T - K)_+\right] \\ &= Z_t(T)[F_t(P, T)N(d_+) - KN(d_-)],\end{aligned} \tag{2.54}$$

where

$$d_\pm = \frac{\log(F_t(P, T)/K) \pm \frac{1}{2}\bar{\sigma}^2(T - t)}{\bar{\sigma}\sqrt{T - t}}, \tag{2.55}$$

$N(\cdot)$ is the cumulative standard normal distribution function, and $\bar{\sigma}$ is defined as in equation (2.38), with time average taken over the squared lognormal volatility of the bond forward price. It is important to note that this model is inaccurate when the lifetime of the bond is comparable to the time to maturity, in which case there can be a significant deviation from lognormality due to the *pull to par* effect.

### 2.2.5 Futures–Forward Price Spread

The spread between the futures price $F_t^*(A, T)$ and the forward price $F_t(A, T)$ of an underlying asset $A$, whose spot price at time $t$ is $A_t$, is given by equation (1.330). This difference was demonstrated in Section 1.11 to be zero in the case when interest rates are deterministic functions of time or when the asset price process is statistically independent of the short rate process. Here the numeraire $g_t = B_t$ is the money-market account. Let us now compute the spread assuming that interest rates are generally stochastic. It suffices to compute the expectation

$$E_t^{Q(B)}\left[A_T\right] = E_t^{Q(B)}\left[F_T(A, T)\right]. \tag{2.56}$$

We consider the stochastic differential of the forward price process $F_t(A, T)$,

$$\frac{dF_t(A, T)}{F_t(A, T)} = \mu_t^{F(A,T)} dt + \sigma_t^{F(A,T)} dW_t, \tag{2.57}$$

given that the asset price satisfies

$$\frac{dA_t}{A_t} = \mu_t^A \, dt + \sigma_t^A \, dW_t. \tag{2.58}$$

Using the results in equations (1.324) and (1.325) for the stochastic differential of the quotient $F_t(A, T) = A_t Z_t(T)^{-1}$, we have

$$\sigma_t^{F(A,T)} = \sigma_t^A - \sigma_t^{Z(T)} \tag{2.59}$$

and

$$\mu_t^{F(A,T)} = \sigma_t^{Z(T)} \left(\sigma_t^{Z(T)} - \sigma_t^A\right). \tag{2.60}$$

Here we have used the fact that, under the risk-neutral measure $Q(B)$, the drift of the asset price $A_t$ and the bond price (which is also an asset) are equal, and both are given by the short rate. As was seen in Chapter 1, this follows as a consequence of the important no-arbitrage property, that all assets drift at the instantaneous short rate $r_t$ under the risk-neutral measure with the money-market account as numeraire. We should emphasize here that the formulas throughout this section obviously extend to the case of many base risk factors as well. In such cases the drifts and volatilities are vector quantities with components in the base risk factors.

We now make the simplifying assumption that the volatilities of the asset $A_t$ and the bond $Z_t(T)$ are deterministic functions of time, i.e.,

$$\sigma_t^A = \sigma^A(t), \qquad \sigma_t^{Z(T)} = \sigma^{Z(T)}(t). \tag{2.61}$$

The forward price volatility $\sigma_t^{F(A,T)} = \sigma^F(t)$ and drift $\mu_t^{F(A,T)} = \mu^F(t)$, for fixed $T$ and given asset $A$, are then also deterministic functions of time as given by equations (2.59) and (2.60). This then allows us to obtain a more explicit formula for the futures–forward price spread, as follows.

Under the measure $Q(B)$, the probability density for the forward price attaining a value $F_T(A, T) = F_T$ at time $T$, given $F_t(A, T) = F_t$ at time $t$, has the lognormal form

$$p(F_t, t; F_T, T) = \frac{1}{\bar{\sigma} F_T \sqrt{2\pi(T-t)}} e^{-[\log(F_t/F_T) + (\bar{\mu} - \bar{\sigma}^2/2)(T-t)]^2 / 2\bar{\sigma}^2(T-t)}, \tag{2.62}$$

with time-averaged time-dependent drift and volatility

$$\bar{\mu} = \frac{1}{(T-t)} \int_t^T \mu^F(\tau) d\tau, \qquad \bar{\sigma}^2 = \frac{1}{(T-t)} \int_t^T (\sigma^F(\tau))^2 \, d\tau. \tag{2.63}$$

An expression for the futures price, in terms of the forward price, is now readily obtained from the integral

$$\begin{aligned} F_t^*(A, T) &= E_t^{Q(B)}\left[F_T(A, T)\right] = \int_0^\infty F_T p(F_t, t; F_T, T) dF_T \\ &= F_t(A, T) e^{\bar{\mu}(T-t)}. \end{aligned} \tag{2.64}$$

From equations (2.60), (2.63), and (2.64), the futures–forward price spread is therefore

$$F_t^*(A, T) - F_t(A, T) = F_t(A, T) \left[ \exp\left( \int_t^T (\sigma^{Z(T)}(\tau) - \sigma^A(\tau)) \sigma^{Z(T)}(\tau) d\tau \right) - 1 \right]. \tag{2.65}$$

Finally, note that for given $T$ and asset $A$, equation (2.64) shows that $F_t^*(A, T)/F_t(A, T)$ is a deterministic function and hence the volatility of the futures and forward price are assumed to be the same,

$$\sigma^{F^*}(t) = \sigma^F(t) = \sigma^A(t) - \sigma^{Z(T)}(t). \tag{2.66}$$

### 2.2.6 Bond Futures Options

Consider a European call option on a futures contract on a zero-coupon bond $Z_T(U)$, with option strike price $K$ and maturity date $T$, with $T < U$. Here the underlying asset $A$ is the zero-coupon bond whose maturity date is $U$ [i.e., $A_t = Z_t(U)$ for given bond maturity date $U$, and at the option expiry date $A_T = Z_T(U)$]. The futures price at any time $t \leq T$ is denoted by $F_t^*(Z_t(U), T)$; hence the pay-off at the option's expiry time $T$ can be written as follows:

$$\mathbf{BO}_T = (F_T^*(Z_T(U), T) - K)_+ = (F_T(Z_T(U), T) - K)_+. \tag{2.67}$$

Here we used the property $F_T^*(A, T) = F_T(A, T)$ for any asset $A$. In order to price this option, we will choose as numeraire the zero-coupon bond with maturity $T$, i.e., $g_t = Z_t(T)$. In this measure the forward price $F_t \equiv F_t(Z_t(U), T) = Z_t(U)/Z_t(T)$ is a martingale. We now make the same assumptions as in the previous section and postulate that the lognormal volatility of a zero-coupon bond of given maturity (i.e., for any $T$ and $U$ values) is a deterministic function of time $t$, with values $\sigma^{Z(T)}(t)$ and $\sigma^{Z(U)}(t)$, for maturities $T$ and $U$, respectively. Here, however, we are working in a probability space, with $F_t$ having zero drift. Using equation (2.64), we have $F_t^* = F_t e^{\bar{\mu}(T-t)}$ for the price $F_t^*(Z_t(U), T)$, with

$$\bar{\mu} = \frac{1}{(T-t)} \int_t^T (\sigma^{Z(T)}(\tau) - \sigma^{Z(U)}(\tau))\sigma^{Z(T)}(\tau) d\tau. \tag{2.68}$$

The probability density $p(F_t, t; F_T, T)$ for the forward price attaining a value $F_T$ at time $T$, given a value $F_t$ at time $t$, is given by the lognormal form as in equation (2.37) with zero drift coefficient.

The pricing formula for the call on the bond futures contract then follows from similar steps as in the previous subsections:

$$\begin{aligned}
\mathbf{BO}_t &= Z_t(T) E_t^{Q(Z_t(T))}\left[(F_T^*(Z_T(U), T) - K)_+\right] \\
&= Z_t(T) E_t^{Q(Z_t(T))}\left[(F_T(Z_T(U), T) - K)_+\right] \\
&= Z_t(T)[F_t(Z_t(U), T) N(d_+) - K N(d_-)] \\
&= Z_t(T)\left[e^{-\bar{\mu}(T-t)} F_t^*(Z_t(U), T) N(d_+) - K N(d_-)\right],
\end{aligned} \tag{2.69}$$

where

$$d_\pm = \frac{\log \frac{F_t(Z_t(U), T)}{K} \pm \frac{1}{2}\bar{\sigma}^2(T-t)}{\bar{\sigma}\sqrt{T-t}} = \frac{\log \frac{F_t^*(Z_t(U), T)}{K} + (-\bar{\mu} \pm \frac{1}{2}\bar{\sigma}^2)(T-t)}{\bar{\sigma}\sqrt{T-t}} \tag{2.70}$$

and $N(\cdot)$ is the cumulative standard normal distribution function. Here $\bar{\mu}$ is given by equation (2.68), whereas

$$\bar{\sigma} = \sqrt{\frac{1}{T-t} \int_t^T \left(\sigma^{Z(T)}(\tau) - \sigma^{Z(U)}(\tau)\right)^2 d\tau}. \tag{2.71}$$

The option price can therefore be expressed either in terms of the futures or forward price, as well as the zero-coupon bond volatility for the two maturities $T$ and $U$.

## Problems

**Problem 1.** Demonstrate a put-call parity relation for European options on a futures contract on an underlying zero-coupon bond as described in Section 2.2.6.

**Problem 2.** Derive an option-pricing formula similar to that in Section 2.2.6 for a forward contract on a bond.

**Problem 3.** Derive a Black–Scholes formula for a European bond put option struck at exercise $K$, of maturity $T$. Is put-call parity satisfied with respect to the call price given in Section 2.2.4?

**Problem 4.** A *floorlet* is similar to a caplet, except the floating rate is bounded from below with payoff $(r_K - y_T^{(\tau)}(T+\tau))_+ \tau Z_T(T+\tau)$. Derive a Black–Scholes formula for a floorlet. Is there a relationship between a floorlet and a caplet?

**Problem 5.** *Caps* and *floors* are collections of caplets and floorlets, respectively, applied to periods $[T_j, T_j+\tau]$, $j = 1, \ldots, n$. Show that a model-independent relationship *cap* = *floor* + *swap* exists.

**Problem 6.** Provide a Black–Scholes type of formula for a receiver swaption with payoff $\tau(r_K - r_T^s)_+ \sum_{j=1}^n Z_T(T_j)$.

**Problem 7.** Provide a Black–Scholes type of formula for a European call option with maturity $T$ and strike $K$ and written on a (unit-nominal) zero-coupon bond with maturity $S > T$. Denoting its pricing function by $\mathbf{ZBC}_t(T, S, K)$,

$$\mathbf{ZBC}_t(T, S, K) = E_t^{Q(B)}\left[e^{-\int_t^T r_s ds}(Z_T(S) - K)_+\right].$$

Assume the forward price of the bond $F_t(Z(S), T) = Z_t(S)/Z_t(T)$ follows a zero-drift lognormal process with time-dependent volatility $\sigma(t)$ under the $T$-forward measure $Q(Z(T))$ with $Z_t(T)$ as numeraire asset price.

# .3 One-Factor Models for the Short Rate

## 3.1 *Bond-Pricing Equation*

A possible way of specifying an interest rate process is to assign a stochastic differential equation for the short rate

$$dr_t = \mu^g(r_t, t)dt + \sigma(r_t, t)dW_t. \tag{2.72}$$

Here $g$ is the numeraire asset. The functions $\mu^g$ and $\sigma$ give the drift and volatility, respectively, of the short rate $r_t$ under the measure, with $g$ as numeraire asset. Here we note that the drift and volatility functions in general have an explicit dependence on $r$ and $t$ variables.

**Theorem. (Bond-Pricing Equation)** *If the short rate process described by equation (2.72) is Markovian, then the zero-coupon bond price process $Z_t(T)$ is given by a pricing function $Z(r, t, T)$ so that*

$$Z_t(T) = Z(r_t, t, T). \tag{2.73}$$

- *The function $Z(r, t, T)$ solves the following partial differential equation:*

$$\frac{\partial Z}{\partial t} + r\frac{\partial Z}{\partial r} + \frac{\sigma(r,t)^2}{2}\frac{\partial^2 Z}{\partial r^2} = rZ. \tag{2.74}$$

- *The drift of the short term rate is given by*

$$\mu^g(r,t) = r + \sigma^g(r,t)\sigma(r,t), \tag{2.75}$$

*where $\sigma_t^g = \sigma^g(r,t)$, as a function of $r$ and $t$, denotes the volatility function for the numeraire asset price $g_t$ at calendar time $t$.*

- *Under the risk-neutral measure with choice of numeraire asset as the savings (i.e., money-market) account process, $g_t = B_t = e^{\int_0^t r_s ds}$, the discount function at present time $t$, maturing at time $T$, is given by the conditional expectation under the risk-neutral measure*

$$Z_t(T) = E_t^{Q(B)}\left[e^{-\int_t^T r_s ds}\right] = E^{Q(B)}\left[e^{-\int_t^T r_s ds}\middle| r_t = r\right], \tag{2.76}$$

*i.e., with condition $r_t = r$.*

- *The probability density $P(r,t)$ for the short rate having value $r$ at time $t$, given an initial condition for the density $P(r,0)$ at time $t = 0$, satisfies the equation*

$$\frac{\partial P(r,t)}{\partial t} = \frac{1}{2}\frac{\partial^2}{\partial r^2}\left(\sigma(r,t)^2 P(r,t)\right) - \frac{\partial}{\partial r}\left(\mu^g(r,t)P(r,t)\right). \tag{2.77}$$

**Proof.** The representation in equation (2.73) is due to the Markov assumption for the short rate: In this situation, the price of a zero-coupon bond can only depend on the short rate value $r$ at calendar time $t$ and on calendar time $t$, given a maturity $T$. By using Itô's lemma, where $Z$ is considered explicitly as a function of $r = r_t$ and $t$ variables, one obtains the stochastic differential for $Z \equiv Z(r_t, t, T)$:

$$\begin{aligned} dZ &= \left(\frac{\partial Z}{\partial t} + \mu^g(r,t)\frac{\partial Z}{\partial r} + \frac{\sigma^2(r,t)}{2}\frac{\partial^2 Z}{\partial r^2}\right)dt + \frac{\partial Z}{\partial r}\sigma(r,t)dW_t \\ &\equiv \mu^{Z_t(T)} Z dt + \sigma^{Z_t(T)} Z dW_t, \end{aligned} \tag{2.78}$$

with bond volatility $\sigma^{Z_t(T)} = \sigma^{Z(T)}(r,t)$ as

$$Z\sigma^{Z(T)}(r,t) = \sigma(r,t)\frac{\partial Z(r,t,T)}{\partial r}. \tag{2.79}$$

Here the bond volatility function is denoted explicitly as a function of $r$ and $t$, for given maturity $T$. The Black–Scholes equation for the stochastic differential equation (2.78) gives the pricing function for bonds as satisfying

$$\frac{\partial Z}{\partial t} + \mu^g(r,t)\frac{\partial Z}{\partial r} + \frac{\sigma(r,t)^2}{2}\frac{\partial^2 Z}{\partial r^2} = rZ + q_t^g \sigma(r,t)\frac{\partial Z}{\partial r}, \tag{2.80}$$

with price of risk $q_t^g = \sigma_t^g$. Note that this also follows by taking expectations on both sides of equation (2.78) while using $E_t^{Q(B)}[dZ] = (r + q_t^g \sigma^{Z_t(T)})Z\ dt$, $E_t^{Q(B)}[dW_t] = 0$. This is essentially a special case of the Feynman–Kac result. In the special case where $g_t = B_t$, i.e., the money-market account, the price of risk is zero (hence also giving $\mu^g = r$) and we finally find equation (2.74). Since the pricing function $Z(r, t, T)$ is not dependent on the price of risk, we conlude that the drift $\mu^g$ of the short rate process satisfies equation (2.75). By applying Itô's lemma to the expectation in equation (2.76), one can show that $Z$ satisfies partial differential equation (2.74) with condition $r = r_t$, thus verifying formula (2.76). Another simple proof of the bond-pricing equation is to apply the Feynman–Kac formula to the conditional expectation in equation (2.76), which can be written as $B_t E_t^{Q(B)}[B_T^{-1}]$, where this last expectation satisfies a Feynman–Kac PDE. Finally, equation (2.77) follows from the Fokker–Planck equation (or Kolmogorov forward equation) for the probability density corresponding to the process in equation (2.72). □

## 3.2 *Hull–White, Ho–Lee, and Vasicek Models*

There is empirical evidence that the interest rate process in the real-world measure is mean reverting. The series for the five-year U.S. dollar rate in Figure 2.7 shows this phenomenon visually. The periods with high and low rates alternate, following the expansion and recession cycles of the economy. There is also strong evidence from option prices that the risk-neutral process is mean reverting as well. Notice that this conclusion is not obvious, mathematically, since the price of risk can in principle offset the mean-reverting character of the overall process. Nevertheless, market expectations as they are reflected through cap prices, for instance, reveal that the market expects rates to fluctuate not far from the historical mean on long time scales. A large class of stochastic models with the mean-reversion property can be constructed based on two processes: the Ornstein–Uhlenbeck process and the Cox–Ingersol–Ross process. We construct both models emphasizing both the continuous-time interpretation and the discrete-time recurrence relations they satisfy. This approach has the advantage of clarifying the methodology for statistical estimations using daily or weekly data and to generate Monte Carlo simulations.

In what follows we describe an explicit method of obtaining expectations of stochastic quantities, as well as the discount function, by the use of a discrete stochastic calculus approach combined with a subsequent continuous-time limit. Let us first consider the time interval $[0, t]$ and its discrete subdivision, with the points $\mathcal{T} = \{t_0 = 0, t_1, \ldots t_n = t\}$ making

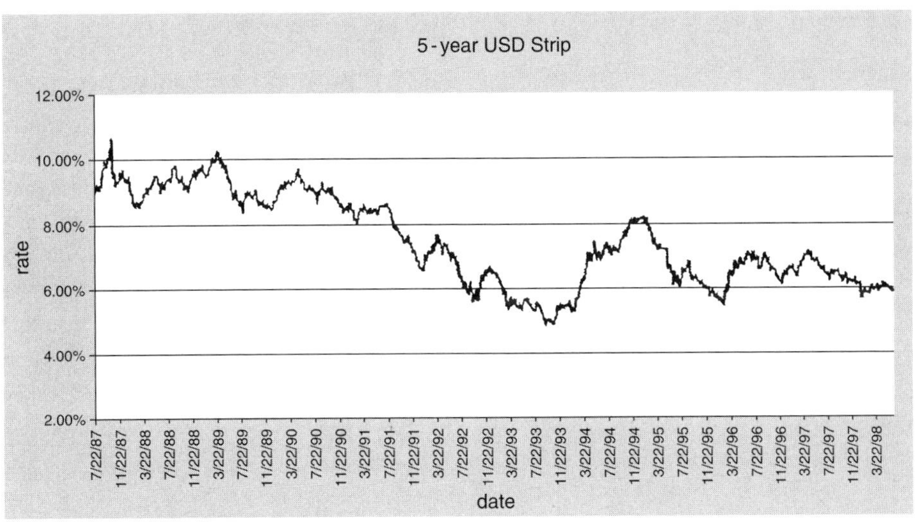


FIGURE 2.7 A time series for the 5-year U.S. dollar (USD) rate.

up $n$ subintervals of length $\delta t_i = t_{i+1} - t_i$. Subinterval paths are defined by means of the recurrence relations

$$r_{t_{i+1}} = e^{-b(t_i)\delta t_i} r_{t_i} + a(t_i)\delta t_i + \sigma(t_i)\delta W_{t_i}, \tag{2.81}$$

for all $i = 0, \ldots, N-1$, where $\delta W_{t_i}$ are uncorrelated Brownian increments such that

$$E_{t_0}\left[\delta W_{t_i}\ \delta W_{t_j}\right] = \delta_{ij}\ \delta t_i. \tag{2.82}$$

The solution to these recurrence relations is readily found by iteration, giving

$$r_{t_n} = e^{-\sum_{i=0}^{n-1} b(t_i)\delta t_i} r_0 + \sum_{i=0}^{n-1} e^{-\sum_{k=i+1}^{n-1} b(t_k)\delta t_k} (a(t_i)\delta t_i + \sigma(t_i)\delta W_{t_i}). \tag{2.83}$$

In the continuous-time limit, as the partition of the interval $[0, t]$ becomes finer and finer, i.e., in the limit $\delta t_i \to 0$ (or $n \to \infty$), this expression for the stochastic process is given by the stochastic integral:

$$r_t = r_0 e^{-\int_0^t b(s)ds} + \int_0^t e^{-\int_s^t b(u)du}(a(s)ds + \sigma(s)dW_s). \tag{2.84}$$

Notice that this expression reduces to equation (2.83) if the functions $a(t)$, $b(t)$, and $\sigma(t)$ are piecewise constant in the intervals $[t_i, t_i + \delta t_i)$. Differentiating this expression with respect to $t$ while using Leibniz's rule for the derivative of the integral on the right gives the stochastic differential equation satisfied by $r_t$ as

$$dr_t = (a(t) - b(t)r_t)dt + \sigma(t)dW_t. \tag{2.85}$$

This model encapsulates both the Hull–White and Vasicek models [HW93, Vas77]. The Hull–White model obtains by setting $b(t) = b$, $\sigma(t) = \sigma$ as constants and keeping $a(t)$ as time dependent. The Vasicek model obtains by also setting $a(t) = a$ as constant. The Ho–Lee model corresponds to setting $b(t) = 0$, $\sigma(t) = \sigma$ as constants and $a(t)$ as generally time dependent. The Black–Karasinski model obtains by replacing the short rate $r_t$ with the logarithm $\log r_t$ in equation (2.85).

From the solution in equation (2.83) one can obtain the expectation, at time $t = 0$, of the random variable $r_{t_n}$ by making use of $E_0[\delta W_{t_i}] = 0$ and then taking the continuous-time limit of the sums, giving

$$E_0[r_t] = e^{-\int_0^t b(s)ds} r_0 + \int_0^t e^{-\int_s^t b(u)du} a(s)ds. \tag{2.86}$$

The reader will also note that this is consistent with taking expectations on both sides of equation (2.84) and using the property of zero expectation for the stochastic integral part, as discussed in Section 1.4. Similarly, the variance can be obtained by considering the following expectation in the continuous-time limit:

$$\begin{aligned} E_0[(r_{t_n} - E_0[r_{t_n}])^2] &= E_0\left[\left(\sum_{i=0}^{n-1} e^{\sum_{k=i+1}^{n-1} b(t_k)\delta t_k} \sigma(t_i)\delta W_{t_i}\right)^2\right] \\ &= \sum_{i=0}^{n-1}\sum_{j=0}^{n-1} e^{\sum_{k=i+1}^{n-1} b(t_k)\delta t_k + \sum_{k=j+1}^{n-1} b(t_k)\delta t_k} \sigma(t_i)\sigma(t_j)E_0[\delta W_{t_i}\delta W_{t_j}] \\ &= \sum_{i=0}^{n-1} e^{\sum_{k=i+1}^{n-1} b(t_k)\delta t_k} \sigma(t_i)^2 \to \int_0^t e^{-2\int_s^t b(u)du} \sigma(s)^2 ds, \end{aligned} \tag{2.87}$$

where the last expression is obtained in the limit $n \to \infty$. The reader will note that equation (2.87) follows also by equations (2.84) and (2.86) after applying the Property (1.106).

In this model for the short rate process we have the useful result that the variable defined by the integral $X_t^T \equiv \int_t^T r_s\, ds$, for any time interval $[t, T]$, is a normal random variable. Hence, the discount function can be obtained in terms of the mean and standard deviation of the random variable $X_t^T$, as is shown next. To compute the mean and standard deviation of $X_t^T$, consider now the interval $[t, T]$ with $n$ subdivisions within time points $t_0 \equiv t, t_1, \ldots, t_n \equiv T$ and, as before, $\delta t_i = t_{i+1} - t_i$. The discretized form of the integral is

$$X_t^T = \sum_{k=0}^{n-1} r_{t_k}\, \delta t_k. \tag{2.88}$$

Taking the expectation, at time $t$, of this sum while using equation (2.83) for $r_{t_k}$ and $E_t[\delta W_{t_k}] = 0$ gives

$$E_t[X_t^T] = r_t \sum_{k=0}^{n-1} e^{-\sum_{i=0}^{k-1} b(t_i)\delta t_i}\, \delta t_k + \sum_{k=0}^{n-1}\sum_{i=0}^{k-1} e^{-\sum_{j=i+1}^{k-1} b(t_j)\delta t_j} a(t_i)\delta t_i\, \delta t_k. \tag{2.89}$$

In the continuous-time limit we have the mean

$$\begin{aligned} \bar{X}_t^T \equiv E_t[X_t^T] &= r_t \int_t^T e^{-\int_t^s b(u)du}\, ds + \int_t^T \int_t^s a(u) e^{-\int_u^s b(v)dv}\, du\, ds \\ &\equiv r_t n(t, T) + m_0(t, T), \end{aligned} \tag{2.90}$$

where the functions $n(t, T)$, $m_0(t, T)$ have been defined through the integrals. The reader can also verify that this result obtains by applying Property (1.105) together with (2.84), after a time shift. The variance follows from the expectation:

$$\begin{aligned} \mathrm{var}[X_t^T] &\equiv E_t[(X_t^T - E_t[X_t^T])^2] = E_t\left[\left(\sum_{k=0}^{n-1}\sum_{i=0}^{k-1} e^{-\sum_{j=i+1}^{k-1} b(t_j)\delta t_j} \sigma(t_i)\delta W_{t_i}\, \delta t_k\right)^2\right] \\ &= \sum_{k=0}^{n-1}\sum_{i=0}^{k-1}\sum_{k'=0}^{n-1}\sum_{i'=0}^{k'-1} e^{-\sum_{j=i+1}^{k-1} b(t_j)\delta t_j - \sum_{j=i'+1}^{k'-1} b(t_j)\delta t_j} \sigma(t_i)\sigma(t_{i'})\delta t_k\, \delta t_{k'} E_t[\delta W_{t_i}\, \delta W_{t_{i'}}] \\ &= \left[\sum_{k=0}^{n-1}\sum_{i=0}^{k-1}\sum_{k'=k}^{n-1} + \sum_{k=0}^{n-1}\sum_{k'=0}^{k-1}\sum_{i=0}^{k'-1}\right] e^{-\sum_{j=i+1}^{k-1} b(t_j)\delta t_j - \sum_{j=i+1}^{k'-1} b(t_j)\delta t_j} \sigma(t_i)^2 \delta t_k\, \delta t_{k'}\, \delta t_i \\ &\to \left[\int_t^T ds \int_t^s d\tau \int_s^T du + \int_t^T ds \int_t^s du \int_t^u d\tau\right] \sigma(\tau)^2 e^{-\int_\tau^s b(v)dv - \int_\tau^u b(v)dv}, \end{aligned} \tag{2.91}$$

where the last expression obtains in the limit $n \to \infty$. By reversing the order of integration in these integrals one can write the expression as one integral term, giving:

$$\begin{aligned} \mathrm{var}[X_t^T] = \int_t^T \sigma(\tau)^2 \left(\int_\tau^T e^{-\int_\tau^s b(u)du} ds\right)^2 d\tau &= \int_t^T \sigma(\tau)^2 n(\tau, T)^2 d\tau \\ &\equiv m_1(t, T). \end{aligned} \tag{2.92}$$

Having obtained $\bar{X}_t^T$ and $\mathrm{var}[X_t^T]$, we therefore have the probability density for the normal random variable $X_t^T \sim N(\bar{X}_t^T, \mathrm{var}[X_t^T])$ taking on a value $y$, as viewed at time $t$, given by a Gaussian:

$$p(y) = \frac{1}{\sqrt{2\pi\, \mathrm{var}[X_t^T]}} \exp\left(-\frac{(y - \bar{X}_t^T)^2}{2\, \mathrm{var}[X_t^T]}\right). \tag{2.93}$$

The discount function $Z_t(T) = Z(r_t, t, T)$ is finally obtained in terms of the expectation

$$Z_t(T) = E_t[e^{-X_t^T}] = \int_{-\infty}^{\infty} p(y)e^{-y}\, dy = e^{\frac{1}{2}var[X_t^T] - \bar{X}_t^T} = e^{m(t,T)-n(t,T)r_t}, \tag{2.94}$$

where

$$m(t, T) = \frac{1}{2}m_1(t, T) - m_0(t, T). \tag{2.95}$$

Note that this discount function can also be derived by using the method discussed in the next section. There, the solution for $Z(r_t, t, T)$ in the form of an exponential of an affine function in $r = r_t$ [see equation (2.94) or (2.116)], is obtained by simply plugging the expression into bond-pricing equation, where the volatility function is independent of the short rate. The functions $m(t, T)$ and $n(t, T)$ are readily shown to satisfy a system of first-order equations,

$$\frac{\partial n}{\partial t} = bn - 1, \tag{2.96}$$

and

$$\frac{\partial m}{\partial t} - an + \frac{1}{2}\sigma(t)^2 n^2 = 0, \tag{2.97}$$

with final time conditions $m(T, T) = n(T, T) = 0$. For these models, this system is exactly integrable, giving the same integral expressions as before.

For purposes of yield curve fitting, it is of interest to consider the formulas for the discount function in terms of the zero-coupon yields. In particular, the foregoing solution reads

$$y_t(T) = (T - t)^{-1}(n(t, T)r_t - m(t, T)). \tag{2.98}$$

The interpretation of this equation is that, for one-factor models having discount functions as exponentials of affine functions of the short rate, the shocks due to changes in the short rate are the only ones to affect the shape of the yield curve, which moves parallel to itself, according to equation (2.98).

The function $n(t, T)$ is linked to the term structure of volatility at calendar time $t$. In fact, by taking the stochastic differential of $y_t(T)$ in equation (2.98) while using equation (2.85), the yield is shown to have volatility

$$\sigma_t^{y(T)} = \frac{n(t, T)\sigma(t)}{T - t}. \tag{2.99}$$

The variance of the differential of the yield hence has a quadratic form given by

$$\text{var}(dy_t(T)) = (\sigma_t^{y(T)})^2\, dt = \frac{n(t, T)^2\sigma^2(t)}{(T - t)^2} dt. \tag{2.100}$$

The foregoing yield volatility equation allows one to fit the function $b(t)$ in terms of the current term structure of volatility. Indeed, since

$$n(t, T) = \int_t^T e^{-\int_t^s b(u)du}\, ds, \tag{2.101}$$

by differentiating with respect to $T$ we have

$$b(T) = -\frac{\partial}{\partial T}\left(\log \frac{\partial}{\partial T} n(t, T)\right). \tag{2.102}$$

Note that we can rewrite this equation by changing variable names, letting maturity $T \to t$ and present time $t \to t_0$, giving

$$b(t) = -\frac{\partial}{\partial t}\left(\log \frac{\partial}{\partial t} n(t_0, t)\right). \tag{2.103}$$

Given the fitted function $b(t)$, one can then fit (or retrieve) the function $a(t)$ from the discount function or using equation (2.98). Moreover, for the case of the Vasicek, Hull–White, and Ho–Lee models, all of the preceding integral expressions are readily worked out exactly in terms of exponential functions. Let us specifically work out the formulas for the case of the Hull–White model. Since $b$, $\sigma$ are constants, equation (2.101) is integrated to give

$$n(t, T) = \frac{1}{b}(1 - e^{-b(T-t)}). \tag{2.104}$$

And for $m(t, T)$ we have

$$m(t, T) = \frac{1}{b}\int_t^T [e^{-b(T-\tau)} - 1]a(\tau)d\tau + \frac{\sigma^2}{2b^2}\int_t^T (1 - e^{-b(T-\tau)})^2\, d\tau. \tag{2.105}$$

Taking logarithms of equation (2.94) gives

$$\log Z_t(T) = m(t, T) - n(t, T)r_t. \tag{2.106}$$

Differentiating this equation with respect to $T$ while using equations (2.105) and (2.104) gives

$$\frac{\partial}{\partial T}\log Z_t(T) = \frac{\sigma^2}{2b^2}\left[1 - 2e^{-b(T-t)} + e^{-2b(T-t)}\right] - re^{-b(T-t)} - \int_t^T a(\tau)e^{-b(T-\tau)}\, d\tau. \tag{2.107}$$

Differentiating again while using equation (2.107) then gives

$$a(T) = -\frac{\partial^2}{\partial T^2}\log Z_t(T) - b\frac{\partial}{\partial T}\log Z_t(T) + \frac{\sigma^2}{2b}(1 - e^{-2b(T-t)}). \tag{2.108}$$

Changing the variable name $T$ to $t$ and taking the initial time as zero gives

$$a(t) = -\frac{\partial^2}{\partial t^2}\log Z_0(t) - b\frac{\partial}{\partial t}\log Z_0(t) + \frac{\sigma^2}{2b}(1 - e^{-2bt)}). \tag{2.109}$$

This last equation gives us a useful relationship between the drift function and the zero-coupon bond prices, as a function of the maturity. In particular, one can rewrite this in terms of the instantaneous continuously compounded forward rates [these are defined in a later section; see equation (2.153)]:

$$a(t) = \frac{\partial}{\partial t}f_0(t) + bf_0(t) + \frac{\sigma^2}{2b}(1 - e^{-2bt)}). \tag{2.110}$$

Lastly, notice that the option-pricing formulas in the previous section, obtained under the forward measure, can be applied as the log-normal volatility of a zero-coupon bond forward given by

$$\left(\frac{dF_t(T, T')}{F_t(T, T')}\right)^2 = \sigma^2(n(t, T') - n(t, T))^2 \, dt. \tag{2.111}$$

The pricing formulas in the previous section, however, require the bond-forward measure. Examples are the formulas for swaptions and options on coupon bonds, which are not applicable here because the resulting volatility is not a deterministic function of time.

The foregoing short-rate models are among the popular models used for pricing interest rate options. In particular, lattice methods are useful for calibration and pricing. For an actual implementation of binomial and trinomial lattice trees within the Ho–Lee, Black–Derman–Toy, Hull–White, and Black–Karasinski models, the reader is referred to the project on interest rate trees in Part II of this book. The project contains an elaborate discussion of the various implementation steps for calibrating binomial and trinomial short-rate lattices, and for numerically pricing interest rate derivatives within these four models.

### 2.3.3 Cox–Ingersoll–Ross Model

The stationary Cox–Ingersoll–Ross (CIR) model for the short-rate process is generally defined as follows under the risk-neutral measure:

$$dr_t = (a - br_t)dt + \sigma\sqrt{r_t}\, dW_t. \tag{2.112}$$

According to the foregoing theorem, the bond-pricing PDE for this process is:

$$\frac{\partial Z}{\partial t} + (a - br)\frac{\partial Z}{\partial r} + \frac{\sigma^2 r}{2}\frac{\partial^2 Z}{\partial r^2} - rZ = 0. \tag{2.113}$$

The stochastic differential equation satisfied by $Z = Z(r, t, T)$, where $r = r_t$, is

$$dZ = rZ\, dt + \sigma_Z Z\, dW_t, \tag{2.114}$$

where

$$\sigma_Z = \frac{\sigma\sqrt{r}}{Z}\frac{\partial Z}{\partial r} \tag{2.115}$$

Note that the CIR model is sometimes written so that the risk-neutral drift term has the form $\kappa(\theta - r)$, where the constants $\kappa$ and $\theta$ correspond to the rate of reversion and mean level, respectively. In our convention, this simply corresponds to setting $\theta = a/b$ and $\kappa = b$.

As with the Vasicek model in the previous section, the discount function for the CIR model takes the form of an exponential of an affine function in $r$:

$$Z(r, t, T) = \exp\left(m(t, T) - n(t, T)r\right). \tag{2.116}$$

Direct substitution leads to the equations

$$\frac{\partial n}{\partial t} - \frac{1}{2}\sigma^2 n^2 - bn + 1 = 0 \tag{2.117}$$

and

$$\frac{\partial m}{\partial t} = an. \tag{2.118}$$

The final-time condition $Z(r, T, T) = 1$ gives $m(T, T) = n(T, T) = 0$. Note the difference between these equations and equations (2.96) and (2.97), obtained for the models considered in the previous section. Again, exact expressions for $m(t, T)$ and $n(t, T)$ are readily obtained by integrating equation (2.117) and subsequently equation (2.118), giving

$$m(t, T) = \frac{2a}{\sigma^2} \log \left[ \frac{\gamma e^{\frac{b\tau}{2}}}{\gamma \cosh \gamma\tau + \frac{1}{2} b \sinh \gamma\tau} \right] \tag{2.119}$$

and

$$n(t, T) = \frac{\sinh \gamma\tau}{\gamma \cosh \gamma\tau + \frac{1}{2} b \sinh \gamma\tau}, \tag{2.120}$$

where $\tau = T - t$ is the time to maturity and $\gamma = \frac{1}{2}\sqrt{b^2 + 2\sigma^2}$.

The Fokker–Planck equation for the risk-neutral probability density of the spot rate is

$$\frac{\partial p(r, t)}{\partial t} = bp(r, t) - (a - br)\frac{\partial}{\partial r} p(r, t) + \frac{1}{2}\frac{\partial^2}{\partial r^2}(\sigma^2 r p(r, t)). \tag{2.121}$$

In the long time limit $t \to \infty$ the distribution approaches a steady state with $\partial p/\partial t \to 0$. As one can verify by direct substitution into the right-hand side of equation (2.121), the stationary probability distribution, denoted by $p_\infty(r)$, is

$$p_\infty(r) = \frac{\left(2b/\sigma^2\right)^{2a/\sigma^2}}{\Gamma(2a/\sigma^2)} r^{(2a/\sigma^2)-1} e^{-(2b/\sigma^2)r}, \qquad b > 0, \tag{2.122}$$

where $\Gamma(\cdot)$ is the gamma function. Notice that when $a > \sigma^2/2$, $p_\infty(r) \to 0$ as $r \to 0$, i.e., gives zero probability of attaining zero interest rates. Otherwise, the stationary probability distribution diverges in the limit $r \to 0$ when $a < \sigma^2/2$.

In particular, the distribution integrates to unity for $a > 0$, has an integrable singularity at $r = 0$ for values $0 < a < \sigma^2/2$, and is nonintegrable for $a \leq 0$. For $a \in (0, \sigma^2/2]$ the origin is reflective. These same conclusions also apply to the time dependent density just below.

An exact analytical solution of the time-dependent Fokker–Planck equation (2.121) for the distribution function $p(r, t) = p(r, r_0; t)$, subject to the initial-time condition $p(r, t = 0) = \delta(r - r_0)$, can be shown to take the form $(a, b > 0)$

$$p(r, r_0; t) = c_t \left(\frac{re^{bt}}{r_0}\right)^{q/2} \exp\left(-c_t(r_0 e^{-bt} + r)\right) I_q\left(2c_t(r_0 r e^{-bt})^{1/2}\right), \tag{2.123}$$

where $c_t \equiv 2b/(\sigma^2(1 - e^{-bt}))$, $q \equiv (2a/\sigma^2) - 1$, and $I_q(\cdot)$ is the modified Bessel function of the first kind of order $q$. Useful properties of the Bessel functions are contained in Appendix C of Chapter 3. Further properties of this density are given as problems at the end of this section. By using the series expansion of the modified Bessel function, the distribution function in equation (2.123) can be shown to be related to the noncentral chi-squared function $f_{\chi^2}(x, \nu, \lambda)$, since

$$f_{\chi^2}(x, \nu, \lambda) = \frac{1}{2}\left(\frac{x}{\lambda}\right)^{(\frac{\nu}{2}-1)/2} e^{-(x+\lambda)/2} I_{\frac{\nu}{2}-1}(\sqrt{\lambda x}), \tag{2.124}$$

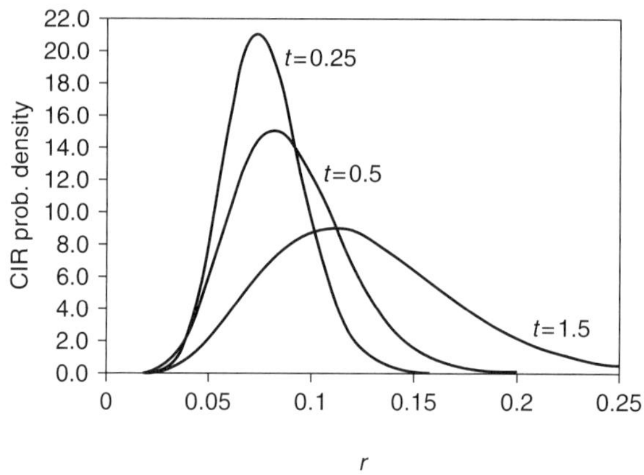


FIGURE 2.8 Plots of the CIR risk-neutral transition probability density as a function of the short rate, at three different chosen times.

where $\nu$ and $\lambda$ are the number of degrees of freedom and the noncentrality parameter, respectively. In particular, for the CIR model under the risk-neutral measure, the spot rate $r = r_t$ at time $t$ is a random variable generated by

$$r_t = \frac{\sigma^2(1 - e^{-tb})}{4b}\rho, \tag{2.125}$$

where $\rho$ is a noncentral chi-squared random variable with $2(q+1) = 4a/\sigma^2$ degrees of freedom and time-dependent noncentrality parameter equal to $2c_t r_0 e^{-bt}$. Figure 2.8 gives a plot of the foregoing risk-neutral density for different time values $t = 0.25$, 0.5, and 1.5 and with choice of parameters $a = 0.075$, $b = 0.35$, $\sigma = 0.15$, $r_0 = 0.065$ (all units are on a yearly basis). With this choice of parameters, the steady-state distribution is nearly attained at values of $t \sim 20$.

Under the forward measure with numeraire $Z_t(T)$, the equation for $Z \equiv Z_t(T) = Z(r_t, t, T)$ is

$$dZ = (r_t + \sigma_Z^2)Z\,dt + \sigma_Z Z\,dW_t^T, \tag{2.126}$$

where $dW_t^T$ is the Brownian increment under that measure. Assuming that under the forward measure the short rate evolves as

$$dr_t = \mu(r_t)dt + \sigma\sqrt{r_t}\,dW_t^T, \tag{2.127}$$

this implies, due to Itô's lemma, and from equation (2.113),

$$E_t[dZ] = \left(\frac{\partial Z}{\partial t} + \frac{\partial Z}{\partial r}\mu(r) + \frac{\sigma^2 r}{2}\frac{\partial^2 Z}{\partial r^2}\right)dt = \left(rZ + \frac{\partial Z}{\partial r}(\mu(r) - a + br)\right)dt \tag{2.128}$$

where $r_t = r$. Hence, the drift obtains as

$$(\mu(r) - a + br) = \sigma_Z^2 Z\left(\frac{\partial Z}{\partial r}\right)^{-1} = \frac{\sigma^2 r}{Z}\frac{\partial Z}{\partial r}, \tag{2.129}$$

giving

$$dr_t = \left(a - br_t + \frac{\sigma^2 r_t}{Z}\frac{\partial Z}{\partial r_t}\right)Z\,dt + \sigma\sqrt{r_t}\,dW_t^T. \tag{2.130}$$

Under this forward measure, one can also solve the Fokker–Planck equation for the process defined by the corresponding stochastic differential equation, giving a slightly more algebraically involved analytical expression for the density, yet again in terms of the modified Bessel function. This follows from the fact that $Z = Z(r, t, T) = e^{m(t,T)-n(t,T)r}$, so $(\partial Z/\partial r)/Z = -n(t, T)$ (independent of $r$). Hence the foregoing SDE has the same structure as the original SDE for the CIR process in the risk-neutral measure, except for an additional time dependence introduced into the mean-reversion coefficient. The solution follows by applying appropriate transformations.[1] In particular, it can be shown that a random variable for the short rate $r = r_\tau$ at any intermediate time $\tau$ with $0 \le t < \tau \le T$ has the form

$$r_\tau = \frac{\sigma^2(n(\tau, T) - n(t, T))}{4\partial n(\tau, T)/\partial \tau} \bar{\rho}, \tag{2.131}$$

where $\bar{\rho}$ is a noncentral chi-squared random variable with $4a/\sigma^2$ degrees of freedom and noncentrality parameter given by

$$\frac{4\partial n(t, T)/\partial t}{\sigma^2(n(\tau, T) - n(t, T))} r_t. \tag{2.132}$$

Note that a simplification arises with the choice of time parameters $t = 0$ and $\tau = T$. We refer to the literature on the CIR model [CIR85] for a derivation of these results. The more advanced material in Chapter 3 that deals with Green's function methods for the Fokker–Planck equation actually provides the reader with the mathematical tools for deriving analytically exact transition probability densities for the short-rate process within the CIR and other models from first principles. Such transition densities, or formulas of the type just given, allow one to price most European-style interest rate derivatives and to generate exact scenarios for the short rate under the CIR model.

## Problems

**Problem 1.** Show that the transition probability density function $p(r, r_0; t)$ in equation (2.123) satisfies the Fokker–Planck equation (2.121), with initial condition $p(r, r_0; t = 0) = \delta(r - r_0)$. Hint: After inserting the solution into the Fokker–Planck equation, differentiating and collecting terms, arrive at a second-order ordinary differential equation for the modified Bessel functions; i.e., show that this gives the modified Bessel equation of the form (see Appendix C in Chapter 3)

$$\frac{d^2}{dx^2} I_\nu(x) + \frac{1}{x}\frac{d}{dx} I_\nu(x) - \big(1 + \frac{\nu^2}{x^2}\big) I_\nu(x) = 0,$$

where $\nu$ is the order.

**Problem 2.** Verify that the CIR density in equation (2.123) where $a, b > 0$ gives

$$\int_0^\infty p(r, r_0; t) dr = 1, \tag{2.133}$$

[1] Let $P(r, r_0; t)$ be the transition density for the process $dr_t = (a - b(t) r_t) dt + \sigma \sqrt{r_t} dW_t$, with deterministic time dependent coefficient $b(t)$ and define the respective scale and time changes: $\lambda(t) = e^{\int_0^t b(u)du}$ and $\tau(t) = \frac{\sigma^2}{4}\int_0^t \lambda(u) du$. Then $P(r, r_0; t) = \lambda(t) u(\lambda(t) r, r_0; \tau(t))$, where $u$ is the density for the Bessel process as given in equation (3.215) with Bessel order $\mu = (2a/\sigma^2) - 1$. Note: when $b(t) = b$ is constant this corresponds to the density in equation (2.123).

hence demonstrating that short rates are never negative, i.e., that any short-rate path starting at time $t = 0$ at any finite positive value $r_0$ will end up in the positive axis with probability 1 at any finite later time $t > 0$. Hint: Use the Bessel integral property (3.357) in Appendix C of Chapter 3.

**Problem 3.** Show that the CIR density in equation (2.123) satisfies the Chapman–Kolmogorov equation

$$\int_0^\infty p(r_T, r; T-t)p(r, r_0; t)dr = p(r_T, r_0; T). \tag{2.134}$$

Hint: Use an appropriate Bessel integral property from Appendix C of Chapter 3.

**Problem 4.** The integrated form of equation (2.112) from time $s$ to time $t$ gives

$$r_t = r_s + \int_s^t (a - br_\tau)d\tau + \int_s^t \sigma\sqrt{r_\tau}\, dW_\tau. \tag{2.135}$$

(a) Show that $E_s[r_t] = E[r_t | r_{t=s} = r_s]$ satisfies a first-order ODE in time $t \geq s$, with initial condition $E_s[r_s] = r_s$ at $t = s$. Solve the initial-value problem and thereby obtain an exact expression for the conditional mean $E_s[r_t]$.
(b) Obtain an exact expression for the conditional variance $\text{Var}(r_t | r_{t=s} = r_s) = E_s[(r_t)^2] - (E_s[r_t])^2$.

**Problem 5.** Assume the short rate satisfies SDE (2.85).

(a) Find an expression for the auto-correlation function

$$\text{Corr}(r_s, r_t) = \text{Cov}(r_s, r_t)/\sqrt{\text{Var}(r_s)\text{Var}(r_t)}$$

for $s < t$.
(b) Find an exact closed-form expression for $\text{Corr}(r_s, r_t)$ by considering $b(t) = b$, $\sigma(t) = \sigma$ as constants. Explain your answer in terms of the mean-reversion parameter and what it represents in the limit $b \to 0$.

**Problem 6.** Consider the European call option on a zero-coupon bond as stated in Problem 7 at the end of Section 2.2. Find a closed-form analytical expression for this option price $\mathbf{ZBC}_t(T, S, K)$ in:

(a) The Hull–White model with constant mean-reversion coefficient $b$ and constant volatility $\sigma$
(b) The CIR model described in this last section

Hint: For part (b) choose $g_t = Z_t(T)$ as numeraire (i.e., use the $T$-forward measure for taking expectations) and use the formulas at the end of this section. In particular, use the appropriate transition density for the short rate (within the $T$-forward measure), and obtain your final result as a sum of two terms involving the cumulative chi-squared density.

### 2.3.4 Flesaker–Hughston Model

The Flesaker–Hughston (FH) model is based on the original idea of defining a numeraire asset process without a direct financial meaning. Interest in this model stems from the fact that it is possible to derive analytical closed-form solutions for both caps and swaptions.

The numeraire process in FH models is defined as follows:

$$g_t = \frac{1}{f(t) + g(t)x_t}, \tag{2.136}$$

where $f(t)$ and $g(t)$ are deterministic and strictly decreasing positive functions of calendar time $t$, and $x_t$ is a positive definite martingale. A zero-drift geometric Brownian motion gives a possible definition of $x_t$, i.e.,

$$dx_t = \sigma(t)x_t\, dW_t, \tag{2.137}$$

with some chosen initial condition $x_0 = 1$. Notice that in this model, $\log x_t$ follows a simple Wiener process with drift $-(\sigma(t))^2/2$ and diffusion $\sigma(t)$. An alternative definition of $x_t$ is the variance-gamma process. Within the FH model one readily arrives at an arbitrage-free price at time $t$ of a zero-coupon bond of unit worth at maturity time $T$ as

$$Z_t(T) = g_t E_t^{Q(g)}\left[\frac{1}{g_T}\right] = \frac{f(T) + g(T)x_t}{f(t) + g(t)x_t}. \tag{2.138}$$

Here we have used the martingale condition $E_t[x_T] = x_t$. The instantaneous short rate also has a simple expression since $f_t(t) = r_t$, as discussed in Section 2.4; hence,

$$r_t = -\frac{\partial}{\partial T}\log Z_t(T)\bigg|_{T=t} = -\frac{f'(t) + g'(t)x_t}{f(t) + g(t)x_t}. \tag{2.139}$$

Simply compounded (time-$t$) forward LIBOR rates $L_t(T)$ with settlement date $T$, tenor $\tau$, and given compounding period $\tau$ solve the equation

$$1 + \tau L_t(T) = \frac{Z_t(T)}{Z_t(T+\tau)} = \frac{f(T) + g(T)x_t}{f(T+\tau) + g(T+\tau)x_t} \tag{2.140}$$

and are thus given by

$$L_t(T) = \frac{1}{\tau}\left[\frac{f(T) + g(T)x_t}{f(T+\tau) + g(T+\tau)x_t} - 1\right]. \tag{2.141}$$

Using $g_t$ as numeraire and following the pricing methodology as in the worked-out examples of Section 2.2, a caplet struck at rate $\kappa$ and maturity $T$ is hence priced as follows:

$$\begin{aligned}\mathbf{Cpl}_t(\kappa, T) &= g_t E_t^{Q(g)}\left[\frac{\tau Z_T(T+\tau)(L_T(T+\tau) - \kappa)_+}{g_T}\right] \\ &= g_t E_t^{Q(g)}\left[\left(a_0(\kappa, T) + b_0(\kappa, T)x_T\right)_+\right],\end{aligned}$$

where

$$a_0(\kappa, T) \equiv f(T) - (1 + \kappa\tau)f(T+\tau), \qquad b_0(\kappa, T) \equiv g(T) - (1 + \kappa\tau)g(T+\tau). \tag{2.142}$$

By using the lognormal probability density function for $x_T$, which results from the process in equation (2.137), this expectation integral gives rise to an exact pricing formula:

$$\mathbf{Cpl}_t(\kappa, T) = g_t\big[a_0(\kappa, T)N(h^0_-(t, T, \kappa)) + b_0(\kappa, T)x_t N(h^0_+(t, T, \kappa))\big], \tag{2.143}$$

where

$$h^0_\pm(t, T, \kappa) = \frac{\log\left(-\frac{b_0(\kappa,T)x_t}{a_0(\kappa,T)}\right) \pm \frac{1}{2}\bar{\sigma}^2(T-t)}{\bar{\sigma}\sqrt{T-t}}, \tag{2.144}$$

the time-averaged volatility is

$$\bar{\sigma}^2 = \frac{1}{(T-t)}\int_t^T (\sigma(u))^2\, du, \tag{2.145}$$

and $N(\cdot)$ is the cumulative standard normal distribution function. This formula is valid for cases in which $b_0(\kappa, T)/a_0(\kappa, T) < 0$. Deriving a similar pricing formula for the case $b_0/a_0 > 0$ is left as an exercise for the reader.

A payer's swaption was considered in Section 2.2.2, with payoff

$$\mathbf{PSO}_T = \tau\left(r^s_T - \kappa\right)_+ \sum_{j=1}^{n} Z_T(T_j), \tag{2.146}$$

where the swap rate $r^s_t$ at time $t$ and the strike rate $\kappa$ are in units of an interest rate (i.e., time$^{-1}$). Assuming $n$ payments and a swap rate of the form

$$r^s_t = \frac{1 - Z_t(T_n)}{\tau\sum_{j=1}^n Z_t(T_j)}, \tag{2.147}$$

we can write the price of a payer's swaption maturing at time $T$ as

$$\begin{aligned}\mathbf{PSO}_t(\kappa, T) &= g_t E_t^{Q(g)}\left[\frac{\left(1 - Z_T(T_n) - \kappa\tau\sum_{j=1}^n Z_T(T_j)\right)_+}{g_T}\right]\\ &= g_t E_t^{Q(g)}\left[\left(a_n(\kappa, T) + b_n(\kappa, T)x_T\right)_+\right].\end{aligned}$$

In the last equation we have used the identity [see equation (2.138)]

$$Z_T(T_j)[f(T) + g(T)x_T] = f(T_j) + g(T_j)x_T, \tag{2.148}$$

giving

$$a_n(\kappa, T) = f(T) - f(T_n) - \kappa\tau\sum_{j=1}^n f(T_j), \qquad b_n(\kappa, T) = g(T) - g(T_n) - \kappa\tau\sum_{j=1}^n g(T_j). \tag{2.149}$$

As before, by using the lognormal probability density function for $x_T$, the expectation integral gives rise to an exact pricing formula:

$$\mathbf{PSO}_t(\kappa, T) = g_t\big[a_n(\kappa, T)N(h^n_-(t, T, \kappa)) + b_n(\kappa, T)x_t N(h^n_+(t, T, \kappa))\big], \tag{2.150}$$

where

$$h^n_\pm(t, T, \kappa) = \frac{\log\left(-\frac{b_n(\kappa,T)x_t}{a_n(\kappa,T)}\right) \pm \frac{1}{2}\bar{\sigma}^2(T-t)}{\bar{\sigma}\sqrt{T-t}}, \tag{2.151}$$

the time-averaged volatility is given by equation (2.145), and $N(\cdot)$ is the cumulative standard normal distribution function. This formula is valid for cases in which $b_n(\kappa, T)/a_n(\kappa, T) < 0$. Deriving a similar pricing formula for the case $b_n/a_n > 0$ is left as an exercise.

## 2.4 Multifactor Models

Multifactor models make use of the observed-yield curve, and this in turn can be described either as a collection of zero-coupon bonds (i.e., discount bonds) of various maturities $T$ with respect to an arbitrary calendar time $t$ with price $Z_t(T)$ or by the instantaneous forward rates. In what follows we denote present (today's) calendar time as $t = 0$, whereas time $t \geq 0$ generally stands for any time in the future or today. It is useful at this point to review very briefly the connection between these quantities and their relation to the instantaneous short rate. Let us recall the continuously compounded time-$t$ forward rate for a future finite time interval $[T, T+\tau]$ as given by

$$f_t(T, T+\tau) = -\frac{\log Z_t(T+\tau) - \log Z_t(T)}{\tau}. \tag{2.152}$$

In the limit $\tau \to 0$ this defines the *instantaneous forward rate* $f_t(T)$ as

$$f_t(T) = -\frac{\partial}{\partial T} \log Z_t(T). \tag{2.153}$$

Hence, forward rates and discount bond prices are also linked by

$$Z_t(T) = \exp\left( - \int_t^T f_t(s) ds \right). \tag{2.154}$$

This simple expression can be directly contrasted to that of the discount bond price given in terms of the risk-neutral expectation involving the instantaneous short rate $r_t$,

$$Z_t(T) = E_t^{Q(B)}\left[e^{-\int_t^T r_s ds}\right]. \tag{2.155}$$

The bond price is therefore related to a *path-integral* of the stochastic variable $r_t$ rather than to a simple (nonstochastic) integral as in the case of the forward rates. This path-integral expectation shows that if the short rate is stochastic, then $f_t(T) \neq r_T$ $(t < T)$, whereas when $r_t$ is deterministic the expectation is simply a regular integral and we have $f_t(T) = r_T$ for all $t \leq T$. In the HJM treatment described shortly, one is directly modeling the forward rates as local stochastic (i.e., Markov) processes. In view of the path-integral relationship between the short rate and the forward rates, one anticipates a generally non-Markovian theory for the short rate. A simple result of the formulation is that for generally stochastic short-rate processes we have

$$f_t(t) = r_t. \tag{2.156}$$

This obtains by equating the right-hand sides of equations (2.154) and (2.155), with $T = t + \epsilon$ $(\epsilon > 0)$, and differentiating with respect to $\epsilon$, giving

$$E_t^{Q(B)}\left[e^{-\int_t^{t+\epsilon} r_s ds} r_{t+\epsilon}\right] = f_t(t+\epsilon) e^{-\int_t^{t+\epsilon} f_t(s) ds}. \tag{2.157}$$

Taking the limit $\epsilon \to 0$ gives equation (2.156), since $E_t[r_t] = r_t$, i.e., the value of the instantaneous short rate at time $t$.

### 2.4.1 Heath–Jarrow–Morton with No-Arbitrage Constraints

An arbitrage-free dynamics of the yield curve in a diffusion model must satisfy constraints that take up various forms, depending on the modeling framework. In this section, we review the Heath–Jarrow–Morton (HJM) constraint for models of instantaneous forward rates [HJM92]. In the next section, we discuss the Brace–Gatarek–Musiela–Jamshidian (BGMJ) condition, where one models LIBOR rates instead. We present formulas in the context of one independent risk factor; however, the multifactor extension follows in an obvious manner, and we leave the derivation as an exercise problem.

Consider an interest rate stochastic process specified through the short rate

$$dr_t = \mu_t^g(r_t, t)dt + \sigma^r(r_t, t)dW_t^g, \tag{2.158}$$

in a suitable measure $Q(g)$. When working within the risk-neutral measure, recall that all assets drift at the instantaneous short rate $r_t$. In particular, all discount bonds of any maturity $T$ are assets, and hence

$$dZ_t(T) = r_t Z_t(T)dt + \sigma_t^{Z(T)} Z_t(T)dW_t^g, \tag{2.159}$$

under the risk-neutral measure with numeraire $g_t = B_t$ and $dW_t$ as Brownian increment in $Q(B)$. Notice that if one chooses a numeraire other than the money-market account $B_t$, then, in accordance with the asset pricing theorem in Chapter 1, the drift for any asset (including any discount bond) will have an extra term added to $r_t$ to account for the price of risk. We use shorthand notation to denote $\sigma_t^{Z(T)} \equiv \sigma(t, T, Z_t(T))$ as the time-$t$ volatility of the bond price. It is important to observe that in general, the bond price volatility is allowed to be a function of calendar time $t$, maturity time $T$, and the (stochastic) bond price $Z_t(T)$ at time $t$.

Thanks to Itô's lemma, the logarithm of the discount function obeys the following stochastic differential equation:

$$d[\log Z_t(T)] = \left[r_t - \frac{1}{2}(\sigma_t^{Z(T)})^2\right]dt + \sigma_t^{Z(T)}\, dW_t. \tag{2.160}$$

Since this equation applies for any value of $T$, we can use it for maturity $T$ and $T+\tau$. Combining this with equation (2.152) gives the stochastic differential of the rate $f_t(T, T+\tau)$:

$$d[f_t(T, T+\tau)] = \frac{(\sigma_t^{Z(T+\tau)})^2 - (\sigma_t^{Z(T)})^2}{2\tau} dt - \frac{\sigma_t^{Z(T+\tau)} - \sigma_t^{Z(T)}}{\tau} dW_t. \tag{2.161}$$

The stochastic differential of the instantaneous forward rate in the risk-neutral measure now obtains in the limit $\tau \to 0$:

$$\begin{aligned} df_t(T) &= \sigma_t^{Z(T)} \sigma_t^{\prime Z(T)}\, dt - \sigma_t^{\prime Z(T)}\, dW_t, \\ &\equiv \mu_t^{f(T)}\, dt + \sigma_t^{f(T)}\, dW_t. \end{aligned} \tag{2.162}$$

The last equation defines the drift $\mu_t^{f(T)} = \mu^f(t, T, f_t(T))$ and volatility $\sigma_t^{f(T)} = \sigma^f(t, T, f_t(T))$ of the instantaneous forward rate. The superscript prime is used to denote differentiation with respect to $T$, i.e., $\sigma_t^{\prime Z(T)} \equiv \partial \sigma_t^{Z(T)} / \partial T$. It turns out that one can relate the drift with the volatility of $f_t(T)$, since a simple integration of the bond price volatility derivative, with respect to maturity time, gives

$$\int_t^T \sigma_t^{\prime Z(\tau)} d\tau = \sigma_t^{Z(T)}. \tag{2.163}$$

In this equation we have used the fact that $\sigma_t^{Z(t)} = 0$, which says that the bond price has zero volatility with known unit value when $t = T$. Then, using the earlier relations for the drift and volatility of $f_t(T)$ in terms of the bond price volatility, we arrive at

$$\mu_t^{f(T)} = \sigma_t^{f(T)} \int_t^T \sigma_t^{f(\tau)} d\tau. \tag{2.164}$$

This result shows that the drift of $f_t(T)$ is linked to its volatility and the volatilities of all forward rates $f_t(\tau)$ between times $\tau = t$ and $\tau = T$. The link between the drift and volatility of the instantaneous forward rate was first noted by Heath, Jarrow, and Morton.

From this treatment one can arrive at the risk-neutral process for the short rate stated in equation (2.158). Using equation (2.162) rewritten in the form

$$df_\tau(t) = \sigma_\tau^{Z(t)} \sigma_\tau'^{Z(t)} \; d\tau + \sigma_\tau'^{Z(t)} \; dW_\tau, \tag{2.165}$$

integrating and using equation (2.156), we find

$$r_t = f_0(t) + \int_0^t \sigma_\tau^{Z(t)} \sigma_\tau'^{Z(t)} \; d\tau + \int_0^t \sigma_\tau'^{Z(t)} \; dW_\tau. \tag{2.166}$$

At this point one can apply the rule for differentiating an Itô integral,

$$\frac{\partial}{\partial t}\left[\int_0^t h(\tau, t) dW_\tau\right] = h(t, t) + \int_0^t \frac{\partial h(\tau, t)}{\partial t} dW_\tau, \tag{2.167}$$

where $h(\tau, t)$ is any smooth function. By differentiating the integral expression for $r_t$ and again using $\sigma_t^{Z(t)} = 0$, we obtain the stochastic process for the short rate as:

$$dr_t = \left\{ f_0'(t) + \left[ \int_0^t \left[\sigma_\tau^{Z(t)} \sigma_\tau''^{Z(t)} + (\sigma_\tau'^{Z(t)})^2\right] d\tau + \int_0^t \sigma_\tau''^{Z(t)} dW_\tau \right] \right\} dt$$
$$+ \sigma_t'^{Z(t)} dW_t. \tag{2.168}$$

The risk-neutral drift for the short rate is, therefore, non-Markovian, since it has a dependence on stochastic variables for times earlier than $t$, as given by the integral and stochastic integral over all times $\tau = 0$ to $\tau = t$ of factors involving the bond volatilities and their derivatives.

## Problems

**Problem 1.** Suppose we have $n$ independent risk factors. The instantaneous forward-rate process of equation (2.162) then takes the form

$$df_t(T) = \mu_t^{f(T)} \; dt + \sum_{j=1}^{n} \sigma_{t,j}^{f(T)} \; dW_t^j, \tag{2.169}$$

where $\sigma_{t,j}^{f(T)}$ are volatilities corresponding to the $j$th risk factor. Show that equation (2.164) generalizes to

$$\mu_t^{f(T)} = \sum_{j=1}^{n} \sigma_{t,j}^{f(T)} \int_t^T \sigma_{t,j}^{f(\tau)} d\tau. \tag{2.170}$$

**Problem 2.** Using the result of Problem 1, obtain the multifactor extension for the short-rate process given by equation (2.168).

## 2.4.2 Brace–Gatarek–Musiela–Jamshidian with No-Arbitrage Constraints

The reader can observe that in all previously presented treatments of the yield curve, including HJM, the theory has made use of either a continuum of discount bonds (i.e., of any maturity) or a continuum of instantaneous forward rates. Such continua provide a basis for the description of points on the yield curve lying on just discrete time intervals, as in LIBOR-based instruments. In contrast, in this section we briefly present the BGMJ (after Brace, Gatarek, Musiela, and Jamshidian), which models discrete market quantities, namely, the LIBOR rates [BGM97].

Within BGMJ one considers a situation with a lattice of $n$ maturities $T_i = T_1 + i\tau$, $i = 0, 1, 2, \ldots, n-1$, and the corresponding simply compounded forward rates $f_t^{(\tau)}(T_i, T_{i+1})$ for a finite period $\tau$ (e.g., 1 month, 3 months, 6 months). Recall the formula for the forward rate in terms of discount bond price ratios,

$$1 + \tau f_t^{(\tau)}(T_i, T_{i+1}) = \frac{Z_t(T_i)}{Z_t(T_{i+1})}. \tag{2.171}$$

To keep the notation simple, we now introduce the symbol (for given $\tau$)

$$L_t(T_i) = f_t^{(\tau)}(T_i, T_{i+1}). \tag{2.172}$$

Moreover, we present the treatment within a one-factor notation, although the extension to many independent risk factors readily follows, and we leave this as an exercise. We now proceed by assuming that each LIBOR rate is a random variable obeying an SDE of the form

$$\frac{dL_t(T_i)}{L_t(T_i)} = \mu_t^{L(T_i)}\, dt + \sigma_t^{L(T_i)}\, dW_t; \tag{2.173}$$

similarly, for each maturity one writes an SDE for each discount bond price process as

$$\frac{dZ_t(T_i)}{Z_t(T_i)} = \mu_t^{Z(T_i)}\, dt + \sigma_t^{Z(T_i)}\, dW_t. \tag{2.174}$$

Here we have used shorthand notation to denote the drifts and volatilities, which can generally be functions of $t$, $T_i$, and the underlying rate or bond price:

$$\begin{aligned} \mu_t^{L(T_i)} &= \mu^{L_i}(t, T_i, L_t(T_i)), & \sigma_t^{L(T_i)} &= \sigma^{L_i}(t, T_i, L_t(T_i)), \\ \mu_t^{Z(T_i)} &= \mu^{Z_i}(t, T_i, Z_t(T_i)), & \sigma_t^{Z(T_i)} &= \sigma^{Z_i}(t, T_i, Z_t(T_i)). \end{aligned} \tag{2.175}$$

Also, we assume to be in a basis of risk factors with no correlations among the LIBOR rates and bond prices. The addition of correlations along with the multifactor extension of the formulas is fairly straightforward and will be left as an exercise.

Taking the stochastic time-$t$ differential of equation (2.171) on both sides and using Itô's lemma in the form of equation (1.137), one finds

$$\begin{aligned} \tau L_t(T_i)\big(\mu_t^{L(T_i)} dt + \sigma_t^{L(T_i)} dW_t\big) = \frac{Z_t(T_i)}{Z_t(T_{i+1})}\Big[&(q_t - \sigma_t^{Z(T_i)})(\sigma_t^{Z(T_i)} - \sigma_t^{Z(T_{i+1})})dt \\ &+ (\sigma_t^{Z(T_i)} - \sigma^{Z(T_{i+1})})dW_t\Big]. \end{aligned} \tag{2.176}$$

Notice that here $q_t$ is a price of risk, which is generally nonzero since the underlying measure is not necessarily assumed to be the forward-neutral measure, wherein forward rates are martingales. Typically, the measure can be chosen to be the so-called spot-LIBOR measure, under which $Z_t(T_i)^{-1}$ [and not $Z_t(T_{i+1})^{-1}$] is a martingale. In this case the forward rates are not martingales.

Using equations (2.171) and (2.172) within the last equation and equating coefficients in $dW_t$ gives a recurrence relation among the bond volatilities at the different maturities:

$$\sigma_t^{Z(T_{i+1})} = \sigma_t^{Z(T_i)} - \frac{\tau L_t(T_i)\sigma_t^{L(T_i)}}{1+\tau L_t(T_i)}. \tag{2.177}$$

This is easily iterated to give

$$\sigma_t^{Z(T_{i+1})} = \sigma_t^{Z(T_1)} - \sum_{k=1}^{i} \frac{\tau L_t(T_k)\sigma_t^{L(T_k)}}{1+\tau L_t(T_k)}, \qquad i \geq 1. \tag{2.178}$$

On the other hand, the drift of the LIBOR forward rates is given by equating coefficients in $dt$ in the preceding SDE while using equation (2.177); hence,

$$\begin{aligned}\mu_t^{L(T_i)} &= \sigma_t^{L(T_i)}(q_t - \sigma_t^{Z(T_{i+1})}) \\ &= \sigma_t^{L(T_i)}\left(q_t - \sigma_t^{Z(T_1)} + \sum_{k=1}^{i} \frac{\tau L_t(T_k)\sigma_t^{L(T_k)}}{1+\tau L_t(T_k)}\right).\end{aligned} \tag{2.179}$$

Possible specifications of the volatility are of the form

$$\sigma_t^{L(T_i)} = L_t(T_i)^{\beta}\sigma(t, T_i), \tag{2.180}$$

where $\beta = 1$ corresponds to lognormal models and $\beta = \frac{1}{2}$ to square-root models.

We conclude this section by providing a pricing formula for the special case of the lognormal model with $\beta = 1$. In particular, the pricing formula for caplets of tenor $\tau$ and with settlement at one of the maturities $T_i$ can be computed in analytical closed form. Using similar methods as discussed in previous sections, one can arrive at a Black–Scholes type of pricing formula for a caplet struck at rate $\kappa$ and tenor $\tau$:

$$\mathbf{Cpl}_t(T_i, \kappa) = \tau Z_t(T_i + \tau)\big[L_t(T_i)N(d_+(t, T_i, \kappa)) - \kappa N(d_-(t, T_i, \kappa))\big], \tag{2.181}$$

where

$$d_\pm(t, T, \kappa) = \frac{\log\left(\frac{L_t(T)}{\kappa}\right) \pm \frac{1}{2}\bar{\sigma}(t, T)^2}{\bar{\sigma}(t, T)}, \tag{2.182}$$

$N(\cdot)$ is the cumulative standard normal distribution function, and the unnormalized average LIBOR rate volatility is given by

$$\bar{\sigma}(t, T)^2 = \int_t^T \sigma(s, T)^2\, ds. \tag{2.183}$$

Swaptions are more problematic. Pricing a swaption struck at rate $\kappa$ requires an evaluation of an expectation under the measure with $Z_t(T_{i+1})$ as numeraire,

$$\mathbf{PSO}_t(T, \kappa) = \tau \sum_{i=1}^{n} Z_t(T_i) E_t^{Q(Z(T_{i+1}))}\big[(L_T(T_i) - \kappa)1_D\big], \tag{2.184}$$

where $1_D$ is the indicator function of the set $D$ of paths for which the payer's swaption ends up in the money, i.e., the set

$$D = \left\{ (r_T^s - \kappa) > 0 \right\}, \tag{2.185}$$

where the swap rate $r_t^s$ is given by equation (2.23). The random variables $L_T(T_j)$, $j = 1, \ldots, n$, are also assumed to be correlated.

## 2.5 Real-World Interest Rate Models

Modeling the real-world evolution of interest rate curves over long time periods is interesting for applications in risk management for assessing overnight risk. In corporate finance as well, these models are used to assess the risk exposure over a time horizon of several years for portfolios of interest-sensitive assets. Acceptable models should ensure that all forward interest rates are positive at all times and will involve some sort of principal-component analysis. In this section we discuss a simple model with the salient features.

An acceptable model meeting the no-arbitrage condition for forward rates is conveniently formulated in terms of the logarithms of forward rates. Consider a situation with a finite number of key rates for maturity times $T_1, T_2, \ldots, T_N$. A possible choice of $T_j$, following *RiskMetrics*™ [Mor96b], is to select the terms 1 m, 2 m, 3 m, 6 m, 9 m, 12 m, 2 y, 3 y, 4 y, 5 y, 7 y, 10 y, 15 y, 20 y, 25 y, and 30 y. Consider the logarithms of the time-$t$ forward rates for the intervals $[T_i, T_{i+1}]$:

$$x_t(i) = \log f_t(T_i, T_{i+1}). \tag{2.186}$$

There are several different ways to go about performing statistical estimations.

If short-term scenarios over time horizons of 1–10 days are sought, one can study the log-returns over the desired period of the time series $x_t(i)$,

$$\delta x_t(i) = x_t(i) - x_{t-1}(i), \tag{2.187}$$

and estimate the covariance matrix as the historical expectation

$$C_{ij} = E[\delta x(i) \cdot \delta x(j)] = \frac{1}{M} \sum_{t=1}^{M} \delta x_t(i) \delta x_t(j). \tag{2.188}$$

Here we assume a return time series of length $M$. Over short time horizons, the fat-tailed character of return distributions is an important feature to take into account. It appears that the degree of kurtosis depends on the term, with shorter maturities being more sensitive to shocks caused by changes of Central Bank policies. In this case, a possible approach is to estimate each term separately. Another approach is to perform portfolio-dependent estimations. The latter method is more accurate but less general.

If long-term scenarios are sought, it is appropriate to compute a singular-value decomposition of the rectangular matrix $Y$ made up of the mean subtracted-time series

$$y_t(i) = x_t(i) - E[x_t(i)] = x_t(i) - \frac{1}{M} \sum_{t=1}^{M} x_t(i). \tag{2.189}$$

Here the expectation is again computed by taking historical averages. $M$ is the number of historical data points, and $N$ is the number of forward dates. The matrix $Y$ has $M$ rows and $N$ columns, and its singular-value decomposition

$$Y = U \cdot S \cdot V' \tag{2.190}$$

involves an $M \times M$ matrix $U$, an $N \times N$ matrix $V$ and an $M \times N$ diagonal matrix S of singular values. The columns of the matrix $V$ are the principal components, denoted with $\mathbf{u}^{\alpha}$ (see Figure 2.9). If one projects the time series $\mathbf{y}_t$ along the principal components, one finds times series for the component scores:

$$\eta_t^{\alpha} = \sum_{i=1}^{N} y_t(i) u^{\alpha}(i), \qquad \alpha = 1, \ldots, N. \tag{2.191}$$

The component scores show a clear tendency to follow a mean reverting process. A statistical model can be built by first finding the auto-regression coefficients $m(\alpha)$ such that

$$\eta_t^{\alpha} - \eta_{t-1}^{\alpha} = -m(\alpha)\eta_{t-1}^{\alpha} + \sigma_{\alpha}\epsilon_t^{\alpha}. \tag{2.192}$$

Here, the coefficients $m(\alpha)$ are computed by solving a least-squares problem. Second, one can postulate that the residuals $\epsilon_t^{\alpha}$ are normally distributed and estimate the covariance matrix as

$$C^{\alpha\beta} = E\left[\eta^{\alpha}\eta^{\beta}\right] = \frac{1}{M}\sum_{t=1}^{M} \eta_t^{\alpha}\eta_t^{\beta}. \tag{2.193}$$

It is common for portfolios to be sensitive to rates in one currency as well as to interest rates in foreign currencies and on the exchange rates as well. Hence, one can consider the case of $R$ interest rate discount curves $Z_t^i(T)$, $i = 0, \ldots, R-1$, and $R$ currencies, giving rise to $(R-1)$ independent exchange rates $X^i$, $i = 1, \ldots, R-1$, giving the worth of one unit of the $i$th currency in the base currency with $i = 0$. In this case, long-term statistical estimations

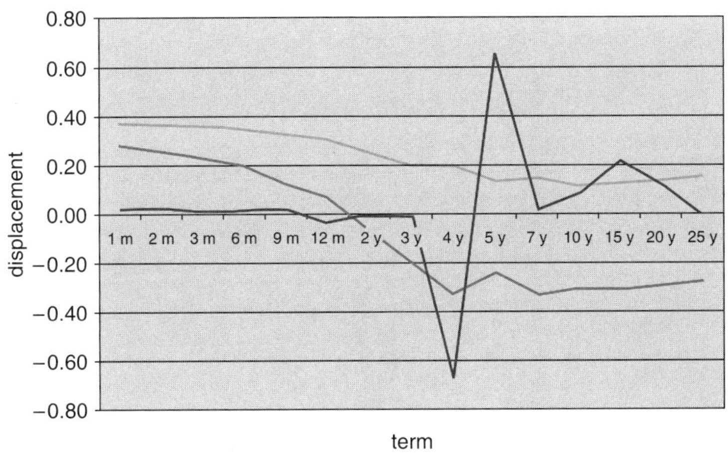


FIGURE 2.9 Three typical principal components for the forward curve as a function of the key maturity dates (i.e., the term) using a time series of U.S. Treasury curves.

must also account for the arbitrage condition, yielding forward exchange rates in terms of the spot exchange rates and interest rate curves. Namely,

$$F_t(X^i, T) = X_t^i \frac{Z_t^0(T)}{Z_t^i(T)}. \tag{2.194}$$

For a long-term statistical analysis, one can still accomplish a principal-component analysis and estimate the mean reversion rates for interest rates along the same lines. In addition, one needs a model for the spot foreign exchange rates, which, jointly with the no-arbitrage constraint in equation (2.194), yields all of the foreign exchange curves.

CHAPTER • 3

# Advanced Topics in Pricing Theory: Exotic Options and State-Dependent Models

*Exotic options* is a term used to describe derivative securities having cash flow or payoff structures that are more intricate and more complex than standard contracts such as plain-vanilla calls and puts. One main reason for trading, and hence pricing, such contracts is that they permit a much larger degree of flexibility for use in risk management and speculation. The payoff structure of these contracts can be fabricated to provide a higher leverage from an investor's viewpoint. Examples of this arise in so-called barrier options, the pricing of which is presented in great detail in this chapter. The theoretical pricing and hedging of exotic as well as standard derivatives depends largely on the stochastic model employed for the underlying asset price processes. The study of various models for the underling asset price process is therefore of importance to pricing theory as a whole.

This chapter is largely devoted to the development and application of exact solution methodologies for pricing derivatives under state-dependent asset price processes. A fairly general mathematical framework is presented for obtaining pricing kernels satisfying various boundary conditions. The kernels are then used to obtain new families of analytically exact closed-form pricing formulas for standard as well as barrier-style European options under various types of multiparameter state-dependent volatility models. The approach we take for tackling state-dependent models is of a general nature whereby we solve for the most fundamental quantities: the pricing kernels or transition probability density functions. This, in turn, is achieved by introducing a new and special type of "mapping" of the original state-dependent diffusion problem onto a related, yet simpler, diffusion problem corresponding to an appropriately chosen, simpler underlying process. The original diffusion problem is essentially reduced to a simpler diffusion for which exact pricing kernels are obtained by means of more standard methods. Once a kernel for the simpler underlying diffusion process is obtained, pricing kernels for a family of more complicated state-dependent volatility models are generated by direct substitution into a formula that provides an exact relationship between any two kernels — one for the simple diffusion and the other belonging to the family of kernels for the original state-dependent volatility model. The derivation of this useful formula is discussed at length in this chapter. Throughout this chapter we refer to the underlying (simpler) diffusion process as the so-called *x-space* process, while the price process of interest (i.e., the more complex process we wish to describe for pricing) is referred to as the *F-space* process. The process $F_t$ can be used to denote either an asset price or a forward price at time $t$.

Two particularly useful choices of underlying $x$-space processes are (i) the Wiener process and (ii) the Bessel process. We present exact solution methods for the transition density functions (i.e., the $x$-space kernels) for the Wiener and Bessel processes, separately, subject to nonabsorbing as well as all types of absorbing boundary conditions that correspond to either single- or double-barrier cases. The single- and double-barrier pricing kernels in the forward (or asset price) space of *interest are* then immediately generated by direct substitutions via our main formula. We shall see that the $F$-space pricing kernels for the linear and quadratic volatility models with two distinct roots can be generated simply from the standard Wiener densities. More complex and more abundant state-dependent pricing kernels arise from underlying densities for the Bessel process. In particular, a considerably larger family of analytically exact ($F$-space) pricing kernels containing as many as six adjustable parameters, which we shall refer to as the *Bessel family*, is generated from the underlying Bessel process. The Bessel family of solutions involves Bessel functions, as the name naturally suggests. This family is quite elaborate in structure because it is also shown to represent the exact solutions to most of the popular pricing models, including the linear, quadratic, and constant-elasticity-of-variance (CEV) volatility models as special cases. Some applications of the Bessel family of pricing kernels to option pricing are discussed in this chapter.

The first section introduces barrier options. The mathematical framework for obtaining probability densities for a process involving absorption at a barrier is then introduced in Section 3.2, where the simplest case is considered: a single-barrier Wiener process. The method of images is used to obtain the Wiener density for one absorbing barrier. Building on the results of Section 3.2, exact pricing kernels as well as single-barrier option formulas for the affine (linear volatility or lognormal model) and quadratic diffusion models are presented in Section 3.5. The method of Green's functions is then presented in Section 3.6 for solving the Kolmogorov partial differential equations for the kernel. In particular, we consider an underlying $x$-space diffusion process and show how analytical formulas for the time-dependent transition probability density for (barrier-type) absorbing boundary conditions

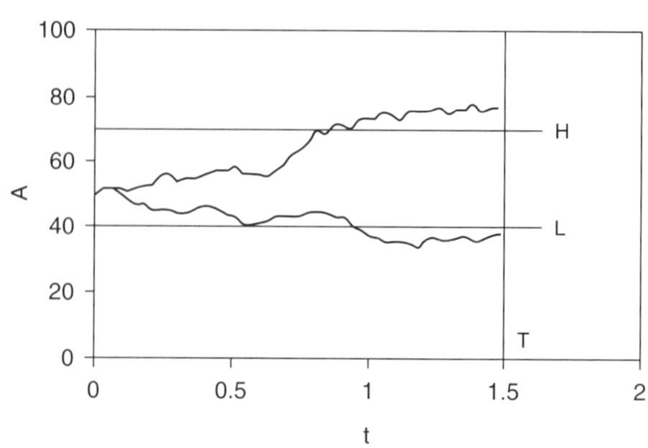


FIGURE 3.1 Sample asset price paths hitting a lower or upper barrier.

as well as nonabsorbing (barrier-free) conditions are generated via the (time-independent) Green's functions. In doing so, we also briefly present the basic important features of the Sturm–Liouville theory of ordinary differential equations for obtaining Green's functions. The Green's functions are obtained in two forms: (i) as special functions and (ii) as eigenfunction expansions. Green's functions of the first form lead to exact closed-form solutions for the transition density, generally in terms of special functions, whereas Green's functions of the second form give analytical series expansions for the kernel. The Green's functions formulas are then used in the subsequent sections to obtain transition densities for the Bessel process via complex variable contour integration methods. We then show how these densities can be used to directly generate new pricing kernels and European option pricing formulas for new families of diffusion models. Formulas are presented for: barrier free, single barriers, and double barriers. A discussion on the hierarchy of state-dependent models is also presented in light of the Bessel family as providing a model that recovers solutions to a class of popular models.

## .1 Introduction to Barrier Options

A barrier option is a particular kind of exotic option because it is to some extent path dependent. That is, the option's pay-off and hence value depends on the realized underlying asset path via the level attained any time before a given maturity time $T$. That is, if one considers an asset of price $A_t$ (e.g., a stock price), then a barrier for an option contract is generally given by a time-dependent price threshold $H_t$, $t \leq T$, on which the pay-off depends. [Note: As seen later, most standard barrier option contracts are structured as having a fixed (i.e., time-independent) barrier level or levels for a chosen underlying asset price.] Barrier options can be conveniently characterized in terms of stopping times. Let us denote $\tau(A, H)$ as the minimum time $\tau \in [t_0, T]$ for which the asset price $A_t$, starting at $A_0$ at current (initial) time $t = t_0$, first crosses or hits the barrier at level $H_\tau$, i.e., the first time $\tau$ for which $A_\tau \geq H_\tau$. Note that the stopping time is dependent on the complete path $A_t$ and the barrier level $H_t$ at all times $t \in [t_0, T]$.

There are two basic types of single-barrier options: (i) knockout options, which have a nonzero pay-off only if a level $H$ is not attained, and (ii) knock-in options, which have a nonzero pay-off only if the level $H$ is attained before or at maturity time $T$. There are then different flavors of these corresponding to whether the barrier level $H$ is placed above (single upper-barrier option) or below (single lower-barrier option) or both above and below (double-barrier option) the initial asset price. We refer the reader to the project in Part II of this book for further details on these contracts and how one can go about hedging them with plain-vanilla puts and calls. These and other examples of elementary single-barrier options and their corresponding payoff structures can be characterized in terms of stopping times, as follows.

(i) **Knockout options** with pay-off at time $T$:

$$\phi(A_T)\left(1 - \mathbf{1}_{\tau<T}\right), \tag{3.1}$$

and **knock-in options** with pay-off at time $T$:

$$\phi(A_T)\mathbf{1}_{\tau<T}, \tag{3.2}$$

with single-barrier level $H$. Here $\phi$ is a certain payoff function [i.e., $\phi(A) = (A - K)_+$ for a call struck at $K$], $\tau = \tau(A, H)$ is the stopping time for barrier level $H$, and $\mathbf{1}_{\mathcal{B}}$ is the indicator function taking on value 1 or 0 if event $\mathcal{B}$ occurs or not, respectively. For double-barrier knock-in/knockout options with lower level $L$ and upper level $H > L$, the pay-off is of the same form, where the indicator function in the foregoing two expressions is now replaced by $\mathbf{1}_{\min(\tau_L,\tau_H)<T}$ with $\tau_L$, $\tau_H$ as stopping times for hitting levels $L$, $H$, respectively.

(ii) **Corridor options** with two barrier levels $H^{(1)} < H^{(2)}$ and pay-off at time $T$:

$$\phi(A_T)\mathbf{1}_{\tau_1<T}\mathbf{1}_{\tau_2<T}, \tag{3.3}$$

where $\tau_1 = \tau(A, H^{(1)})$, $\tau_2 = \tau(A, H^{(2)})$ are stopping times for hitting the two respective levels. Corridor options hence have a nonzero pay-off only if the asset price hits both levels before time $T$.

(iii) **Pay-at-hit one-touch options** with pay-off at time $\tau$:

$$\phi(A_\tau)\mathbf{1}_{\tau<T}. \tag{3.4}$$

In contrast to the previous contracts, here the pay-off occurs *at* the stopping time rather than at maturity $T$, which is given by $\tau = \tau(A, H)$ in the case of a single level $H$.

(iv) **Upper-wall options**, with payoff

$$\frac{1}{T-t_0}\int_{t_0}^{T} \phi(A_t)\mathbf{1}_{A_t>H_t}\, dt, \tag{3.5}$$

and **lower-wall options**, with payoff

$$\frac{1}{T-t_0}\int_{t_0}^{T} \phi(A_t)\mathbf{1}_{A_t<H_t}\, dt. \tag{3.6}$$

The pay-offs of these contracts are given by the time average of a certain pay-off over all time intervals for which the asset price is above or below the barrier level $H_t$.

These elementary pay-offs can be engineered together to create more complex structures. These options are path-dependent securities and their price is affected by the dynamics of the implied volatility surface. From the modeling point of view it is often convenient to work in the space of the forward price process $F_t = F_t(A, T)$.

## 3.2 Single-Barrier Kernels for the Simplest Model: The Wiener Process

### 3.2.1 *Driftless Case*

Recall equation (1.86), which is the probability density for free Brownian motion with drift and no barriers (i.e., with nonabsorbing homogeneous zero-boundary conditions imposed at $\pm\infty$). Setting the drift to zero gives the transition probability density for a pure Wiener process $x_t$, with constant volatility. Let us reconsider the Wiener process $x_t$, obeying the SDE: $dx_t = \nu(x)dW_t$, with constant volatility function[1] $\nu(x) = \sqrt{2}$, zero drift, and focus now on solving the corresponding forward and backward Kolmogorov partial differential equations:

$$\frac{\partial}{\partial t}u(x, t; x_0, t_0) = \frac{\partial^2}{\partial x^2}u(x, t; x_0, t_0) \tag{3.7}$$

[1] This choice of volatility proves convenient because solutions for arbitrary constant volatility $\nu(x) = \sigma = const$ obtain by a simple time scale change, i.e., by the replacement $t \to \frac{1}{2}\sigma^2 t$, $t_0 \to \frac{1}{2}\sigma^2 t_0$ within the solutions for $\nu(x) = \sqrt{2}$.

and

$$\frac{\partial}{\partial t_0}u(x, t; x_0, t_0) + \frac{\partial^2}{\partial x_0^2}u(x, t; x_0, t_0) = 0, \tag{3.8}$$

subject to delta function initial (or final) time condition in the case of the forward (backward) equation: $\lim_{t \to t_0} u(x, t; x_0, t_0) = \delta(x - x_0)$, $t - t_0 > 0$. More formal methods for solving equation (3.7) or (3.8), in the case of general time independent volatility and drift functions, by application of Laplace transform and Green's functions techniques, are discussed in Section 3.6. In this particularly simple example, however, we simply make use of the solution for the barrier-less case obtained in Chapter 1. Namely, the solution $u(x, t; x_0, t_0) = g_0(x, x_0; \tau)$ for the infinite domain $x, x_0 \in (-\infty, \infty)$, allowing paths to attain any finite value, is simply

$$g_0(x, x_0; \tau) \equiv \frac{e^{-(x-x_0)^2/4\tau}}{2\sqrt{\pi\tau}}. \tag{3.9}$$

Note: Throughout this section we define $\tau \equiv t - t_0$. In most of what follows we shall work in terms of this time quantity, since the drift and volatility terms are not explicitly time dependent, hence giving rise to time-homogeneous solutions dependent on $\tau$. The boundary conditions are homogeneous: $\lim_{x \to \pm\infty} g_0(x, x_0; \tau) = 0$, given any $x_0$, and for the backward equation $\lim_{x_0 \to \pm\infty} g_0(x, x_0; \tau) = 0$, given any $x$, and finite time $\tau$. This so-called elementary solution can be used to obtain the solution to any other initial-value problem satisfying equation (3.7) [or (3.8)] and obeying homogeneous boundary conditions on the infinite domain. Indeed, the solution to the forward-time equation (3.7) for an initial distribution condition $u(x, t = t_0) = f(x)$ is given by the integral

$$u(x, t) = \int_{-\infty}^{\infty} f(x_0) g_0(x, x_0; \tau) dx_0. \tag{3.10}$$

The function $g_0(x, x_0; \tau)$ is also referred to as a *time-dependent Green's function* or *kernel* or *fundamental solution* for the preceding diffusion process. Physically, this corresponds to the transition probability density of the random variable $x_t$ having value $x_0$ at an initial time $t_0(\tau = 0)$ and taking on the value $x$ at a later time $t$. For any time value $\tau > 0$ and any fixed initial value $x_0$, one readily verifies that this Gaussian-shaped density integrates to unity exactly over $x \in (-\infty, \infty)$. In the limit $\tau \to 0$ the kernel is the delta function, thereby also integrating to unity, as required. This kernel hence corresponds to the case of no absorption outside the entire region; i.e., probability is conserved in the entire region $x \in (-\infty, \infty)$.

Let us now consider a solution to the forward-time equation (3.7) by imposing a zero boundary condition at a finite upper-barrier value $x = x_H$, i.e., $u(x_H, t; x_0, t_0) = 0$, with solution region of interest defined by $x_0, x \leq x_H$. As is seen shortly, this gives rise to absorption of paths (at $x = x_H$) into the region outside the interval $(-\infty, x_H)$. We will now demonstrate the use of the so-called method of images. In this technique the exact solution to the forward-time Kolmogorov equation, for arbitrary initial condition $u(x, t = t_0) = f(x)$, is obtained by extending the ("physical") region $x \leq x_H$ to include the ("nonphysical") region $x > x_H$ via the definition

$$\bar{f}(x) = \begin{cases} f(x), & x \leq x_H \\ -f(2x_H - x), & x > x_H. \end{cases} \tag{3.11}$$

This function is antisymmetric about the point $x = x_H$: $\bar{f}(x_H - \epsilon) = -\bar{f}(x_H + \epsilon)$ for any $\epsilon > 0$. Then using the solution in the form

$$u(x,t) = \int_{-\infty}^{\infty} \bar{f}(x_0) g_0(x, x_0; \tau) dx_0 \tag{3.12}$$

with $g_0$ given by equation (3.9) one can easily show by a change of integration variables that $u(x = x_H, t) = 0$. This is a consequence of the antisymmetric property. By splitting this integral into the regions $(-\infty, x_H]$ and $(x_H, \infty)$, using equation 3.11, and changing integration variables in one of the integrals, one finally has the solution to the initial-value problem on the interval $x \in (-\infty, x_H]$ satisfying the forward-time PDE of the form in equation (3.7), with $u(x, t = t_0) = f(x)$ and zero-boundary condition $u(x = x_H, t) = 0$:

$$u(x,t) = \int_{-\infty}^{x_H} f(x_0) g^u(x_H, x, x_0; \tau) dx_0, \tag{3.13}$$

where

$$\begin{aligned} g^u(x_H, x, x_0; \tau) &= g_0(x, x_0; \tau) - g_0(x, 2x_H - x_0; \tau) \\ &= g_0(x, x_0; \tau) - g_0(2x_H - x, x_0; \tau) \\ &= \frac{1}{2\sqrt{\pi\tau}} \left( e^{-(x-x_0)^2/4\tau} - e^{-(x+x_0-2x_H)^2/4\tau} \right). \end{aligned} \tag{3.14}$$

This last quantity is hence the time-dependent Green's function or kernel $u(x, t; x_0, t_0) = g^u(x_H, x, x_0; \tau)$ for the Wiener process in the region $x_0, x \leq x_H$, with the condition that there is absorption at the barrier level $x = x_H$. The fact that absorption occurs when imposing a zero boundary condition on the solution $u$ at a finite level is examined more precisely later, where we also show explicitly why $g^u(x_H, x, x_0; \tau)$ is considered a probability density for Wiener (Brownian) paths starting from $x_0 < x_H$ and ending at any point $x \leq x_H$ in time $\tau$, conditional on absorption of all paths crossing the barrier level $x_H$. Note that $g^u$ is given by subtracting the original (i.e., no-barrier) density $g_0$ centered at $x_0$ with the same density centered at $2x_H - x_0$ within the nonphysical region $x \in (x_H, \infty)$ (see Figure 3.2). This is essentially the reflection principle arising from the method of images, where the image source is a sink at the point $2x_H - x_0$. Since $g^u$ is a linear combination of two solutions to the Kolmogorov equations (which are linear partial differential equations), $g^u$ as given by equation (3.14) is then also a solution to the Kolmogorov equations and, moreover, is readily seen to satisfy the required zero-boundary condition at the barrier, $g^u(x_H, x = x_H, x_0; \tau) = 0$, as well as $g^u(x_H, x = -\infty, x_0; \tau) = 0$. Using the delta function definition, we have $\lim_{\tau \to 0} g^u = \delta(x - x_0) - \delta(x - (2x_H - x_0))$. Hence from the integral property of the delta function, the solution given by equation (3.13) is indeed shown to satisfy the required initial condition. Note that the second delta function does not contribute to the integral, for it is centered in the nonphysical region and is precisely the term that acts as a so-called sink (or negative point source), as mentioned earlier.

The foregoing method applies in identical fashion if we are interested in obtaining solutions within the upper half-line region $x_0, x \geq x_L$, where $x_L$ is now any finite lower-absorption boundary point with $u(x = x_L, t) = 0$. In this case the kernel $u(x, t; x_0, t_0) = g^l(x_L, x, x_0; \tau)$ for the Wiener process in the region $x_0, x \geq x_L$, given the absorption condition at the lower barrier level $x = x_L$ is given by $g^l(x_L, x, x_0; \tau) = g_0(x, x_0; \tau) - g_0(x, 2x_L - x_0; \tau)$, and equation (3.13) is replaced by

$$u(x,t) = \int_{x_L}^{\infty} f(x_0) g^l(x_L, x, x_0; \tau) dx_0. \tag{3.15}$$

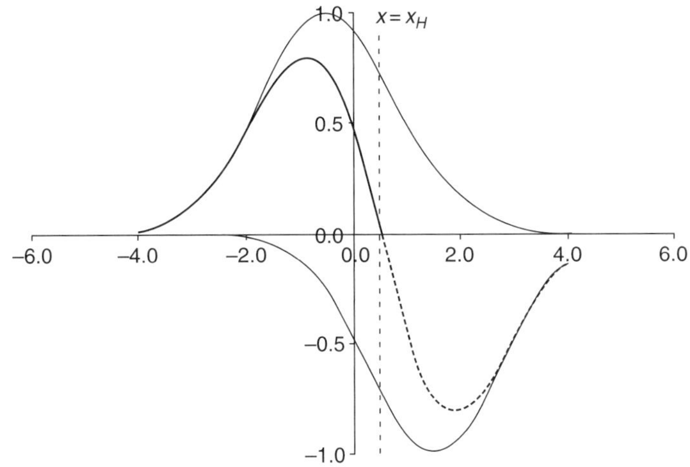


FIGURE 3.2 A sample plot of the kernel $g^u(x_H, x, x_0; \tau)$ for absorption at an upper barrier with parameter choices $x_H = 0.5$, $x_0 = -0.5$, $\tau = 0.75$. The thicker solid line gives $g^u$ in the physical solution region, while the dashed line extends into the nonphysical region. The plot of $g^u$ is obtained by subtracting two barrier-free kernels (i.e., summing the two thin solid lines): $g_0(x, x_0; \tau) - g_0(x, 2x_H - x_0; \tau)$, where $2x_H - x_0 = 1.5$.

This therefore gives the solution to the initial-value problem on the interval $x \in [x_L, \infty)$ satisfying the forward-time Kolmogorov PDE with arbitrary initial condition $u(x, t_0) = f(x)$ and zero-boundary conditions $u(x_L, t) = 0$. We note that if $f(x)$ is integrable over the entire solution domain, then $u(\infty, t) = 0$ also. Due to the symmetry of the Wiener process, we also have $g^l(x_b, x, x_0; \tau) = g^u(x_b, x, x_0; \tau)$ for any real barrier value $x_b$. This follows from the symmetry of the Green's function $g_0(x, x_0; \tau) = g_0(x_0, x; \tau)$.

It is important to observe that our analysis can be applied similarly to solve the backward-time Kolmogorov PDE, where $t_0 = t$ now corresponds to a *final-time* condition instead of an initial-time condition. The foregoing transition density function $g_0$ also satisfies the backward PDE with zero-(homogeneous)-boundary conditions at infinity, $\lim_{x_0 \to \pm\infty} g_0(x, x_0; \tau) = 0$, given any $x$. If a zero-boundary condition is placed at some upper level $x_0 = x_H$, then the solution kernel for equation (3.8) on the interval $x, x_0 \in (-\infty, x_H]$ is again given by $u(x, t; x_0, t_0) = g^u(x_H, x, x_0; \tau)$ since expression (3.14) satisfies the backward PDE and $g^u(x_H, x, x_0 = x_H; \tau) = g^u(x_H, x, x_0 = -\infty; \tau) = 0$ for any fixed $x$. In general, the solution to the backward PDE with arbitrary final-time condition $u(x_0, t_0 = t) = \phi(x_0)$ and kernel $u(x, t; x_0, t_0)$ can be represented as

$$u(x_0, t_0) = \int_{\mathcal{D}} \phi(x) u(x, t; x_0, t_0) dx, \tag{3.16}$$

where the integral is over the appropriate solution interval $\mathcal{D}$ and $u$ is the kernel with appropriate boundary conditions imposed at two endpoints. In particular, the solution with zero-boundary condition imposed at the endpoint $x_0 = x_H$ is given by the integral

$$u(x_0, t_0) = \int_{-\infty}^{x_H} \phi(x) g^u(x_H, x, x_0; \tau) dx, \tag{3.17}$$

while for zero boundary condition at a lower endpoint $x_0 = x_L$

$$u(x_0, t_0) = \int_{x_L}^{\infty} \phi(x) g^l(x_L, x, x_0; \tau) dx. \tag{3.18}$$

If $\phi$ is further assumed to be a compact integrable function over the entire solution domain, then $u(x_0, t_0)$ will also have zero-boundary condition as $x_0 \to \pm\infty$ accordingly.

It is instructive to reconsider the preceding absorbing barrier problem from a different point of view using purely probabilistic arguments and basic properties of Brownian paths. In particular, let $x_t$ denote the Brownian motion starting at $x_0 < x_H$ at initial time $t_0$ with upper absorbing barrier at $x = x_H$. Let $\tilde{x}_t$ denote the same Brownian motion but with no barrier, i.e., the standard Brownian (or Wiener) process with transition density $g_0(\tilde{x}_t, x_0; \tau)$, $\tau = t - t_0$. Let us focus on the case of an upper barrier (the derivation for the case of a lower barrier is similar; see Problem 4 of this section) and set out to compute the probability that a path $x_s$, $t_0 \le s \le t$, has the value of $X$ or less at time $t$, where $X < x_H$:

$$P\{x_t \le X\} = P\{\tilde{x}_t \le X, \sup_{t_0 \le s \le t} \tilde{x}_s < x_H\}. \tag{3.19}$$

This expression follows from the fact that if a free Brownian path $\tilde{x}_s$ crosses the barrier, $x_s$ will be absorbed and hence would never attain a value below $x_H$. Now, from first principles the total probability for the event

$$\{\tilde{x}_t \le X\} = \{\tilde{x}_t \le X, \sup_{t_0 \le s \le t} \tilde{x}_s < x_H\} \cup \{\tilde{x}_t \le X, \sup_{t_0 \le s \le t} \tilde{x}_s \ge x_H\}$$

is given by the sum of the probabilities of the two mutually exclusive events:

$$P\{\tilde{x}_t \le X\} = P\{\tilde{x}_t \le X, \sup_{t_0 \le s \le t} \tilde{x}_s < x_H\} + P\{\tilde{x}_t \le X, \sup_{t_0 \le s \le t} \tilde{x}_s \ge x_H\}. \tag{3.20}$$

Any path contributing to the second term must therefore cross the barrier. The density for the $\tilde{x}_t$ motion is given by $g_0(\tilde{x}_t, x_0; \tau)$, so $\tilde{x}_t$ follows a symmetric random walk in time. In particular, if we let $t_H < t$ denote the time at which a path first hits $x_H$, then the probability density that a Brownian path at $x_H$ at time $t_H$ subsequently attains the value $X$ at terminal time $t$ is the same as that for a (reflected) path starting at $x_H$ at time $t_H$ and attaining a value $2x_H - X$ at time $t$ (see Figure 3.3). Indeed, for both paths this probability density is

$$g_0(X, x_H; t - t_H) = g_0(2x_H - X, x_H; t - t_H) = \frac{e^{-(X - x_H)^2/4(t - t_H)}}{2\sqrt{\pi(t - t_H)}}. \tag{3.21}$$

Using this, the second term in equation (3.20) becomes

$$\begin{aligned} P\{\tilde{x}_t \le X, \sup_{t_0 \le s \le t} \tilde{x}_s \ge x_H\} &= P\{\tilde{x}_t \ge 2x_H - X, \sup_{t_0 \le s \le t} \tilde{x}_s \ge x_H\} \\ &= P\{\tilde{x}_t \ge 2x_H - X\}, \end{aligned} \tag{3.22}$$

where the last term follows because the supremum condition is redundant. Substituting this result into equation (3.20) and using equation (3.19) gives

$$P\{x_t \le X\} = P\{\tilde{x}_t \le X\} - P\{\tilde{x}_t \ge 2x_H - X\} \tag{3.23}$$

for all $X < x_H$.

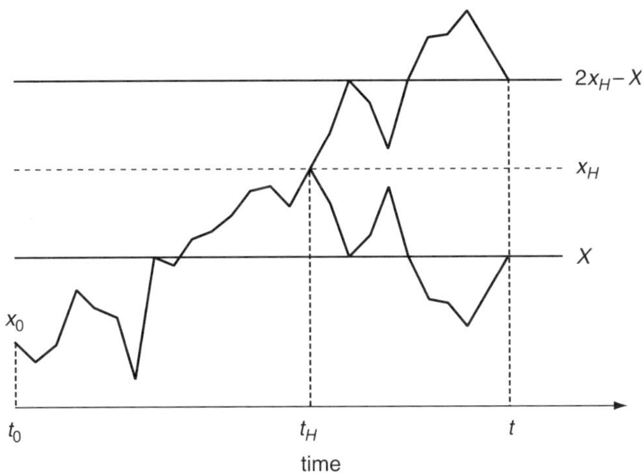


FIGURE 3.3 The reflection principle for Brownian paths.

Placing the density $g_0$ into equation (3.23) hence gives the probability of any path initiating below the barrier at $x_0 < x_H$ and attaining any value $x_t \le X < x_H$ within a time interval $\tau$, with the condition of paths being absorbed if the barrier level $x_H$ is crossed, as:

$$
\begin{aligned}
P\{x_t \le X\} &= \int_{-\infty}^{X} g_0(x, x_0; \tau)dx - \int_{2x_H - X}^{\infty} g_0(x, x_0; \tau)dx \\
&= \int_{-\infty}^{X} g^u(x_H, x, x_0; \tau)dx,
\end{aligned}
\tag{3.24}
$$

where the last expression is obtained by a change of variable in the second integral. Since the density is obtained by differentiating the cumulative probability function (or by the standard definition of a cumulative density function) we conclude that the kernel $g^u(x_H, x, x_0; \tau)$ in equation (3.14), as derived earlier by the method of images, is indeed the transition probability density for Brownian motion $x_t$ on the interval $x, x_0 \in (-\infty, x_H]$ with an absorbing barrier at $x_H$. The probability in the last equation is readily evaluated as the difference of two cumulative normal functions:

$$
P\{x_t \le X\} = N\left(\frac{X - x_0}{\sqrt{2\tau}}\right) - N\left(\frac{X + x_0 - 2x_H}{\sqrt{2\tau}}\right), \tag{3.25}
$$

$\tau = t - t_0$. The absorption of paths crossing the barrier can then be quantified precisely as follows. Let $P(\tau)$ denote the probability of any path initiating at $x_0 < x_H$ and terminating within time $\tau$ in the interval $x \in (-\infty, x_H]$, conditional on absorption at $x_H$. Then $P(\tau) = P\{x_t \le x_H\}$, where the conditional probability is computed using the density $g^u$:

$$
P(\tau) = N\left(\frac{x_H - x_0}{\sqrt{2\tau}}\right) - N\left(-\frac{x_H - x_0}{\sqrt{2\tau}}\right). \tag{3.26}
$$

Hence the probability does not integrate to unity and is in fact time dependent with $P(\tau) < 1$, implying absorption with $1 - P(\tau)$ giving the probability of absorption. Moreover, $P(\tau) \to 1$ as $\tau \to 0$ and $P(\tau) \to 0$ as $\tau \to \infty$. One can also compute the rate of absorption $R(\tau) = -dP(\tau)/d\tau$ or flux across the barrier (i.e., the rate at which probability leaks). From equation (3.26) (and the analogous formula for the case of a lower barrier, wherein $x_H - x_0$ is replaced by $x_0 - x_L$), we generally have

$$R(\tau) = \frac{|x_b - x_0|}{2\sqrt{\pi\tau^3}} e^{-(x_b-x_0)^2/4\tau}, \tag{3.27}$$

where $x_b$ is either a lower or an upper barrier and $x_0$ is above or below the barrier, respectively.

### 3.2.2 Brownian Motion with Drift

The analysis of the previous section is readily extended to the case of a constant drift $\mu$ and constant volatility $\sigma$, i.e., drifted Brownian motion $x_t$ with stochastic increment $dx_t = \mu\, dt + \sigma\, dW_t$. The transition density for this process with no barrier [recall equation (1.86)] is

$$g_{0,\mu}(x, x_0; \tau) \equiv \frac{e^{-(x-x_0-\mu\tau)^2/2\sigma^2\tau}}{\sigma\sqrt{2\pi\tau}}. \tag{3.28}$$

Rewriting gives

$$g_{0,\mu}(x, x_0; \tau) = e^{\frac{\mu}{\sigma^2}(x-x_0)-\frac{\mu^2}{2\sigma^2}\tau} g_0(x, x_0; \tau), \tag{3.29}$$

where

$$g_0(x, x_0; \tau) = \frac{e^{-(x-x_0)^2/2\sigma^2\tau}}{\sigma\sqrt{2\pi\tau}} \tag{3.30}$$

is the corresponding density for zero drift and no barrier [i.e., the density in equation (3.9) with $\tau \to \frac{1}{2}\sigma^2\tau$]. A transition probability density function for the drifted process, denoted by $u_\mu = u_\mu(x, x_0; \tau)$, is a fundamental solution to the forward and backward time-homogeneous Kolmogorov equations, which can be respectively written as

$$\frac{\partial u_\mu}{\partial \tau} = \frac{1}{2}\sigma^2\frac{\partial^2 u_\mu}{\partial x^2} - \mu\frac{\partial u_\mu}{\partial x} \tag{3.31}$$

and

$$\frac{\partial u_\mu}{\partial \tau} = \frac{1}{2}\sigma^2\frac{\partial^2 u_\mu}{\partial x_0^2} + \mu\frac{\partial u_\mu}{\partial x_0}, \tag{3.32}$$

with delta function condition $\lim_{\tau\to 0} u_\mu(x, x_0; \tau) = \delta(x - x_0)$. For the case of free motion on the entire infinite domain, we have $u_\mu = g_{0,\mu}$, since this kernel solves equations (3.31) and (3.32) with zero-boundary conditions at $x, x_0 \to \pm\infty$ and $\lim_{\tau\to 0} g_{0,\mu}(x, x_0; \tau) = \delta(x - x_0)$. As in the case of zero-drift, we are interested in further obtaining kernels satisfying zero-boundary conditions at any specified finite barrier level. For this purpose, relation (3.29) points to the following generally useful result.

**Proposition 3.1.** *Let $u_\mu(x, x_0; \tau)$ be a fundamental solution to the Kolmogorov equations (3.31) and (3.32) for drifted Brownian motion and satisfying homogeneous zero-boundary conditions (in $x$ or $x_0$) at any two endpoints of a finite, infinite, or semi-infinite solution domain. Assume the corresponding fundamental solution for zero drift ($\mu = 0$) is given by $u_0(x, x_0; \tau) \equiv u(x, x_0; \tau)$ and that this solution satisfies the same endpoint zero-boundary conditions, we have the relation*

$$u_\mu(x, x_0; \tau) = e^{\frac{\mu}{\sigma^2}(x-x_0)-\frac{\mu^2}{2\sigma^2}\tau} u(x, x_0; \tau). \tag{3.33}$$

The solution in equation (3.33) is verified by directly substituting into equations (3.31) and (3.32), differentiating, and using the fact that $u(x, x_0; \tau)$ solves the same forward and backward Kolmogorov equations for $\mu = 0$. In the limit $\tau \to 0$, $u_\mu$ obviously approaches the delta function since $u$ does. Moreover, note that the exponential term in equation (3.33) is bounded for all finite values of $x$, $x_0$, and grows only with linear exponent at infinite absolute values of $x$ or $x_0$. Hence, any zero-boundary condition on $u$ (placed at a finite or infinite point in $x$ or $x_0$) is automatically also satisfied by $u_\mu$ at the same point.

Based on the foregoing proposition, the barrier kernels for the drifted Wiener process are automatically obtained from those for zero drift. Although in this section we are explicitly discussing only the single-barrier case, the reader should realize that the proposition also applies directly to the case of the double-barrier kernels. Using equation (3.14) (with the replacement $\tau \to \frac{1}{2}\sigma^2\tau$) for the case of an upper absorbing barrier at $x = x_H$, the transition density denoted by $u_\mu = g_\mu^u$ on the domain $x, x_0 \in (-\infty, x_H]$ is then equivalently given by

$$\begin{aligned} g_\mu^u(x_H, x, x_0; \tau) &= \frac{e^{\frac{\mu}{\sigma^2}(x-x_0) - \frac{\mu^2}{2\sigma^2}\tau}}{\sigma\sqrt{2\pi\tau}}\left(e^{-(x-x_0)^2/2\sigma^2\tau} - e^{-(x+x_0-2x_H)^2/2\sigma^2\tau}\right) \\ &= g_{0,\mu}(x, x_0; \tau) - e^{\frac{2\mu}{\sigma^2}(x_H - x_0)} g_{0,\mu}(x, 2x_H - x_0; \tau) \\ &= g_{0,\mu}(x, x_0; \tau)\left[1 - e^{-2(x_H^2 + x x_0 - x_H(x+x_0))/\sigma^2\tau}\right], \end{aligned} \tag{3.34}$$

where the function $g_{0,\mu}$ is defined by equation (3.28). This density satisfies zero-boundary conditions at the barrier level $x, x_0 = x_H$ as well as at $x, x_0 \to -\infty$, as required. The kernel for the case of a lower barrier at $x = x_L$ is identical with transition density for $x, x_0 \in [x_L, \infty)$ given by $g_\mu^l(x_L, x, x_0; \tau) = g_\mu^u(x_L, x, x_0; \tau)$, with zero-boundary condition at $x, x_0 = x_L$ and at $x, x_0 \to \infty$. It is easy to verify by comparison of the relative magnitudes of the exponents that these densities are indeed strictly nonnegative on their respective semi-infinite solution domains.

These kernels can be used to provide analogous probability formulas to those in the previous section. For example, the kernel $g^l$ can be used to compute the probability that a drifted Brownian path initiating at any point above the barrier at $x_0 > x_L$, at time $t_0$, and attaining any value $x_t \geq X$, for $X \geq x_L$, within a time interval $t - t_0 = \tau$, conditional on the path being absorbed if it crosses below the barrier level $x_L$, as

$$\begin{aligned} P\{x_t \geq X \geq x_L | x_0 > x_L\} &= \int_X^\infty g_\mu^l(x_L, x, x_0; \tau)dx \\ &= N\left(\frac{x_0 - X + \mu\tau}{\sigma\sqrt{\tau}}\right) - e^{\frac{2\mu}{\sigma^2}(x_L - x_0)} N\left(\frac{2x_L - x_0 - X + \mu\tau}{\sigma\sqrt{\tau}}\right). \end{aligned} \tag{3.35}$$

The analogous probability for the case of an upper barrier at $x_H$ is

$$\begin{aligned} P\{x_t \leq X \leq x_H | x_0 < x_H\} &= \int_{-\infty}^X g_\mu^u(x_H, x, x_0; \tau)dx \\ &= N\left(\frac{X - x_0 - \mu\tau}{\sigma\sqrt{\tau}}\right) - e^{\frac{2\mu}{\sigma^2}(x_H - x_0)} N\left(\frac{X + x_0 - 2x_H - \mu\tau}{\sigma\sqrt{\tau}}\right). \end{aligned} \tag{3.36}$$

## Problems

**Problem 1.** Consider the Wiener process with lower absorbing barrier as discussed in Section 3.2.1. Obtain analogues of equations (3.19) through equation (3.27). Provide an analogous plot to the one in Figure 3.2 for the kernel $g^l(x_L, x, x_0; \tau)$.

**Problem 2.** What are the limiting values of $P(\tau)$ and $R(\tau)$ in equations (3.26) and (3.27) as $x_H \to \infty$? Explain.

**Problem 3.** Obtain formulas for $P(\tau)$ and $R(\tau)$ for the case of a driftless Wiener process with constant volatility $\sigma$. Explain the dependence of $P(\tau)$ and $R(\tau)$ on volatility. What are the limiting values as $\sigma \to \infty$ and $\sigma \to 0$?

**Problem 4.** Consider driftless Brownian motion with constant volatility $\nu(x) = \sigma$ and absorption at a lower barrier $x_L$. Using steps similar to those in equations (3.19) to (3.25), show that a path $x_s$, $t_0 \le s \le t$, conditional on starting at $x_0 > x_L$ at time $t_0$, has value $x_t \ge X$ at time $t$, where $X \ge x_L$, with probability given by

$$P\{x_t \ge X\} = N\left(\frac{X + x_0 - 2x_L}{\sigma\sqrt{\tau}}\right) - N\left(\frac{X - x_0}{\sigma\sqrt{\tau}}\right), \tag{3.37}$$

where $\tau = t - t_0$. Show that this result is consistent with equation (3.35) when $\mu = 0$.

**Problem 5.** By using equations (3.35) and (3.36) with $X = x_L$ and $X = x_H$, respectively, derive an expression for the rate of absorption across a barrier. Explain the particular dependence on the drift rate $\mu$.

## 3.3 Pricing Kernels and European Barrier Option Formulas for Geometric Brownian Motion

The kernels for the drifted Brownian motion obtained in the previous section can be used to provide exact pricing kernels and hence pricing formulas for which the underlying asset price process $S_t$ at time $t$ is assumed to obey a linear volatility and linear drift model (i.e., geometric Brownian motion or the standard Black–Scholes model):

$$dS_t = \mu S_t\, dt + \sigma S_t\, dW_t, \qquad S_t > 0.$$

Let us begin by defining the variable transformation $x = X(S) \equiv \log(S)$, with inverse $S = e^x$, mapping the domains $x \in (-\infty, \infty)$ and $S \in (0, \infty)$ into one another. From Itô's lemma, the process $x_t = \log S_t$ has SDE

$$dx_t = \left(\mu - \frac{\sigma^2}{2}\right) dt + \sigma\, dW_t.$$

Hence, the transition density for the random variable $\log S_t$ is given by the transition density for the simple Brownian motion $x_t$ with constant drift $\mu - \frac{1}{2}\sigma^2$ and volatility $\sigma$. Changing variables with Jacobian $d\log S/dS = 1/S$ therefore gives a general relationship between the $S$-space and the $x$-space densities:

$$U(S, S_0; \tau) = \frac{1}{S} u_{\mu - \frac{1}{2}\sigma^2}(X(S), X(S_0); \tau), \tag{3.38}$$

for all $S, S_0 > 0$. Here the notation $u_\mu$ refers to a kernel for simple Brownian motion with drift $\mu$, as discussed in the previous section. It is also readily shown by direct substitution, using equations (3.31) and (3.32), that the density $U$ satisfies the appropriate forward and backward Kolmogorov equations in $S$, $S_0$ (i.e., the Kolmogorov equations for lognormal diffusion with linear drift and volatility functions $\mu S$ and $\sigma S$, respectively, as discussed in Section 1.13).

Relation (3.38) holds true for any homogeneous zero-boundary conditions. The case of zero-boundary conditions imposed on the pricing kernel $U$ at $S, S_0 \to 0$ and $S, S_0 \to \infty$ corresponds to imposing zero-boundary conditions on the kernel $u$ at $x, x_0 \to \pm\infty$. Such boundary conditions give free geometric Brownian motion on the entire half-line $S, S_0 \in (0, \infty)$. The pricing kernel for the case of no barriers, denoted by $U_0$, is then obtained via equation (3.38) by substituting the barrier-free solution for drifted Brownian motion of the previous section $u_{\mu-\frac{1}{2}\sigma^2} = g_{0,\mu-\frac{1}{2}\sigma^2}(x, x_0; \tau)$, with $x = X(S) = \log S$, $x_0 = X(S_0) = \log S_0$, giving

$$U_0(S, S_0; \tau) = \frac{1}{S} g_{0,\mu-\frac{1}{2}\sigma^2}(\log S, \log S_0; \tau)$$
$$= \frac{1}{\sigma S\sqrt{2\pi\tau}} e^{-[\log(S/S_0)-(\mu-\sigma^2/2)\tau]^2/2\sigma^2\tau}. \qquad (3.39)$$

This is the familiar lognormal density for the Black–Scholes model discussed in Chapter 1. However, here we arrived at this density from a different perspective, one that allows us to readily derive pricing kernels subject to different boundary conditions. To obtain the pricing kernel for the case of a single absorbing barrier at $S = H$, the barrier points in the two spaces are related by $x_H = X(H) = \log H$. Then by simply substituting the appropriate single-barrier $x$-space kernel of equation (3.34) into equation (3.38) we obtain the equivalent forms:

$$\begin{aligned} U(H, S, S_0; \tau) &= \frac{1}{S} g^u_{\mu-\frac{1}{2}\sigma^2}(\log H, \log S, \log S_0; \tau) \\ &= \frac{1}{S}\Big[g_{0,\mu-\frac{1}{2}\sigma^2}(\log S, \log S_0; \tau) \\ &\quad - (H/S_0)^{\frac{2\mu}{\sigma^2}-1} g_{0,\mu-\frac{1}{2}\sigma^2}(\log S, \log(H^2/S_0); \tau)\Big] \\ &= U_0(S, S_0; \tau) - (H/S_0)^{\frac{2\mu}{\sigma^2}-1} U_0(S, H^2/S_0; \tau) \\ &= U_0(S, S_0; \tau)\left[1 - \exp\left[-\frac{\log(S/H)\log(S_0/H)}{\frac{1}{2}\sigma^2\tau}\right]\right], \end{aligned} \qquad (3.40)$$

where $U_0$ is given by equation (3.39). This single-barrier kernel hence satisfies zero-boundary conditions at the barrier value for both $S = H$ and $S_0 = H$, as well as approaching zero as $S, S_0 \to 0$ and as $S, S_0 \to \infty$. Kernel (3.40) is therefore valid as a single-barrier kernel (transition probability density) for either the lower domain, $S, S_0 \in (0, H]$, or the upper domain, $S, S_0 \in [H, \infty)$, with level $H$ being an upper barrier or lower barrier, respectively. The price level $H$ therefore plays the role of either upper or lower barrier in the respective solution domains.

Pricing kernel (3.40) can be used to obtain exact analytical formulas for various types of single-barrier European-style options under the Black–Scholes model where $\mu = r$, the assumed interest rate. If the underlying asset has constant dividend yield $q$, then $\mu = r - q$. Without loss in generality, in what follows we derive explicit formulas for $q = 0$.[2] Given an arbitrary payoff function $\Lambda(S)$ at maturity time $T$, the fair value at current time $t_0$ and

[2] The pricing formulas for $q \neq 0$ obtain trivially from the $q = 0$ formulas. Indeed, let $V(S_0, r, q, \tau)$ represent any option-pricing function for the case of a constant dividend $q$. Then from the discounted risk-neutral pricing integrals we directly have $V(S_0, r, q, \tau) = e^{-q\tau} V(S_0, r - q, \tau)$, where the latter is the corresponding option-pricing function $V(S_0, r, \tau)$ derived for zero dividend but with subsequent drift replacement $r \to r - q$.

spot price $S_0 > H$ of a down-and-out option with barrier level $H$ is given by the discounted risk-neutral expectation over the domain above the barrier:

$$V^{DO}(S_0, \tau) = e^{-r\tau} \int_H^\infty U(H, S, S_0; \tau)\Lambda(S)dS, \tag{3.41}$$

where the option price is considered a function of $\tau = T - t_0$, the time to maturity. Recall from previous contract definitions that this option automatically expires worthless if the stock or asset price $S_t$ attains or falls below the barrier price level $H$ for any time before maturity. The value of the corresponding up-and-out option with spot price $S_0 < H$ is given by the discounted risk-neutral expectation over the domain below the barrier:

$$V^{UO}(S_0, \tau) = e^{-r\tau} \int_0^H U(H, S, S_0; \tau)\Lambda(S)dS. \tag{3.42}$$

The values of the knock-in barrier options (i.e., the up-and-in and down-and-in options) follow simply by (knock-in)-(knockout) symmetry:

$$V^{UI} + V^{UO} = V^{DI} + V^{DO} = V,$$

where

$$V(S_0, \tau) = e^{-r\tau} \int_0^\infty U_0(S, S_0; \tau)\Lambda(S)dS \tag{3.43}$$

is the value of the plain European option. [Note that these integral solutions are consistent with the fact that $V$, $V^{UI}$, $V^{DI}$, $V^{UO}$, and $V^{DO}$ all satisfy the usual time-homogeneous Black–Scholes partial differential equation (BSPDE) in the variables $S_0$, $\tau$ with appropriate boundary values in $S_0$ and whose value at zero time to maturity is determined uniquely by the pay-off (and the barrier level with respect to $S_0$ in the case of the barrier options). This follows, since one can interchange the order of taking partial derivatives in $S_0$, $\tau$ with integrating over $S$, and using the fact that $e^{-r\tau}U_0$ and $e^{-r\tau}U$ solve the BSPDE in $S_0$, $\tau$ (for fixed $S$) with appropriate boundary conditions in $S_0$ and delta function value at zero time to maturity.]

Recall from contract definitions that the knock-in options have zero value unless the asset price $S_t$ attains the barrier at a time before maturity time $T$, upon which the option immediately becomes the plain European. The foregoing symmetry relation follows from the fact that the knock-in solution is expressible as a linear combination of the knockout and barrier-free solutions. The unique combination then follows by satisfying boundary conditions. In particular, $V^{DI} = V - V^{DO}$ since at the barrier the knock-in must have the same value as the plain option: $V^{DI}(S_0 = H, \tau) = V(S_0 = H, \tau)$ for all nonzero times to maturity. Also, at the other boundary, $S_0 = 0$, the two option prices must both equal zero. Finally, at maturity, $V^{DI}(S_0, \tau = 0) = V(S_0, \tau = 0) - V^{DO}(S_0, \tau = 0) = 0$ since $V$ and $V^{DO}$ are equal for all $S_0 > H$, at zero time to maturity. This last property (i.e., the initial condition $\tau = 0$) must be satisfied since the asset price starts above the barrier and stays there, hence the barrier is never attained, giving zero value for the knock-in. A similar argument applied to the up-and-in option also leads to the foregoing symmetry.

We now provide the derivation of exact pricing formulas for single-barrier European calls and puts. As a first example, we consider a down-and-out call with strike $K$ and barrier level $H$. In this case $\Lambda(S) = (S - K)_+$ and equation (3.41) gives

$$C^{DO}(H, S_0, K, \tau) = e^{-r\tau} \int_B^\infty U(H, S, S_0; \tau)(S - K)dS, \tag{3.44}$$

where $B = H$ if $H \geq K$ and $B = K$ if $H \leq K$. In general, it proves useful to evaluate the two integrals defined by

$$\phi(B) \equiv \int_B^\infty U(H, S, S_0; \tau) dS \tag{3.45}$$

and

$$\bar{\phi}(B) \equiv \int_B^\infty U(H, S, S_0; \tau) S dS \tag{3.46}$$

for any $B \geq 0$. Using equation (3.40) with $\mu = r$, and changing integration variable $S = e^x$, we have

$$\phi(B) = \int_{\log B}^\infty \left[ g_{0,r-\frac{1}{2}\sigma^2}(x, \log S_0; \tau) - (H/S_0)^{\frac{2r}{\sigma^2}-1} g_{0,r-\frac{1}{2}\sigma^2}(x, \log(H^2/S_0); \tau) \right] dx.$$

This integral is evaluated using steps similar to those in previous derivations of the Black–Scholes formula for a plain call. In particular, using equation (3.28) gives

$$\begin{aligned}\int_{\log B}^\infty g_{0,r-\frac{1}{2}\sigma^2}(x, \log S_0; \tau) dx &= \frac{1}{\sigma\sqrt{2\pi\tau}} \int_{\log B}^\infty e^{-(x-\log S_0-(r-\frac{1}{2}\sigma^2)\tau)^2/2\sigma^2\tau} dx \\ &= \frac{1}{\sqrt{2\pi}} \int_{-\infty}^{\frac{\log(S_0/B)+(r-\frac{1}{2}\sigma^2)\tau}{\sigma\sqrt{\tau}}} e^{-\frac{1}{2}y^2} dy \\ &= N\left(d_-\left(\frac{S_0}{B}\right)\right),\end{aligned} \tag{3.47}$$

where here and throughout we define

$$d_\pm(x) \equiv \frac{\log x + (r \pm \frac{1}{2}\sigma^2)\tau}{\sigma\sqrt{\tau}} \tag{3.48}$$

where $d_-(x) = d_+(x) - \sigma\sqrt{\tau}$. The second line in equation (3.47) follows simply by a linear change of variables $x = \log S_0 + (r - \frac{1}{2}\sigma^2)\tau - \sigma\sqrt{\tau}y$. The second term in $\phi(B)$ is integrated in identical fashion, with $S_0$ replaced by $H^2/S_0$, and combining gives

$$\phi(B) = N\left(d_-\left(\frac{S_0}{B}\right)\right) - \left(\frac{H}{S_0}\right)^{\frac{2r}{\sigma^2}-1} N\left(d_-\left(\frac{H^2}{S_0 B}\right)\right). \tag{3.49}$$

The integrand for the $\bar{\phi}(B)$ integral is similar, except for an extra $e^x$ factor. Upon completing the squares in the integrand exponents and using similar steps as before, one readily obtains

$$\bar{\phi}(B) = e^{r\tau}\left[ S_0 N\left(d_+\left(\frac{S_0}{B}\right)\right) - S_0\left(\frac{H}{S_0}\right)^{\frac{2r}{\sigma^2}+1} N\left(d_+\left(\frac{H^2}{S_0 B}\right)\right)\right]. \tag{3.50}$$

From equations (3.44), (3.45), and (3.46), $C^{DO}(H, S_0, K, \tau) = e^{-r\tau}[\bar{\phi}(B) - K\phi(B)]$. Hence plugging the value $B = H$ if $H \geq K$ ($B = K$ if $H \leq K$) gives the exact pricing formula for the down-and-out option in terms of cumulative normal density functions:

$$\begin{aligned}C^{DO}(H, S_0, K, \tau) = {} & S_0 N\left(d_+\left(\frac{S_0}{H}\right)\right) - S_0\left(\frac{H}{S_0}\right)^{\frac{2r}{\sigma^2}+1} N\left(d_+\left(\frac{H}{S_0}\right)\right) \\ & - Ke^{-r\tau} N\left(d_-\left(\frac{S_0}{H}\right)\right) + Ke^{-r\tau}\left(\frac{H}{S_0}\right)^{\frac{2r}{\sigma^2}-1} N\left(d_-\left(\frac{H}{S_0}\right)\right)\end{aligned} \tag{3.51}$$

for $H \geq K$, and

$$C^{DO}(H, S_0, K, \tau) = S_0 N\left(d_+\left(\frac{S_0}{K}\right)\right) - S_0\left(\frac{H}{S_0}\right)^{\frac{2r}{\sigma^2}+1} N\left(d_+\left(\frac{H^2}{S_0 K}\right)\right)$$
$$- Ke^{-r\tau} N\left(d_-\left(\frac{S_0}{K}\right)\right) + Ke^{-r\tau}\left(\frac{H}{S_0}\right)^{\frac{2r}{\sigma^2}-1} N\left(d_-\left(\frac{H^2}{S_0 K}\right)\right) \tag{3.52}$$

for $H \leq K$. Note that for the case $H \leq K$, one also has the compact form in terms of plain calls:

$$C^{DO}(H, S_0, K, \tau) = C(S_0, K, \tau) - (H/S_0)^{\frac{2r}{\sigma^2}-1} C(H^2/S_0, K, \tau). \tag{3.53}$$

From the symmetry $C^{DI} + C^{DO} = C$, this expression gives the down-and-in value $C^{DI}$ explicitly. Rearranging equation (3.51) we can also extract an exact expression for $C^{DI}$ when $H \geq K$.

The down-and-out put value $P^{DO} = 0$ for $H \geq K$ since the put payoff $(K - S)_+$ is zero in this trivial case. Symmetry then gives $P^{DI} = P$, the plain European put value. In contrast, the case $H \leq K$ gives

$$\begin{aligned}
P^{DO}(H, S_0, K, \tau) &= e^{-r\tau} \int_H^K U(H, S, S_0; \tau)(K - S)dS \\
&= e^{-r\tau}\left[K(\phi(H) - \phi(K)) + \bar{\phi}(K) - \bar{\phi}(H)\right] \\
&= Ke^{-r\tau}\Bigg[N\left(d_-\left(\frac{S_0}{H}\right)\right) - \left(\frac{H}{S_0}\right)^{\frac{2r}{\sigma^2}-1} N\left(d_-\left(\frac{H}{S_0}\right)\right) \\
&\quad -N\left(d_-\left(\frac{S_0}{K}\right)\right) + \left(\frac{H}{S_0}\right)^{\frac{2r}{\sigma^2}-1} N\left(d_-\left(\frac{H^2}{S_0 K}\right)\right)\Bigg] \\
&\quad +S_0 N\left(d_+\left(\frac{S_0}{K}\right)\right) - S_0\left(\frac{H}{S_0}\right)^{\frac{2r}{\sigma^2}+1} N\left(d_+\left(\frac{H^2}{S_0 K}\right)\right) \\
&\quad -S_0 N\left(d_+\left(\frac{S_0}{H}\right)\right) + S_0\left(\frac{H}{S_0}\right)^{\frac{2r}{\sigma^2}+1} N\left(d_+\left(\frac{H}{S_0}\right)\right)
\end{aligned} \tag{3.54}$$

By using $C(S_0, K, \tau) = S_0 N(d_+(S_0/K)) - Ke^{-r\tau} N(d_-(S_0/K))$, the property $N(d_\pm(S_0/H)) = 1 - N(-d_\pm(S_0/H))$, and put-call parity for the plain call and put option price, this result is also expressible as

$$P^{DO}(H, S_0, K, \tau) = P(S_0, K, \tau) - P^{DI}(H, S_0, K, \tau), \tag{3.55}$$

where

$$\begin{aligned}
P^{DI}(H, S_0, K, \tau) &= -S_0 N\left(-d_+\left(\frac{S_0}{H}\right)\right) + Ke^{-r\tau} N\left(-d_-\left(\frac{S_0}{H}\right)\right) \\
&\quad + S_0\left(\frac{H}{S_0}\right)^{\frac{2r}{\sigma^2}+1}\left[N\left(d_+\left(\frac{H^2}{S_0 K}\right)\right) - N\left(d_+\left(\frac{H}{S_0}\right)\right)\right] \\
&\quad - Ke^{-r\tau}\left(\frac{H}{S_0}\right)^{\frac{2r}{\sigma^2}-1}\left[N\left(d_-\left(\frac{H^2}{S_0 K}\right)\right) - N\left(d_-\left(\frac{H}{S_0}\right)\right)\right]
\end{aligned} \tag{3.56}$$

is the value of the down-and-in put. Note that for $H = K$ these expressions give $P^{DI}(H, S_0, K, \tau) = P(S_0, K, \tau)$, the plain put value, and $P^{DO} = 0$, as required.

Up-and-out calls and puts are obtained using equation (3.42). For a put we have

$$
\begin{aligned}
P^{UO}(H, S_0, K, \tau) &= e^{-r\tau} \int_0^B U(H, S, S_0; \tau)(K - S)dS \\
&= Ke^{-r\tau} \int_0^B U\, dS - e^{-r\tau} \int_0^B US\, dS \\
&= Ke^{-r\tau}[\phi(0) - \phi(B)] + [\bar{\phi}(B) - \bar{\phi}(0)]e^{-r\tau} \\
&= -S_0\left[N\left(-d_+\left(\frac{S_0}{B}\right)\right) - \left(\frac{H}{S_0}\right)^{\frac{2r}{\sigma^2}+1} N\left(-d_+\left(\frac{H^2}{S_0 B}\right)\right)\right] \\
&\quad + Ke^{-r\tau}\left[N\left(-d_-\left(\frac{S_0}{B}\right)\right) - \left(\frac{H}{S_0}\right)^{\frac{2r}{\sigma^2}-1} N\left(-d_-\left(\frac{H^2}{S_0 B}\right)\right)\right],
\end{aligned}
$$

where $B = H$ for $H \leq K$ and $B = K$ for $H \geq K$. Here we have used the properties $N(d_\pm(\infty)) = N(\infty) = 1$ and $1 - N(x) = N(-x)$. Substituting $B = H$ or $B = K$ then gives the exact expressions for the up-and-out put:

$$
\begin{aligned}
P^{UO}(H, S_0, K, \tau) &= -S_0 N\left(-d_+\left(\frac{S_0}{H}\right)\right) + Ke^{-r\tau} N\left(-d_-\left(\frac{S_0}{H}\right)\right) \\
&\quad + S_0\left(\frac{H}{S_0}\right)^{\frac{2r}{\sigma^2}+1} N\left(-d_+\left(\frac{H}{S_0}\right)\right) \\
&\quad - Ke^{-r\tau}\left(\frac{H}{S_0}\right)^{\frac{2r}{\sigma^2}-1} N\left(-d_-\left(\frac{H}{S_0}\right)\right)
\end{aligned}
\tag{3.57}
$$

for $H \leq K$, and

$$
\begin{aligned}
P^{UO}(H, S_0, K, \tau) &= P(S_0, K, \tau) + S_0\left(\frac{H}{S_0}\right)^{\frac{2r}{\sigma^2}+1} N\left(-d_+\left(\frac{H^2}{S_0 K}\right)\right) \\
&\quad - Ke^{-r\tau}\left(\frac{H}{S_0}\right)^{\frac{2r}{\sigma^2}-1} N\left(-d_-\left(\frac{H^2}{S_0 K}\right)\right)
\end{aligned}
\tag{3.58}
$$

for $H \geq K$. The exact expressions for the up-and-in put follow simply by symmetry, $P^{UI} = P - P^{UO}$.

From equation (3.42), the up-and-out call is given by

$$
C^{UO}(H, S_0, K, \tau) = e^{-r\tau} \int_0^H U(H, S, S_0; \tau)(S - K)_+\, dS. \tag{3.59}
$$

Since the payoff function is zero for $S \leq K$, $C^{UO} = 0$ for $H \leq K$. For $H \geq K$ the option value can be rewritten as

$$
C^{UO}(H, S_0, K, \tau) = e^{-r\tau} \int_H^K U(H, S, S_0; \tau)(K - S)dS. \tag{3.60}
$$

As observed from equation (3.54), this is precisely the value of the down-and-out put option for $H \leq K$. By extracting out the plain call value $C(S_0, K, \tau)$ from the last expression on the right-hand side of (3.54), the result can be recast as

$$
C^{UO}(H, S_0, K, \tau) = C(S_0, K, \tau) - C^{UI}(H, S_0, K, \tau), \tag{3.61}
$$

with up-and-in call option value

$$C^{UI}(H, S_0, K, \tau) = S_0 N\left(d_+\left(\frac{S_0}{H}\right)\right) - Ke^{-r\tau}N\left(d_-\left(\frac{S_0}{H}\right)\right)$$
$$- S_0\left(\frac{H}{S_0}\right)^{\frac{2r}{\sigma^2}+1}\left[N\left(d_+\left(\frac{H}{S_0}\right)\right) - N\left(d_+\left(\frac{H^2}{S_0 K}\right)\right)\right]$$
$$+ Ke^{-r\tau}\left(\frac{H}{S_0}\right)^{\frac{2r}{\sigma^2}-1}\left[N\left(d_-\left(\frac{H}{S_0}\right)\right) - N\left(d_-\left(\frac{H^2}{S_0 K}\right)\right)\right] \quad (3.62)$$

for $H \geq K$. For $H = K$ we have $C^{UI} = C$, the plain call value, and $C^{UO} = 0$, as required.

All of the preceding analytical pricing formulas for geometric Brownian motion are also readily extended to the case of a time-dependent barrier that has an exponential form $H(\tau) = He^{-\alpha\tau}$, with $\alpha$, $H$ as constants. Assuming the choice $\alpha > 0$, the barrier boundary is an increasing function of calendar time (or decreasing function of time to maturity $\tau$). For a given $\tau$, the solution domain for the underlying asset price is $[H(\tau), \infty)$ for a down-and-out and $(0, H(\tau)]$ for an up-and-out. Assuming geometric Brownian motion as before with constant drift $\mu$ and volatility $\sigma$, the single-barrier kernel for this exponentially shaped barrier with zero-boundary condition at the $\tau$-dependent boundary level $S_0 = H(\tau)$ (and at the other endpoint $S_0 = 0$ or $S_0 = \infty$) is denoted by $U^{H(\tau)}(S, S_0, \mu; \tau)$. [Note that we use a notation involving the explicit functional dependence on the drift parameter needed to precisely clarify the arguments that follow.] It can be readily shown (see Problem 3) that this kernel is given by the constant barrier kernel in equation (3.40), now denoted by $U(H, S, S_0, \mu; \tau)$, where we replace the arguments $S_0 \to S_0 e^{\alpha\tau}$ and $\mu \to \mu - \alpha$. That is,

$$U^{H(\tau)}(S, S_0, \mu; \tau) = U(H, S, S_0 e^{\alpha\tau}, \mu - \alpha; \tau). \quad (3.63)$$

The risk-neutral pricing kernel for the exponential barrier with lognormal drift $\mu = r$ (the assumed constant interest rate) is then explicitly given by

$$U^{H(\tau)}(S, S_0, r; \tau) = \frac{1}{\sigma S\sqrt{2\pi\tau}}\left[e^{-[\log\frac{S}{S_0} - (r - \frac{\sigma^2}{2})\tau]^2/2\sigma^2\tau}\right.$$
$$\left. - \left(\frac{H(\tau)}{S_0}\right)^{\frac{2(r-\alpha)}{\sigma^2}-1} e^{-[\log\frac{S}{S_0} - (2\log\frac{H(\tau)}{S_0} + (r - \frac{\sigma^2}{2})\tau)]^2/2\sigma^2\tau}\right]. \quad (3.64)$$

Setting $\alpha = 0$ obviously recovers the previous risk-neutral density for the case with constant barrier.

Exact pricing formulas for European knockouts and knock-ins for exponential barriers can be obtained by integrating the density given by equation (3.64) and following similar steps as were used earlier for the case of a constant barrier. However, a straightforward approach is to make use of relation (3.63) directly in the risk-neutral pricing formula. Consider a down-and-out with payoff $\Lambda(S)$: The risk-neutral price is

$$V^{DO}(H(\tau), S_0, r, \tau) = e^{-r\tau}\int_{H(\tau)}^{\infty} U^{H(\tau)}(S, S_0, r; \tau)\Lambda(S)dS$$
$$= e^{-\alpha\tau}e^{-(r-\alpha)\tau}\int_{H}^{\infty} U(H, S, S_0 e^{\alpha\tau}, r - \alpha; \tau)\Lambda(S)dS$$
$$= e^{-\alpha\tau}V^{DO}(S_0 e^{\alpha\tau}, H, r - \alpha, \tau), \quad (3.65)$$

where $V^{DO}(S_0 e^{\alpha\tau}, H, r-\alpha, \tau)$ is the value of the down-and-out with spot $S_0 e^{\alpha\tau}$, constant barrier level at $H$, effective interest rate $r-\alpha$, and time to maturity $\tau$. Similarly, an up-and-out has value

$$V^{UO}(H(\tau), S_0, r, \tau) = e^{-\alpha\tau} V^{UO}(S_0 e^{\alpha\tau}, H, r-\alpha, \tau). \tag{3.66}$$

The corresponding prices of the knock-ins obtain simply from knock-in/knockout symmetry. Since the barrier-free pricing kernel (3.39) satisfies the invariance relation $U_0(S, S_0, r; \tau) = U_0(S, S_0 e^{\alpha\tau}, r-\alpha; \tau)$, the plain-vanilla price satisfies

$$V(S_0, r, \tau) = e^{-\alpha\tau} V(S_0 e^{\alpha\tau}, r-\alpha, \tau). \tag{3.67}$$

Given a pricing formula for the constant barrier case, the corresponding pricing formula for the exponentially shaped barrier follows from equation (3.65) or (3.66). For example, applying equation (3.65) to equation (3.53) gives the exact price of a down-and-out call with exponential barrier for $H(\tau) \leq K$ as

$$C^{DO}(H(\tau), S_0, K, \tau) = C(S_0, K, \tau) - \left(\frac{H(\tau)}{S_0}\right)^{\frac{2(r-\alpha)}{\sigma^2}-1} C\left(\frac{H(\tau)^2}{S_0}, K, \tau\right), \tag{3.68}$$

where equation (3.67) has been used on the two plain calls. Analogous formulas for the other types of knock-in and knockout barrier options discussed follow in similar fashion.

## Problems

**Problem 1.** Show that the function $V(S_0, \tau) = S_0^{\alpha} \overline{V}(a S_0^{\beta}, \tau)$ satisfies the Black–Scholes equation

$$\frac{\partial V}{\partial \tau} = \frac{1}{2}\sigma^2 S_0^2 \frac{\partial^2 V}{\partial S_0^2} + r S_0 \frac{\partial V}{\partial S_0} - rV, \tag{3.69}$$

where $\overline{V}(\overline{S}_0, \tau)$ is assumed to satisfy the same equation in the $(\overline{S}_0, \tau)$ variables, $\overline{S}_0 \equiv a S_0^{\beta}$, and provided we make the parameter choice $\alpha = 1 - 2r/\sigma^2$, $\beta = -1$, for arbitrary nonzero constant $a$. Then consider expressing the price of a down-and-out call struck at $K$, with constant barrier at $H \leq K$, as a linear combination of two solutions using plain calls

$$C^{DO} = C(S_0, K, \tau) + b S_0^{1-\frac{2r}{\sigma^2}} C(a/S_0, K, \tau). \tag{3.70}$$

Determine the constants $a$ and $b$ by satisfying the zero-boundary condition at the barrier $S_0 = H$ and the initial condition $C^{DO} \to (S_0 - K)_+$ as $\tau \to 0$, hence arriving at (3.53).

**Problem 2.** Derive the greeks $\Delta$, $\Gamma$, $\Theta$ (as defined in Chapter 1) for the down-and-out call, with value $V = C^{DO}$ given by equation (3.53). Is the relationship $\Theta = \frac{1}{2}\sigma^2 S_0^2 \Gamma + r(S_0\Delta - V)$ satisfied?

**Problem 3.** Consider the exponential barrier $H(\tau) = He^{-\alpha\tau}$, with $H$ and $\alpha$ as constants. Let $\tilde{U}(S, S_0; \tau) = U^{H(\tau)}(S, S_0; \tau)$ be the pricing kernel solving

$$\frac{\partial \tilde{U}}{\partial \tau} = \frac{1}{2}\sigma^2 S_0^2 \frac{\partial^2 \tilde{U}}{\partial S_0^2} + \mu S_0 \frac{\partial \tilde{U}}{\partial S_0} \tag{3.71}$$

and satisfying $\tilde{U}(S, S_0 = H(\tau); \tau) = 0$ and $\tilde{U}(S, S_0 = \infty; \tau) = 0$ for the fundamental solution in the upper domain $H(\tau) \leq S_0 < \infty$ or $\tilde{U}(S, S_0 = 0; \tau) = 0$ for the case of the lower solution domain $0 < S_0 \leq H(\tau)$. Let $\tilde{U}(S, S_0; \tau) = \bar{U}(S, \bar{S}; \tau)$, $\bar{S} = S_0 e^{\alpha\tau}$, and show that $\bar{U}$ solves

$$\frac{\partial \bar{U}}{\partial \tau} = \frac{1}{2}\sigma^2 \bar{S}^2 \frac{\partial^2 \bar{U}}{\partial \bar{S}^2} + (\mu - \alpha)\bar{S}\frac{\partial \bar{U}}{\partial \bar{S}}, \tag{3.72}$$

with $\bar{U}(S, \bar{S} = H; \tau) = 0$ and $\bar{U}(S, \bar{S} = \infty; \tau) = 0$ for the fundamental solution in the upper domain $H \leq \bar{S} < \infty$ or $\bar{U}(S, \bar{S} = 0; \tau) = 0$ for the case of the lower domain $0 < \bar{S} \leq H$. Hence $\bar{U}(S, \bar{S}; \tau) = U(H, S, S_0 e^{\alpha\tau}, \mu - \alpha; \tau)$, with the function $U$ given by equation (3.40) for constant barrier level $H$ and drift $\mu - \alpha$ (in the place of $\mu$), and conclude that the kernel $U^{H(\tau)}$ for the time-dependent exponential barrier with drift $\mu$ is given by equation (3.63), while setting $\mu = r$ gives equation (3.64).

## 3.4 First-Passage Time

When pricing exotic barrier options it is useful to consider the first-passage time of a diffusion process, i.e., the first time at which a process achieves a particular value or enters (exits) a region. In particular, for the sake of pricing, we are interested in the first-passage time for an asset price process crossing a specified constant barrier level $H > 0$. We hence consider calculating the probability distribution for the first-passage time, the time taken to attain the absorbing barrier. Consider the case of an upper barrier with current asset price $S_0 < H$, and let $t - t_0 = \tau > 0$ be the amount of time spent from current time $t_0$ until the barrier is first attained at time $t$. Then

$$\Phi(H, S_0, \tau) = 1 - \int_0^H U(H, S, S_0; \tau)dS \tag{3.73}$$

represents the probability (cumulative in the passage time $\tau$) that the asset price process has attained the upper barrier $H$ and has been absorbed. Indeed, this is just 1 minus the probability that the asset price remains below the barrier, or, equivalently, $\Phi$ is the probability of absorption. If we denote $\tau_p = \min\{\tau; S_t \geq H, S_0 < H\}$ as the first-passage time random variable, then $\Phi(H, S_0, \tau)$ is the probability $P\{\tau_p \leq \tau\}$. The function $U(H, S, S_0; \tau)$ is the kernel for the solution region $[0, H]$ with absorbing boundary condition at the barrier. [Note that although we are considering a time-homogeneous process, with state-dependent drift and volatility functions, the formal theory extends in the obvious manner for the general case of a time-inhomogeneous process, where we would consider a kernel $U(H, S, t; S_0, t_0)$ having explicit dependence on $t$ and $t_0$ rather than $\tau = t - t_0$.] As $\tau \to 0$, the integrand gives a Dirac delta function contribution $\delta(S - S_0)$ in the region $[0, H]$ and hence integrates to unity; therefore $\Phi(H, S_0, \tau = 0) = 0$. Since $U(H, S, S_0; \tau)$ is identically zero for $S_0 = H$, $\Phi$ has boundary condition $\Phi(H, S_0 = H, \tau) = 1$. Moreover, $U$ is a kernel and hence obviously solves both forward and backward Kolmogorov equations for the asset price diffusion process. Since partial derivatives with respect to $S_0$ and $\tau$ can be taken inside the integral, the cumulative probability density for the first passage time, $\Phi$, is therefore a solution of the time-homogeneous backward (and not the forward) Kolmogorov partial differential equation in $S_0$, $\tau$ subject to the foregoing conditions.

The other case, where $H$ is a lower barrier with current asset price $S_0 > H$, is similar, with equation (3.73) replaced by

$$\Phi(H, S_0, \tau) = 1 - \int_H^\infty U(H, S, S_0; \tau)dS, \tag{3.74}$$

which is the cumulative probability that the asset price process has attained the lower barrier and has been absorbed, where $U$ is the kernel for the solution region $[H, \infty)$ with absorbing boundary condition at the barrier. The first passage time random variable is now the stopping time $\tau_p = \min\{\tau;\ S_t \leq H, S_0 > H\}$. From similar arguments as before, one again obtains that $\Phi$ solves the same backward Kolmogorov equation with unit boundary condition at the barrier, $\Phi(H, S_0 = H, \tau) = 1$, and zero initial condition $\Phi(H, S_0, \tau = 0) = 0$.

In both cases, the function $\Phi$ can be obtained by solving the backward Kolmogorov equation subject to the stated conditions. However, given the kernel $U$, $\Phi$ is simply determined by an integration via equation (3.73) [or (3.74)]. If $\Phi$ is a cumulative function, the probability density function $f$ for the first passage time must be given by differentiation:

$$f(H, S_0, \tau) = \frac{\partial}{\partial \tau} P\{\tau_p \leq \tau\} = \frac{\partial \Phi(H, S_0, \tau)}{\partial \tau}. \tag{3.75}$$

For $f$ to be a bona fide probability density, it must be strictly nonnegative and must integrate to unity over all positive $\tau$. Integrating

$$\int_0^\infty f(H, S_0, \tau) d\tau = \int_0^\infty \frac{\partial \Phi(H, S_0, \tau)}{\partial \tau} d\tau = \Phi(H, S_0, \infty) \tag{3.76}$$

hence gives $\Phi(H, S_0, \infty) = 1$ as the latter condition. This is not generally satisfied, as we shall see next for the specific case of geometric Brownian motion. Since the integral in equation (3.76) gives the probability that (given any amount of time) a path starting at $S_0$ will eventually be absorbed at the barrier, this quantity is generally less than or equal to 1. The condition of nonnegativity of $f$, however, can be shown to follow for quite general processes (see Problem 1).

For geometric Brownian motion with drift $r$ and volatility $\sigma$ it is a simple matter to obtain exact formulas for the first-passage densities based on the exact kernel in equation (3.40). In particular, the integrals in equations (3.73) and (3.74) are given by direct use of equation (3.49), giving

$$\begin{aligned}
\Phi(H, S_0, \tau) &= N\left(-d_-\left(\frac{S_0}{H}\right)\right) + \left(\frac{H}{S_0}\right)^{\frac{2r}{\sigma^2}-1} N\left(d_-\left(\frac{H}{S_0}\right)\right) \\
&= 1 - N\left(\frac{\log \frac{S_0}{H} + (r - \frac{1}{2}\sigma^2)\tau}{\sigma\sqrt{\tau}}\right) \\
&\quad + \left(\frac{H}{S_0}\right)^{\frac{2r}{\sigma^2}-1} N\left(\frac{-\log \frac{S_0}{H} + (r - \frac{1}{2}\sigma^2)\tau}{\sigma\sqrt{\tau}}\right)
\end{aligned} \tag{3.77}$$

for $S_0 > H$ and

$$\begin{aligned}
\Phi(H, S_0, \tau) &= N\left(d_-\left(\frac{S_0}{H}\right)\right) + \left(\frac{H}{S_0}\right)^{\frac{2r}{\sigma^2}-1} N\left(-d_-\left(\frac{H}{S_0}\right)\right) \\
&= N\left(\frac{\log \frac{S_0}{H} + (r - \frac{1}{2}\sigma^2)\tau}{\sigma\sqrt{\tau}}\right) \\
&\quad + \left(\frac{H}{S_0}\right)^{\frac{2r}{\sigma^2}-1} N\left(\frac{\log \frac{S_0}{H} - (r - \frac{1}{2}\sigma^2)\tau}{\sigma\sqrt{\tau}}\right)
\end{aligned} \tag{3.78}$$

for $S_0 < H$. Hence, for $S_0 > H$ we obtain the limiting value of equation (3.77):

$$\Phi(H, S_0, \infty) = \begin{cases} 1, & r \leq \frac{1}{2}\sigma^2 \\ \left(\frac{H}{S_0}\right)^{\frac{2r}{\sigma^2}-1}, & r > \frac{1}{2}\sigma^2. \end{cases} \tag{3.79}$$

upon using $N(\infty) = 1$, $N(-\infty) = 0$. Hence, if $r \leq \frac{1}{2}\sigma^2$, the cumulative density approaches unity in the infinite-passage time limit so that, with probability 1, absorption eventually occurs. On the other hand, if $r > \frac{1}{2}\sigma^2$, the cumulative density approaches a number strictly between 0 and 1, since $H/S_0 < 1$ and $\frac{2r}{\sigma^2} - 1 > 0$, so the probability of eventual absorption is less than 1. In contrast, taking the infinite time limit of equations (3.78) gives, for $S_0 < H$,

$$\Phi(H, S_0, \infty) = \begin{cases} 1, & r \geq \frac{1}{2}\sigma^2 \\ \left(\frac{H}{S_0}\right)^{\frac{2r}{\sigma^2}-1}, & r < \frac{1}{2}\sigma^2. \end{cases} \tag{3.80}$$

In this case, the reverse is observed, whereby the density approaches unity only if $r \geq \frac{1}{2}\sigma^2$ and otherwise approaches a number strictly between 0 and 1. A basic interpretation of this is that eventual absorption will take place with certainty only if the effective drift, which is given by $r - \frac{1}{2}\sigma^2$, is not positive (or not negative) if the process starts above (or below) the barrier. By differentiating equations (3.77) and (3.78) and combining, the exact first-passage time density can be written as a single expression:

$$f(H, S_0, \tau) = \frac{|\log \frac{S_0}{H}|}{\sigma\tau^{3/2}\sqrt{2\pi}} e^{-[\log(S_0/H)+(r-\frac{1}{2}\sigma^2)\tau]^2/2\sigma^2\tau}, \tag{3.81}$$

for all $S_0, H > 0$.

The first-passage time density is useful when pricing "pay-at-hit one touch" type of options or for pricing barrier options that also provide a rebate payment to the holder once the barrier is hit. In the case of a down-and-out option with a rebate, equation (3.41) becomes

$$V^{DO}(S_0, \tau) = e^{-r\tau} \int_H^\infty U(H, S, S_0; \tau)\Lambda(S)dS + \int_0^\tau e^{-rt} R(\tau - t) f(H, S_0, t)dt.$$

The time integral term is just the expected present value of the rebate, whereby discounted payments occurring at an elapsed time $t$ in the future from the present are weighted with the first-passage time density for hitting the barrier after time $t$. The time-dependent rebate function is here assumed to be a function of the time remaining to maturtiy.

The first-passage time is also a very useful tool for computing options whose prices depend on stopping times that can be interpreted as *first hitting times*. Nice examples of such options are the *American digitals*. We have already discussed the payoff structure of European digitals. The Black–Scholes price of European digitals is simple to obtain (see Problem 8 in Section 1.8). The pay-off of an American digital is similar — the holder of an American digital receives one dollar if, and at the first time that, the underlying stock price hits the fixed strike level $K$. Since the option expires with a pay-off to the holder at the instant the spot hits the strike level, the early-exercise boundary, as such, is trivially fixed at the strike $K$. The time optionality in this case is simpler than in the standard American contracts (e.g., a put or call with dividend, etc.) studied in Chapter 1, where the early-exercise boundary is moving with time. The optimal stopping time is, in this case, just the first hitting time $\tau$ such that $S_\tau = K$. Once a hitting time $\tau$ occurs, the contract expires, paying one dollar at time $\tau$, and the value of that cash flow is the discounted value of one dollar, i.e., $e^{-r\tau}$. Given the probability density $f(H, S_0, \tau)$ for the first hitting time as provided by equation (3.75), the fair price of the American digital at time $t_0 = 0$, with maturity $T$, spot $S_0$, and strike $K$, reduces to a time integral:

$$V(S_0, T) = \int_0^T e^{-r\tau} f(K, S_0, \tau) d\tau = \int_0^T e^{-r\tau} \frac{\partial \Phi(K, S_0, \tau)}{\partial \tau} d\tau. \tag{3.82}$$

Closed-form analytical expressions can therefore be derived assuming a geometric Brownian motion model (see Problem 4).

## Problems

**Problem 1.** Consider a process with state-dependent drift $\mu(S)$ and volatility $\sigma(S)$. Argue that the first-passage time density for either lower or upper barrier case is strictly nonnegative. In developing your argument, consider the derivative with respect to $\tau$ of $\Phi$ defined via equation (3.73) [and (3.74) separately] and make use of the forward equation for the single barrier density $U = U(H, S, S_0; \tau)$:

$$\frac{\partial U}{\partial \tau} = \frac{1}{2} \frac{\partial^2}{\partial S^2} \left( \sigma^2(S) U \right) - \frac{\partial}{\partial S} \left( \mu(S) U \right). \tag{3.83}$$

Integrating over $S$ and assuming $\mu(S)U$ and $\sigma^2(S)U$ satisfy zero-boundary conditions at the endpoints, arrive at the expressions

$$f(H, S_0, \tau) = \pm \frac{1}{2} \sigma^2(H) \left. \frac{\partial U}{\partial S} \right|_{S=H},$$

where the plus sign is for $S_0 > H$ and the minus sign is for $S_0 < H$. Using the fact that the kernel is a positive differentiable function of $S$ within either solution interval and has zero value at the barrier endpoint, further argue that

$$f(H, S_0, \tau) = \frac{1}{2} \sigma^2(H) \left| \left. \frac{\partial U}{\partial S} \right|_{S=H} \right|, \tag{3.84}$$

which is hence strictly nonnegative.

**Problem 2.** Using the kernel in equation (3.40), give an explicit verification that equation (3.84) gives the exact first-passage time density in equation (3.81) for geometric Brownian motion where $\mu(S) = rS$, $\sigma(S) = \sigma S$.

**Problem 3.** Assume the exponentially time-dependent barrier of the previous section, $H(\tau) = He^{-\alpha\tau}$, $\alpha > 0$. Show that the first-passage time density for $S_0 > H$ is given by

$$f(H, S_0, \tau) = \frac{\log \frac{S_0}{H(\tau)}}{\sigma \tau^{3/2} \sqrt{2\pi}} e^{-[\log(S_0/H(\tau)) + (r - \alpha - \frac{1}{2}\sigma^2)\tau]^2 / 2\sigma^2 \tau}. \tag{3.85}$$

**Problem 4.** Obtain an analytical pricing formula for an American digital within the geometric Brownian motion model.

## 3.5 Pricing Kernels and Barrier Option Formulas for Linear and Quadratic Volatility Models

The kernels for the Wiener process obtained in the previous sections are readily used as a basis for providing other exact kernels and barrier-pricing formulas for affine and quadratic volatility models. The formulas follow as a simple consequence of a more general method, which we coin as the *diffusion canonical mapping reduction methodology*. This mathematical framework is presented in detail later in this chapter. In particular, it provides a precise relationship between the transition probability density or pricing kernel $U(F, F_0; \tau)$ for the space of a process $F_t$ and the transition density $u(x, x_0; \tau)$ for a process $x_t$ under a simpler diffusion. In this section it suffices to consider $x_t$ as the pure Wiener process. The process $F_t$ represents an underlying asset price, such as a forward price at time $t$. Hence, given an exact kernel for the simpler $x$-space process, we show how the mapping reduction method automatically provides the desired exact pricing kernel for the more complicated $F$-space process. Moreover, the desired boundary conditions in $F$-space (e.g., in the desired asset price space) are satisfied by mapping onto the corresponding boundary conditions in $x$-space.

### 3.5.1 *Linear Volatility Models Revisited*

Although we have already dealt with the linear volatility model (i.e., the standard Black–Scholes model) in great detail in previous sections, it is instructive to see how the solutions to the linear volatility model also arise as a very special case of the diffusion canonical mapping reduction method, wherein the underlying $x$-space process is the simple Wiener process. In particular, assume the two processes satisfy $dx_t = \sqrt{2}\, dW_t$ and $dF_t = \sigma(F_t)dW_t$, under appropriate respective measures, where the $F_t$ process is considered to have zero drift and linear volatility function $\sigma(F) = \sigma F$, $\sigma = \text{const}$. The (forward price) space of $F$ values is mapped one to one onto the entire space of the Wiener process with the variable transformation

$$x = X(F) = (\sqrt{2}/\sigma)\log F \tag{3.86}$$

with inverse $F = F(x) = e^{\sigma x/\sqrt{2}}$. Since $\frac{dx}{dF} = \frac{\sqrt{2}}{\sigma F}$, the transformation reduction equation (3.259), of Lemma 3.1 to be derived in Section 3.8.1, specializes to give

$$\begin{aligned} U(F, F_0; \tau) &= \frac{\sqrt{2}}{\sigma F}\exp\left[\frac{1}{2}\log\frac{F_0}{F} - \frac{\sigma^2}{8}\tau\right]u(X(F), X(F_0); \tau) \\ &= \frac{\sqrt{2}}{\sigma}\sqrt{\frac{F_0}{F^3}}e^{-\frac{\sigma^2}{8}\tau}u(X(F), X(F_0); \tau). \end{aligned} \tag{3.87}$$

Here we have used $\alpha = \alpha_{x\to F} = -\sigma^2/8$, which results from equation (3.257) while substituting for the $x$-space volatility function (as constant) $\nu(x) = \sqrt{2}$ and drift $\lambda(x) = 0$. At this point the reader should note that the two transition probability densities $U$ and $u$ are not just simply related by a change of variables (i.e., the two functions are *not* the same probability densities expressed in terms of two different variables), but rather also involve the exponential multiplicative term due essentially to a measure change. This point will become clear later in this chapter when we come to discuss the mapping reduction framework in general. The mapping $x = X(F)$ and its inverse is monotonically increasing, with domain $x \in (-\infty, \infty)$ mapped onto $F \in (0, \infty)$. By direct substitution, while changing variables of differentiation and using equations (3.7) and (3.8), the reader can readily verify that $U = U(F, F_0; \tau)$ in equation (3.87) indeed satisfies both forward and backward equations: $\partial U/\partial\tau = \frac{\sigma^2}{2}\partial^2(F^2U)/\partial F^2$ and $\partial U/\partial\tau = \frac{1}{2}\sigma^2 F_0^2\, \partial^2 U/\partial F_0^2$.

Equation (3.87) gives an exact relationship between a kernel $U$ for the linear volatility model and a kernel $u$ for the Wiener process. The unique pricing kernels for the barrier-free case as well as for the case of single and double barriers then follow automatically by substitution of the particular kernel $u$ that satisfies the appropriate boundary conditions. For the barrier-free case, the zero-boundary conditions $U(F = 0, F_0; \tau) = U(F = \infty, F_0; \tau) = 0$ (with the same conditions also holding in $F_0$) are satisfied by substituting the solution $u(x, x_0; \tau) = g_0(x, x_0; \tau)$ of equation (3.9) into equation (3.87). Upon using equation (3.86) and completing the square in the exponent, one obtains the zero-drift lognormal density

$$U(F, F_0; \tau) = \frac{1}{\sigma F\sqrt{2\pi\tau}} \exp\left[ -\left(\log\frac{F_0}{F} - \frac{\sigma^2}{2}\tau\right)^2 /2\sigma^2\tau\right]. \tag{3.88}$$

As required, this formula is consistent with equation (3.39), where $S_\tau = e^{\mu\tau}F_\tau$ or, alternatively, with the case of zero drift $\mu = 0$, with $S = F$, $S_0 = F_0$. A barrier level at $F = H$ (or $F_0 = H$) corresponds to $H = F(x_H) = e^{\sigma x_H/\sqrt{2}}$, so $x_H = X(H) = (\sqrt{2}/\sigma)\log H$. Hence the lower-region $F, F_0 \in (0, H]$ maps onto $x, x_0 \in (-\infty, x_H]$, whereas the upper-region $F, F_0 \in [H, \infty)$ maps onto $x, x_0 \in [x_H, \infty)$. The density for a single absorbing barrier at $F, F_0 = H$ is hence obtained by simply substituting the kernel $u(X(F), X(F_0); \tau) = g^u(X(H), X(F), X(F_0); \tau)$ of equation (3.14) into relation (3.87), giving:

$$\begin{aligned} U(H, F, F_0; \tau) &= \frac{\sqrt{2}}{\sigma F}\exp\left[\frac{1}{2}\log\frac{F_0}{F} - \frac{\sigma^2}{8}\tau\right] g^u(X(H), X(F), X(F_0); \tau) \\ &= U(F, F_0; \tau)\left[1 - \exp\left[-\frac{\log(F/H)\log(F_0/H)}{\sigma^2\tau/2}\right]\right], \end{aligned} \tag{3.89}$$

with $U(F, F_0; \tau)$ given by equation (3.88). Note that this gives (absorbing) zero-boundary conditions $U(H, F = H, F_0; \tau) = U(H, F, F_0 = H; \tau) = 0$ and that equation (3.89) is exactly consistent with equation (3.40) when $\mu = 0$.

Exact analytical expressions for single-barrier options follow from the kernel in equation (3.89). Ignoring discounting,[3] an up-and-out European-style option expiring worthless if the upper forward price barrier $F = H$ is crossed before a time to maturity $\tau$, with current (forward) price level $F_0 \in (0, H)$, has a price given by [in direct analogy with equation (3.42)]

$$V^{UO}(F_0, \tau) = \int_0^H U(H, F, F_0; \tau)\Lambda(F)dF, \tag{3.90}$$

where $\Lambda(F)$ is an assumed payoff function. The corresponding down-and-out option with $F_0 > H$ has price [in direct analogy with equation (3.41)]

$$V^{DO}(F_0, \tau) = \int_H^\infty U(H, F, F_0; \tau)\Lambda(F)dF. \tag{3.91}$$

The knock-in barrier option prices are obtained from (knock-in)-(knockout) symmetry as discussed in Section 3.3. The plain-vanilla option price follows by integrating the barrier-free kernel (3.88) against the payoff function:

$$V(F_0, \tau) = \int_0^\infty U(F, F_0; \tau)\Lambda(F)dF. \tag{3.92}$$

[3] Throughout Section 3.5 we shall simply omit the overall discount factor in all the option-pricing formulas.

We shall not repeat the explicit intermediate steps in the derivations of the single-barrier European call and put pricing formulas since the procedure follows in exactly the same manner as discussed in Section 3.3. Call and put pay-offs with forward price struck at $K$ are assumed to have payoffs $\Lambda(F) = (F - K)_+$ and $\Lambda(F) = (K - F)_+$, respectively. The integrals in equations (3.90) and (3.91) are then readily evaluated by considering the analogues of equations (3.45) and (3.46); now defined by:

$$\phi(B) \equiv \int_B^\infty U(H, F, F_0; \tau) dF, \tag{3.93}$$

$$\bar{\phi}(B) \equiv \int_B^\infty U(H, F, F_0; \tau) F dF \tag{3.94}$$

for any $B \geq 0$, with $U(H, F, F_0; \tau)$ given by equation (3.89). In particular, the price of a down-and-out call on the underlying forward price struck at $K$ with single barrier at forward price level $H$ is given by (ignoring discounting)

$$C^{DO}(H, F_0, K, \tau) = \bar{\phi}(B) - K\phi(B), \tag{3.95}$$

where $B = H$ if $H \geq K$ and $B = K$ if $H \leq K$. Exact expressions for $\phi(B)$ and $\bar{\phi}(B)$ follow from equations (3.49) and (3.50) with $r = 0$ and $S_0 = F_0$:

$$\phi(B) = N\left(d_-\left(\frac{F_0}{B}\right)\right) - \frac{F_0}{H} N\left(d_-\left(\frac{H^2}{F_0 B}\right)\right), \tag{3.96}$$

$$\bar{\phi}(B) = F_0 N\left(d_+\left(\frac{F_0}{B}\right)\right) - HN\left(d_+\left(\frac{H^2}{F_0 B}\right)\right), \tag{3.97}$$

where

$$d_\pm(x) = \frac{\log x \pm \frac{1}{2}\sigma^2\tau}{\sigma\sqrt{\tau}}. \tag{3.98}$$

Hence setting $B = H$ if $H \geq K$ (and $B = K$ if $H \leq K$) gives the exact pricing formula for the down-and-out call in terms of cumulative normal density functions:

$$\begin{aligned} C^{DO}(H, F_0, K, \tau) = {} & F_0 N\left(d_+\left(\frac{F_0}{H}\right)\right) - HN\left(d_+\left(\frac{H}{F_0}\right)\right) \\ & - KN\left(d_-\left(\frac{F_0}{H}\right)\right) + \left(\frac{KF_0}{H}\right) N\left(d_-\left(\frac{H}{F_0}\right)\right) \end{aligned} \tag{3.99}$$

for $H \geq K$; and for $H \leq K$,

$$\begin{aligned} C^{DO}(H, F_0, K, \tau) = {} & F_0 N\left(d_+\left(\frac{F_0}{K}\right)\right) - HN\left(d_+\left(\frac{H^2}{F_0 K}\right)\right) \\ & - KN\left(d_-\left(\frac{F_0}{K}\right)\right) + \frac{KF_0}{H} N\left(d_-\left(\frac{H^2}{F_0 K}\right)\right) \end{aligned} \tag{3.100}$$

All other cases of single-barrier calls and puts are derived in the same manner, as described in detail in Section 3.3. The exact pricing formulas are obtained by simply setting $r = 0$ and $S_0 = F_0$ in all of the option-pricing expressions in Section 3.3. This is indeed not surprising, since discounting is ignored and the underlying is a forward price, as is the barrier level.

To complete this section we consider the problem of pricing a European *double-knockout*-barrier option on the underlying asset price process $F_t$. This option expires worthless if at any time before maturity the underlying price attains either barrier at $L$ or $H$ with $L < H$. The transition density in this case must have absorbing (i.e., zero) boundary conditions at both finite barrier endpoints. These points are mapped into the $x$-space endpoints: $x_H = X(H) = (\sqrt{2}/\sigma)\log H$, $x_L = X(L) = (\sqrt{2}/\sigma)\log L$. In virtue of the general relationship given by equation (3.87), the $F$-space density follows by simply substituting the $x$-space transition density satisfying zero-boundary conditions: $u(x = x_L, x_0; \tau) = u(x = x_H, x_0; \tau) = 0$. The problem is hence again reduced to finding $u(x, x_0; \tau)$. This density is readily obtained as an exact series expansion in sine functions via the method of eigenfunction expansions. This method and its relation to the Laplace transform technique for solving the Kolmogorov equations subject to different types of boundary conditions is generally described later in this chapter, i.e., where the method of Green's functions is discussed. Here we simply state the result (see Problem 1 of this section for an alternate derivation):

$$u(x, x_0; \tau) = \frac{2}{x_H - x_L}\sum_{n=1}^{\infty} e^{-\bar{\rho}_n \tau} \sin\frac{n\pi(x_0 - x_L)}{x_H - x_L}\sin\frac{n\pi(x - x_L)}{x_H - x_L}, \tag{3.101}$$

for $x, x_0 \in [x_L, x_H]$, where $\bar{\rho}_n = n^2\pi^2/(x_H - x_L)^2$. [In mathematical physics, this is the well-known Fourier sine series solution to the simple heat conduction problem for an initial point source of heat diffusing on a finite one-dimensional domain (e.g., a rod) with insulation at both endpoints.] This series converges for all positive $\tau$ and gives a representation of the Dirac delta function $\delta(x - x_0)$ for the finite domain $[x_L, x_H]$ when $\tau = 0$. Inserting equation (3.101) into equation (3.87) while using equation (3.86) gives the transition density satisfying double-barrier zero-boundary conditions at $L$ and $H$, denoted by $U^{DB}$, for the linear volatility model as an exact series:

$$U^{DB}(F, F_0; \tau) = \frac{2}{\log\frac{H}{L}}\sqrt{\frac{F_0}{F^3}}\sum_{n=1}^{\infty} e^{-\rho_n \tau}\sin\left(n\pi\gamma(F_0)\right)\sin\left(n\pi\gamma(F)\right) \tag{3.102}$$

for $F_0, F \in [L, H]$, where

$$\gamma(F) \equiv \frac{X(F) - X(L)}{X(H) - X(L)} = \frac{\log\frac{F}{L}}{\log\frac{H}{L}}, \tag{3.103}$$

$$\rho_n = \frac{\sigma^2}{8} + \bar{\rho}_n = \frac{\sigma^2}{8} + \frac{n^2\pi^2\sigma^2}{2\log^2\frac{H}{L}}. \tag{3.104}$$

This series is easily shown to converge for all positive $\tau$ values and gives a representation of the Dirac delta function $\delta(F - F_0)$ for the finite domain $F, F_0 \in [L, H]$ when $\tau = 0$. [Note: $\delta(F - F_0) = \frac{dx}{dF}\delta(X(F) - X(F_0)) = \frac{\sqrt{2}}{\sigma F}\delta(X(F) - X(F_0))$.] Of practical importance is the fact that the convergence of the series is fairly rapid since the eigenvalues $\rho_n$ grow as $n^2$ as $n$ increases; i.e., contributions from the higher-frequency sine functions are diminished by the dominant factor $e^{-\rho_n \tau}$, which decreases rapidly as a Gaussian function in $n$. Although more terms are required to achieve the same level of accuracy as the time to maturity is decreased, a uniformly high level of accuracy (and positivity in the density) can be achieved by retaining a relatively small number of terms in the sum (see Figure 3.4). Moreover, similar expressions (as demonstrated next) for pricing double-barrier options require a substantially smaller number of terms for high accuracy. A double knockout European-style option maturing in time $\tau$ is

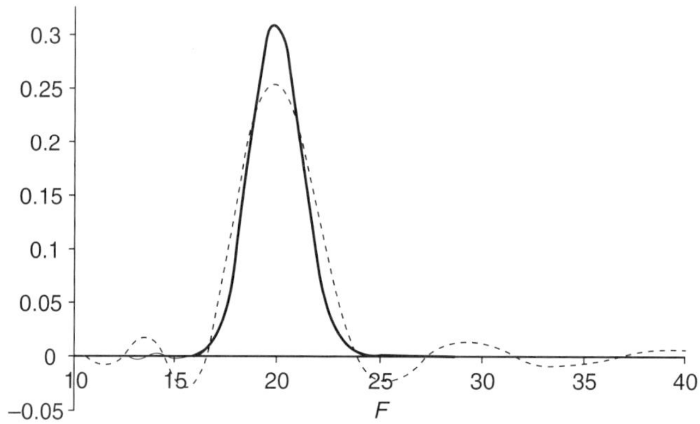


FIGURE 3.4 Uniform convergence of the density given by equation (3.102) for $L = 10$, $H = 50$, $F_0 = 20$, $\sigma = 0.2$, $\tau = 0.1$. The three curves correspond to using the first 10 (dashed line), 20, and 30 terms (the thick solid line) in the series sum.

then priced by taking the expectation of the payoff $\Lambda(F)$ over the allowable region (ignoring discounting):

$$V^{DB}(F_0, \tau) = \int_L^H U^{DB}(F, F_0; \tau)\Lambda(F)dF. \tag{3.105}$$

For example, a double knockout European call struck at $K$ is priced by inserting $U^{DB}(F, F_0; \tau)$ with $\Lambda(F) = (F - K)_+$ and integrating, term by term, in the series to obtain exact analytical series expressions for the option value. In carrying out the integration it is very convenient to change integration variables $F \to x$ as defined by the original variable transformation $F = F(x) = e^{\sigma x/\sqrt{2}}$, i.e., using the $F$-space density $F'(x)U^{DB}(F(x), F_0; \tau)$ expressed as a function of the $x$ variable (see Problem 2). Two separate formulas arise accounting for whether $K \le L$ or $K \ge L$ (in both cases $K < H$; otherwise the strike is above the upper barrier and the option is worthless). For $K \ge L$ the formula for the call is

$$C^{DB}(F_0, K, \tau) = \frac{\sigma^2}{\log\frac{H}{L}}\sqrt{F_0}\sum_{n=1}^{\infty}\frac{e^{-\rho_n\tau}}{\rho_n}\sin\left(n\pi\gamma(F_0)\right) \times\left[\frac{n\pi}{\log\frac{H}{L}}\frac{K-H}{\sqrt{H}}(-1)^n - \sqrt{K}\sin\left(n\pi\gamma(K)\right)\right] \tag{3.106}$$

and for the case $K \le L$ is

$$C^{DB}(F_0, K, \tau) = \frac{\pi\sigma^2}{\log^2\frac{H}{L}}\sqrt{F_0}\sum_{n=1}^{\infty}n\frac{e^{-\rho_n\tau}}{\rho_n}\sin\left(n\pi\gamma(F_0)\right) \times\left[\frac{K-H}{\sqrt{H}}(-1)^n - \frac{K-L}{\sqrt{L}}\right], \tag{3.107}$$

where $\gamma(\cdot)$ and $\rho_n$ are given by equations (3.103) and (3.104). Note that the two expressions are equivalent when $K = L$. Figure 3.5 displays a typical convergence when applying equation (3.106).

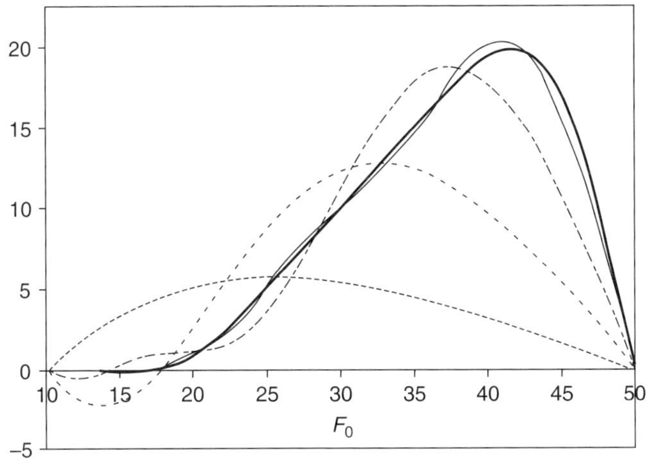


FIGURE 3.5 Uniform rapid convergence of the value of the double knockout call as more terms are used in the series formula for $C^{DB}$, where $L = 10$, $H = 50$, $K = 20$, $\sigma = 0.2$, $\tau = 0.25$. The five separate curves correspond to the truncated series sum in the first 1, 2, 4, 8, and 12 (solid line) terms of equation (3.106).

## Problems

**Problem 1.** We wish to solve

$$\frac{\partial u}{\partial \tau} = \frac{\partial^2 u}{\partial x^2},$$

subject to $u(x = x_L, \tau) = u(x = x_H, \tau) = 0$ and initial condition $u(x, \tau = 0) = u_0(x)$. Since the solution must vanish at the endpoints of the interval $[x_L, x_H]$, one method is to express $u$ as a Fourier sine series:

$$u(x, \tau) = \sum_{n=1}^{\infty} b_n(\tau) \sin \frac{n\pi(x - x_L)}{x_H - x_L}, \tag{3.108}$$

with coefficients $b_n(\tau)$ depending only on $\tau$. Using direct substitution and by satisfying the initial condition, show that

$$b_n(\tau) = a_n e^{-\bar{\rho}_n \tau}, \tag{3.109}$$

where $\bar{\rho}_n = n^2\pi^2/(x_H - x_L)^2$ and where $a_n = b_n(0)$ is

$$a_n = \frac{2}{x_H - x_L} \int_{x_L}^{x_H} u_0(x) \sin \frac{n\pi(x - x_L)}{x_H - x_L} dx. \tag{3.110}$$

Hence, recover equation (3.101) when $u_0(x) = \delta(x - x_0)$.

**Problem 2.** From equation (3.105) we see that the double-barrier call option can be derived by computing the integrals

$$\phi(K) = \int_K^H U^{DB}(F, F_0; \tau) dF, \tag{3.111}$$

$$\bar{\phi}(K) = \int_K^H U^{DB}(F, F_0; \tau) F dF, \tag{3.112}$$

for $L \leq K < H$. By using equation (3.102) and a change-of-integration variable $F \to x$ defined by $F = e^{\sigma x/\sqrt{2}}$, show that the latter integral is

$$\bar{\phi}(K) = \frac{2\sqrt{F_0}}{\frac{\sqrt{2}}{\sigma}\log\frac{H}{L}} \sum_{n=1}^{\infty} e^{-\rho_n \tau} \sin\left(n\pi\gamma(F_0)\right) \int_{X(K)}^{X(H)} e^{\frac{\sigma}{2\sqrt{2}}x} \sin\left[\frac{n\pi(x - X(L))}{X(H) - X(L)}\right] dx,$$

where $X(\cdot)$ is defined by equation (3.86). Apply the indefinite integral identity $\int e^{ax}\sin bx \; dx = e^{ax}[a\sin bx - b\cos bx]/(a^2+b^2) + c$, where $a$, $b$, $c$ are any constants, and obtain

$$\bar{\phi}(K) = \frac{\pi\sigma^2\sqrt{F_0}}{\log^2\frac{H}{L}} \sum_{n=1}^{\infty} \frac{e^{-\rho_n\tau}}{\rho_n} \sin\left(n\pi\gamma(F_0)\right)\Bigg[-n(-1)^n\sqrt{H} + \sqrt{K}\left(n\cos\left(n\pi\gamma(K)\right) - \frac{\log\frac{H}{L}}{2\pi}\sin\left(n\pi\gamma(K)\right)\right)\Bigg]. \tag{3.113}$$

Using a similar procedure, obtain

$$\phi(K) = \frac{\pi\sigma^2\sqrt{F_0}}{\log^2\frac{H}{L}} \sum_{n=1}^{\infty} \frac{e^{-\rho_n\tau}}{\rho_n} \sin\left(n\pi\gamma(F_0)\right)\Bigg[-\frac{n(-1)^n}{\sqrt{H}} + \frac{1}{\sqrt{K}}\left(n\cos\left(n\pi\gamma(K)\right) + \frac{\log\frac{H}{L}}{2\pi}\sin\left(n\pi\gamma(K)\right)\right)\Bigg]. \tag{3.114}$$

Using $C^{DB} = \bar{\phi}(K) - K\phi(K)$ for $K \geq L$ and $C^{DB} = \bar{\phi}(L) - K\phi(L)$ for $K \leq L$, obtain equations (3.106) and (3.107).

**Problem 3.** By computing the integrals

$$\Phi(K) = \int_L^K U^{DB} dF \tag{3.115}$$

and

$$\bar{\Phi}(K) = \int_L^K U^{DB} F dF \tag{3.116}$$

and using steps similar to those in Problem 2, derive an exact series expression for the corresponding double-barrier put option value $P^{DB}(F_0, K, \tau)$ for strike $K$, $L < K \leq H$.

## 3.5.2 *Quadratic Volatility Models*

We now consider the problem of pricing European options, including barriers, for the more complex quadratic volatility model with two *distinct roots*:[4]

$$\sigma(F) = \frac{\sigma}{(\bar{\bar{F}} - \bar{F})}(F - \bar{F})(\bar{\bar{F}} - F). \tag{3.117}$$

[4] A quadratic volatility function is generally of the form $\sigma(F) = \sigma_0(F - \bar{F})(\bar{\bar{F}} - F)$, where $\bar{F}$, $\bar{\bar{F}}$ are two real roots. Here, we find it useful to express the nonzero parameter as a ratio $\sigma_0 = \sigma/(\bar{\bar{F}} - \bar{F})$. In this way, the parameter $\sigma$ corresponds to the volatility parameter in the linear model in the limit $\bar{\bar{F}} \to \infty$.

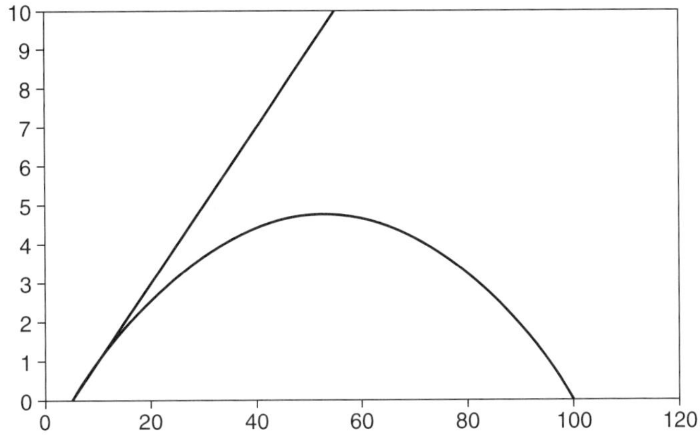


FIGURE 3.6 Example of a quadratic volatility function with two distinct roots $\bar{F} = 5$, $\bar{\bar{F}} = 100$, $\sigma = 0.2$. The linear volatility function for given parameter $\sigma$ is drawn for direct comparison. The linear model obtains in the limit $\bar{\bar{F}} \to \infty$.

Figure 3.6 depicts the shape of a quadratic volatility function in comparison with an affine (linear) model $\sigma(F) = \sigma(F - \bar{F})$, for given volatility parameter $\sigma$. Without loss in generality, throughout we assume $\bar{F} < \bar{\bar{F}}$, with underlying asset price $F \in [\bar{F}, \bar{\bar{F}}]$. We note that the separate case of the single double-root quadratic model is discussed later in this chapter. This model is a special case of the constant-elasticity-of-variance (CEV) model, which itself is shown to obtain as a special case of a more general Bessel family of solutions. These more general families of exact solutions are discussed later in this chapter. Here, we consider obtaining solutions to the model in equation (3.117) by mapping the (forward) price space $F$ onto the $x$-space of the Wiener process. That is, the zero-drift Wiener process with constant volatility $\nu(x) = \sqrt{2}$ can again be chosen as underlying process in $x$-space and thereby ultimately provide exact solutions to the quadratic volatility model in $F$-space. As in the linear model, we have the constant $\alpha = -\sigma^2/8$. (Note: This is the constant $\alpha_{x \to F}$ corresponding to the diffusion canonical transformation $x \to F$ described later in the chapter.) The transformation of $x = X(F)$ is defined by equating the Jacobian of the transformation to the ratio of the volatility functions in both spaces:

$$\frac{dX(F)}{dF} = \frac{\nu(X(F))}{\sigma(F)} = \frac{\sqrt{2}(\bar{F} - \bar{\bar{F}})}{\sigma(F - \bar{F})(F - \bar{\bar{F}})}. \tag{3.118}$$

This implies the following monotonically increasing map:

$$X(F) = \frac{\sqrt{2}}{\sigma} \log \left| \frac{F - \bar{F}}{F - \bar{\bar{F}}} \right| = \frac{\sqrt{2}}{\sigma} \log \frac{F - \bar{F}}{\bar{\bar{F}} - F}. \tag{3.119}$$

This is a one-to-one map of the domain $x \in (-\infty, +\infty)$ into the domain $F \in (\bar{F}, \bar{\bar{F}})$, with inverse relation

$$F = F(x) = \frac{\bar{F} + \bar{\bar{F}} e^{\frac{\sigma}{\sqrt{2}} x}}{1 + e^{\frac{\sigma}{\sqrt{2}} x}}. \tag{3.120}$$

That is, $X(\bar{F}) = -\infty$, $X(\bar{\bar{F}}) = \infty$, $F(-\infty) = \bar{F}$, $F(\infty) = \bar{\bar{F}}$. As shown later in this section, the form in equation (3.117) is convenient for showing that the solutions to the quadratic (double-root) model directly recover the corresponding known solutions to the linear volatility (i.e., affine or lognormal model) by simply taking the limit $\bar{\bar{F}} \to \infty$ in all expressions. This is in fact a mathematical consistency requirement of the theory.

By specializing equation (3.259), the exact relationship between a transition probability density function, or pricing kernel, $U$ for the quadratic volatility model and a kernel $u$ for the Wiener process is given by [i.e., the analogue of equation (3.87)]

$$U(F, F_0; \tau) = \frac{\sqrt{2}}{\sigma}(\bar{\bar{F}} - \bar{F})\sqrt{\frac{(F_0 - \bar{F})(\bar{\bar{F}} - F_0)}{(F - \bar{F})^3(\bar{\bar{F}} - F)^3}}\, e^{-\sigma^2\tau/8} u(X(F); X(F_0); \tau), \tag{3.121}$$

$F, F_0 \in (\bar{F}, \bar{\bar{F}})$. This equation relates the density for the quadratic model to that of the simple Wiener model. By direct substitution and by using equations (3.7) and (3.8), one can verify that $U(F, F_0; \tau)$ in equation (3.121) satisfies both forward and backward time-homogeneous Kolmogorov equations in $F, F_0$ for the zero-drift function and volatility function given by equation (3.117). Later in the chapter, the reader will learn to derive this relation based on the canonical diffusion mapping methodology.

Following a similar procedure to that in the previous section, the pricing kernels for the barrier-free case as well as for single and double barriers arise by direct substitution of the $x$-space kernel $u$ satisfying the appropriate boundary conditions. In particular, zero-boundary conditions, $U(F = 0, F_0; \tau) = U(F = \infty, F_0; \tau) = 0$ (with the same conditions also holding in $F_0$), are satisfied by substituting the solution $u(x, x_0; \tau) = g_0(x, x_0; \tau)$ of equation (3.9) into equation (3.121). Upon using equation (3.119) and rearranging logarithmic terms, the *barrier-free* kernel is given in exact form:

$$\begin{aligned} U(F, F_0; \tau) = {} & \frac{(\bar{\bar{F}} - \bar{F})}{\sigma\sqrt{2\pi\tau}}\sqrt{\frac{(F_0 - \bar{F})(\bar{\bar{F}} - F_0)}{(F - \bar{F})^3(\bar{\bar{F}} - F)^3}}\, e^{-\sigma^2\tau/8} \\ & \times \exp\left[-\frac{1}{2\sigma^2\tau}\log^2\frac{(F - \bar{F})(\bar{\bar{F}} - F_0)}{(\bar{\bar{F}} - F)(F_0 - \bar{F})}\right]. \end{aligned} \tag{3.122}$$

This kernel may be compared to the zero-drift lognormal density kernel in equation (3.88), which obtains as a simpler case in the limit $\bar{\bar{F}} \to \infty$, $\bar{F} = 0$. For computing integral expectations (i.e., for pricing purposes) it is convenient to work in terms of the $x$ variable. Using equation (3.120), $F = F(x)$, and equation (3.122) gives the density (see Problem 1)

$$U(F(x), F_0; \tau)\frac{dF}{dx} = \frac{e^{-\frac{\sigma^2}{8}\tau}}{2\sqrt{\pi\tau}}\frac{\cosh(\sigma x/2\sqrt{2})}{\cosh(\sigma x_0/2\sqrt{2})}e^{-(x-x_0)^2/4\tau}, \tag{3.123}$$

$x_0 = X(F_0)$. From this it readily follows that the barrier-free kernel conserves probability (see Problem 1). The price of a plain European-style option maturing in time $\tau$ is given by (ignoring discounting)

$$V(F_0, \tau) = \int_{\bar{F}}^{\bar{\bar{F}}} U(F, F_0; \tau)\Lambda(F)dF. \tag{3.124}$$

Calls or puts on the forward price struck at $K$ with payoff $\Lambda(F) = (F-K)_+$ or $\Lambda(F) = (K-F)_+$, respectively, are readily priced. In particular, the price of a call with $\bar{F} \leq K < \bar{\bar{F}}$ takes the form (ignoring discounting)

$$C(F_0, K, \tau) = \bar{\phi}(K) - K\phi(K), \tag{3.125}$$

where

$$\phi(K) = \int_K^{\bar{\bar{F}}} U(F, F_0; \tau)dF, \qquad \bar{\phi}(K) = \int_K^{\bar{\bar{F}}} U(F, F_0; \tau)FdF. \tag{3.126}$$

By changing integration variable $F \to x = X(F)$ and using equation (3.123),

$$\phi(K) = \frac{e^{-\frac{\sigma^2}{8}\tau}}{2\sqrt{\pi\tau}} \frac{1}{\cosh\left(\frac{\sigma x_0}{2\sqrt{2}}\right)} \int_{X(K)}^{\infty} \cosh\left(\tfrac{\sigma x}{2\sqrt{2}}\right) e^{-(x-x_0)^2/4\tau} dx, \tag{3.127}$$

with $X(K) = \frac{\sqrt{2}}{\sigma} \log[(K-\bar{F})/(\bar{\bar{F}}-K)]$. This integral is readily evaluated via the identity (see Problem 3)

$$\int_{X(K)}^{\infty} e^{\pm\frac{\sigma x}{2\sqrt{2}}} e^{-(x-x_0)^2/4\tau} dx = 2\sqrt{\pi\tau} e^{\pm\frac{\sigma x_0}{2\sqrt{2}}} e^{\frac{\sigma^2}{8}\tau} N\big(d_{\pm}(\mathcal{X})\big), \tag{3.128}$$

where

$$\mathcal{X} = \frac{(\bar{\bar{F}}-K)(F_0-\bar{F})}{(K-\bar{F})(\bar{\bar{F}}-F_0)} \tag{3.129}$$

and $N(\cdot)$ is the standard cumulative normal density function. Throughout this section we define

$$d_{\pm}(x) \equiv \frac{\log x \pm \frac{1}{2}\sigma^2\tau}{\sigma\sqrt{\tau}}. \tag{3.130}$$

From equation (3.128) and using $x_0 = X(F_0)$, namely, $e^{\frac{\sigma x_0}{2\sqrt{2}}} = [(F_0-\bar{F})/(\bar{\bar{F}}-F_0)]^{\frac{1}{2}}$ and $\cosh(\frac{\sigma x_0}{2\sqrt{2}}) = \frac{1}{2}(\bar{\bar{F}}-\bar{F})[(F_0-\bar{F})(\bar{\bar{F}}-F_0)]^{-\frac{1}{2}}$, we hence obtain

$$\phi(K) = (\bar{\bar{F}}-\bar{F})^{-1}\big[(F_0-\bar{F})N\big(d_+(\mathcal{X})\big) + (\bar{\bar{F}}-F_0)N\big(d_-(\mathcal{X})\big)\big]. \tag{3.131}$$

The second integral in equation (3.126) is evaluated in similar fashion, namely, by changing integration variable $F \to x = X(F)$, using the identity

$$F(x)\cosh(\sigma x/2\sqrt{2}) = \frac{1}{2}[\bar{F}e^{-\sigma x/2\sqrt{2}} + \bar{\bar{F}}e^{\sigma x/2\sqrt{2}}], \tag{3.132}$$

which follows from equation (3.120), and integrating with the use of equation (3.128),

$$\bar{\phi}(K) = (\bar{\bar{F}}-\bar{F})^{-1}\big[\bar{\bar{F}}(F_0-\bar{F})N\big(d_+(\mathcal{X})\big) + \bar{F}(\bar{\bar{F}}-F_0)N\big(d_-(\mathcal{X})\big)\big]. \tag{3.133}$$

Combining equations (3.131) and (3.133) finally gives the call price:

$$\begin{aligned} C(F_0, K, \tau) = (\bar{\bar{F}}-\bar{F})^{-1}\big[&(\bar{\bar{F}}-K)(F_0-\bar{F})N\big(d_+(\mathcal{X})\big) \\ &- (K-\bar{F})(\bar{\bar{F}}-F_0)N\big(d_-(\mathcal{X})\big)\big]. \end{aligned} \tag{3.134}$$

The put price is derived in similar fashion (see Problem 4).

For a single barrier with absorption at level $H$, $\bar{F} < H < \bar{\bar{F}}$, the pricing kernel is obtained in exact form by simply substituting the kernel $u(X(F), X(F_0); \tau) = g^u(X(H), X(F), X(F_0); \tau)$ of equation (3.14) into equation (3.121), giving

$$U(H, F, F_0; \tau) = \frac{(\bar{\bar{F}} - \bar{F})}{\sigma\sqrt{2\pi\tau}} \sqrt{\frac{(F_0 - \bar{F})(\bar{\bar{F}} - F_0)}{(F - \bar{F})^3(\bar{\bar{F}} - F)^3}}\, e^{-\sigma^2\tau/8} \times \left\{ \exp\left[ -\frac{1}{2\sigma^2\tau} \log^2 \frac{(F - \bar{F})(\bar{\bar{F}} - F_0)}{(\bar{\bar{F}} - F)(F_0 - \bar{F})} \right] - \exp\left[ -\frac{1}{2\sigma^2\tau} \log^2 \frac{(F - \bar{F})(F_0 - \bar{F})(\bar{\bar{F}} - H)^2}{(\bar{\bar{F}} - F)(\bar{\bar{F}} - F_0)(H - \bar{F})^2} \right] \right\}. \tag{3.135}$$

The boundary conditions $U(H, F = H, F_0; \tau) = 0$, $U(H, F, F_0 = H; \tau) = 0$ are obviously satisfied. This kernel is hence useful for pricing single-barrier options for the quadratic (double-root) volatility models. Note that the kernel in equation (3.89) obtains in the limit $\bar{\bar{F}} \to \infty$, $\bar{F} = 0$. Exact formulas for single-barrier knock-in and knockout calls/puts are most readily derived by changing variables of integration, as was done in the earlier barrier-free case. In particular, using equation (3.120), $F = F(x)$, and equation (3.135) gives the analogue of equation (3.123):

$$U(H, F(x), F_0; \tau)\frac{dF}{dx} = \frac{e^{-\frac{\sigma^2}{8}\tau} \cosh(\frac{\sigma x}{2\sqrt{2}})}{2\sqrt{\pi\tau} \cosh(\frac{\sigma x_0}{2\sqrt{2}})}\left[e^{-\frac{(x-x_0)^2}{4\tau}} - e^{-\frac{(x+x_0-2x_H)^2}{4\tau}}\right], \tag{3.136}$$

$x_0 = X(F_0)$, $x_H = X(H)$.

European-style single-barrier knock-in and knockout option price formulas are then derived by integrating the density in equation (3.136) against the pay-off in the appropriate domain. In what follows we derive the knockout option prices as the knock-in prices follow simply from (knock-in)–(knockout) symmetry. A down-and-out call option, expiring worthless if the barrier $F = H$ is crossed before a time to maturity $\tau$ with current (forward) price $F_0 > H$, $\bar{F} \leq H$, $K < \bar{\bar{F}}$, has value (ignoring discounting throughout)

$$C^{DO}(H, F_0, K, \tau) = \int_H^{\bar{\bar{F}}} U(H, F, F_0; \tau)(F - K)_+ dF = \begin{cases} \bar{\phi}(K) - K\phi(K), & K \geq H \\ \bar{\phi}(H) - K\phi(H), & K \leq H, \end{cases} \tag{3.137}$$

where $\phi(\cdot)$ and $\bar{\phi}(\cdot)$ are defined by

$$\phi(B) = \int_B^{\bar{\bar{F}}} U(H, F, F_0; \tau)dF, \qquad \bar{\phi}(B) = \int_B^{\bar{\bar{F}}} U(H, F, F_0; \tau)FdF \tag{3.138}$$

any real value $B$ such that $\bar{F} \leq B \leq \bar{\bar{F}}$. Following similar steps as earlier, these integrals are reduced to standard cumulative normal functions. Changing variables $F \to x = X(F)$ and using equation (3.136),

$$\phi(K) = \frac{e^{-\frac{\sigma^2}{8}\tau}}{2\sqrt{\pi\tau}} \frac{1}{\cosh\left(\frac{\sigma x_0}{2\sqrt{2}}\right)} \int_{X(K)}^{\infty} \cosh\left(\frac{\sigma x}{2\sqrt{2}}\right)\left[e^{-\frac{(x-x_0)^2}{4\tau}} - e^{-\frac{(x+x_0-2x_H)^2}{4\tau}}\right]dx.$$

This integral is evaluated in two parts. The first exponential integral [i.e., identically as in equation (3.127)] is given by expression (3.131). The second term is integrated in identical fashion by using the same integral identity (3.128) with the replacement $x_0 \to 2x_H - x_0$ and then using $2x_H - x_0 = 2X(H) - X(F_0) = \frac{\sqrt{2}}{\sigma} \log \frac{(H-\bar{F})^2(\bar{\bar{F}}-F_0)}{(\bar{\bar{F}}-H)^2(F_0-\bar{F})}$,

$$e^{\sigma(2x_H - x_0)/2\sqrt{2}} = [(H - \bar{F})/(\bar{\bar{F}} - H)][(\bar{\bar{F}} - F_0)/(F_0 - \bar{F})]^{\frac{1}{2}}. \tag{3.139}$$

Combining the results of the two integrals:

$$\begin{aligned}\phi(K) = (\bar{\bar{F}} - \bar{F})^{-1}\Bigg[&(F_0 - \bar{F})N\big(d_+(\mathcal{X})\big) + (\bar{\bar{F}} - F_0)N\big(d_-(\mathcal{X})\big)\\ &-\left(\frac{H - \bar{F}}{\bar{\bar{F}} - H}\right)(\bar{\bar{F}} - F_0)N\big(d_+(\mathcal{Y})\big) - \left(\frac{\bar{\bar{F}} - H}{H - \bar{F}}\right)(F_0 - \bar{F})N\big(d_-(\mathcal{Y})\big)\Bigg],\end{aligned} \tag{3.140}$$

where $d_\pm(\cdot)$ and $\mathcal{X}$ are given by equations (3.130) and (3.129) and

$$\mathcal{Y} = \frac{(\bar{\bar{F}} - K)(\bar{\bar{F}} - F_0)(H - \bar{F})^2}{(K - \bar{F})(F_0 - \bar{F})(\bar{\bar{F}} - H)^2}. \tag{3.141}$$

The second integral in equation (3.138) for $B = K$ is evaluated in similar fashion, namely, by using equation (3.123) and identity (3.132):

$$\begin{aligned}\bar{\phi}(K) = \frac{e^{-\frac{\sigma^2}{8}\tau}}{2\sqrt{\pi\tau}} \frac{1}{\cosh\left(\frac{\sigma x_0}{2\sqrt{2}}\right)}\Bigg[&\frac{\bar{\bar{F}}}{2}\int_{X(K)}^{\infty} e^{-\frac{\sigma x}{2\sqrt{2}}}[e^{-\frac{(x-x_0)^2}{4\tau}} - e^{-\frac{(x+x_0-2x_H)^2}{4\tau}}]dx\\ &+\frac{\bar{F}}{2}\int_{X(K)}^{\infty} e^{\frac{\sigma x}{2\sqrt{2}}}[e^{-\frac{(x-x_0)^2}{4\tau}} - e^{-\frac{(x+x_0-2x_H)^2}{4\tau}}]dx\Bigg].\end{aligned}$$

Applying equation (3.128) on each of the four terms and expressing the result in terms of $F_0$, $H$, $K$ while using equation (3.139) and simplifying gives

$$\begin{aligned}\bar{\phi}(K) = (\bar{\bar{F}} - \bar{F})^{-1}\Bigg[&\bar{\bar{F}}(F_0 - \bar{F})N\big(d_+(\mathcal{X})\big) + \bar{F}(\bar{\bar{F}} - F_0)N\big(d_-(\mathcal{X})\big)\\ &-\bar{\bar{F}}\left(\frac{H - \bar{F}}{\bar{\bar{F}} - H}\right)(\bar{\bar{F}} - F_0)N\big(d_+(\mathcal{Y})\big) - \bar{F}\left(\frac{\bar{\bar{F}} - H}{H - \bar{F}}\right)(F_0 - \bar{F})N\big(d_-(\mathcal{Y})\big)\Bigg].\end{aligned} \tag{3.142}$$

Combining equations (3.140) and (3.142) in equation (3.137) hence gives the exact down-and-out call price for $K \geq H$:

$$\begin{aligned}C^{DO}(H, F_0, K, \tau) = (\bar{\bar{F}} - \bar{F})^{-1}\Bigg[&(\bar{\bar{F}} - K)(F_0 - \bar{F})N\big(d_+(\mathcal{X})\big) + (\bar{F} - K)(\bar{\bar{F}} - F_0)N\big(d_-(\mathcal{X})\big)\\ &+(K - \bar{\bar{F}})(\bar{\bar{F}} - F_0)\left(\frac{H - \bar{F}}{\bar{\bar{F}} - H}\right)N\big(d_+(\mathcal{Y})\big)\\ &+(K - \bar{F})(F_0 - \bar{F})\left(\frac{\bar{\bar{F}} - H}{H - \bar{F}}\right)N\big(d_-(\mathcal{Y})\big)\Bigg].\end{aligned} \tag{3.143}$$

By taking the limit $\bar{\bar{F}} \to \infty$ of this expression, the reader can easily verify that the exact formula for the down-and-out price for the affine linear volatility model is obtained and that in particular by also setting $\bar{F} = 0$, equation (3.100) (i.e., the price assuming a lognormal density model) is exactly recovered, as required.

For $K \leq H$ the down-and-out call price is obtained by evaluating $\phi(H)$ and $\bar{\phi}(H)$. These quantities are derived by replacing the lower integration value $X(K)$ by $X(H)$ in the foregoing integrals for $\phi(K)$ and $\bar{\phi}(K)$. This is equivalent to setting $K \to H$ in equations (3.129) and (3.141), giving

$$\phi(H) = (\bar{\bar{F}} - \bar{F})^{-1}\Bigg[(F_0 - \bar{F})N\big(d_+(\mathcal{A})\big) + (\bar{\bar{F}} - F_0)N\big(d_-(\mathcal{A})\big) - \left(\frac{H - \bar{F}}{\bar{\bar{F}} - H}\right)(\bar{\bar{F}} - F_0)N\big(d_+(\mathcal{B})\big) - \left(\frac{\bar{\bar{F}} - H}{H - \bar{F}}\right)(F_0 - \bar{F})N\big(d_-(\mathcal{B})\big)\Bigg], \tag{3.144}$$

$$\bar{\phi}(H) = (\bar{\bar{F}} - \bar{F})^{-1}\Bigg[\bar{\bar{F}}(F_0 - \bar{F})N\big(d_+(\mathcal{A})\big) + \bar{F}(\bar{\bar{F}} - F_0)N\big(d_-(\mathcal{A})\big) - \bar{\bar{F}}\left(\frac{H - \bar{F}}{\bar{\bar{F}} - H}\right)(\bar{\bar{F}} - F_0)N\big(d_+(\mathcal{B})\big) - \bar{F}\left(\frac{\bar{\bar{F}} - H}{H - \bar{F}}\right)(F_0 - \bar{F})N\big(d_-(\mathcal{B})\big)\Bigg] \tag{3.145}$$

where

$$\mathcal{A} = \frac{(\bar{\bar{F}} - H)(F_0 - \bar{F})}{(H - \bar{F})(\bar{\bar{F}} - F_0)}, \qquad \mathcal{B} = \frac{(\bar{\bar{F}} - F_0)(H - \bar{F})}{(F_0 - \bar{F})(\bar{\bar{F}} - H)} \tag{3.146}$$

and $d_\pm(\cdot)$ is defined by equation (3.130). Combining these expressions gives the analytically exact down-and-out call price for $K \leq H$:

$$\begin{aligned} C^{DO}(H, F_0, K, \tau) = (\bar{\bar{F}} - \bar{F})^{-1}\Bigg[&(\bar{\bar{F}} - K)(F_0 - \bar{F})N\big(d_+(\mathcal{A})\big) + (\bar{F} - K)(\bar{\bar{F}} - F_0)N\big(d_-(\mathcal{A})\big) \\ &+(K - \bar{\bar{F}})(\bar{\bar{F}} - F_0)\left(\frac{H - \bar{F}}{\bar{\bar{F}} - H}\right)N\big(d_+(\mathcal{B})\big) \\ &+ (K - \bar{F})(F_0 - \bar{F})\left(\frac{\bar{\bar{F}} - H}{H - \bar{F}}\right)N\big(d_-(\mathcal{B})\big)\Bigg]. \end{aligned} \tag{3.147}$$

The limit $\bar{\bar{F}} \to \infty$ of this expression reduces to the exact formula for the down-and-out price for the affine linear volatility model; and further, by setting $\bar{F} = 0$, equation (3.99) is also recovered. The price of a down-and-out put is derived in similar fashion (see Problem 5).

An up-and-out call option, expiring worthless if the upper barrier $F = H$ is crossed before a time to maturity $\tau$ with current (forward) price $F_0 < H$ and $\bar{F} \leq K < H \leq \bar{\bar{F}}$, has value (ignoring discounting)

$$\begin{aligned} C^{UO}(H, F_0, K, \tau) &= \int_K^H U(H, F, F_0; \tau)(F - K)dF \\ &= \bar{\phi}(K) - K\phi(K) - [\bar{\phi}(H) - K\phi(H)] \end{aligned} \tag{3.148}$$

and value zero for $K \geq H$. The second expression obtains by writing the integral as the difference of two integrals [one from $K$ to $\bar{\bar{F}}$ and the other from $H$ to $\bar{\bar{F}}$ within equation (3.138)].

Using equations (3.140), (3.142), (3.144), and (3.145) (or taking the difference between the two down-and-out call prices) gives the up-and-out call price (excluding discounting)

$$
\begin{aligned}
C^{UO}(H, F_0, K, \tau) = {} & (\bar{\bar{F}} - \bar{F})^{-1}\Big[(\bar{\bar{F}} - K)(F_0 - \bar{F})[N(d_+(\mathcal{X})) - N(d_+(\mathcal{A}))] \\
& + (\bar{F} - K)(\bar{\bar{F}} - F_0)[N(d_-(\mathcal{X})) - N(d_-(\mathcal{A}))] \\
& + (K - \bar{\bar{F}})(\bar{\bar{F}} - F_0)\left(\frac{H - \bar{F}}{\bar{\bar{F}} - H}\right)[N(d_+(\mathcal{Y})) - N(d_+(\mathcal{B}))] \\
& + (K - \bar{F})(F_0 - \bar{F})\left(\frac{\bar{\bar{F}} - H}{H - \bar{F}}\right)[N(d_-(\mathcal{Y})) - N(d_-(\mathcal{B}))]\Big]. \quad (3.149)
\end{aligned}
$$

The limit $\bar{\bar{F}} \to \infty$ of this expression reduces to the exact formula for the up-and-out price for the affine linear volatility model; and for $\bar{F} = 0$, equation (3.61) is recovered for $r = 0$ and $S_0 = F_0$. The price of an up-and-out put can be derived in similar fashion (see Problem 6).

We now present the valuation of European-style double-knockout-barrier options with underlying asset price $F_t$ satisfying the driftless process, with quadratic volatility function as in equation (3.117). Although the mapping and the functional relationship between the kernels in $F$-space and $x$-space differ, the procedure is similar to the one employed for the linear volatility. In particular, we impose zero boundary conditions at both barrier endpoints $L$ (lower barrier) and $H$ (upper barrier) of the double-knockout-barrier pricing kernel, which we denote by $U^{DB}$, where $\bar{F} < L < H < \bar{\bar{F}}$. The values $L$, $H$ are mapped onto the $x$-space endpoints via equation (3.119):

$$
x_H = X(H) = \frac{\sqrt{2}}{\sigma}\log\frac{H - \bar{F}}{\bar{\bar{F}} - H}, \qquad x_L = X(L) = \frac{\sqrt{2}}{\sigma}\log\frac{L - \bar{F}}{\bar{\bar{F}} - L} \quad (3.150)
$$

From equation (3.121), an exact series for the $F$-space density follows by simply substituting the $x$-space transition density satisfying zero-boundary conditions; i.e., we substitute equation (3.101) into equation (3.121) while using equation (3.119), giving

$$
\begin{aligned}
U^{DB}(F, F_0; \tau) = {} & \frac{2(\bar{\bar{F}} - \bar{F})}{\log\frac{(H-\bar{F})(\bar{\bar{F}}-L)}{(L-\bar{F})(\bar{\bar{F}}-H)}}\sqrt{\frac{(F_0 - \bar{F})(\bar{\bar{F}} - F_0)}{(F - \bar{F})^3(\bar{\bar{F}} - F)^3}} \\
& \times \sum_{n=1}^{\infty} e^{-\rho_n \tau}\sin\big(n\pi\gamma(F_0)\big)\sin\big(n\pi\gamma(F)\big), \quad (3.151)
\end{aligned}
$$

where $F, F_0 \in [L, H]$ and

$$
\gamma(F) = \frac{X(F) - X(L)}{X(H) - X(L)} = \frac{\log\frac{(F-\bar{F})(\bar{\bar{F}}-L)}{(L-\bar{F})(\bar{\bar{F}}-F)}}{\log\frac{(H-\bar{F})(\bar{\bar{F}}-L)}{(L-\bar{F})(\bar{\bar{F}}-H)}}, \quad (3.152)
$$

$$
\rho_n = \frac{\sigma^2}{8} + \frac{n^2\pi^2\sigma^2}{2\log^2\frac{(H-\bar{F})(\bar{\bar{F}}-L)}{(L-\bar{F})(\bar{\bar{F}}-H)}}. \quad (3.153)
$$

This series possesses the same rapid convergence properties as equation (3.102) for positive $\tau$ and also represents the Dirac delta function $\delta(F - F_0)$ for the finite domain $[L, H]$ when $\tau = 0$. A double knockout option maturing in time $\tau$ with payoff $\Lambda(F)$ is priced using equation (3.105), with $U^{DB}$ now given by equation (3.151). In particular, the value of a double knockout call is given by (excluding an overall discount factor)

$$C^{DB}(F_0, K, \tau) = \int_L^H U^{DB}(F, F_0; \tau)(F-K)_+ \, dF = \begin{cases} \bar{\phi}(K) - K\phi(K), & K \geq L \\ \bar{\phi}(L) - K\phi(L), & K \leq L, \end{cases} \tag{3.154}$$

where $\phi(\cdot)$ and $\bar{\phi}(\cdot)$ are defined by

$$\phi(B) = \int_B^H U^{DB}(F, F_0; \tau) dF, \quad \bar{\phi}(B) = \int_B^H U^{DB}(F, F_0; \tau) F dF, \tag{3.155}$$

for any real value $B \in [L, H]$. As in the single-barrier case, these integrals are most readily evaluated by changing variables using equation (3.120). From equation (3.151),

$$U^{DB}(F(x), F_0; \tau)\frac{dF}{dx} = \frac{2\sqrt{2}\sigma}{\bar{\bar{F}} - \bar{F}} \frac{[(F_0 - \bar{F})(\bar{\bar{F}} - F_0)]^{\frac{1}{2}}}{\log \frac{(H-\bar{F})(\bar{\bar{F}}-L)}{(L-\bar{F})(\bar{\bar{F}}-H)}} \cosh \tfrac{\sigma x}{2\sqrt{2}} \times \sum_{n=1}^{\infty} e^{-\rho_n \tau} \sin\left(n\pi\gamma(F_0)\right) \sin \tfrac{n\pi(x-x_L)}{x_H - x_L}, \tag{3.156}$$

and the integrals in equation (3.155) give

$$\phi(B) = \frac{2\sqrt{2}\sigma}{\bar{\bar{F}} - \bar{F}} \frac{[(F_0 - \bar{F})(\bar{\bar{F}} - F_0)]^{\frac{1}{2}}}{\log \frac{(H-\bar{F})(\bar{\bar{F}}-L)}{(L-\bar{F})(\bar{\bar{F}}-H)}} \sum_{n=1}^{\infty} e^{-\rho_n \tau} \sin\left(n\pi\gamma(F_0)\right) I_n(B), \tag{3.157}$$

$$\bar{\phi}(B) = \frac{2\sqrt{2}\sigma}{\bar{\bar{F}} - \bar{F}} \frac{[(F_0 - \bar{F})(\bar{\bar{F}} - F_0)]^{\frac{1}{2}}}{\log \frac{(H-\bar{F})(\bar{\bar{F}}-L)}{(L-\bar{F})(\bar{\bar{F}}-H)}} \sum_{n=1}^{\infty} e^{-\rho_n \tau} \sin\left(n\pi\gamma(F_0)\right) \bar{I}_n(B), \tag{3.158}$$

where

$$I_n(B) = \int_{x_B}^{x_H} \cosh \tfrac{\sigma x}{2\sqrt{2}} \sin \tfrac{n\pi(x-x_L)}{x_H - x_L} dx, \tag{3.159}$$

$$\bar{I}_n(B) = \int_{x_B}^{x_H} F(x) \cosh \tfrac{\sigma x}{2\sqrt{2}} \sin \tfrac{n\pi(x-x_L)}{x_H - x_L} dx, \tag{3.160}$$

$x_B = X(B) = \frac{\sqrt{2}}{\sigma} \log \frac{B-\bar{F}}{\bar{\bar{F}}-B}$. These integrals are readily evaluated in exact closed form (see Problem 7):

$$\begin{aligned} I_n(B) = \frac{\pi\sigma}{2\sqrt{2}\rho_n \log \frac{(H-\bar{F})(\bar{\bar{F}}-L)}{(L-\bar{F})(\bar{\bar{F}}-H)}} \Bigg\{ & -n(-1)^n \left[ \sqrt{\frac{\bar{\bar{F}} - H}{H - \bar{F}}} + \sqrt{\frac{H - \bar{F}}{\bar{\bar{F}} - H}} \right] \\ & + \frac{1}{2\pi} \log \frac{(H-\bar{F})(\bar{\bar{F}}-L)}{(L-\bar{F})(\bar{\bar{F}}-H)} \sin\left(n\pi\gamma(B)\right) \left[ \sqrt{\frac{\bar{\bar{F}} - B}{B - \bar{F}}} - \sqrt{\frac{B - \bar{F}}{\bar{\bar{F}} - B}} \right] \\ & + n \cos\left(n\pi\gamma(B)\right) \left[ \sqrt{\frac{\bar{\bar{F}} - B}{B - \bar{F}}} + \sqrt{\frac{B - \bar{F}}{\bar{\bar{F}} - B}} \right] \Bigg\} \end{aligned} \tag{3.161}$$

$$\bar{I}_n(B) = \frac{\pi\sigma}{2\sqrt{2}\rho_n \log\frac{(H-\bar{F})(\bar{\bar{F}}-L)}{(L-\bar{F})(\bar{\bar{F}}-H)}} \left\{ -n(-1)^n \left[ \bar{F}\sqrt{\frac{\bar{\bar{F}}-H}{H-\bar{F}}} + \bar{\bar{F}}\sqrt{\frac{H-\bar{F}}{\bar{\bar{F}}-H}} \right] \right.$$
$$+ \frac{1}{2\pi}\log\frac{(H-\bar{F})(\bar{\bar{F}}-L)}{(L-\bar{F})(\bar{\bar{F}}-H)} \sin\left(n\pi\gamma(B)\right) \left[ \bar{F}\sqrt{\frac{\bar{\bar{F}}-B}{B-\bar{F}}} - \bar{\bar{F}}\sqrt{\frac{B-\bar{F}}{\bar{\bar{F}}-B}} \right]$$
$$\left. + \; n\cos\left(n\pi\gamma(B)\right) \left[ \bar{F}\sqrt{\frac{\bar{\bar{F}}-B}{B-\bar{F}}} + \bar{\bar{F}}\sqrt{\frac{B-\bar{F}}{\bar{\bar{F}}-B}} \right] \right\}. \tag{3.162}$$

Finally, from equation (3.154) we see that using these expressions within equations (3.157) and (3.158) for $B = K$ and then separately for $B = L$ and simplifying gives:

$$C^{DB}(F_0, K, \tau) = \frac{\sigma^2\sqrt{(F_0-\bar{F})(\bar{\bar{F}}-F_0)}}{\log\frac{(H-\bar{F})(\bar{\bar{F}}-L)}{(L-\bar{F})(\bar{\bar{F}}-H)}} \sum_{n=1}^{\infty} \frac{e^{-\rho_n\tau}}{\rho_n} \sin\left(n\pi\gamma(F_0)\right)$$
$$\times \left[ \frac{n\pi(-1)^n(K-H)}{\log\frac{(H-\bar{F})(\bar{\bar{F}}-L)}{(L-\bar{F})(\bar{\bar{F}}-H)}\sqrt{(\bar{\bar{F}}-H)(H-\bar{F})}} - \frac{\sqrt{(K-\bar{F})(\bar{\bar{F}}-K)}}{(\bar{\bar{F}}-\bar{F})} \sin\left(n\pi\gamma(K)\right) \right] \tag{3.163}$$

for $K \geq L$, and

$$C^{DB}(F_0, K, \tau) = \frac{\pi\sigma^2\sqrt{(F_0-\bar{F})(\bar{\bar{F}}-F_0)}}{\log^2\frac{(H-\bar{F})(\bar{\bar{F}}-L)}{(L-\bar{F})(\bar{\bar{F}}-H)}} \sum_{n=1}^{\infty} n\frac{e^{-\rho_n\tau}}{\rho_n} \sin\left(n\pi\gamma(F_0)\right)$$
$$\times \left[ \frac{(-1)^n(K-H)}{\sqrt{(\bar{\bar{F}}-H)(H-\bar{F})}} + \frac{L-K}{\sqrt{(\bar{\bar{F}}-L)(L-\bar{F})}} \right] \tag{3.164}$$

for $K \leq L$. This last expression has a simpler form since $\gamma(L) = 0$. Note that the two formulas are identical when $K = L$. An example of the rapid convergence of these series solutions is given in Figure 3.7. The limit $\bar{\bar{F}} \to \infty$ of these expressions gives exact formulas for the case of an affine linear volatility model; further, by setting $\bar{F} = 0$, we also exactly recover equations (3.106) and (3.107), respectively. Figure 3.8 demonstrates this explicitly. Indeed for a given double-barrier call option contract, one observes uniform agreement of the option prices for the quadratic model with those of the linear model, as the quadratic volatility function is made to coincide more and more closely with that of the corresponding linear volatility function.

## Problems

### Problem 1.

(a) Using equation (3.120) show that

$$[(F(x)-\bar{F})(\bar{\bar{F}}-F(x))]^{-\frac{1}{2}} = 2(\bar{\bar{F}}-\bar{F})^{-1}\cosh(\sigma x/2\sqrt{2}), \tag{3.165}$$

and use this relation and the derivative $F'(x)$ to arrive at equation (3.123).

(b) Use the identity $\int_{-\infty}^{\infty} e^{\sigma x/2\sqrt{2}} e^{-(x-x_0)^2/4\tau}\, dx = 2\sqrt{\pi\tau} e^{\sigma^2\tau/8} e^{\sigma x_0/2\sqrt{2}}$ to show that the barrier-free density satisfies

$$\int_{\bar{F}}^{\bar{\bar{F}}} U(F, F_0; \tau) dF = \int_{-\infty}^{\infty} U(F(x), F_0; \tau) \frac{dF}{dx} dx = 1. \tag{3.166}$$

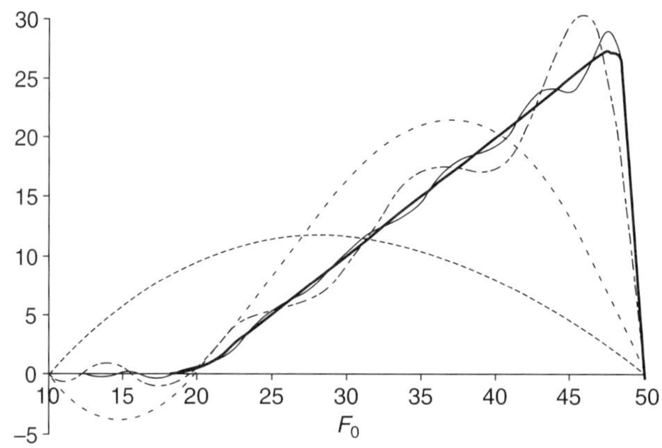


FIGURE 3.7 Rapid convergence of the double knockout call price across the full range of spot $F_0$ as one includes only the first 1, 2, 8, 16, and 32 (thick solid line) terms in the series (3.163), where $L = 10$, $H = 50$, $K = 20$, $\sigma = 0.2$, $\tau = 0.25$.

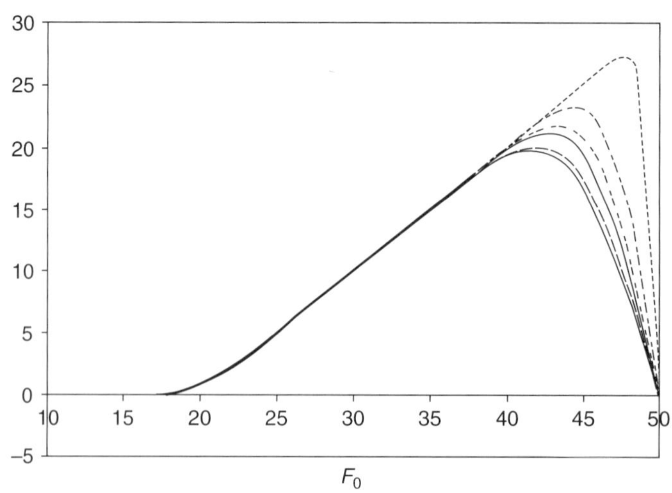


FIGURE 3.8 Uniform approach of the double knockout call price for the quadratic model [given by equation (3.163)] to that of the linear model given by equation (3.106), as $\bar{\bar{F}}$ is pushed to larger values. The five thinner curves represent the converged price [i.e., using equation (3.163)] for the quadratic model for the separate cases of $\bar{\bar{F}} = 60$, 120, 240, 480, and 3200. The curve for $\bar{\bar{F}} = 3200$ is very close to the thick solid line representing the price given by the linear model for the same parameter choice: $L = 10$, $H = 50$, $K = 20$, $\sigma = 0.2$, $\tau = 0.25$.

**Problem 2.** Using parts of Problem 1, show that equation (3.123) leads to the martingale property:

$$E_0[F_\tau] = E\big[F_\tau \big| F_{\tau=0} = F_0\big] = \int_{\bar{F}}^{\bar{\bar{F}}} U(F, F_0; \tau) F dF = F_0. \tag{3.167}$$

**Problem 3.** Derive equation (3.128) by completing the square in the exponent. Note that the identity $d_\pm(1/x) = -d_\mp(x)$ obtained from equation (3.130) is useful in the manipulation of expressions.

**Problem 4.** By following a similar procedure as was used to derive equation (3.134), derive the exact formula for the corresponding put value. Is a put-call parity relation satisfied?

**Problem 5.** Derive an exact formula for the down-and-out put value.

**Problem 6.** Obtain an exact formula for the up-and-out put value for $K \le H$ and for $K \ge H$.

**Problem 7.** The integrals in equations (3.159) and (3.160) can be evaluated by rewriting them as a sum of integrals of the form

$$\int_{x_B}^{x_H} e^{\pm\sigma x/2\sqrt{2}} \sin \tfrac{n\pi(x-x_L)}{x_H - x_L} dx.$$

Use the antiderivative $\int e^{ax} \sin bx dx = e^{ax}[a \sin bx - b \cos bx]/(a^2 + b^2) + c$, where $a$, $b$, $c$ are any constants, and then recast the variables $x_B$, $x_H$, $x_L$ in the resulting integrations in terms of the $F$-space variables $B$, $H$, $L$ and arrive at equations (3.161) and (3.162).

## 3.6 Green's Functions Method for Diffusion Kernels

In this section we present a standard Green's function framework for finding solutions for the $x$-space kernel subject to homogeneous boundary conditions. Throughout this section we shall assume one-dimensional diffusions, i.e., a diffusion process $x_t$ obeying

$$dx_t = \lambda(x_t)dt + \nu(x_t)dW_t, \tag{3.168}$$

with $W_t$ as the standard Wiener process. This process is assumed to have a differentiable drift function $\lambda(x)$ and a twice differentiable diffusion function or volatility function $\nu(x)$. The goal is to solve for the kernel or density $u(x, x_0; \tau)$, subject to appropriate boundary conditions.

Since the drift and volatility functions are assumed to have no explicit time dependence, the kernel $u = u(x, x_0; \tau)$ satisfies the time-homogeneous forward Kolmogorov equation

$$\frac{\partial u}{\partial \tau} = \frac{1}{2}\frac{\partial^2}{\partial x^2}\Big(\nu(x)^2 u\Big) - \frac{\partial}{\partial x}\Big(\lambda(x) u\Big) \equiv \mathcal{L}_x u \tag{3.169}$$

and the corresponding backward equation

$$\frac{\partial u}{\partial \tau} = \frac{1}{2}\nu(x_0)^2 \frac{\partial^2 u}{\partial x_0^2} + \lambda(x_0)\frac{\partial u}{\partial x_0} \equiv \tilde{\mathcal{L}}_{x_0} u, \tag{3.170}$$

subject to the initial condition $u(x, x_0; 0) = \delta(x - x_0)$. As in Chapter 1, we have defined the Fokker–Planck differential operator $\mathcal{L}_x$ that acts on the variable $x$ and its formal Lagrange adjoint $\tilde{\mathcal{L}}_{x_0}$ acting on $x_0$. One technical point to note is that the differential operator $\mathcal{L}$ is generally not self-adjoint, i.e., $\tilde{\mathcal{L}} \neq \mathcal{L}$, and the solution for the transition density is generally not symmetric with respect to interchanging $x$ and $x_0$. However, as is seen from the transformations provided next, the corresponding time-independent Green's function technique for solving either forward or backward equations can be treated within a common footing.

In developing a solution framework for $u(x, x_0; \tau)$, we consider the corresponding *time-independent* Green's function $G(x, x_0; s)$, which is defined via the Laplace transform with respect to time:

$$G(x, x_0; s) = L\big[u(x, x_0; \tau)\big][s] \equiv \int_0^\infty e^{-s\tau} u(x, x_0; \tau) d\tau. \tag{3.171}$$

[Without loss in generality, we shall assume that $u$ is absolutely integrable with respect to $\tau$ on any interval $0 \leq \tau \leq T$ and that $G(x, x_0; s)$ exists for some real value of $s = a$. Then from the theory of Laplace transforms it can be shown that $G(x, x_0; s)$ is an analytic function on the complex $s$-plane for $\mathrm{Re}\, s > a$. As will be seen, what is important to keep in mind for the discussion at hand is that the function $G(x, x_0; s)$ is uniquely determined by satisfying appropriate boundary conditions in $x$, for $\mathrm{Re}\, s > a$.] Taking Laplace transforms with respect to time $\tau$ on both sides of forward equation (3.169) while making use of the well-known identity for the Laplace transform of the derivative of a function and the initial delta function condition on $u$ gives a *nonhomogeneous ordinary differential equation* for the Green's function $G \equiv G(x, x_0; s)$,

$$\frac{1}{2}\frac{d^2}{dx^2}\Big(\nu(x)^2 G\Big) - \frac{d}{dx}\Big(\lambda(x) G\Big) - sG \equiv \mathcal{L}_x G - sG = -\delta(x - x_0). \tag{3.172}$$

Note here that the partial derivatives have been replaced by ordinary derivatives, where one is holding $x_0$ (and $s$) fixed in the Green's function. In similar fashion, by taking Laplace transforms on both sides of backward equation (3.170), one also obtains the adjoint equation to equation (3.172):

$$\frac{1}{2}\nu(x_0)^2 \frac{d^2 G}{dx_0^2} + \lambda(x_0)\frac{dG}{dx_0} - sG \equiv \tilde{\mathcal{L}}_{x_0} G - sG = -\delta(x - x_0). \tag{3.173}$$

Again, the partial derivatives have been replaced by ordinary derivatives, where one is now holding $x$ (and $s$) fixed in the Green's function.

Using either of these equations, the objective is now to solve the ordinary differential equation (i.e., with delta function as the inhomogeneous source term) for the function $G(x, x_0; s)$, subject to the same homogeneous boundary conditions that are imposed on the function $u(x, x_0; \tau)$. Hence either one can solve equation (3.172) with imposed boundary conditions in $x$, or one solves the corresponding adjoint equation (3.173) with boundary conditions imposed in $x_0$. Upon unique determination of $G(x, x_0; s)$, one then has the desired unique solution for the kernel $u(x, x_0; \tau)$ (which satisfies the same desired homogeneous boundary conditions) via the Laplace inversion

$$u(x, x_0; \tau) = L^{-1}\big[G(x, x_0; s)\big][\tau] = \frac{1}{2\pi i}\int_{\gamma - i\infty}^{\gamma + i\infty} e^{s\tau} G(x, x_0; s) ds. \tag{3.174}$$

We shall use $L^{-1}[F(s)][t]$ to denote the inverse Laplace transform of a function $F(s)$ evaluated at $t$. This inversion formula, which can generally be used to compute the inverse Laplace transform, is the Bromwich contour integral or the Mellin integral arising in the theory of

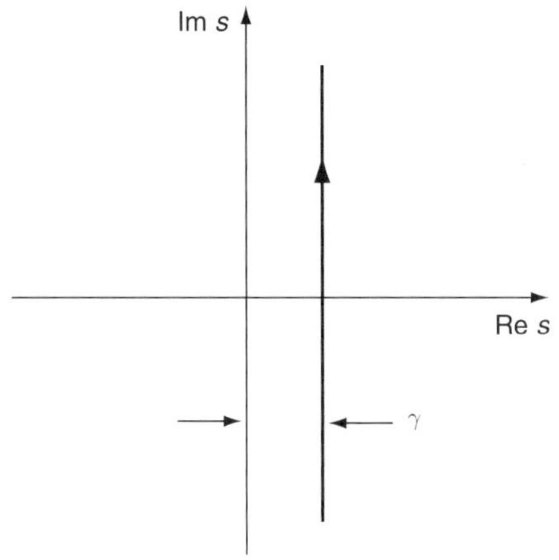


FIGURE 3.9 The Bromwich contour extends from $\gamma - i\infty$ to $\gamma + i\infty$.

Laplace and other integral transforms. This contour, depicted in Figure 3.9, is the infinite line $\mathrm{Re}\, s = \gamma$ on the complex $s$-plane parametrized by $s = \gamma + ir$, with real parameter $r$ running from $-\infty$ to $\infty$. Here $\gamma$ is any real number such that all singularities of $G(x, x_0; s)$ (now considered as a complex-valued function of $s$ for any fixed real values $x$,$x_0$) lie to the left of the line $\mathrm{Re}\, s = \gamma$ on the complex $s$-plane. Throughout, $i$ denotes the usual complex number, with $z = \mathrm{Re}\, z + i\mathrm{Im}\, z$, where Re and Im denote the real and imaginary parts, respectively. Later we also make use of polar coordinate form of a complex number, $z = re^{i\theta}$, where $r = |z|$ is the modulus and $\theta = \arg z$ is the argument of $z$.

Once one has obtained $G(x, x_0; s)$ analytically, the integral in equation (3.174) is then in itself an exact integral representation for the transition density $u(x, x_0; \tau)$. This is partly the reason for sometimes also referring to $G(x, x_0; s)$ as the *resolvent kernel*. As shown shortly, in the analytical evaluation of the Laplace inverse, it proves useful to extend the Bromwich contour to form a *closed contour* integral enclosing the negative real half of the complex $s$-plane. A simple application of the infamous *residue theorem* of complex analysis then further allows us to evaluate the integral either as an exact series or in terms of exact closed-form special functions. A rather general procedure for achieving this purpose is to try to close the contour in such a manner that the Green's function is an analytic function of the complex variable $s$ everywhere on the closed contour. Yet inside the contour, $G$ may either be analytic or have a finite number of isolated simple poles (i.e., simple singularities). After justifying the equivalence and hence replacement of the Bromwich integral with the *closed contour*, or *loop integral*, we then subsequently apply the standard residue theorem to compute the result. In more general applications the Green's function may have a *branch point* (e.g., due to factors such as $\sqrt{s}$) that gives rise to a *branch cut* on the complex $s$-plane. From complex analysis we know that the residue theorem cannot be used to evaluate a contour integral that encloses a branch cut. However, there is a standard technique that can be used in such a case. This is sometimes referred to as "shrinking the contour onto the branch cut." This method is generally best described by example. Later we give concrete examples of this Laplace inversion of $G$, for the simple case of the Wiener process and also for the more complex case of the Bessel process. As we shall see, for the case of finite double barriers one obtains a rapidly converging analytical infinite series representation for the transition density $u$. This is the *eigenfunction* expansion solution for the transition density. Such expansions, as discussed briefly in Section 3.6.1, also follow from the spectral theory of eigenfunction expansions.

The Green's function methodology we now present is based on the Sturm–Liouville theory of linear ordinary differential equations [MF53, But80, Duf01b, Dav02]. However, we specialize the theory to the diffusion equation relevant to pricing theory. In what follows it is convenient to make direct use of equation (3.172). That is, we consider solving the nonhomogeneous equation (3.172); i.e., we now build the Green's function $G(x, x_0; s)$ by considering solutions $y(x; s)$ to the corresponding *homogeneous* equation

$$\mathcal{L}_x y(x; s) - sy(x; s) = 0, \tag{3.175}$$

subject to appropriate boundary conditions. Note: For shorthand we shall also simply write $y(x)$ to mean $y(x; s)$, because $s$ is a fixed parameter in the differential equation. In order to make use of established results from Sturm–Liouville theory, we shall first transform the original equation (3.175) into one in *standard Sturm–Liouville form*. This is accomplished via a transformation to a new function defined by

$$\bar{y}(x) \equiv \frac{\nu(x)}{\nu(x_0)} \exp\left( - \int_{x_0}^{x} \frac{\lambda(u)}{\nu(u)^2} du \right) y(x). \tag{3.176}$$

Using this definition we can show by direct differentiation that

$$\mathcal{L}_x y(x) = \frac{\nu(x_0)}{\nu(x)} \exp\left( \int_{x_0}^{x} \frac{\lambda(u)}{\nu(u)^2} du \right) \bar{\mathcal{L}}_x \bar{y}(x), \tag{3.177}$$

where the new differential operator $\bar{\mathcal{L}}_x$ is defined by

$$\bar{\mathcal{L}}_x f(x) \equiv \frac{d}{dx}\left( p(x) \frac{df(x)}{dx} \right) - q(x) f(x) \tag{3.178}$$

for any arbitrary twice differentiable function $f(x)$. Here the functions $p(x)$ and $q(x)$ are given in terms of the drift and volatility functions:

$$p(x) = \frac{1}{2}\nu(x)^2, \tag{3.179}$$

$$q(x) = \frac{1}{2}\left[ \lambda'(x) + \left( \frac{\lambda(x)}{\nu(x)} \right)^2 - 2\lambda(x)\frac{\nu'(x)}{\nu(x)} - \nu(x)\nu''(x) \right]. \tag{3.180}$$

(Prime is used to denote differentiation.) The operator $\bar{\mathcal{L}}$ is now in standard Sturm–Liouville form and is hence also self-adjoint. One should note here that the Green's function methodology may also be directly applied to the original nonself-adjoint problem. However, transforming the equations into the standard self-adjoint Sturm–Liouville form and then solving and transforming back proves very convenient, as the whole following analysis shows.

From equation (3.176), it follows that a related (new or modified) Green's function $\bar{G}$ is similarly defined as

$$\bar{G}(x, x_0; s) \equiv \frac{\nu(x)}{\nu(x_0)} \exp\left( - \int_{x_0}^{x} \frac{\lambda(u)}{\nu(u)^2} du \right) G(x, x_0; s), \tag{3.181}$$

leading to the transformed nonhomogeneous equation in Sturm–Liouville form:

$$\bar{\mathcal{L}}_x \bar{G}(x, x_0; s) - s\bar{G}(x, x_0; s) = -\delta(x - x_0) \tag{3.182}$$

Note that the inhomogeneous term again contains only the Dirac delta function since the property $\frac{f(x)}{f(x_0)}\delta(x-x_0)=\delta(x-x_0)$ has been used where the ratio $\frac{f(x)}{f(x_0)}$ is nonsingular [i.e., $f(x)=\nu(x)\exp(\int^x(\lambda(u)/\nu(u)^2)du)$] within the allowable solution region, and evaluates to unity when $x=x_0$. Equation (3.182) is now the desired standard form, which may be solved subject to various homogeneous boundary conditions. Upon solving for $\bar{G}$ we then simply invert equation (3.181), giving $G$, as shown next. By using standard textbook methods of solution for nonhomogeneous second-order ordinary differential equations (e.g., the method of variation of parameters), $\bar{G}(x, x_0; s)$ is readily obtained from the solutions $\bar{y}(x) \equiv \bar{y}(x; s)$ to the corresponding homogeneous equation [i.e., the homogeneous counterpart of equation (3.182)]:

$$\bar{\mathcal{L}}_x\bar{y}(x) - s\bar{y}(x) = 0. \tag{3.183}$$

Generally, if $\bar{y}_1$ and $\bar{y}_2$ are two *linearly independent* solutions to equation (3.183), then the Green's function is readily shown to take the form

$$\bar{G}(x, x_0; s) = -\begin{cases} \dfrac{\bar{y}_1(x)\bar{y}_2(x_0)}{p\bar{W}}, & x \le x_0 \\[2ex] \dfrac{\bar{y}_2(x)\bar{y}_1(x_0)}{p\bar{W}}, & x \ge x_0. \end{cases} \tag{3.184}$$

Here $p\bar{W} = p(x_0)\bar{W}(x_0) = p(x)\bar{W}(x)$ is a constant independent of $x$ and $x_0$ (not constant w.r.t. $s$), as can be shown from the properties of the Wronskian of any two solutions to equation (3.183): $\bar{W}(x) \equiv W[\bar{y}_1(x), \bar{y}_2(x)] \equiv \bar{y}_1(x)\bar{y}_2'(x) - \bar{y}_1'(x)\bar{y}_2(x)$. The boundary conditions for the Green's function are matched by the choice of the two solutions $\bar{y}_1$, $\bar{y}_2$. The reader should also note that, since equation (3.184) involves a ratio of the product of two independent solutions divided by their Wronskian, the Green's function is still uniquely determined if we multiply any of the two solutions by an arbitrary nonzero constant. The symmetry $\bar{G}(x, x_0; s) = \bar{G}(x_0, x; s)$ with respect to interchanging $x$ and $x_0$ is also a useful property, following from the fact that the Sturm–Liouville operator is self-adjoint. The solution $\bar{y}_1$ is chosen to match the boundary condition at the lower region, while $\bar{y}_2$ is chosen to match the boundary at the upper region. For example, if one requires zero-boundary conditions at two points $x = x_L$ and $x = x_H$ ($x_L < x_H$) with $\bar{G}(x = x_L, x_0; s) = 0$ and $\bar{G}(x = x_H, x_0; s) = 0$, then a linear combination of independent solutions to equation (3.183) must be formed to give $\bar{y}_1(x_L) = \bar{y}_2(x_H) = 0$. Inserting the two solutions and their Wronskian into equation (3.184) gives $\bar{G}$.

From equation (3.181), the Green's function to the *original* problem (3.172) is then obtained as

$$G(x, x_0; s) = -\frac{e^{\int_{x_0}^{x}\frac{\lambda(u)}{\nu(u)^2}du}}{\nu(x)/\nu(x_0)}\begin{cases} \dfrac{\bar{y}_1(x)\bar{y}_2(x_0)}{p\bar{W}}, & x_L \le x \le x_0 \\[2ex] \dfrac{\bar{y}_2(x)\bar{y}_1(x_0)}{p\bar{W}}, & x_0 \le x \le x_H. \end{cases} \tag{3.185}$$

Here we have assumed that the multiplicative factor to the left of the curly bracket in equation (3.185) is finite at the solution endpoints. In the special case of a singular multiplicative factor at an endpoint, we assume that either $\bar{y}_1$ or $\bar{y}_2$ approaches zero more rapidly at the endpoint so that $G$ satisfies the same zero-boundary condition as $\bar{G}$. The foregoing expression is applicable to all cases of homogeneous boundary conditions that we shall encounter. [It is noted that this approach can also handle boundary conditions of a mixed kind to accommodate for other types of solutions, such as relection at a boundary. However, throughout we are only concerned with zero-(i.e., Dirichlet)-boundary conditions for the purpose of pricing barrier as well as barrier-free European options for state-dependent volatility models to follow.] The points $x_L$, $x_H$ can be finite, or either point can be taken in the infinite limit, depending on the allowable solution space. Note that, in contrast to $\bar{G}$, $G$ is generally not symmetric with respect to interchanging $x$ and $x_0$. However, by direct inspection we see that $G$ in equation (3.185) is a product of functions in $x$ and $x_0$ and hence automatically provides us with solutions to the homogeneous equation (3.175) and its adjoint equation where the functions $\nu(x_0)\exp\big(-\int^{x_0}\frac{\lambda(u)}{\nu(u)^2}du\big)\bar{y}_1(x_0)$ and $\nu(x_0)\exp\big(-\int^{x_0}\frac{\lambda(u)}{\nu(u)^2}du\big)\bar{y}_2(x_0)$ form two linearly independent solutions to the homogeneous version of equation (3.173); i.e., these form two linearly independent solutions to $\tilde{\mathcal{L}}_{x_0}\tilde{y}-s\tilde{y}=0$. Renaming variables $x_0\to x$ and $s\to\rho$ hence gives a general solution to this homogeneous ordinary differential equation rewritten in terms of $x$ in equation (3.272), which we obtain simply by inspection of the Green's function $G$. We shall denote this solution by $\hat{u}(x;\rho)$, where

$$\hat{u}(x;\rho)=\nu(x)e^{-\int^{x}\frac{\lambda(u)}{\nu(u)^2}du}[q_1\bar{y}_1(x;\rho)+q_2\bar{y}_2(x;\rho)] \tag{3.186}$$

and $q_1$,$q_2$ are arbitrary constants. The function $\hat{u}(x;\rho)$ (in Section 3.8.1 it is referred to as a *generating function*) will turn out to play an important role in generating new pricing kernels for an $F$-space process from known $x$-space kernels, as is discussed later in this chapter.

In closing this section, we demonstrate the Green's function procedure with a standard example covering the different cases of boundary conditions.

**Example 5. The Wiener Process.**

Let's consider the process $dx_t=\sqrt{2}\,dW_t$, where $\nu(x)=\sqrt{2}$ (constant volatility) and $\lambda(x)=0$ (zero drift). From equations (3.179) and (3.180), the functions $p(x)=1$ and $q(x)=0$ are trivial. In this special case $\tilde{\mathcal{L}}=\mathcal{L}=\bar{\mathcal{L}}$, and $G=\bar{G}$, which satisfies

$$\frac{d^2}{dx^2}G-sG=-\delta(x-x_0), \tag{3.187}$$

where $\bar{y}=y$ satisfies the corresponding homogeneous equation [i.e., equation (3.183) or (3.175)] $y''-sy=0$. Two independent solutions of this equation are $e^{\sqrt{s}x}$ and $e^{-\sqrt{s}x}$. If we seek barrier-free kernel solutions, then we impose zero-boundary conditions at $x\to\pm\infty$ (e.g., $x_L=-\infty$ and $x_H=\infty$ in the earlier notation). Therefore we let $\bar{y}_1=e^{\sqrt{s}x}$, $\bar{y}_2=e^{-\sqrt{s}x}$ since $e^{\pm\sqrt{s}x}\to 0$ as $x\to\mp\infty$ for real values of $s>0$ (also generally true for Re $s>0$). The Wronskian of these two solutions gives $p\bar{W}=-2\sqrt{s}$. Using equation (3.185), where the multiplicative factor is just unity, gives the Green's function for the barrier-free case:

$$G(x,x_0;s)=\begin{cases} e^{\sqrt{s}(x-x_0)}/2\sqrt{s}, & x\le x_0\\ e^{-\sqrt{s}(x-x_0)}/2\sqrt{s}, & x\ge x_0\end{cases} \tag{3.188}$$

Obtaining the kernel $u(x,x_0;\tau)$ is now just a matter of Laplace-inverting this function from $s$ back into the time $\tau$ domain. Note that we may rewrite $G=e^{-k\sqrt{s}}/2\sqrt{s}$, $k=x^>-x^<$ $(k\ge 0)$, where $x^>$ $(x^<)$ stands for the greater (smaller) of the two real numbers $x$, $x_0$. The Laplace transform can in some cases be found directly with the use of tables. For instance, in this case one can look up a table of transforms to find $L^{-1}[e^{-k\sqrt{s}}/\sqrt{s}][\tau]=e^{-k^2/4\tau}/\sqrt{\pi\tau}$. Since $k^2=(x-x_0)^2$ (regardless of the relative magnitudes of $x$ and $x_0$), we then recover the known solution $u(x, x_0; \tau) = g_0(x, x_0; \tau)$ exactly as in equation (3.9), and the problem of obtaining the barrier-free kernel has been completely solved.

It is now instructive to show how the inverse transform is computed using standard techniques of complex analysis (without the use of tables because this is particularly important for handling nonelementary Green's functions, as discussed later). $G$ is an analytic function on the complex $s$-plane, except for a branch point at $s = 0$ due to the $\sqrt{s}$ factor. For this purpose a branch cut must be introduced. In order to apply the residue theorem, we consider the closed contour in Figure 3.10 (see Section 3.7.1) with branch cut chosen as the negative real line $\arg s = \pi$, with complex $s$-plane $|\arg s| < \pi$, i.e., the principal branch. The Bromwich contour corresponds to the line segment $MN$. Since $G$ is analytic everywhere (i.e., no singularities) on and inside the entire region within the closed contour for all values of the semicircular radius $R > 0$ as well as for positive parameters $\rho$, $\delta$, and $\gamma$ taken arbitrarily close to zero, Cauchy's integral formula gives a value of zero for the complete loop integral. Hence the Bromwich integral is equal to the negative of the sum of all the other contour integrals that make up the closed loop. From this fact, the kernel is

$$\begin{aligned} u(x, x_0; \tau) &= \frac{1}{2\pi i} \int_{\gamma - i\infty}^{\gamma + i\infty} e^{s\tau} G(x, x_0; s) ds \\ &= -\frac{1}{2\pi i} \left[ \int_{C_R^+} + \int_{C_R^-} + \int_P^Q + \int_{Q'}^{P'} + \int_{C_\rho} \right] e^{s\tau} \frac{e^{-k\sqrt{s}}}{2\sqrt{s}} ds. \end{aligned} \tag{3.189}$$

This sum of integrals is dramatically reduced using standard arguments as follows. Taking limits $R \to \infty$ and $\gamma, \rho, \delta \to 0$, the $C_R^\pm$ integrals vanish, since along the semicircular contours $s = \mathrm{R}e^{i\theta}$, $\frac{\pi}{2} < |\theta| < \pi$, hence $\cos\theta < 0$, so the modulus of the integrand $|Ge^{s\tau}| = e^{R\tau\cos\theta} e^{-k\sqrt{R}\cos(\frac{\theta}{2})}/2\sqrt{R} \to 0$ as $R \to \infty$. The $C_\rho$ integral for the circular segment $QQ'$ also vanishes since $s = \rho e^{i\theta}$, $-\pi < \theta < \pi$, so the modulus (as $\rho \to 0$) of this integral has value $\leq \sqrt{\rho} \times$ const., which goes to zero in the limit $\rho \to 0$. The only nonzero integrals are along the branch cut corresponding to the $PQ$ and $Q'P'$ segments, where $s = re^{i\pi}$ ($\sqrt{s} = i\sqrt{r}$) and $s = re^{-i\pi}$ ($\sqrt{s} = -i\sqrt{r}$), respectively, in the limit $\delta \to 0$, with $\rho \leq r \leq R$. In the limits $\rho \to 0$ and $R \to \infty$ the two integrals are combined to give the real-valued integral

$$u(x, x_0; \tau) = \frac{1}{2\pi} \int_0^\infty e^{-r\tau} \frac{\cos(k\sqrt{r})}{\sqrt{r}} dr = \frac{1}{2\sqrt{\pi\tau}} e^{-k^2/4\tau}, \tag{3.190}$$

where the last result is $g_0(x, x_0; \tau)$, as before, and was obtained by a change of integration variables resulting in the cosine transform of a Gaussian function giving a Gaussian in $k$.

Barrier kernels for the Wiener process are also readily obtained. The Green's functions provide solutions that relate directly to the method of images, discussed partly in Section 3.2.1. In particular, let's reconsider the problem of finding the kernel in the domain $x, x_0 \in (-\infty, x_H]$ for a single upper barrier at level $x_H$. [The steps for the case of a lower barrier are the same.] Since we wish to impose zero-boundary conditions for the kernel at $x_L = -\infty$ and at $x_H$, we form a linear combination of $e^{\sqrt{s}x}$ and $e^{-\sqrt{s}x}$ to set $\bar{y}_2(x) = \sinh\sqrt{s}(x - x_H)$ and set $\bar{y}_1(x) = e^{\sqrt{s}x}$. Hence $\bar{y}_2(x_H) = 0$, $\bar{y}_1(-\infty) = 0$ for $\mathrm{Re}\, s > 0$, as needed. In this case $p\bar{W} = \sqrt{s}e^{\sqrt{s}x_H}$ and the Green's function is

$$G(x, x_0; s) = \frac{1}{2\sqrt{s}} \left[ e^{-(x^> - x^<)\sqrt{s}} - e^{-(2x_H - x - x_0)\sqrt{s}} \right] \tag{3.191}$$

since $x^> + x^< = x + x_0$. This involves the difference of two expressions of the same functional form in $s$ as in the barrier-free Green's function. Hence, Laplace-inverting gives precisely the kernel $u(x, x_0; \tau) = g^u(x_H, x, x_0; \tau)$ of equation (3.14). This result was previously derived by the method of images.

The last case of interest is the kernel having zero value at two finite endpoints $x_L$, $x_H$ (i.e., the double-barrier) with solution domain $x, x_0 \in [x_L, x_H]$. Following similar steps as before we have

$$G(x, x_0; s) = -\frac{\sinh\sqrt{s}(x^< - x_L)\sinh\sqrt{s}(x^> - x_H)}{\sqrt{s}\sinh\sqrt{s}(x_H - x_L)}. \tag{3.192}$$

Note that this function is zero at both endpoints. This Green's function leads to two separate types of exact series expansions for the kernel. The first type is an eigenfunction expansion. The relation between eigenfunction expansions for diffusion kernels and Green's functions is discussed in the next section. Here we show explicitly how such an expansion arises from the Laplace inversion of equation (3.192). Observe that $G$ is a ratio of two analytic functions of complex $s$, despite the appearance of the $\sqrt{s}$ factor. Indeed this can be seen by a direct Taylor expansion of the hyperbolic sine in both numerator and denominator. The only singularities of $G$ are *isolated simple poles* along the negative real axis. In fact, using the identity $\sinh(ix) = i\sin(x)$ and letting $s = -|\epsilon|$, the denominator of $G$ along the negative real axis is $\sqrt{s}\sinh\sqrt{s}(x_H - x_L) = -\sqrt{|\epsilon|}\sin\sqrt{|\epsilon|}(x_H - x_L)$. Therefore the zeros of the sine function give the simple poles of $G$ at positions $s = \epsilon_n \equiv -n^2\pi^2/(x_H - x_L)^2$, $n = 1, \ldots$. Note that $s = 0$ is a removable singularity in this case, as is shown by a Taylor expansion of the denominator about $s = 0$. The Bromwich integral can therefore be closed by joining a single semicircular contour $C_R$ enclosing the negative real half of the complex $s$-plane, as long as the contour does not coincide with any of the isolated poles. Since the modulus of the integrand in the $C_R$ integral approaches zero as $R \to \infty$, the residue theorem gives

$$u(x, x_0; \tau) = \sum_{n=1}^{\infty} e^{\epsilon_n \tau} Res G(x, x_0; s = \epsilon_n). \tag{3.193}$$

Since the Green's function is a ratio of two analytic functions, e.g., $G(s) = P(s)/Q(s)$, where $Q(s)$ has simple zeros at $s = \epsilon_n$, then from complex analysis we know that the residue at each pole is given by $\operatorname{Res} G(s = \epsilon_n) = P(\epsilon_n)/Q'(\epsilon_n)$. Evaluating the derivative of the denominator in equation (3.192) and the numerator at each pole while making use of the identity $\sinh(ix) = i\sin(x)$ and recasting one of the resulting sine functions in the numerator, we obtain

$$\operatorname{Res} G(x, x_0; s = \epsilon_n) = \frac{2}{x_H - x_L}\sin\frac{n\pi(x - x_L)}{x_H - x_L}\sin\frac{n\pi(x_0 - x_L)}{x_H - x_L}, \tag{3.194}$$

which is valid regardless of the relative magnitude of $x$ and $x_0$. Substituting into equation (3.193) therefore recovers the kernel (3.101). Recall that this kernel was used in Section 3.5 to generate rapidly convergent exact series solutions for the affine and quadratic (with two distinct roots) volatility models.

Green's function (3.192) can also be used to generate a second type of exact infinite expansion for the kernel, which is not based on eigenfunctions but rather gives exactly what one would obtain by applying the method of infinite images. The idea is to reexpress $G$ in a Taylor expansion involving an infinite sum of exponential terms, which upon Laplace inversion gives rise to an infinite sum of kernels of the barrier-free type that will be centered at the image points located at a sequence of increasing distances from either side of the solution domain. We leave this as an exercise for the interested reader (see Problem 1).

## Problems

**Problem 1.** By using the Taylor expansion identity $1/\sinh x = 2\sum_{n=0}^{\infty} e^{-(2n+1)x}$, show that the Green's function in equation (3.192) is given by

$$G = \frac{1}{2\sqrt{s}}\sum_{n=0}^{\infty} e^{-2n(x_H - x_L)\sqrt{s}}\Big[e^{-\sqrt{s}(2(x_H-x_L)-(x^> - x^<))} + e^{-\sqrt{s}(x^> - x^<)} - e^{-\sqrt{s}(x^> + x^< - 2x_L)} - e^{-\sqrt{s}(2x_H - x^< - x^>)}\Big]. \quad (3.195)$$

By taking the Laplace inverse of this series, obtain an infinite series for the kernel.

**Problem 2.** Verify that

$$u(x, x_0; \tau) = \sum_{n=-\infty}^{\infty}\Big[g_0(x, x_0 + 2nL; \tau) - g_0(x, 2nL - x_0; \tau)\Big], \quad (3.196)$$

where $g_0$ is defined by equation (3.9), $\tau = t - t_0$, is a solution to equations (3.7) and (3.8) in the finite domain $0 < x, x_0 < L$. Determine the boundary conditions at the endpoints.

## 3.6.1 *Eigenfunction Expansions for the Green's Function and the Transition Density*

Green's functions are intimately tied to the eigenvalue-eigenfunction problem of the corresponding homogeneous equation. Here it suffices to give only the most basic and brief discussion of this useful aspect of the theory. In particular, as an alternative to the closed-form expressions of the previous section, it is sometimes useful to consider Green's function solutions directly in terms of eigenfunction expansions when possible. Let us again consider equation (3.183). This equation, together with the imposed boundary conditions, constitutes an eigenvalue problem of the Sturm–Liouville type. For the case of zero-homogeneous boundary conditions at two finite boundaries it follows from *regular* Sturm–Liouville theory that if the functions $p(x)$ and $q(x)$ in equations (3.179) and (3.180) are well behaved (i.e., $p(x) > 0$ and $p$, $p'$, $q$ are continuous in a finite solution domain $[x_L, x_H]$), then the Green's function admits a *spectral resolution* of the form

$$\bar{G}(x, x_0; s) = \sum_{n=1}^{\infty} \frac{\phi_n(x)\phi_n(x_0)}{s - \epsilon_n}, \quad (3.197)$$

where the eigenfunctions $\phi_n(x)$ satisfy the eigenvalue equation

$$\bar{\mathcal{L}}_x \phi_n(x) = \epsilon_n \phi_n(x) \quad (3.198)$$

with eigenvalue $\epsilon_n$ and boundary conditions $\phi_n(x_L) = \phi_n(x_H) = 0$. The expression in equation (3.197) is readily verified to satisfy equation (3.182) by differentiating, term by term, in the sum and using upcoming equation (3.200). Also from Sturm–Liouville theory we have that the eigenvalue spectrum $\epsilon_n = -|\epsilon_n|$, $n = 1, 2, \ldots, \infty$, for a regular problem is real and discrete (infinitely countable) where $|\epsilon_n|$ form an increasing sequence. The corresponding eigenfunctions $\phi_n(x)$ form a complete orthonormal basis set with

$$(\phi_m, \phi_n) \equiv \int_{x_L}^{x_H} \phi_m(x)\phi_n(x)dx = \delta_{m,n}, \quad (3.199)$$

where $\delta_{m,n} = 1$ for $m = n$ and is otherwise zero. Note that completeness of the functions also gives

$$\sum_{n=1}^{\infty} \phi_n(x)\phi_n(x_0) = \delta(x - x_0), \tag{3.200}$$

so any smooth function $f(x)$ admits an eigenfunction expansion

$$f(x) = \sum_{n=1}^{\infty} a_n \phi_n(x) \tag{3.201}$$

with coefficients $a_n = (f, \phi_n)$. Assuming we have determined the eigenfunctions $\phi(x)$, the original Green's function $G(x, x_0; s)$ is then given by equations (3.197) and (3.181). Substituting this form into equation (3.174) and taking the inverse Laplace transform operation inside the summation gives a formal eigenfunction series solution representation for the kernel:

$$\begin{aligned} u(x, x_0; \tau) &= \frac{e^{\int_{x_0}^{x} \frac{\lambda(x')}{\nu(x')^2} dx'}}{\nu(x)/\nu(x_0)} \sum_{n=1}^{\infty} \phi_n(x)\phi_n(x_0) L^{-1}\left[\frac{1}{s + |\epsilon_n|}\right][\tau] \\ &= \frac{e^{\int_{x_0}^{x} \frac{\lambda(x')}{\nu(x')^2} dx'}}{\nu(x)/\nu(x_0)} \sum_{n=1}^{\infty} e^{-|\epsilon_n|\tau} \phi_n(x)\phi_n(x_0). \end{aligned} \tag{3.202}$$

Note that in the last step the Laplace transform is trivially known and one does not really need to resort to the residue theorem to compute the Laplace inverse transform. This result also follows, though, from a straightforward application of the residue theorem by closing the Bromwich contour with an infinite semicircular portion to the left and thereby picking up the contributions from the residues occurring at the simple poles of $\bar{G}$ that lie along the negative real axis.

Equation (3.202) is a generic series solution for the kernel when $\lambda(x)$, $\nu(x)$, the solution interval being considered, and the imposed boundary conditions all combined are such that one indeed has a regular Sturm–Liouville problem at hand, i.e., if it is true that the Green's function ($G$ or $\bar{G}$) has the assumed discrete eigenfunction-eigenvalue expansion. In many applications, however, the Sturm–Liouville problem of interest may *not* be of regular type but, rather, of so-called *singular* Sturm–Liouville type. This situation occurs in a variety of cases, such as when $p(x)$ in equation (3.178) attains a zero value at either solution endpoint or the functions $p$, $q$ become unbounded or the solution interval is unbounded (e.g., $x \in [0, \infty)$, $(-\infty, \infty)$, etc.). The eigenvalues may not be discrete in such cases, and the problem may have a continuous or a mixed eigenvalue spectrum, in which cases the generic formulas are generally not valid. Even in singular Sturm–Liouville problems for which the spectrum is discrete, the convergence of the eigenfunction expansions must also be examined on an individual basis. However, a substantial class of important singular Sturm–Liouville boundary value problems involving the so-called hypergeometric and confluent hypergeometric equations (such as Bessel's equation for which an in-depth Green's function development is given in the next section) can still be treated within the earlier eigenfunction formulation. This class of problems will generally admit a spectral resolution (or decomposition) of the Green's function $G$ as well as the kernel $u$ as a sum of a discrete and a continuous eigenvalue-eigenfunction portion. In closing this section, we emphasize that the complex contour integral framework of the previous section has a general applicability. In particular, it is applicable to most singular Sturm–Liouville problems of interest and can be shown to recover the spectral decomposition formulas. In fact the approach of the previous section is used in the next section to arrive at analytically closed-form kernels for the Bessel process involving Bessel's equation. The procedure for extracting the kernels analytically is then basically an advanced exercise in the application of the residue theorem of complex analysis.

## 3.7 Kernels for the Bessel Process

In this section we apply the Green's function methodoloy of Section 3.6 to the so-called Bessel process and obtain exact analytical solutions for the kernel $u(x, x_0; \tau)$ for all cases of interest: (1) no absorption (barrier free), (2) absorption at two finite endpoints (double barrier), (3) absorption at a single upper endpoint (single upper barrier), and (4) absorption at a single lower endpoint (single lower barrier).

The Bessel process is characterized by a square root volatility[5] function $\nu(x) = 2\sqrt{x}$, and drift $\lambda(x) = \lambda = \text{const.}$:

$$dx_t = \lambda\, dt + 2\sqrt{x_t}\, dW_t. \tag{3.203}$$

Moreover, throughout we consider $\lambda > 2$, where all path values are strictly positive $x_t > 0$. The allowable domain for the kernel is hence $x > 0$. The corresponding Sturm–Liouville operator in equation (3.178) has $p(x) = 2x$ and $q(x) = \mu^2/2x$, where $\mu \equiv \frac{\lambda}{2} - 1 > 0$ and equation (3.183) takes the form

$$\bar{y}''(x) + \frac{1}{x}\bar{y}'(x) - \left(\frac{s}{2x} + \frac{\mu^2}{4x^2}\right)\bar{y} = 0. \tag{3.204}$$

By a change of variable this equation leads to the modified Bessel's equation [see equation (3.374) in Appendix C to this chapter], as one can readily verify. Two linearly independent solutions to equation (3.204) are $\bar{y}_1(x) = I_\mu(\sqrt{2sx})$ and $\bar{y}_2(x) = K_\mu(\sqrt{2sx})$. Here $I_\mu$ and $K_\mu$ are the modified (i.e., hyperbolic) Bessel functions of the first and second kinds, respectively, of order $\mu > 0$. These functions are also commonly called the *Macdonald functions* (see, for example, [AS64]). For convenient reference, some common useful properties of the Bessel and modified Bessel functions, are given in this chapter's Appendix C. These functions are linearly independent for all values of $\mu$; hence linear combinations of these two solutions can be used to satisfy the appropriate boundary conditions for the Green's function $\bar{G}$ (and $G$) and hence for the kernel $u(x, x_0; \tau)$.

### 3.7.1 *The Barrier-Free Kernel: No Absorption*

Let us consider the case of homogeneous boundary conditions at the endpoints of the entire positive region $(0, \infty)$. The exact kernel is now readily obtained in analytically closed form. To begin with, the density must satisfy zero-boundary conditions

$$\lim_{x\to\infty} u(x, x_0; \tau) = \lim_{x\to 0} u(x, x_0; \tau) = 0. \tag{3.205}$$

Hence, the Green's function corresponding to equation (3.185), with $x_L \to 0$ and $x_H \to \infty$, obtains by the choice $\bar{y}_1(x; s) = I_\mu(\sqrt{2sx})$ and $\bar{y}_2(x; s) = K_\mu(\sqrt{2sx})$, since (for positive

[5] The Bessel process obeying $dx_t = \lambda dt + \upsilon_0\sqrt{x_t}dW_t$ with arbitrary nonzero constant parameter $\upsilon_0$ is obtained from equation (3.203) by making a scale change in the order and in the time: $\lambda \to \lambda/\alpha$, $t \to \alpha t$, where $\alpha \equiv \nu_0^2/4$. In particular, by simply changing $\lambda \to 4\lambda/\nu_0^2$ and $t \to \nu_0^2 t/4$, all the formulas for the Bessel process with parameter $\nu_0$ follow from those explicitly given for the process obeying equation (3.203) where $\nu_0 = 2$.

order $\mu$) $I_\mu(z) \to 0$ as $|z| \to 0$ and $K_\mu(z) \to 0$ as $|z| \to \infty$ for generally complex $z$. In particular, $I_\mu(\sqrt{2sx}) \to 0$ as $x \to 0$ and $K_\mu(\sqrt{2sx}) \to 0$ as $x \to \infty$, for any value of $s$. The Wronskian of these two functions is $\bar{W}(x) = -1/2x$, so $p\bar{W} = -1$. Combining this into equation (3.184) gives

$$\bar{G}(x, x_0; s) = \begin{cases} I_\mu(\sqrt{2sx})K_\mu(\sqrt{2sx_0}), & x \leq x_0 \\ \\ K_\mu(\sqrt{2sx})I_\mu(\sqrt{2sx_0}), & x_0 \leq x. \end{cases} \tag{3.206}$$

Note that this function has been constructed to match the zero-boundary conditions at $x = 0$ and $x = \infty$. For the Bessel process the multiplicative factor in equation (3.185) is simply $(x/x_0)^{\mu/2}$; hence equation (3.185) reduces to

$$G(x, x_0; s) = \left(\frac{x}{x_0}\right)^{\mu/2} \bar{G}(x, x_0; s), \tag{3.207}$$

giving

$$G(x, x_0; s) = \left(\frac{x}{x_0}\right)^{\mu/2} \begin{cases} I_\mu(\sqrt{2sx})K_\mu(\sqrt{2sx_0}), & 0 < x \leq x_0 \\ \\ K_\mu(\sqrt{2sx})I_\mu(\sqrt{2sx_0}), & x_0 \leq x < \infty. \end{cases} \tag{3.208}$$

Observe from equation (3.206) that the symmetry property $\bar{G}(x, x_0; s) = \bar{G}(x_0, x; s)$ is evident by interchanging $x$ with $x_0$. This is consistent with the fact that the Sturm–Liouville operator $\bar{\mathcal{L}}$ is self-adjoint. Note that this symmetry property is not true for the original Green's function $G$ in equation (3.208), as expected since the Fokker–Planck operator $\mathcal{L}$ in equation (3.169) is not self-adjoint in this case.

From the theory of Section 3.6 we know that the inverse Laplace transform (with respect to $s$) of this function will yield the density according to equation (3.174). We now proceed to compute the Bromwich integral analytically using standard techniques of complex analysis.

In proceeding further, we use a known fact that $I_\mu K_\mu$ (for all $x \leq x_0$ or $x_0 \leq x$) within $G(x, x_0; s)$ is *analytic* on the complex $s$-plane, with the exception of a (square root) branch point at $s = 0$. For this reason we need to introduce a branch cut along some branch or ray emanating from the origin of the complex $s$-plane. It is convenient to choose the principal branch cut defined by $\arg s = \pi$ along the negative real axis and to consider points on the complex $s$-plane with $|\arg s| < \pi$. We therefore extend the Bromwich contour to that of a closed contour that bypasses the branch cut, as in Figure 3.10. Note that this same contour was used earlier for the Wiener process.

The Bromwich integral in equation (3.174) corresponds to the line segment $MN$. Since $G(x, x_0; s)$ is analytic everywhere (i.e., no singularities) on and inside the entire region within the closed contour for all values of the radius $R > 0$ as well as for positive parameters $\rho$, $\delta$, and $\gamma$ taken arbitrarily close to zero, Cauchy's integral formula gives zero for the loop integral. Hence the Bromwich integral is equal to the negative of the sum of all the other contour integrals that make up the closed loop. From this fact, the kernel is then given as the negative sum of such integrals:

$$\begin{aligned} u(x, x_0; \tau) &= \left(\frac{x}{x_0}\right)^{\mu/2} \frac{1}{2\pi i} \int_{\gamma - i\infty}^{\gamma + i\infty} e^{s\tau} \bar{G}(x, x_0; s) ds \\ &= -\left(\frac{x}{x_0}\right)^{\mu/2} \frac{1}{2\pi i} \left[ \int_{C_R^+} + \int_{C_R^-} + \int_P^Q + \int_{Q'}^{P'} + \int_{C_\rho} \right] e^{s\tau} \bar{G}(x, x_0; s) ds. \end{aligned} \tag{3.209}$$

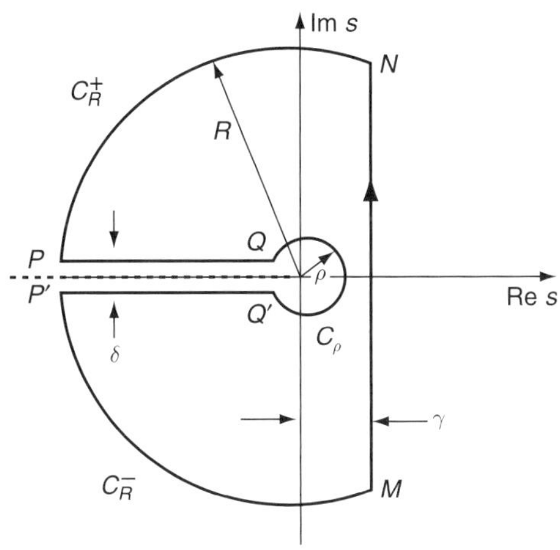


FIGURE 3.10 The closed contour integral for the Laplace inversion of the Green's function for the barrier-free case with a branch cut.

Although this integrand involves nonelementary special functions, the steps that follow actually make use of standard techniques to reduce this sum of seemingly complicated integrals to an analytically tractable form. Taking limits $R \to \infty$ and $\gamma, \rho, \delta \to 0$, it readily follows that the $C_R^{\pm}$ integrals vanish, since along the semicircular contours $s = Re^{i\theta}$, $\frac{\pi}{2} < |\theta| < \pi$, hence $|e^{s\tau}| = e^{R\tau\cos\theta} \to 0$ as $R \to \infty$ with $\cos\theta < 0$. The integrand therefore approaches zero as $R \to \infty$, since $I_\mu(z)K_\mu(z) \sim 1/2z$ as $|z| \to \infty$ from the leading-order asymptotic expansions of the modified Bessel functions. The $C_\rho$ integral for the segment $QQ'$ with $s = \rho e^{i\theta}$, $-\pi < \theta < \pi$, also vanishes as $\rho \to 0$. In particular for $x \leq x_0$,

$$\left| \int_{C_\rho} ds e^{s\tau} \bar{G}(x, x_0; s) \right| \leq \frac{\rho}{2\pi} \int_{-\pi}^{\pi} |I_\mu(\sqrt{2\rho x}e^{i\theta/2}) K_\mu(\sqrt{2\rho x_0}e^{i\theta/2})| d\theta. \tag{3.210}$$

Since $\mu > 0$, $|I_\mu(\sqrt{2\rho x}e^{i\theta/2}) K_\mu(\sqrt{2\rho x_0}e^{i\theta/2})| \to$ const. (independent of $\rho$) in the limit $\rho \to 0$. The same result applies when $x \geq x_0$; hence the $C_\rho$ integral vanishes in the limit $\rho \to 0$. The only nonzero integrals are along the branch cut corresponding to the $PQ$ and $Q'P'$ segments, where $s = re^{i\pi}$ ($\sqrt{s} = i\sqrt{r}$) and $s = re^{-i\pi}$ ($\sqrt{s} = -i\sqrt{r}$), respectively, in the limit $\delta \to 0$, with $\rho \leq r \leq R$. In the limits $\rho \to 0$ and $R \to \infty$ the two integrals are combined to give the real-valued integral

$$u(x, x_0; \tau) = \frac{1}{2}\left(\frac{x}{x_0}\right)^{\frac{\mu}{2}} \int_0^\infty e^{-r\tau} \frac{[\bar{G}(x, x_0; e^{-i\pi} r) - \bar{G}(x, x_0; e^{i\pi} r)]}{\pi i} dr, \tag{3.211}$$

where

$$\bar{G}(x, x_0; e^{-i\pi} r) = \begin{cases} I_\mu(-i\sqrt{2xr}) K_\mu(-i\sqrt{2x_0 r}), & x \leq x_0 \\ K_\mu(-i\sqrt{2xr}) I_\mu(-i\sqrt{2x_0 r}), & x_0 \leq x \end{cases} \tag{3.212}$$

and $\bar{G}(x, x_0; e^{i\pi} r)$ is given by the complex conjugate expression. We note that the integral involves the value of the branch cut discontinuity (or jump discontinuity) of the Green's function along the entire cut. This is typical of Green's functions for barrier-free kernels, as we have seen in the simpler case of the Wiener process.

The integrand in equation (3.211) is readily simplified by computing the jump discontinuity by use of the identity

$$\frac{1}{\pi i}[I_\mu(-ia)K_\mu(-ib) - I_\mu(ia)K_\mu(ib)] = J_\mu(a)J_\mu(b) \tag{3.213}$$

for any real $a$,$b$ and where $J_\mu$ are the ordinary Bessel functions of the first kind. Note that since this expression is symmetric with respect to interchanging $a$ and $b$, it follows that the integral simplifies to

$$u(x, x_0; \tau) = \frac{1}{2}\left(\frac{x}{x_0}\right)^{\mu/2} \int_0^\infty e^{-r\tau} J_\mu(\sqrt{2xr})J_\mu(\sqrt{2x_0 r})dr \tag{3.214}$$

for any $x, x_0 \geq 0$, *irrespective* of the relative magnitude of $x$ and $x_0$. This result is now simplified further by applying the integral identity (3.359) in Appendix C of this chapter with choice $\alpha = \tau$, $\beta = \sqrt{x/2}$, $\gamma = \sqrt{x_0/2}$, finally giving the known exact closed-form expression for the barrier-free kernel:

$$u(x, x_0; \tau) = \frac{1}{2}\left(\frac{x}{x_0}\right)^{\frac{\mu}{2}} \frac{e^{-(x+x_0)/2\tau}}{\tau} I_\mu(\sqrt{x x_0}/\tau). \tag{3.215}$$

## 3.7.2 The Case of Two Finite Barriers with Absorption

Here we consider homogeneous zero-boundary conditions at arbitrary finite endpoints $x_L$ and $x_H$ with $0 < x_L < x_H < \infty$ and thereby obtain the kernel, denoted by $u(x, x_0, x_L, x_H; \tau)$, for two absorbing boundary conditions (i.e., a double barrier) at finite values $x = x_L$ and $x = x_H$. In our notation we explicitly denote the dependence of $u$ on the endpoint values. The boundary conditions imposed on the kernel are

$$u(x = x_L, x_0, x_L, x_H; \tau) = u(x = x_H, x_0, x_L, x_H; \tau) = 0. \tag{3.216}$$

Hence, the Green's function corresponding to equation (3.185) obtains by the choice $\bar{y}_1(x) \equiv \bar{y}_1(x; s) = \varphi_\mu(x_L, x; s)$ and $\bar{y}_2(x) \equiv \bar{y}_2(x; s) = \varphi_\mu(x_H, x; s)$, where we have defined the function

$$\varphi_\mu(a, b; z) \equiv I_\mu(\sqrt{2az})K_\mu(\sqrt{2bz}) - K_\mu(\sqrt{2az})I_\mu(\sqrt{2bz}) \tag{3.217}$$

for generally complex $z$ and real parameters $a$, $b$. The two independent solutions are simply linear combinations of the $I_\mu$ and $K_\mu$ functions satisfying the respective zero-boundary conditions: $\bar{y}_1(x = x_L) = 0$, $\bar{y}_2(x = x_H) = 0$. In this case the Wronskian is shown to give $pW[\bar{y}_1, \bar{y}_2] = \varphi_\mu(x_H, x_L; s) = -\varphi_\mu(x_L, x_H; s)$, and hence the Green's function is given, via equation (3.185), as

$$G(x, x_0; s) = \left(\frac{x}{x_0}\right)^{\frac{\mu}{2}} \begin{cases} \dfrac{\varphi_\mu(x_L, x; s)\varphi_\mu(x_H, x_0; s)}{\varphi_\mu(x_L, x_H; s)}, & x_L \leq x \leq x_0 \\[2ex] \dfrac{\varphi_\mu(x_H, x; s)\varphi_\mu(x_L, x_0; s)}{\varphi_\mu(x_L, x_H; s)}, & x_0 \leq x \leq x_H. \end{cases} \tag{3.218}$$

In order to obtain the transition density we will invert this Green's function again with the use of a closed contour integral while taking into account all singularities of $G$ on the complex $s$-plane. First note that $s = 0$ may be a possible branch point due to the $\sqrt{s}$ argument. Since the functions $\varphi_\mu$ are analytic on the entire $s$-plane (excluding possibly the branch cut), all other singularities of $G$ are the zeros of the denominator $\varphi_\mu(x_L, x_H; s)$ in equation (3.218). From equation (3.217) and using properties of the Bessel functions, we see that the zeros must lie along the negative real axis. Indeed, putting $s = -\epsilon$ for any real $\epsilon > 0$ gives

$$\varphi_\mu(x_L, x_H; -\epsilon) = K_\mu(i\bar{x}_H)I_\mu(i\bar{x}_L) - K_\mu(i\bar{x}_L)I_\mu(i\bar{x}_H). \tag{3.219}$$

To compact notation, we have denoted the real quantities $\bar{x}_L \equiv \sqrt{2\epsilon x_L}$, $\bar{x}_H \equiv \sqrt{2\epsilon x_H}$. Using the properties $I_\mu(ix) = i^\mu J_\mu(x)$ and $K_\mu(iy) = \frac{\pi}{2}[I_{-\mu}(iy) - I_\mu(iy)]/\sin \pi\mu$ for real $x, y$ gives $I_\mu(ix)K_\mu(iy) = \frac{\pi}{2}\csc(\pi\mu)J_\mu(x)[J_{-\mu}(y) - e^{i\pi\mu}J_\mu(y)]$ for any noninteger $\mu$. Using this we obtain the identity

$$K_\mu(ix)I_\mu(iy) - K_\mu(iy)I_\mu(ix) = \frac{\pi}{2}[J_\mu(x)Y_\mu(y) - Y_\mu(x)J_\mu(y)], \tag{3.220}$$

which applies for *all* $\mu$ (integer values included), where the usual limiting procedure (i.e., analytic continuation in $\mu$) is used in the definition of the Bessel $K_\mu$ and $Y_\mu$ functions for the case of integer order $\mu$. The functions $Y_\mu$ are the ordinary Bessel functions of the second kind of order $\mu$ (see, for example, [AS64]). In contrast to the monotonic and positive hyperbolic Bessel functions for real arguments, the ordinary Bessel functions are *oscillatory*. In particular, the functions on the right-hand side of equation (3.220) involving the difference of products of ordinary Bessel functions (these are sometimes referred to as *cylinder functions*) have a countable infinite number of zeros. The zeros of the denominator of the Green's function are hence all real and negative. To simplify notation, we shall denote these zeros by $\epsilon_n \equiv \epsilon_{\mu,n}$, where it is implicitly understood that these are really the $n$th eigenvalues for given $\mu$. The equation determining these zeros (i.e., the eigenvalues of the Sturm–Liouville operator with zero-boundary conditions at two finite endpoints) is therefore $\varphi_\mu(x_L, x_H; s = \epsilon_n) = 0$; i.e., from equations (3.219) and (3.220),

$$J_\mu(\sqrt{2|\epsilon_n|x_H})Y_\mu(\sqrt{2|\epsilon_n|x_L}) - J_\mu(\sqrt{2|\epsilon_n|x_L})Y_\mu(\sqrt{2|\epsilon_n|x_H}) = 0. \tag{3.221}$$

Solving for $|\epsilon_n|$ gives the eigenvalues $\epsilon_n = -|\epsilon_n|$ for all integers $n \geq 1$. The eigenvalues form a sequence of negative values along the entire negative real axis. Note that this is entirely consistent with a regular Sturm–Liouville boundary value problem. These zeros occur in increasing order $|\epsilon_1| < |\epsilon_2| < \ldots$, and are readily obtained by standard numerical procedures.

We are now in a position to compute the Bromwich integral analytically using a similar contour integration procedure as before. However, in contrast to the barrier-free Green's function of the previous section, $G$ has isolated singularities at the zeros of the denominator at $s = -|\epsilon_n|$ along the negative real axis ($\arg s = \pi$). At all other points not lying on the branch cut and for $s \neq \epsilon_n$, $G$ is analytic, since it is a ratio of two analytic functions with nonzero denominator. Although we have freedom in the choice of branch cut, the choice of cut along the negative imaginary axis with $\arg s = \frac{3\pi}{2}$ is convenient. We therefore consider $-\frac{\pi}{2} < \arg s \leq \frac{3\pi}{2}$ and close the Bromwich contour and apply the residue theorem to the

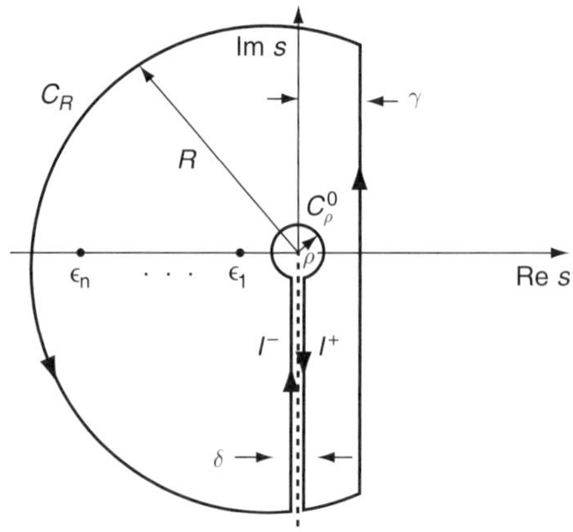


FIGURE 3.11 The contour integral for the Laplace inversion of the Green's function for the case of absorption at two finite endpoints with branch cut along the negative imaginary axis.

loop integral in Figure 3.11. Applying the Cauchy residue formula to the closed contour in Figure 3.11 gives the Laplace inverse of $G$, and hence the kernel, as

$$\begin{aligned} u(x, x_0, x_L, x_H; \tau) &= \frac{1}{2\pi i}\int_{\gamma-i\infty}^{\gamma+i\infty} e^{s\tau} G(x, x_0; s) ds \\ &= \sum_{n=1}^{\infty} e^{-|\epsilon_n|\tau} \text{Res}\, G(x, x_0; s = -|\epsilon_n|) \\ &\quad - \frac{1}{2\pi i}\left[\int_{C_R} + \int_{I^+} + \int_{I^-} + \int_{C_\rho^0}\right] e^{s\tau} G(x, x_0; s) ds, \end{aligned} \tag{3.222}$$

where the first term involves a sum over the *residues* of $G$, as a function of $s$, at all eigenvalues $s = \epsilon_n$, $n = 1, \ldots, \infty$. In this formula the limits $R \to \infty$ and $\gamma, \rho, \delta \to 0$ are implied. Taking such limits it readily follows that the semicircular $C_R$ integral, with $s = Re^{i\theta}$, $\frac{\pi}{2} < \theta < \frac{3\pi}{2}$, approaches zero. This obtains from the property of the $\varphi_\mu$ functions, which in the limit $R \to \infty$ gives $|e^{s\tau} G(x, x_0; s)| \to \kappa e^{-R\tau|\cos\theta|} e^{\beta\sqrt{R}}/\sqrt{R} \to 0$, where $\kappa$, $\beta$ are positive constants dependent on $x$, $x_0$, $x_L$, $x_H$. Then using the property $\lim_{\rho\to 0} G(x, x_0; s = \rho e^{i\theta}) \to$ const., independent of $\rho$, a similar argument as used in the previous section allows us to conclude that the $C_\rho^0$ integral approaches zero. The sum of $I^+$ and $I^-$ integrals in the limits $\rho \to 0$, $R \to \infty$ give

$$\frac{1}{2\pi}\int_0^\infty e^{-ir\tau}[G(x, x_0; re^{i\frac{3\pi}{2}}) - G(x, x_0; re^{-i\frac{\pi}{2}})]dr. \tag{3.223}$$

By completing a circuit around the origin, however, one easily proves the property

$$\varphi_\mu(a, b; e^{2\pi i} z) = \varphi_\mu(a, b; z) \tag{3.224}$$

for any complex $z \neq 0$ and any positive real $a$, $b$. This shows that there is *no* jump discontinuity in the function $\varphi_\mu$ along any choice of branch cut. Since $G$ is a function of a product and a ratio of such functions, $G$ also has no jumps. Indeed, applying the last identity with the particular choice $z = re^{-i\frac{\pi}{2}}$, equation (3.218) gives $G(x, x_0; re^{i\frac{3\pi}{2}}) = G(x, x_0; re^{-i\frac{\pi}{2}})$. Equation (3.222) hence reduces to only the sum of residues:

$$u(x, x_0, x_L, x_H; \tau) = \sum_{n=1}^{\infty} e^{-|\epsilon_n|\tau} \text{Res}\, G(x, x_0; s = -|\epsilon_n|). \tag{3.225}$$

The residues of the Green's function are evaluated analytically as follows. From the analyticity of the $\varphi_\mu$ functions we observe that every point $s = \epsilon_n$ is a simple pole of $G(x, x_0; s)$; i.e., $\varphi_\mu(x_{,0}; s)$ has simple zeros at every $s = \epsilon_n$, as will be shown. Hence we have

$$\operatorname{Res} G(x, x_0; \epsilon_n) = \left(\frac{x}{x_0}\right)^{\frac{\mu}{2}} \begin{cases} \dfrac{\varphi_\mu(x_L, x; \epsilon_n)\varphi_\mu(x_H, x_0; \epsilon_n)}{\frac{\partial \varphi_\mu(x_L, x_H; s)}{\partial s}\Big|_{s=\epsilon_n}}, & x \le x_0 \\ \dfrac{\varphi_\mu(x_H, x; \epsilon_n)\varphi_\mu(x_L, x_0; \epsilon_n)}{\frac{\partial \varphi_\mu(x_L, x_H; s)}{\partial s}\Big|_{s=\epsilon_n}}, & x_0 \le x. \end{cases} \tag{3.226}$$

To compute this residue we use $\frac{\partial \varphi_\mu(x_L, x_H; s)}{\partial s}\Big|_{s=\epsilon_n} = -\frac{\partial \varphi_\mu(x_L, x_H; -\epsilon)}{\partial \epsilon}\Big|_{\epsilon=|\epsilon_n|}$, since $\epsilon_n = -|\epsilon_n|$, and we hence consider

$$\varphi_\mu(x_L, x_H; -\epsilon) = \frac{\pi}{2}[J_\mu(\bar{x}_H)Y_\mu(\bar{x}_L) - J_\mu(\bar{x}_L)Y_\mu(\bar{x}_H)], \tag{3.227}$$

which follows from equations (3.219) and (3.220) for real $\epsilon > 0$. Differentiating this equation at $\epsilon = |\epsilon_n|$ while making use of equation (3.221) and the recurrence relations $J'_\mu(z) = (\mu/z)J_\mu(z) - J_{\mu+1}(z)$, $Y'_\mu(z) = (\mu/z)Y_\mu(z) - Y_{\mu+1}(z)$ gives

$$\begin{aligned} \frac{\partial \varphi_\lambda(x_L, x_H; -\epsilon)}{\partial \epsilon}\Bigg|_{\epsilon=|\epsilon_n|} &= \frac{\pi}{4|\epsilon_n|}\Big[\bar{x}_H[J_\mu(\bar{x}_L)Y_{\mu+1}(\bar{x}_H) - Y_\mu(\bar{x}_L)J_{\mu+1}(\bar{x}_H)] \\ &\quad + \bar{x}_L[J_{\mu+1}(\bar{x}_L)Y_\mu(\bar{x}_H) - Y_{\mu+1}(\bar{x}_L)J_\mu(\bar{x}_H)]\Big] \\ &= \frac{1}{2|\epsilon_n|}\left[\frac{Y_\mu(\bar{x}_H)}{Y_\mu(\bar{x}_L)} - \frac{Y_\mu(\bar{x}_L)}{Y_\mu(\bar{x}_H)}\right], \end{aligned} \tag{3.228}$$

where $\bar{x}_L \equiv \sqrt{2|\epsilon_n|x_L}$, $\bar{x}_H \equiv \sqrt{2|\epsilon_n|x_H}$. The last expression obtains from the identity $J_\mu(z)Y_{\mu+1}(z) - J_{\mu+1}(z)Y_\mu(z) = -2/\pi z$. Note that the expression in equation (3.228) is readily seen to be nonzero, since $x_H > x_L$ and the zeros $|\epsilon_n| \equiv |\epsilon_{\mu,n}|$ of equation (3.221) cannot also be zeros of the $Y_\mu$ functions for given order $\mu$. All poles $s = -|\epsilon_n|$ are therefore simple, justifying our assumption. Substituting the expression in equation (3.228) into equation (3.226) and again using equation (3.221) with some tedious algebraic manipulation gives the closed-form compact formula for the residue:

$$\operatorname{Res} G(x, x_0; s = -|\epsilon_n|) = \left(\frac{x}{x_0}\right)^{\frac{\mu}{2}} \phi_n(x)\phi_n(x_0), \tag{3.229}$$

where $\phi_n(x)$ are (eigenfunctions) given by

$$\phi_n(x) = \mathcal{N}_n\Big[J_\mu(\bar{x}_L)Y_\mu(\bar{x}) - Y_\mu(\bar{x}_L)J_\mu(\bar{x})\Big], \tag{3.230}$$

with normalization factor

$$\mathcal{N}_n = \pi\sqrt{\frac{|\epsilon_n|/2}{[Y_\mu(\bar{x}_L)/Y_\mu(\bar{x}_H)]^2 - 1}}. \tag{3.231}$$

Here we have used the shorthand notation $\bar{z} \equiv \sqrt{2|\epsilon_n|z}$. Note that the result is valid for all $x, x_0 > 0$ values, for it does not actually depend on the relative magnitude of $x, x_0$. As is the case in all eigenfunction expansion solutions for a Sturm–Liouville problem (for $\bar{G}$ in this case), the occurrence of $\phi_n(x)\phi_n(x_0)$ is symmetric with respect to interchanging $x$ and $x_0$. Finally, inserting this residue formula into equation (3.225) gives the kernel for the domain $x_L \leq x \leq x_H$ subject to double-ended zero-boundary conditions at $x = x_L, x_H$ as an exact closed-form eigenfunction series:

$$u(x, x_0, x_L, x_H; \tau) = \left(\frac{x}{x_0}\right)^{\frac{\mu}{2}} \sum_{n=1}^{\infty} e^{-|\epsilon_n|\tau} \phi_n(x)\phi_n(x_0). \tag{3.232}$$

From the distribution of increasing values of $|\epsilon_n|$ with $n$, as can be shown from equation (3.221), this series converges fairly rapidly for finite values of time $\tau$, particularly for large values of $\tau$ relative to the first value $|\epsilon_1|$. It is interesting to remark that the complex analysis approach to the Green's function methodology also automatically guarantees that the eigenfunctions $\phi(x)$ are normalized, since in the limit $\tau \to 0$ the density $u$ must approach the Dirac delta function $\delta(x - x_0)$. From another perspective, this result is also entirely consistent with Sturm–Liouville theory as well as spectral theory for the eigenvalue problem corresponding to the operator $\bar{\mathcal{L}}_x$ defined earlier. A direct, yet algebraically very tedious, proof of the normalization $\int_{x_L}^{x_H} \phi_m(x)\phi_n(x)dx = \delta_{mn}$ also follows by use of appropriate integral properties of products of the Bessel $J$ and $Y$ functions, as provided in this chapter's Appendix C.

## 3.7.3 *The Case of a Single Upper Finite Barrier with Absorption*

This situation corresponds to zero-boundary conditions at $x = 0$ and at a finite upper endpoint $x = x_H$, $0 < x_H < \infty$. We shall denote the kernel for this case by $u_H(x, x_0, x_H; \tau)$. The upper endpoint turns out to be an absorbing-boundary condition (at a single upper barrier). The boundary conditions imposed on the kernel are

$$u_H(x = 0, x_0, x_H; \tau) = u_H(x = x_H, x_0, x_H; \tau) = 0. \tag{3.233}$$

Hence, the Green's function corresponding to equation (3.185) obtains with choice $\bar{y}_1(x; s) = I_\mu(\sqrt{2sx})$ and $\bar{y}_2(x; s) = \varphi_\mu(x_H, x; s)$. The Wronskian of these functions gives $pW[\bar{y}_1, \bar{y}_2] = -I_\mu(\sqrt{2sx_H})$; hence the Green's function is

$$G(x, x_0; s) = \left(\frac{x}{x_0}\right)^{\frac{\mu}{2}} \begin{cases} \dfrac{I_\mu(\bar{x})[I_\mu(\bar{x}_H)K_\mu(\bar{x}_0) - K_\mu(\bar{x}_H)I_\mu(\bar{x}_0)]}{I_\mu(\bar{x}_H)}, & x \leq x_0 \\[2ex] \dfrac{I_\mu(\bar{x}_0)[I_\mu(\bar{x}_H)K_\mu(\bar{x}) - K_\mu(\bar{x}_H)I_\mu(\bar{x})]}{I_\mu(\bar{x}_H)}, & x_0 \leq x, \end{cases} \tag{3.234}$$

where we use shorthand notation $\bar{z} \equiv \sqrt{2sz}$. We can split this into a difference of two functions,

$$G(x, x_0; s) = \left(\frac{x}{x_0}\right)^{\frac{\mu}{2}} \begin{cases} I_\mu(\bar{x})K_\mu(\bar{x}_0), & x \leq x_0 \\ I_\mu(\bar{x}_0)K_\mu(\bar{x}), & x_0 \leq x \end{cases} - g^H(x, x_0; s), \tag{3.235}$$

where the first part corresponds to the barrier-free Green's function of equation (3.208) and the second part is

$$g^H(x, x_0; s) = \left(\frac{x}{x_0}\right)^{\frac{\mu}{2}} \frac{K_\mu(\sqrt{2sx_H})}{I_\mu(\sqrt{2sx_H})} I_\mu(\sqrt{2sx})I_\mu(\sqrt{2sx_0}). \tag{3.236}$$

The inverse Laplace transform is the difference of two Laplace inverses. The first part is exactly $u(x, x_0; \tau)$ of equation (3.215) for $0 \le x, x_0 \le x_H$, while the second inverse Laplace contour integral is computed using exactly the same methods as in the previous section, i.e., using the closed contour integral in Figure 3.11 since $g^H$ is analytic, except for the branch point at $s = 0$ and at simple poles along the negative real $s$-axis. The simple poles of $g^H$ are $s = -|\epsilon_n|$, where $\epsilon_n \equiv \epsilon_{\mu,n}$, $n = 1, \ldots,$ are now simply the zeros of the ordinary Bessel function

$$J_\mu(\sqrt{2|\epsilon_n| x_H}) = 0. \tag{3.237}$$

We note that the value $\epsilon_1$ is the first nonzero root.

Using the residue theorem, the Bromwich contour integral reduces to

$$L^{-1}[g^H(x, x_0; s)][\tau] = \sum_{n=1}^{\infty} e^{-|\epsilon_n|\tau} Res g^H(x, x_0; s = -|\epsilon_n|) + \frac{1}{2}\int_0^\infty e^{-r\tau} \frac{[g^H(x, x_0; re^{-i\pi}) - g^H(x, x_0; re^{i\pi})]}{\pi i} dr. \tag{3.238}$$

The branch cut discontinuity in $g^H$ is readily computed using the properties of the modified Bessel functions for purely complex arguments, namely,

$$\begin{aligned} g^H(x, x_0; re^{-i\pi}) &= \left(\frac{x}{x_0}\right)^{\frac{\mu}{2}} \frac{K_\mu(-i\sqrt{2rx_H})}{I_\mu(-i\sqrt{2rx_H})} I_\mu(-i\sqrt{2rx}) I_\mu(-i\sqrt{2rx_0}) \\ &= \left(\frac{x}{x_0}\right)^{\frac{\mu}{2}} \frac{J_\mu(\sqrt{2rx_0})}{J_\mu(\sqrt{2rx_H})} I_\mu(-i\sqrt{2rx}) K_\mu(-i\sqrt{2rx_H}), \end{aligned} \tag{3.239}$$

giving

$$\frac{[g^H(x, x_0; re^{-i\pi}) - g^H(x, x_0; re^{i\pi})]}{\pi i} = \left(\frac{x}{x_0}\right)^{\frac{\mu}{2}} J_\mu(\sqrt{2rx_0}) J_\mu(\sqrt{2rx}),$$

where the identity in equation (3.213) has been used. Inserting this expression into the integrand shows that the integral term is exactly the barrier-free kernel $u(x, x_0; \tau)$, as in equation (3.214). Taking the difference of Laplace inverses for the two terms in equation (3.235) therefore cancels out the barrier-free portion and we are left with

$$u_H(x, x_0, x_H; \tau) = -\sum_{n=1}^{\infty} e^{-|\epsilon_n|\tau} Res g^H(x, x_0; s = -|\epsilon_n|). \tag{3.240}$$

Let $g^H = (x/x_0)^{\mu/2} \bar{g}_H$; then the residues at the simple poles are given by

$$\text{Res } \bar{g}^H(x, x_0; s = -|\epsilon_n|) = \frac{K_\mu(i\sqrt{2|\epsilon_n| x_H}) I_\mu(i\sqrt{2|\epsilon_n| x}) J_\mu(\sqrt{2|\epsilon_n| x_0})}{-\frac{dJ_\mu(\sqrt{2\epsilon x_H})}{d\epsilon}\Big|_{\epsilon = |\epsilon_n|}}. \tag{3.241}$$

Upon evaluating the derivative and using the relation $I_\mu(ix)K_\mu(iy) = -\frac{\pi}{2} J_\mu(x) Y_\mu(y)$ we have

$$\text{Res } \bar{g}^H(x, x_0; s = -|\epsilon_n|) = \pi\sqrt{\frac{|\epsilon_n|}{2x_H}} \frac{Y_\mu(\sqrt{2|\epsilon_n| x_H})}{J'_\mu(\sqrt{2|\epsilon_n| x_H})} J_\mu(\sqrt{2|\epsilon_n| x}) J_\mu(\sqrt{2|\epsilon_n| x_0}). \tag{3.242}$$

This expression is simplified via the Wronskian property of the ordinary Bessel functions and by making use of equation (3.237); i.e., using

$$Y_\mu(\sqrt{2|\epsilon_n|x_H}) = \frac{\sqrt{2/|\epsilon_n|x_H}}{\pi J_{\mu+1}(\sqrt{2|\epsilon_n|x_H})} \tag{3.243}$$

we obtain

$$\text{Res } g^H(x, x_0; s = -|\epsilon_n|) = -\left(\frac{x}{x_0}\right)^{\frac{\mu}{2}} \frac{J_\mu(\sqrt{2|\epsilon_n|x})J_\mu(\sqrt{2|\epsilon_n|x_0})}{x_H J^2_{\mu+1}(\sqrt{2|\epsilon_n|x_H})}. \tag{3.244}$$

Inserting this expression into equation (3.240) finally gives an exact closed-form eigenfunction series solution for the kernel

$$u_H(x, x_0, x_H; \tau) = \left(\frac{x}{x_0}\right)^{\frac{\mu}{2}} \sum_{n=1}^{\infty} e^{-|\epsilon_n|\tau} \phi_n(x)\phi_n(x_0) \tag{3.245}$$

in terms of the normalized ordinary Bessel eigenfunctions:

$$\phi_n(x) = \frac{J_\mu(\sqrt{2|\epsilon_n|x})}{\sqrt{x_H} J_{\mu+1}(\sqrt{2|\epsilon_n|x_H})} \tag{3.246}$$

In closing this section we note that this result is also readily proven to obtain as the limit $x_L \to 0$ in the double-barrier solution $u(x, x_0, x_L, x_H; \tau)$ of the previous section. We leave it as an exercise for the interested reader.

### 3.7.4 The Case of a Single Lower Finite Barrier with Absorption

This last case corresponds to zero-boundary conditions at a lower finite endpoint $x_L \geq 0$ and at infinity with $0 \leq x_L < \infty$. The domain of the solution is the semi-infinite interval $[x_L, \infty)$. We denote the kernel by $u_L(x, x_0, x_L; \tau)$. The imposed boundary conditions are now

$$u_L(x = x_L, x_0, x_L; \tau) = u_L(x = \infty, x_0, x_L; \tau) = 0. \tag{3.247}$$

For the limiting value $x_L = 0$ the solution is simply that of the barrier-free (no absorption) problem; for $x_L > 0$, $x_L$ is a single lower absorbing barrier. The Green's function corresponding to equation (3.185) obtains with choice $\bar{y}_1(x; s) = \varphi_\mu(x_L, x; s)$ and $\bar{y}_2(x; s) = K_\mu(\sqrt{2sx})$ since $K_\mu(\sqrt{2sx}) \to 0$ as $x \to \infty$. The Wronskian of these functions gives $pW[\bar{y}_1, \bar{y}_2] = K_\mu(\sqrt{2sx_L})$; hence the Green's function is

$$G(x, x_0; s) = \left(\frac{x}{x_0}\right)^{\frac{\mu}{2}} \begin{cases} \dfrac{[K_\mu(\bar{x}_L)I_\mu(\bar{x}) - I_\mu(\bar{x}_L)K_\mu(\bar{x})]K_\mu(\bar{x}_0)}{K_\mu(\bar{x}_L)}, & x \leq x_0 \\[2ex] \dfrac{K_\mu(\bar{x})[K_\mu(\bar{x}_L)I_\mu(\bar{x}_0) - I_\mu(\bar{x}_L)K_\mu(\bar{x}_0)]}{K_\mu(\bar{x}_L)}, & x_0 \leq x. \end{cases} \tag{3.248}$$

Here again we use shorthand notation $\bar{z} \equiv \sqrt{2sz}$. Rewriting, we have

$$G(x, x_0; s) = G^0(x, x_0; s) - g^L(x, x_0; s), \tag{3.249}$$

where $g^L(x, x_0; s) \equiv (x/x_0)^{\frac{\mu}{2}} \bar{g}^L(x, x_0; s)$,

$$\bar{g}^L(x, x_0; s) = \frac{I_\mu(\sqrt{2sx_L})}{K_\mu(\sqrt{2sx_L})} K_\mu(\sqrt{2sx}) K_\mu(\sqrt{2sx_0}), \tag{3.250}$$

and $G^0$ denotes the barrier-free Green's function given by equation (3.208) for $x_L \leq x, x_0 < \infty$.

Laplace-inversion of the first term, $G^0$, gives the barrier-free contribution $u(x, x_0; \tau)$ of equation (3.215). Laplace-inversion of the $g^L$ term follows by using the same contour as in the barrier-free case, i.e., Figure 3.10. With branch cut along the negative real axis, $s = |s|e^{i\theta}$, $-\pi < \theta \leq \pi$, and the function $g^L$ is hence analytic except at the branch point $s = 0$ and cut along arg $s = \pi$. From the properties of the modified Bessel functions of the second kind, we know that $K_\mu(z)$ has no zeros in the region $|\arg z| \leq \frac{\pi}{2}$ for real $\mu$ (see [AS64]). Hence the denominator $K_\mu(\sqrt{2sx_L}) = K_\mu(\sqrt{2|s|x_L}e^{i\theta/2})$ is never zero at every point on and inside the closed contour of Figure 3.10. We therefore deduce that $\bar{g}^L(x, x_0; s)$ has no poles and is analytic on and inside the contour. Using the residue theorem and following similar steps as in the previous cases, the Bromwich contour integral reduces to give

$$L^{-1}[\bar{g}^L(x, x_0; s)][\tau] = \frac{1}{2\pi i} \int_0^\infty e^{-r\tau}[\bar{g}^L(x, x_0; re^{-i\pi}) - \bar{g}^L(x, x_0; re^{i\pi})]dr.$$

The branch cut discontinuity in $g^L$ is readily computed by making use of the properties $I_\mu(-ix) = e^{-i\pi\mu/2} J_\mu(x)$ and

$$e^{\pm i\pi\mu/2} K_\mu(ix) = \mp\frac{\pi i}{2}[J_\mu(x) \mp iY_\mu(x)]$$

for real $\mu$, $x$. After some tedious algebraic manipulation, this gives the imaginary part

$$\begin{aligned}\operatorname{Im} \bar{g}^L(x, x_0; re^{-i\pi}) &= \frac{1}{2i}[\bar{g}^L(x, x_0; re^{-i\pi}) - \bar{g}^L(x, x_0; re^{i\pi})] \\ &= \frac{\pi}{2} \frac{J_\mu(\bar{x}_L)[\phi_\mu^{(1)}(\bar{x}, \bar{x}_0) J_\mu(\bar{x}_L) + \phi_\mu^{(2)}(\bar{x}, \bar{x}_0) Y_\mu(\bar{x}_L)]}{J_\mu^2(\bar{x}_L) + Y_\mu^2(\bar{x}_L)},\end{aligned}$$

where we define new functions

$$\phi_\mu^{(1)}(x, y) \equiv J_\mu(x) J_\mu(y) - Y_\mu(x) Y_\mu(y), \tag{3.251}$$

$$\phi_\mu^{(2)}(x, y) \equiv J_\mu(x) Y_\mu(y) + J_\mu(y) Y_\mu(x) \tag{3.252}$$

with shorthand notation $\bar{z} \equiv \sqrt{2rz}$.

An exact closed-form expression for the kernel is therefore

$$u_L(x, x_0, x_L; \tau) = u(x, x_0; \tau) - \bar{u}(x, x_0, x_L; \tau), \tag{3.253}$$

where $u(x, x_0; \tau)$ is the barrier-free part as given by equation (3.215) and $\bar{u}(x, x_0, x_L; \tau)$ has the integral representation

$$\bar{u}(x, x_0, x_L; \tau) = \frac{1}{2}\left(\frac{x}{x_0}\right)^{\frac{\mu}{2}} \int_0^\infty e^{-r\tau} \frac{J_\mu(\bar{x}_L)[\phi_\mu^{(1)}(\bar{x}, \bar{x}_0) J_\mu(\bar{x}_L) + \phi_\mu^{(2)}(\bar{x}, \bar{x}_0) Y_\mu(\bar{x}_L)]}{J_\mu^2(\bar{x}_L) + Y_\mu^2(\bar{x}_L)} dr,$$

with $\bar{x}_L = \sqrt{2rx_L}$, $\bar{x} = \sqrt{2rx}$, $\bar{x}_0 = \sqrt{2rx_0}$. The zero-boundary condition at $x = x_L$ is readily verified. In particular, setting $x = x_L$ while making use of the functions $\phi_\mu^{(1)}(\bar{x}_L, \bar{x}_0)$ and $\phi_\mu^{(2)}(\bar{x}_L, \bar{x}_0)$, the integrand in our integral representation reduces to $e^{-r\tau} J_\mu(\bar{x}_L) J_\mu(\bar{x}_0)$. From equation (3.214) we arrive at $\bar{u}(x = x_L, x_0, x_L; \tau) = u(x_L, x_0; \tau)$; hence $u_L(x = x_L, x_0, x_L; \tau) = 0$.

## 3.8 New Families of Analytical Pricing Formulas: "From $x$-Space to $F$-Space"

In this section we present a mathematical framework for generating various families of exact analytical pricing kernels for nonlinear state-dependent diffusion processes. We shall refer to this construction as the *diffusion canonical transformation* methodology. The method is a reduction approach that essentially reduces the more complex state-dependent diffusion problem (i.e., the so-called $F$-space problem that we wish to solve) into a simpler underlying diffusion process (i.e., the $x$-space problem). One of the basic ideas of the approach is to consider an $x$-space diffusion process that is analytically tractable, e.g., for which Green's function methods can be used to arrive at a solution. Pricing kernels for $F$-space then arise as a result of having obtained transition kernels for an underlying $x$-space process. As seen next, the technique makes use of a special combination of transformations.

### 3.8.1 *Transformation Reduction Methodology*

Throughout we shall consider time-homogeneous drift and volatility functions having no explicit time dependence. Hence, without loss in generality we set initial time $t_0 = 0$, and in particular for the $x$-space transition probability densities we simply write $u(x, x_0; \tau)$ [or $u(x, x_0; t)$] in place of $u(x, t; x_0, t_0)$, and $U(F, F_0; \tau)$ [or $U(F, F_0; t)$] denotes the $F$-space transition density or pricing kernel. The basis of our reduction methodology arises from Lemma 3.1 and ultimately Theorem 3.1 relating fundamental solutions of the Fokker–Planck (or Kolmogorov) equation under two different stochastic processes.

Consider an underlying diffusion process with SDE

$$dx_t = \lambda(x_t)dt + \nu(x_t)dW_t, \tag{3.254}$$

where $W_t$ is a standard Wiener process. As already mentioned, the term $\nu(x)$ is the $x$-space diffusion function or (generally state-dependent) volatility function, while $\lambda(x)$ is the drift function. The $x$-space kernel $u = u(x, x_0; \tau)$ satisfies the corresponding forward and backward Kolmogorov PDE (3.169) and (3.170). In $F$-space (e.g., forward-price space) we are interested in finding pricing kernels for the corresponding SDE:

$$dF_t = \sigma(F_t)dW_t, \tag{3.255}$$

where $\sigma(F)$ is the $F$-space diffusion function or state-dependent volatility function, and $W_t$ is a standard Wiener process under some new measure. The $F$-space kernel $U(F, F_0; t)$ satisfies a new time-homogeneous forward (and backward) Kolmogorov PDE for the process described by equation (3.255). An important question that arises is: Can we develop new families of solutions $U(F, F_0; t)$, corresponding to new volatility functions $\sigma(F)$, by making use of (known) solutions $u(x, x_0; t)$? The answer is *yes*, and it is specifically contained in what follows.

**Lemma 3.1.** *Let $u = u(x, x_0; t)$ be a fundamental solution to the Fokker–Planck (forward Kolmogorov) equation for the $x$-space stochastic process*

$$\frac{\partial u}{\partial t} = \frac{1}{2}\frac{\partial^2}{\partial x^2}\left(\nu(x)^2 u\right) - \frac{\partial}{\partial x}\left(\lambda(x)u\right), \tag{3.256}$$

*with Dirac delta function initial condition $\lim_{t\to 0} u(x, x_0; t) = \delta(x - x_0)$, with appropriate boundary conditions at the endpoints of an interval that may be finite, semi-infinite, or* *infinite. Let $x = X(F)$ be the invertible transformation with invertible mapping $F = F(x)$ and having positive semidefinite derivative $dX(F)/dF = \nu(x)/\sigma(F)$ on the interval. Assume that the function defined by*

$$\alpha(x, F) = \frac{1}{2}\Big[\lambda'(x) + \frac{\lambda(x)^2}{\nu(x)^2} - 2\lambda\frac{\nu'(x)}{\nu(x)} + \frac{1}{2}\big(\sigma(F)\sigma''(F) - \nu(x)\nu''(x)\big) + \frac{1}{4}\big(\nu'(x)^2 - \sigma'(F)^2\big)\Big] \tag{3.257}$$

*is a constant $\alpha(x, F) \equiv \alpha$, independent of $x$ with $F = F(x)$, hence also independent of $F$ with $x = X(F)$. The related Fokker–Planck (forward Kolmogorov) equation in $F$-space,*

$$\frac{\partial U}{\partial t} = \frac{1}{2}\frac{\partial^2}{\partial F^2}\Big(\sigma(F)^2 U\Big), \tag{3.258}$$

*for the stochastic process defined by equation (3.255) then admits a fundamental solution $U = U(F, F_0; t)$ of the form*

$$U(F, F_0; t) = \frac{\nu(x)}{\sigma(F)}\exp\left[\alpha t + \frac{1}{2}\log\frac{\nu(x)/\sigma(F)}{\nu(x_0)/\sigma(F_0)} - \int_{x_0}^{x}\frac{\lambda(z)}{\nu(z)^2}dz\right]u(x, x_0; t), \tag{3.259}$$

*where $x = X(F)$, $x_0 = X(F_0)$, with corresponding Dirac delta function initial condition $\lim_{t\to 0} U(F, F_0; t) = \delta(F - F_0)$.*

It is important to note that an equivalent result also obtains for the case that the mapping $x = X(F)$ is assumed to be monotonically decreasing with $dX(F)/dF = -\nu(x)/\sigma(F)$, where $\nu(x)$, $\sigma(F)$ are both positive semidefinite functions. Moreover, under fairly general boundary conditions (such as homogeneous conditions) the kernels $u$ and $U$ are also solutions to the corresponding backward time Kolmogorov equations; i.e., an equivalent result of the foregoing is a statement involving the adjoint or backward time equations. Note also that boundary conditions in the $F$-space kernel can be imposed by setting appropriate boundary conditions in the $x$-space kernel via the mapping $x \to F$. In fact, by taking the simple Wiener process as underlying $x$-space process, in Section 3.5 this procedure formed the basis for deriving exact analytical pricing formulas for standard Equropean as well as various barrier options for the linear and quadratic volatility models. Under fairly general situations, unique solutions for $U$ satisfying homogeneous boundary conditions are obtained by simply matching (i.e., uniquely mapping) these homogeneous conditions in $u$. A direct proof of this lemma is contained in Appendix A of this chapter.

It is crucial to note that equation (3.257) implicitly defines a special class of invertible transformations that are used to generate our next main result. It is useful therefore to introduce a formal definition for such a variable transformation, which we shall refer to as a *diffusion canonical transformation*. One definition based on Lemma 3.1 is as follows.

**Definition 3.1.** *Let $\rho$ be an arbitrary constant, and let the (volatility) functions $\nu(x)$ and $\sigma(F)$ be positive semidefinite twice differentiable functions defined on appropriate finite, semi-infinite, or infinite domains of $x$- and $F$-spaces, respectively. Furthermore, let the function $\alpha(x, F)$ be defined by equation (3.257), where $\lambda(x)$ is a differentiable (drift) function of $x$. A diffusion canonical transformation is an invertible transformation $x = X(F)$ such that*

$$\alpha(x, F) = -\rho \quad \text{and} \quad \frac{dx}{dF} = \pm\frac{\nu(x)}{\sigma(F)}.$$

This definition now leads us to an equivalent, yet more directly useful and transparent, definition, as follows. Note that since $\nu(x)/\sigma(F(x))$ is positive (or negative) semidefinite, we can set $\nu(x)/\sigma(F(x)) = c(\psi(x))^2$, with arbitrary constant $c \neq 0$ and twice differentiable function $\psi(x)$. Differentiating w.r.t. $x$, using $F'(x) = (dx/dF)^{-1} = \sigma/\nu$ (note: without loss in generality the map is assumed either monotonically increasing or decreasing), and dividing both sides by $c\psi(x)^2$ gives

$$\nu(x)\frac{\psi'(x)}{\psi(x)} = \frac{1}{2}\big(\nu'(x) - \sigma'(F)\big). \tag{3.260}$$

Squaring gives

$$\nu(x)^2\left(\frac{\psi'(x)}{\psi(x)}\right)^2 = \frac{1}{4}[\nu'(x)^2 - 2\sigma'(F)\nu'(x) + \sigma'(F)^2], \tag{3.261}$$

and multiplying the previous equation by $\nu'(x)$ gives

$$\nu(x)\nu'(x)\frac{\psi'(x)}{\psi(x)} = \frac{1}{2}[\nu'(x)^2 - \nu'(x)\sigma'(F)]. \tag{3.262}$$

Subtracting this last equation from the previous one gives

$$\frac{1}{4}[\nu'(x)^2 - \sigma'(F)^2] = -\nu(x)^2\left(\frac{\psi'(x)}{\psi(x)}\right)^2 + \nu(x)\nu'(x)\frac{\psi'(x)}{\psi(x)}. \tag{3.263}$$

Now differentiating $\psi'(x)/\psi(x)$ using equation (3.263) and multiplying by $\nu(x)^2$ while using the previous expression, we have

$$\frac{1}{2}[\sigma(F)\sigma''(F) - \nu(x)\nu''(x)] = \nu(x)^2\left(\frac{\psi'(x)}{\psi(x)}\right)^2 - \nu(x)\nu'(x)\frac{\psi'(x)}{\psi(x)} - \nu(x)^2\left(\frac{\psi''(x)}{\psi(x)}\right). \tag{3.264}$$

Note that the left-hand side of equations (3.263) and (3.264) are contained in the expression for $\alpha(x, F)$; hence combining equations (3.263) and (3.264) into the expression for $\alpha(x, F) = -\rho$ and simplifying gives

$$-\psi''(x, \rho) + V(x, \rho)\psi(x, \rho) = 0, \tag{3.265}$$

where

$$V(x, \rho) = \frac{1}{\nu(x)^2}\left[\lambda'(x) + \frac{\lambda(x)^2}{\nu(x)^2} - 2\lambda\frac{\nu'(x)}{\nu(x)} + 2\rho\right]. \tag{3.266}$$

Here we have denoted $\psi = \psi(x, \rho)$ to stress the explicit dependence on the constant parameter $\rho$. Equation (3.265) is a homogeneous linear second-order ordinary differential equation.[6]

Based on the development directly preceding, we now present another equivalent, and more transparent and practical, definition for a diffusion canonical transformation.

[6] The reader familiar with quantum mechanics will observe that equation (3.265) is essentially related to a one-dimensional time-independent Schrodinger-like equation.

**Definition 3.2.** *Let $\rho$ be an arbitrary constant and $\nu(x)$ and $\sigma(F)$ be positive semidefinite twice differentiable (volatility) functions defined on some appropriate finite, semi-infinite, or infinite domains of x- and F-spaces, respectively. Furthermore, let $\lambda(x)$ be a differentiable (drift) function of x. A diffusion canonical transformation is an invertible variable transformation $x = X(F)$ such that*

$$\frac{dx}{dF} = \pm \frac{\nu(x)}{\sigma(F)} \tag{3.267}$$

*and*

$$\sigma(F) = \sigma_0 \frac{\nu(x)}{[\psi(x,\rho)]^2}, \tag{3.268}$$

*with arbitrary constant $\sigma_0 \neq 0$ and $\psi(x,\rho)$ satisfying equation (3.265) with $V(x,\rho)$ given by equation (3.266). The inverse transformation $F = F(x)$ follows from $F'(x) = \pm\sigma(F(x))/\nu(x) = \pm\sigma_0/[\psi(x,\rho)]^2$, and integrating gives*

$$F(x) = \bar{F} \pm \sigma_0 \int_{\bar{x}}^{x} \frac{dz}{[\psi(z,\rho)]^2}, \tag{3.269}$$

*with $\bar{F} = F(\bar{x})$ and $\bar{x}$ as an arbitrary constant. The $\pm$ factor allows for two possible branches of either monotonically increasing or decreasing maps.*

In the analysis that follows throughout the rest of this section it is convenient to work with a slightly modified version of $\psi$, by defining

$$\hat{u}(x,\rho) \equiv \psi(x,\rho) \exp\left( -\int \frac{\lambda(x)}{\nu(x)^2} dx \right). \tag{3.270}$$

The integral here is left as indefinite since any choice of definite integration would simply lead to an overall multiplicative factor. From equation (3.268) we therefore conclude that a diffusion canonical transformation is one that relates the two volatility functions via the (generally implicit) relationship

$$\sigma(F) = \frac{\sigma_0 \nu(x) \exp\left( -2\int \frac{\lambda(x)}{\nu(x)^2} dx \right)}{[\hat{u}(x,\rho)]^2}, \tag{3.271}$$

with $x = X(F)$ and where $\hat{u} = \hat{u}(x,\rho)$ is readily shown to satisfy

$$\frac{1}{2}\nu(x)^2 \frac{d^2}{dx^2}\hat{u} + \lambda(x)\frac{d}{dx}\hat{u} - \rho\hat{u} = 0. \tag{3.272}$$

Indeed equation (3.272) follows by direct differentiation and substitution of equation (3.270) into equation (3.265). As we will see, equation (3.271) is rather central to the whole transformation methodology. Equation (3.272) actually turns out to be the homogeneous adjoint equation for the corresponding $x$-space time independent Green's function discussed in Section 3.6, i.e., the homogeneous version of equation (3.173), with $\delta(x - x_0)$ replaced by zero and Laplace transform variable $s = \rho$. A set of two linearly independent solutions for $\hat{u}$ follow immediately from the Green's function, as shown in Section 3.6. Using equations (3.269) and (3.270), the mapping $F = F(x)$ can now also be rewritten explicitly in terms of $\hat{u}$:

$$F(x) = \bar{F} \pm \sigma_0 \int_{\bar{x}}^{x} \frac{e^{-2\int \frac{\lambda(z)}{\nu(z)^2} dz}}{[\hat{u}(z,\rho)]^2} dz, \tag{3.273}$$

with inverse $x = X(F)$ given (generally implicitly) by inverting this relation using either branch (+ sign branch for monotonically increasing or − sign branch for monotonically decreasing).

Using these equations we now summarize the main result into the following important main theorem, which follows as a direct consequence of the preceding lemma.

**Theorem 3.1. (Reduction-Mapping for Pricing Kernels)** *Given an x-space process satisfying equation (3.254), with transition probability function $u(x, x_0; t)$ as fundamental solution to the corresponding Kolmogorov (forward or backward) equation, and an F-space process described by equation (3.255), with transition probability function $U(F, F_0; t)$ as fundamental solution to the corresponding (forward or backward) Kolmogorov equation, the fundamental solutions are related as follows:*

$$U(F, F_0; t) = \frac{\nu(x)}{\sigma(F)} \frac{\hat{u}(x, \rho)}{\hat{u}(x_0, \rho)} e^{-\rho t} u(x, x_0; t), \tag{3.274}$$

*where $x = X(F)$, $x_0 = X(F_0)$ are (implicitly) given by the diffusion canonical invertible variable transformation defined by equation (3.271), or (3.273), and $\hat{u}(x, \rho)$ solves equation (3.272), with $X'(F) = \pm\nu(X(F))/\sigma(F)$.*

**Proof.** One way to verify this is to show that $U$ in equation (3.274) solves equation (3.258) by changing derivatives w.r.t. $F$ to derivatives w.r.t. $x$ with repeated use of the chain rule and using the fact that $u$ satisfies equation (3.256). Although straightforward, this process is tedious. A simpler proof follows directly from the foregoing lemma. Indeed letting $\alpha(x, F) = -\rho$ in equation (3.257) gives the map $x = X(F)$ defined by equations (3.271) and (3.272), as shown earlier. Hence $\alpha = -\rho$ in equation (3.259). Moreover, using equation (3.271) we have

$$\exp\left[\frac{1}{2}\log\frac{\nu(x)/\sigma(F)}{\nu(x_0)/\sigma(F_0)}\right] = \left(\frac{\nu(x)}{\sigma(F)}\right)^{\frac{1}{2}}\left(\frac{\nu(x_0)}{\sigma(F_0)}\right)^{-\frac{1}{2}}$$
$$= \frac{\hat{u}(x, \rho)}{\hat{u}(x_0, \rho)} \exp\left(\int_{x_0}^{x} \frac{\lambda(z)}{\nu(z)^2} dz\right). \tag{3.275}$$

Substituting directly into equation (3.259) eliminates the exponential term, giving equation (3.274), where we assume $\hat{u}(x, \rho)$ is either positive or negative semidefinite. Note also that generally, and without loss in generality, the ratio of the volatility functions $\nu(x)/\sigma(F)$ is assumed to be positive definite; i.e., both volatility functions can be positive or negative semidefinite. Otherwise, one simply takes the absolute value of the Jacobian of the transformation. □

It is important to point out the basic structure of equation (3.274) and how this relates to the asset pricing theory of Chapter 1. That is, the $F$-space transition density $U$ is related to the $x$-space transition density by a combination of two terms. The first factor, $\nu(x)/\sigma(F)$, is simply the Jacobian resulting from the assumed variable transformation $x \to F$. Within the framework of stochastic differentials, equivalent martingale measures, and the continuous-time asset-pricing theorem discussed in Chapter 1, the second term can now actually be identified as a ratio of two numeraires $g_t/g_0$, where the numeraire at time $t$ is $g_t \equiv e^{\rho t}/\hat{u}(x_t, \rho)$ and the $x$-space process at time $t$ denoted by $x_t$ has value $x$ at time $t$ and value $x_0$ at time zero. Recall from Chapter 1 that a transition density corresponds to the current price of an infinitely narrow butterfly spread pay-off (i.e. a delta function pay-off). Hence by assuming

$g_t$ as numeraire, the asset-pricing formula (1.292) allows us to also rewrite equation (3.274) as the conditional expectation at time zero of the delta function pay-off:

$$U(F, F_0; t) = E_0^{Q(g)}\left[\frac{g_0}{g_t}\delta(F(x_t) - F)\right]. \tag{3.276}$$

Note that process $F_t$ is generated from the underlying process $x_t$ via the mapping $F_t = F(x_t)$. For an alternative and instructive "proof" of Theorem 3.1 as it relates to pricing measures, see Appendix B of this chapter.

In summary, the foregoing reduction methodology provides exact analytical relationships among transition probability densities describing continuous diffusion under classes of different stochastic processes (i.e., $x$-space and $F$-space) with different state-dependent volatility and drift functions. Note that throughout we present the theory with the assumptions of no explicit time dependence for all drift and volatility functions; furthermore it is assumed that the drift function multiplying the $dt$ term in the SDE of the $F_t$ processes in $F$-space is zero. [It should be noted, however, that generally this does *not* necessarily imply that $F_t$ is a driftless (i.e., martingale) process in cases of nonlinear volatility functions $\sigma(F)$.] Extensions that further relax some of these assumptions are possible; however, these are not discussed here. As we will show, this result provides the main tool for generating a substantial number of new families of exactly solvable diffusions and hence for obtaining new pricing kernels under multiparameter volatility functions. The fact that $\sigma(F)$ involves multiple parameters can generally be seen from equation (3.271), wherein $\sigma_0$ and $\rho$ are two obvious parameters, while all other parameters can arise from the underlying $x$-space drift and volatility functions $\lambda(x)$ and $\nu(x)$, respectively. As is shown later, two other adjustable parameters arise if one considers arbitrary linear combinations of two linearly independent solutions to equation (3.272). That is, equation (3.272) admits a family of solutions; and since we are at liberty to choose any particular solution, every choice gives us a particular volatility function in $F$-space.

It is now apparent that if an $F_t$ process can be mapped onto an $x_t$ process (in the "diffusion canonical" sense), then solutions for $F$-space transition probability densities (i.e., pricing kernels) can be obtained by solving the $x$-space diffusion problem with appropriately imposed boundary conditions. Consequently, the functions $\hat{u}$ that solve equation (3.272) are the basic building blocks for ultimately deriving the pricing kernels $U(F, F_0; t)$ and hence for constructing solutions for the $F$-space processes. As described earlier, this arises simply from application of the theory of time-dependent and time-independent Green's functions to the underlying $x$-space diffusion problem. For this reason, we also refer to such a function $\hat{u}$ as a *generating function*. By solving $u(x, x_0; t)$ subject to a judicious choice of boundary conditions in $x$-space, one therefore generates the pricing kernel $U(F, F_0; t)$ via equation (3.274) while satisfying required boundary conditions in $F$-space via the inverse transformation $F = F(x)$. The analytical properties of $U$, such as nonnegativity, integrability, and probability conservation, depend upon the $x$-space drift and volatility functions and the choice of $\rho$.

## 3.8.2 Bessel Families of State-Dependent Volatility Models

Based on the exact analysis of a nontrivial underlying $x$-space process and the foregoing mapping reduction method, we are now ready to develop new families of analytically exact pricing kernels for multiparameter classes of diffusion models. In particular, we shall make use of the solutions to the Bessel process obtained in Section 3.7 and arrive at a new family of pricing kernels with corresponding volatility models that can be expressed in terms of the modified Bessel functions. We shall refer to these new models and solution kernels as the *Bessel family* of volatilities and pricing kernels.

The results follow from a straightforward application of equation (3.274) of Theorem 3.1 starting from the exact form of the generating function $\hat{u}(x, \rho)$ in the case that the underlying $x$-space process has volatility function $\nu(x) = 2\sqrt{x}$ and drift $\lambda(x) = \lambda$, $x \in (0, \infty)$. From the discussion in Section 3.7, and in particular from equation (3.186), $\hat{u}(x, \rho)$ obtains from the general solution to the modified Bessel differential equation (3.204), for $s = \rho$. Explicitly, equation (3.186) with $\bar{y}_1(x, \rho) = I_\mu(\sqrt{2\rho x})$ and $\bar{y}_2(x; \rho) = K_\mu(\sqrt{2\rho x})$, as strictly increasing and decreasing nonnegative functions for $\rho > 0$, $\mu \equiv \frac{\lambda}{2} - 1 > 0$, gives

$$\hat{u}(x, \rho) = x^{-\mu/2}[q_1 I_\mu(\sqrt{2\rho x}) + q_2 K_\mu(\sqrt{2\rho x})]. \tag{3.277}$$

Throughout we shall assume the family of solutions with $q_1$, $q_2$ as real constants and $\rho > 0$ such that $\hat{u}$ is nonnegative. In this case the map $x = X(F)$ [and its inverse $F = F(x)$] is strictly monotonic on the entire half-line $x \in [0, \infty)$. [Note: For $\rho < 0$, the general form for the generating function is expressible in terms of ordinary Bessel functions: $\hat{u}(x, \rho) = x^{-\mu/2}\big(q_1 J_\mu(\sqrt{-2\rho x}) + q_2 Y_\mu(\sqrt{-2\rho x})\big)$. In this case, however, invertible maps exist only on finite piecewise segments along the half-line $x \geq 0$ since the $J_\mu$, $Y_\mu$ functions are oscillatory and have multiple zeros.] Substituting $\hat{u}(x, \rho)$ from equation (3.277) into equation (3.273) and applying a change of integration variable gives

$$F(x) = \bar{F} + 2\sigma_0 \int_{z=\sqrt{2\rho\bar{x}}}^{z=\sqrt{2\rho x}} \frac{dz}{z[q_1 I_\mu(z) + q_2 K_\mu(z)]^2}, \tag{3.278}$$

with constant value $\bar{x}$ mapping into $F(\bar{x}) = \bar{F}$, an arbitrary real constant. Here we have used the + branch of equation (3.273) while a similar result follows for the – branch. This integral leads to two dual families of exact analytical expressions for the transformation $F = F(x)$. This follows directly with the use of the identity

$$\frac{d}{dz}\left(\frac{(1/q_2) I_\mu(z)}{q_1 I_\mu(z) + q_2 K_\mu(z)}\right) = \frac{1}{z[q_1 I_\mu(z) + q_2 K_\mu(z)]^2}, \tag{3.279}$$

in the case of the first family, and with the use of

$$\frac{d}{dz}\left(\frac{-(1/q_1) K_\mu(z)}{q_1 I_\mu(z) + q_2 K_\mu(z)}\right) = \frac{1}{z[q_1 I_\mu(z) + q_2 K_\mu(z)]^2}, \tag{3.280}$$

in the case of the second family. These general identities follow from the Wronskian relation $I_\mu(z)K'_\mu(z) - K_\mu(z)I'_\mu(z) = -1/z$. Using equation (3.279) gives

$$F(x) = c_1 + \frac{2\sigma_0/q_1 q_2}{1 + (q_2/q_1)K_\mu(\sqrt{2\rho x})/I_\mu(\sqrt{2\rho x})}, \tag{3.281}$$

with $q_2 \neq 0$, while use of equation (3.280) gives

$$F(x) = c_2 - \frac{2\sigma_0/q_1 q_2}{1 + (q_1/q_2)I_\mu(\sqrt{2\rho x})/K_\mu(\sqrt{2\rho x})}, \tag{3.282}$$

with $q_1 \neq 0$, for the first and second families, respectively. Here the constants $c_1$, $c_2$ are given by $c_1 = \bar{F} - (2\sigma_0/q_1 q_2)/[1 + (q_2/q_1)K_\mu(\sqrt{2\rho\bar{x}})/I_\mu(\sqrt{2\rho\bar{x}})]$ and $c_2 = \bar{F} + (2\sigma_0/q_1 q_2)/[1 + (q_1/q_2)I_\mu(\sqrt{2\rho\bar{x}})/K_\mu(\sqrt{2\rho\bar{x}})]$. These constants are fixed by setting $\bar{x}$. For the first family, for example, by setting $\bar{x} = 0$ we have $F(0) = \bar{F}$ (i.e., $x = 0$ maps onto $F = \bar{F}$). In the limit

$\bar{x} \to 0$, $K_\mu(\sqrt{2\rho\bar{x}})/I_\mu(\sqrt{2\rho\bar{x}}) \to \infty$, hence $c_1 = \bar{F}$. For the second family, we can choose $\bar{x} = \infty$, giving $c_2 = \bar{F}$, where $x = \infty$ maps onto $\bar{F}$. For the first family, we then have

$$F(x) = \bar{F} + \frac{2\sigma_0/q_1 q_2}{1 + (q_2/q_1)K_\mu(\sqrt{2\rho x})/I_\mu(\sqrt{2\rho x})}, \tag{3.283}$$

with $q_2 \neq 0$. By considering the asymptotic limits $I_\mu(\sqrt{2\rho x})/K_\mu(\sqrt{2\rho x}) \to 0$ as $x \to 0$ and $I_\mu(\sqrt{2\rho x})/K_\mu(\sqrt{2\rho x}) \to \infty$ as $x \to \infty$, we observe that the interval $x \in (0, \infty)$ maps one to one onto $F \in (\bar{F}, \bar{F} + 2\sigma_0/q_1 q_2)$ in this first family. Letting $\bar{x} \to \infty$ in the second family gives an alternate map as

$$F(x) = \bar{F} - \frac{2\sigma_0/q_1 q_2}{1 + (q_1/q_2)I_\mu(\sqrt{2\rho x})/K_\mu(\sqrt{2\rho x})}, \tag{3.284}$$

where $x \in (0, \infty)$ now maps one to one onto $F \in (\bar{F} - 2\sigma_0/q_1 q_2, \bar{F})$.

Applying the foregoing theorem, the volatility for the $F_t$ process is hence given by equation (3.271), which upon inserting the generating function in equation (3.277) gives

$$\sigma(F) = \frac{2\sigma_0}{\sqrt{X(F)}\left[q_1 I_\mu(\sqrt{2\rho X(F)}) + q_2 K_\mu(\sqrt{2\rho X(F)})\right]^2}. \tag{3.285}$$

Solving for $x = X(F)$ using either equation (3.283) or equation (3.284) and inserting into equation (3.285), we observe that the $F$-space volatility function generally involves as many as six adjustable parameters: $\sigma_0$, $\mu$, $\rho$, $q_1$, $q_2$, and $\bar{F}$. It can be seen from the transformations, however, that the effective number of independent parameters reduces to five: $\sigma_0/q_1 q_2$, $q_2/q_1$, $\mu$, $\rho$, $\bar{F}$.

Further properties of these variable transformations lead to other useful subfamilies of volatility models, as follows. As can be seen directly from equation (3.278), the function $F = F(x)$ is monotonically increasing, assuming $\sigma_0 > 0$. By considering the limit $q_1 \to 0$ (for fixed nonzero $q_2$), the first family reduces to a four-parameter subfamily,

$$F(x) = \bar{F} + a\frac{I_\mu(\sqrt{2\rho x})}{K_\mu(\sqrt{2\rho x})}, \tag{3.286}$$

where $x \in (0, \infty)$ maps onto $F \in (\bar{F}, \infty)$ (for constant $a \equiv 2\sigma_0/q_2^2 > 0$), with volatility function

$$\sigma(F) = \frac{a}{\sqrt{X(F)}K_\mu^2(\sqrt{2\rho X(F)})}. \tag{3.287}$$

Similarly, by considering the limit $q_2 \to 0$ (for fixed nonzero $q_1$), the second family, with $F(x)$ defined by equation (3.284), admits another (dual) four-parameter subfamily of solutions with

$$F(x) = \bar{F} - a\frac{K_\mu(\sqrt{2\rho x})}{I_\mu(\sqrt{2\rho x})}, \tag{3.288}$$

where $x \in (0, \infty)$ maps onto $F \in (-\infty, \bar{F})$ (for $a \equiv 2\sigma_0/q_1^2 > 0$), with volatility function

$$\sigma(F) = \frac{a}{\sqrt{X(F)}I_\mu^2(\sqrt{2\rho X(F)})}. \tag{3.289}$$

For option-pricing purposes, it is useful to consider the related family [obtained via the – branch of equation (3.278) and using equation (3.280) for $q_2 = 0$] that maps $x \in (0, \infty)$ onto $F \in [\bar{F}, \infty)$ (for $a \equiv 2\sigma_0/q_1^2 > 0$):

$$F(x) = \bar{F} + a\frac{K_\mu(\sqrt{2\rho x})}{I_\mu(\sqrt{2\rho x})}. \tag{3.290}$$

This family has the same volatility function (3.289) and defines a strictly monotonically decreasing function $F = F(x)$. Indeed by differentiating equation (3.290) w.r.t. $x$ and using the Wronskian property, we find the derivative

$$F'(x) = -\frac{a/2}{xI_\mu^2(\sqrt{2\rho x})}. \tag{3.291}$$

By combining the generating function (3.277), the $F$-space volatility function (3.285), and the $x$-space volatility function $\nu(x) = 2\sqrt{x}$ into our main equation (3.274) of Theorem 3.1, we obtain the relationship between a pricing kernel $U$ for the general (dual) six-parameter Bessel family and a kernel $u$ for the Bessel process:

$$U(F, F_0; t) = \frac{x^{1-\frac{\mu}{2}}[q_1 I_\mu(\sqrt{2\rho x}) + q_2 K_\mu(\sqrt{2\rho x})]^3}{\sigma_0 x_0^{-\frac{\mu}{2}}[q_1 I_\mu(\sqrt{2\rho x_0}) + q_2 K_\mu(\sqrt{2\rho x_0})]} e^{-\rho t} u(x, x_0; t), \tag{3.292}$$

where $x = X(F)$ and $x_0 = X(F_0)$ are given by inverting either equation (3.283) for the first family or equation (3.284) for the second family of solutions. Here $u(x, x_0; t)$ is an $x$-space kernel for the Bessel process, as given in Section 3.7. The particular solution used for $u$ depends on what set of boundary conditions we require $U$ to satisfy. For instance, one uses either the kernel in equation (3.215), (3.232), (3.245), or (3.253), depending on the specific boundary conditions one wishes to impose. We point out that among the general possible Bessel families of pricing kernels given by equation (3.292), only a subclass of solutions with $q_2 = 0$ can provide pricing kernels with no absorption in $F$-space. This important class of solutions is discussed in detail in the next section. For a technical discussion concerning the general question of determining whether or not a given kernel represents a transition density that conserves probability over a solution domain (i.e., whether or not absorption occurs), see Section 3.8.4. It turns out that for nonzero $q_2$ the kernel $U$ in equation (3.292) always gives rise to probability leakage or absorption at an $F$-space endpoint, even in the case where equation (3.215) is used for the $x$-space kernel $u$. Partly because of this property and the added flexibility of the parameter space, the full six-parameter Bessel model is a good candidate for modeling credit-rating migration and default risk and for pricing under a credit setting [ACCZ03].

### 3.8.3 The Four-Parameter Subfamily of Bessel Models

In this section, we specialize the general Bessel family of solutions and consider a subfamily of models containing up to four parameters. The pricing of standard European-style options is considered under this model. Moreover, we show that special cases of this four-parameter Bessel subfamily correspond to other known exact solutions in the literature, such as the CEV (constant-elasticity-of-variance), quadratic, and affine volatility models.

In particular, let us consider the model mentioned in the previous section, with zero-drift function and state-dependent volatility function $\sigma(F)$ given by equation (3.289) and where the inverse map $x = X(F)$ is defined uniquely (and generally implicitly) by inverting equation (3.290). As previously seen, this family is obtained by a one-to-one monotonically decreasing map of the underlying Bessel process space $x \in (0, \infty)$ onto the (asset price) space $F \in (\bar{F}, \infty)$. Figure 3.12 illustrates this map for a particular choice of model parameters. Figure 3.13 gives an illustration of some of the typical local volatility plots obtained within this family of models. In this family, $a$ and $\rho$ are positive parameters, $\bar{F}$ is an arbitrary parameter because it corresponds to a lower bound of the $F_t$ process, and $\mu \equiv \frac{\lambda}{2} - 1 > 0$ since $\lambda > 2$ is chosen so as to guarantee probability conservation for the pricing kernel $U(F, F_0; t)$ in the case of unrestricted barrier-free motion, with process $F_t$ attaining any value in $(\bar{F}, \infty)$.

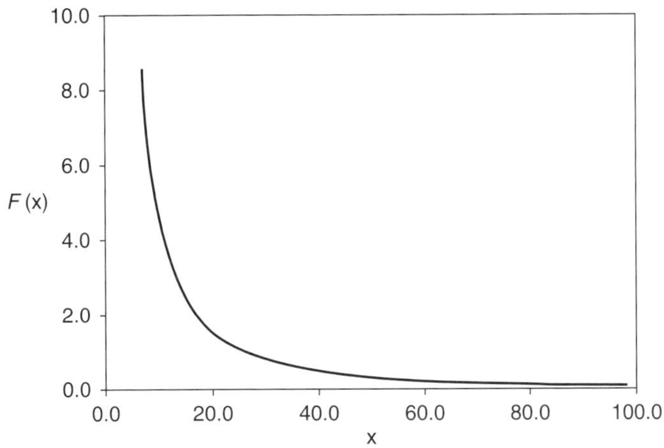


FIGURE 3.12 Plot of $F = F(x)$ using equation (3.290) for $a = 0.1$, $\rho = 0.01$, $\mu = 1.5$, $\bar{F} = 0$.

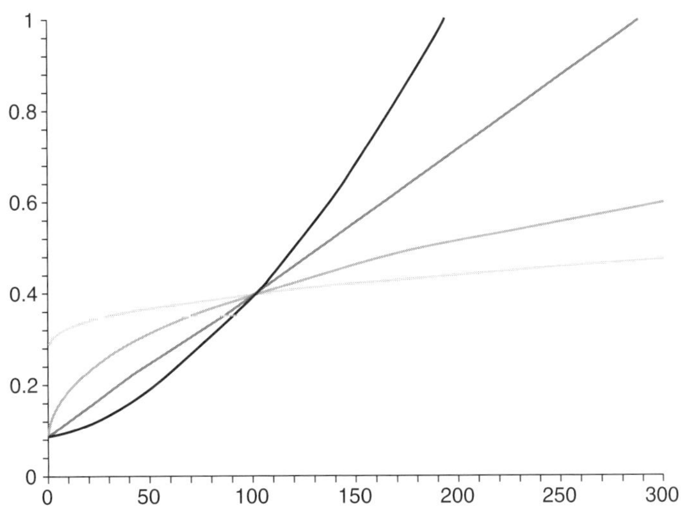


FIGURE 3.13 Local volatility plots of $\sigma(F)/F$ versus $F$, for four sets of choices of model parameters: $(\rho, a, \mu)$ = (0.001, 16, 0.25), (0.001, 9, 0.5), (0.001, 1.7, 1.25), (0.01, 150, 1.25). These choices correspond to most rapidly increasing to least increasing with fixed local volatility at $F = 100$ and the choice $\bar{F} = 0$.

In this case, $q_2 = 0$, and formula (3.292) reduces to

$$U(F, F_0; t) = \frac{2}{a}\frac{x^{1-\frac{\mu}{2}}}{x_0^{-\frac{\mu}{2}}}\frac{I_\mu^3(\sqrt{2\rho x})}{I_\mu(\sqrt{2\rho x_0})}e^{-\rho t}u(x, x_0; t), \tag{3.293}$$

with $x = X(F)$, $x_0 = X(F_0)$ via equation (3.290) and where $u(x, x_0; t)$ is an $x$-space kernel for the Bessel process, as given in the previous sections. This formula hence provides a general link between a pricing kernel for the underlying Bessel process and that for the four-parameter Bessel family. As such it can be used to generate exact analytical pricing kernels for the case of barriers (which are useful for pricing barrier options analytically under the four-parameter Bessel model), or we can simply use it to generate barrier-free pricing kernels.

In this section we focus on the case of barrier-free solutions. Specifically, by inserting equation (3.215) into equation (3.293) we obtain the barrier-free analytical pricing kernel for the four-parameter family in terms of the modified Bessel function of the first kind:

$$U(F, F_0; t) = \frac{e^{-\rho t-(X(F)+X(F_0))/2t}}{at}\frac{X(F)I_\mu^3(\sqrt{2\rho X(F)})}{I_\mu(\sqrt{2\rho X(F_0)})}I_\mu\left(\frac{\sqrt{X(F)X(F_0)}}{t}\right). \tag{3.294}$$

Typical densities are shown in Figure 3.14. As can be observed for the particular choice of model parameters, the densities are significantly skewed, particularly for larger values of time $t$. This pronounced tail feature becomes apparent when comparing the cumulative densities of a four-parameter model with that of the lognormal model while choosing model parameters such that the two transition densities have similar spreads about the spot $F_0$ (i.e., the local volatility at $F_0$ is set to the lognormal volatility). Figure 3.15 gives a relative comparison of the cumulative densities. Note: Given a transition probability density $U(F, F_0; t)$, the cumulative density is defined in the usual manner by $\Phi(F, F_0, t) \equiv \int_{\bar{F}}^{F} U(F', F_0; t)dF'$.

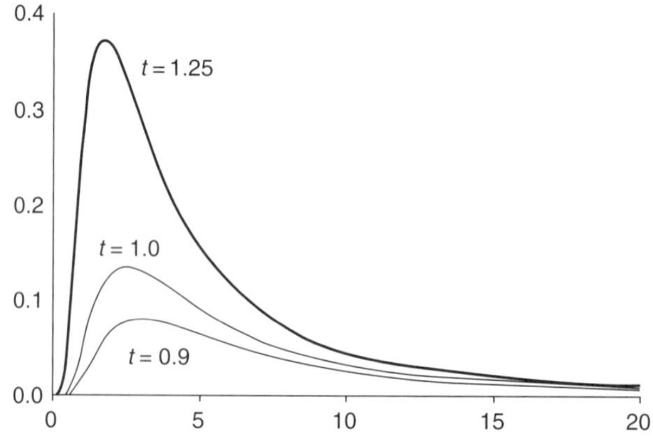


FIGURE 3.14 Plots of the transition probability density (3.294) for $a = 0.1$, $\rho = 0.01$, $\mu = 1.5$, $\bar{F} = 0$, $F_0 = 14.15$.

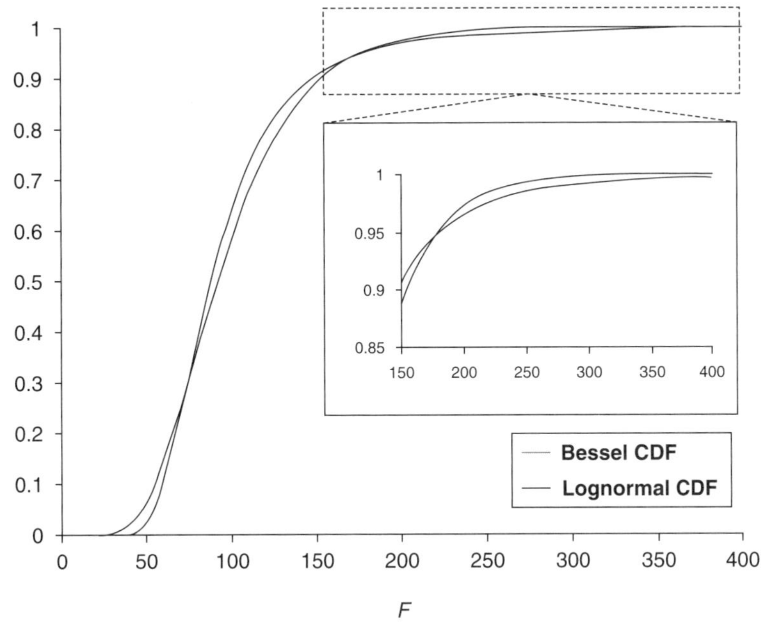


FIGURE 3.15 A relative comparison of cumulative density functions for that of a lognormal transition density (linear model) versus that for a typical four-parameter Bessel family kernel. The parameters are chosen so that the local volatility $\frac{\sigma(F_0)}{F_0}$, at spot $F_0 = 100$, equals the lognormal volatility parameter.

For option-pricing purposes it is useful to consider a change of variables $F \to x$ while using equation (3.291):

$$\left|\frac{dF}{dx}\right| U(F(x), F_0; t) = \frac{e^{-\rho t-(x+x_0)/2t}}{2tI_\mu(\sqrt{2\rho x_0})} I_\mu(\sqrt{2\rho x}) I_\mu\left(\frac{\sqrt{xx_0}}{t}\right), \tag{3.295}$$

$x_0 = X(F_0)$. As function of $x$, this form is now simply a product of two Bessel functions times a decaying exponential factor. Integral identities for such functions are now useful. For instance, using property (3.258), it is easy to verify that the density given by equation (3.294) conserves probability over the allowable path space $F_t \in (\bar{F}, \infty)$:

$$\begin{aligned}\int_{\bar{F}}^{\infty} U(F, F_0; t)dF &= \int_0^{\infty} \left|\frac{dF}{dx}\right| U(F(x), F_0; t)dx \\ &= \frac{e^{-\rho t - x_0/2t}}{2tI_\mu(\sqrt{2\rho x_0})} \int_0^{\infty} e^{-x/2t} I_\mu(\sqrt{2\rho x}) I_\mu\left(\frac{\sqrt{xx_0}}{t}\right) dx \\ &= 1\end{aligned} \tag{3.296}$$

A European-style option with assumed payoff $\Lambda(F)$, given a time to maturity $t$, can then be priced as an expectation integral (ignoring a discount factor throughout):

$$\begin{aligned}V(F_0, t) &= \int_{\bar{F}}^{\infty} U(F, F_0; t)\Lambda(F)dF \\ &= \frac{e^{-\rho t - X(F_0)/2t}}{2tI_\mu(\sqrt{2\rho X(F_0)})} \int_0^{\infty} e^{-x/2t} I_\mu(\sqrt{2\rho x}) I_\mu\left(\frac{\sqrt{xX(F_0)}}{t}\right) \Lambda(F(x))dx.\end{aligned}$$

Notice that expectation integrals are more readily computed by expressing the pay-off in terms of the $x$ variable. In this manner the implicit inversion step from $x$ to $F$ is mainly avoided. A European call written on the (forward) price $F_0$, maturing in time $t$, strike $K \geq \bar{F}$, with payoff $\Lambda(F) = (F - K)_+$ can be priced exactly in terms of Bessel integrals:

$$C(F_0, K, t) = \frac{e^{-\rho t - X(F_0)/2t}}{2tI_\mu(\sqrt{2\rho X(F_0)})}\left[(\bar{F} - K)f^{(1)} + af^{(2)}\right], \tag{3.297}$$

where

$$f^{(1)} \equiv f^{(1)}(F_0, K, t) = \int_0^{X(K)} e^{-x/2t} I_\mu(\sqrt{2\rho x}) I_\mu\left(\frac{\sqrt{xX(F_0)}}{t}\right) dx, \tag{3.298}$$

$$f^{(2)} \equiv f^{(2)}(F_0, K, t) = \int_0^{X(K)} e^{-x/2t} K_\mu(\sqrt{2\rho x}) I_\mu\left(\frac{\sqrt{xX(F_0)}}{t}\right) dx. \tag{3.299}$$

Equation (3.297) is derived by using equation (3.290) within the call pay-off of the expectation integral. The corresponding put option price can be derived in similar fashion (see Problem 3). These integrals are efficiently computed by numerical routines. Figure 3.16 displays some exact numerical call prices by application of equation (3.297).

### 3.8.3.1 Recovering the Constant-Elasticity-of-Variance Model

One way to recover the constant-elasticity-of-variance (CEV) model is to consider the limiting case where $\rho \to 0$ within the foregoing four-parameter Bessel family. For this purpose it is convenient to define a parameter $\theta > 0$ such that $\mu = (2\theta)^{-1}$, i.e., $\lambda = \theta^{-1} + 2$. Using the leading-order small-argument properties of the modified Bessel $I_\mu$ and $K_\mu$ functions with positive order $\mu$, we have the limiting form of the map, equation (3.290), as $\rho \to 0$:

$$\begin{aligned} F(x) &\sim \bar{F} + a\frac{2^{\mu-1}\Gamma(\mu)(2\rho x)^{-\mu/2}}{(2\rho x)^{\mu/2}/(2^\mu\Gamma(\mu+1))} \\ &\sim \bar{F} + (a/2)(\rho/2)^{-\mu}\Gamma(\mu)\Gamma(\mu+1)x^{-\mu} \\ &\sim \bar{F} + Cx^{-\mu}, \end{aligned} \tag{3.300}$$

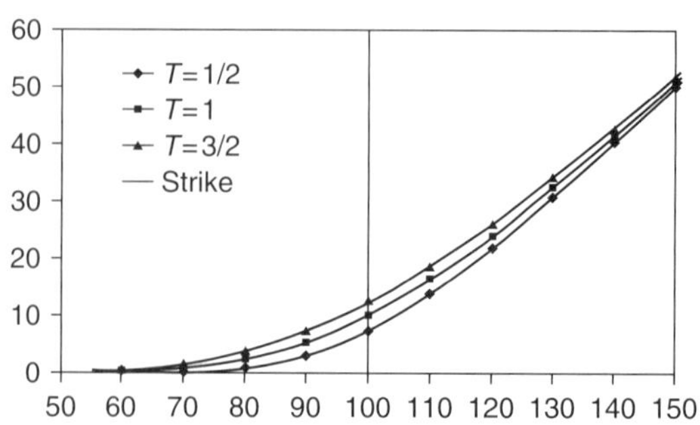


FIGURE 3.16 European call prices as functions of spot $F_0$ for various maturities. The parameters $a = 5.06$, $\rho = 0.001$, $\lambda = 5$ ($\mu = 1.5$), $K = 100$ were chosen such that the local volatility at the strike is $\sigma(K)/K = 0.25$.

where the constant is defined by $C \equiv (a/2)(\rho/2)^{-\mu}\Gamma(\mu)\Gamma(\mu+1)$ and $\Gamma(\cdot)$ is the gamma function. Note: The limiting procedure we are considering is such that $\rho \to 0$ while $a\rho^{-\mu}$ is *kept constant*; i.e., we set the parameter $a = \text{const.} \times \rho^{\mu}$. Expressions are further simplified by defining a positive constant $\sigma_0$ by $C \equiv \sigma_0^{-\theta^{-1}}$. Using $\mu = (2\theta)^{-1}$ within the last expression in equation (3.300) hence gives the limiting form of the map $x \to F$ in terms of $\sigma_0$:

$$F(x) = \bar{F} + (\sigma_0^2 x)^{-(2\theta)^{-1}} \tag{3.301}$$

with inverse

$$x = X(F) = \sigma_0^{-2}(F - \bar{F})^{-2\theta}, \tag{3.302}$$

for any constant $\bar{F}$. Taking the same limit $\rho \to 0$ in equation (3.289) and using equation (3.302) gives

$$\begin{aligned} \sigma(F) &\sim \frac{a}{\sqrt{X(F)}} \frac{[2^{\mu}\Gamma(\mu+1)]^2}{(2\rho X(F))^{\mu}} \\ &\sim 2\sigma_0 \frac{\Gamma(\mu+1)}{\Gamma(\mu)} (F - \bar{F})^{1+\theta}. \end{aligned} \tag{3.303}$$

Now, using the gamma function property $\Gamma(z+1) = z\Gamma(z)$, $\Gamma(\mu+1)/\Gamma(\mu) = \mu = 1/2\theta$, and the volatility function for this model then reduces to the expression[7]

$$\sigma(F) = \frac{\sigma_0}{\theta}(F - \bar{F})^{1+\theta}. \tag{3.304}$$

The exact barrier-free pricing kernel for the CEV volatility model (3.304) is then obtained by taking the same limit $\rho \to 0$ and using the small-argument leading order of the Bessel $I_\mu$ in equation (3.294):

$$U(F, F_0; t) \sim \frac{e^{-(X(F)+X(F_0))/2t}}{at} \frac{(X(F))^{1+\frac{3\mu}{2}}(X(F_0))^{-\frac{\mu}{2}}}{(\rho/2)^{-\mu}[\Gamma(\mu+1)]^2} I_\mu\left(\frac{\sqrt{X(F)X(F_0)}}{t}\right).$$

This expression is further reduced by making use of the map (3.302), substituting $\mu = (2\theta)^{-1}$, using the earlier definition $(a/2)(\rho/2)^{-\mu}\Gamma(\mu)\Gamma(\mu+1) = \sigma_0^{-\theta^{-1}}$ and the property $\Gamma(\mu+1)/\Gamma(\mu) = \frac{1}{2\theta}$. Again we arrive at the barrier-free pricing kernel for the CEV model with volatility given by equation (3.304), and zero-drift function:

$$\begin{aligned} U(F, F_0; t) &= \frac{\theta}{\sigma_0^2 t} \frac{(F_0 - \bar{F})^{\frac{1}{2}}}{(F - \bar{F})^{\frac{3}{2}+2\theta}} e^{-\left((F-\bar{F})^{-2\theta}+(F_0-\bar{F})^{-2\theta}\right)/2\sigma_0^2 t} \\ &\times I_{\frac{1}{2\theta}}\left(\frac{\left((F-\bar{F})(F_0-\bar{F})\right)^{-\theta}}{\sigma_0^2 t}\right), \end{aligned} \tag{3.305}$$

where $F, F_0 \in (\bar{F}, \infty)$. It is important to point out that this result can also be obtained independent of any consideration of the more general four-parameter Bessel family of solutions. In particular, this pricing kernel can be derived using equation (3.259) of Lemma 3.1, where the CEV process is directly mapped onto the underlying $x$-space Bessel process (see Problem 4). Solution (3.305) can also be extended to the case of a linear deterministic drift term (see Problem 5).

[7] The CEV model is usually defined with volatility function $\sigma(F) = \delta(F - \bar{F})^{1+\theta}$. This simply corresponds to setting $\sigma_0 = \delta\theta$ in all our formulas.

Note that this result was derived in the case $\theta > 0$, for which the lower bound of the process, $F_t = \bar{F}$, is not attained. From equation (3.296) or by use of equation (3.357), the density is easily shown to integrate to unity (i.e., no absorption occurs and the density also vanishes at the endpoint $F \to \bar{F}$ and as $F \to \infty$). By replacing $\sigma_0/\theta \to \sigma_0/|\theta|$ in equation (3.304) and considering the kernel defined by equation (3.305) but with the slight modification $\frac{\theta}{\sigma_0^2 t} \to \frac{|\theta|}{\sigma_0^2 t}$, we obtain solutions for the CEV model for $\theta < 0$. Indeed one can verify that this modified pricing kernel is a solution. That is, by direct substitution the kernel is shown to satisfy the forward and backward Kolmogorov PDE. In the range $\theta < 0$, however, the properties of this pricing kernel are more subtle. In particular, one can show that the density integrates to unity for all values $\theta < -\frac{1}{2}$, hence no absorption occurs for $\theta \in (-\infty, -\frac{1}{2})$. The boundary conditions for the density can be shown to be vanishing at $F \to \bar{F}$ (i.e., paths do not attain the lower endpoint) for all $\theta < -1$. In contrast, for $\theta \in (-1, -\frac{1}{2})$ the density becomes singular at the lower endpoint, $F = \bar{F}$ (hence this corresponds to the case where the density has an integrable singularity for which paths can also attain the lower endpoint but are not absorbed). For the special case of $\theta = -\frac{1}{2}$, the formula gives rise to absorption. [Note that for the range $\theta \in (-\frac{1}{2}, 0)$ the assumed pricing kernel is not useful, since it gives rise to a density that has a nonintegrable singularity at $F = \bar{F}$, except for certain fractional values of $\theta$. For $\theta < 0$, however, another solution that is integrable is obtained by only replacing the order $(2\theta)^{-1}$ by $-(2\theta)^{-1}$ in the Bessel function. The latter solution for the density does not integrate to unity and hence gives rise to absorption, whereby the lower finite endpoint $\bar{F}$ is an exit boundary.] The special case of $\theta = -1$ gives a nonzero constant value at the lower endpoint and recovers the Wiener process with reflection at $F = \bar{F}$ and no absorption on the interval $[\bar{F}, \infty)$, with

$$U(F, F_0; t) = \frac{1}{\sigma_0\sqrt{2\pi t}}\left(e^{-(F-F_0)^2/2\sigma_0^2 t} + e^{-(F+F_0-2\bar{F})^2/2\sigma_0^2 t}\right). \tag{3.306}$$

In the limit $\bar{F} \to -\infty$ this gives back the kernel for the pure Wiener process on the entire real line $F \in (-\infty, \infty)$, with $U = e^{-(F-F_0)^2/2\sigma_0^2 t}/\sigma_0\sqrt{2\pi t}$.

### 3.8.3.2 Recovering Quadratic Models

We have already seen, in Section 3.5.2, that the Wiener process constitutes a useful underlying $x$-space process for generating exact $F$-space pricing kernels for the quadratic volatility model of the form in equation (3.117) with two distinct roots. In fact, in Section 3.5.2 we employed the diffusion canonical reduction transformation methodology and thereby generated various exact pricing kernels for this quadratic volatility model by specifically mapping the process onto the constant-volatility Wiener process. It is now instructive to show that the quadratic model with one *double root* (i.e., one root of order 2) at the lower limit, $\bar{F}$, obtains as a special case of the four-parameter Bessel family. For this we simply consider the CEV model with choice $\theta = 1$. From equation (3.304) the volatility function is then

$$\sigma(F) = \sigma_0(F - \bar{F})^2. \tag{3.307}$$

Using the Bessel function $I_{\frac{1}{2}}(z) = \sqrt{2/\pi z}\sinh z$, equation (3.305) gives the exact barrier-free pricing kernel for this model:

$$U(F, F_0; t) = \frac{2}{\sigma_0\sqrt{2\pi t}}\frac{(F_0 - \bar{F})}{(F - \bar{F})^3} e^{-\left((F-\bar{F})^{-2}+(F_0-\bar{F})^{-2}\right)/2\sigma_0^2 t} \times \sinh\left(\frac{(F-\bar{F})^{-1}(F_0-\bar{F})^{-1}}{\sigma_0^2 t}\right), \tag{3.308}$$

where $F, F_0 \in (\bar{F}, \infty)$. This density integrates to unity exactly (see Problem 2).

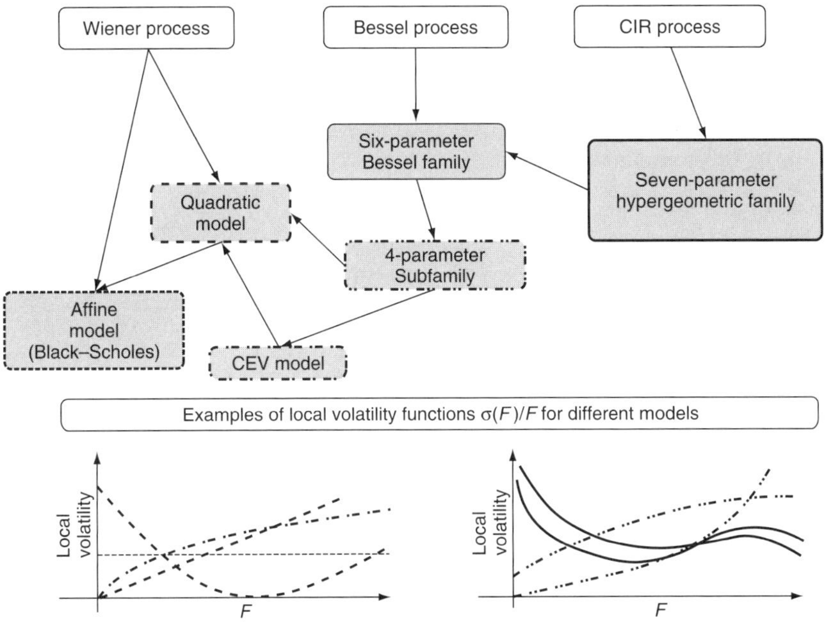


FIGURE 3.17 A hierarchy of analytically solvable state-dependent models with examples of their corresponding typical local volatility curves. The popular linear (Black–Scholes) model gives only the flat-line local volatility shape.

It is interesting to observe that the foregoing exactly solvable state-dependent multi-parameter volatility models form a kind of model hierarchy that can be summarized in a flowchart, as depicted in Figure 3.17. At the top are the underlying ($x$-space) processes that are used to generate the various pricing ($F$-space) models. Most of the models depicted are subsets of the Bessel family. However, extensions to other models are also possible by means of the techniques presented in this chapter. For example, one can enlarge the family of exact pricing kernels by considering the CIR process as an underlying $x$-space process. As seen in Chapter 2, the CIR process has the linear-drift function $\lambda(x) = \lambda_0 + \lambda_1 x$ and hence has one extra parameter as compared to the Bessel process. This gives rise to the family of confluent hypergeometric functions (e.g., Whittaker and Kummer functions), for which the Bessel functions form a special subset, as depicted in Figure 3.17. The so-called confluent hypergeometric family can be shown to contain a total of seven adjustable parameters. We refer the interested reader to some recent literature on this topic [ACCL01, Lip03]. Other extensions are also possible. The search for new families of analytical solutions to complex state-dependent models and their applications to pricing is a topic of current and ongoing research in financial mathematics. For recent works on pricing path-dependent options using new families of state-dependent volatility models see [CaM04a, CaM04b, CaM05].

## Problems

**Problem 1.** Show that the density in equation (3.305) integrates to unity for all $t > 0$ by a change of variables using equation (3.302) and an appropriate Bessel integral identity. Show that in the limit $t \to 0$ the density represents a Dirac delta function.

**Problem 2.** Show that the density in equation (3.308) integrates to unity for all $t > 0$. In doing so, do not employ any Bessel integral identity. Hint: Change variables to $x = \sigma_0^{-2}(F - \bar{F})^{-2}$ and rewrite the integral so as to make use of the identity $\int_{-\infty}^{\infty} ye^{-ay^2+by}\, dy = \sqrt{\frac{\pi}{a}}\left(\frac{b}{2a}\right)e^{b^2/4a}$.

**Problem 3.** Derive the European put option formula analogous to equation (3.297) for the four-parameter Bessel model.

**Problem 4.** Let $\lambda(x) = \lambda = \theta^{-1} + 2$, $\nu(x) = 2\sqrt{x}$, $\sigma(F) = \frac{\sigma_0}{\theta}(F - \bar{F})^{1+\theta}$. Using relation (3.257) with the choice $\rho = 0$ [i.e., $\alpha(x, F) = 0$], show that the mapping in equation (3.302) obtains. By substituting the kernel $u(X(F), X(F_0); t)$ of equation (3.215) into equation (3.259) of Lemma 3.1, arrive at the kernel in equation (3.305).

**Problem 5.** Show that the kernel defined by $U_\mu(F, F_0; t) \equiv e^{-\mu t}U(e^{-\mu t}F, F_0; T(t))$, where $U$ solves the CEV process $dF_t = \delta F_t^{1+\theta} dW_t$, [i.e. as in equation (3.305) with $\bar{F} = 0$, $\sigma_0 = \delta\theta$], is a solution to the corresponding CEV process with an added drift function: $dF_t = \mu F_t dt + \delta F_t^{1+\theta} dW_t$, for arbitrary drift parameter $\mu$. In doing so, arrive at $T(t) = (e^{2\theta\mu t} - 1)/2\theta\mu$ and hence derive the barrier-free kernel

$$U_\mu(F, F_0; t) = \frac{2\mu F_0^{\frac{1}{2}} F^{-\frac{3}{2}-2\theta} e^{\mu t/2}}{sgn(\theta)\delta^2(1 - e^{-2\mu\theta t})} \exp\left(-\frac{\mu}{\theta\delta^2}\frac{F_0^{-2\theta}e^{-2\mu\theta t} + F^{-2\theta}}{(1 - e^{-2\mu\theta t})}\right) \times I_{\frac{1}{2\theta}}\left(\frac{\mu}{\theta\delta^2}\frac{(FF_0)^{-\theta}}{\sinh(\mu\theta t)}\right). \tag{3.309}$$

### 3.8.4 Conditions for Absorption, or Probability Conservation

Consider an $x$-space kernel solving equation (3.256) and a given fixed interval $x \in (a, b)$. Given an initial interior point $x_0 \in (a, b)$ at time $t = 0$, the probability $p(a, b|x_0, t)$ that a sample path $x_\tau$, $0 \le \tau \le t$, will have terminal value $x_t = x \in (a, b)$ within the fixed interval at time $t \ge 0$ is then

$$p(a, b|x_0, t) = \int_a^b u(x, x_0; t)dx. \tag{3.310}$$

The rate of absorption into the interval, or the rate of probability increase, denoted by $r(a, b|x_0, t)$, is then given by $\partial p/\partial t$. Taking the time derivative inside the integral while making use of the forward equation (3.256) and integrating gives

$$r(a, b|x_0, t) = \left[\frac{1}{2}\frac{\partial}{\partial x}\left(\nu(x)^2 u(x, x_0; t)\right) - \lambda(x)u(x, x_0; t)\right]_{x=a}^{x=b}. \tag{3.311}$$

If $r(a, b|x_0, t) = 0$ for any $t$, then no absorption occurs over time; otherwise absorption occurs inside (or outside) the interval $x \in (a, b)$.

Of interest is whether kernels with imposed homogeneous-(zero-)-boundary conditions give rise to absorption or not. In this case we take $a = x_L$ and $b = x_H$ as, respectively, the lower and upper endpoints of the entire solution space and generally assume solutions such that $\lambda(x)u(x, x_0; t) \to 0$ at both endpoints.[8] This is certainly the case for all the $x$-space kernels considered throughout this chapter, as can be verified. It hence follows from equation (3.311) that the kernel gives no absorption if

$$\lim_{x \to x_L^+} \frac{\partial}{\partial x}\left(\nu(x)^2 u(x, x_0; t)\right) = 0 \tag{3.312}$$

[8] Depending on the solution interval, a lower (upper) endpoint $x_L$ ($x_H$) takes on either a finite value or $-\infty$ ($\infty$).

and

$$\lim_{x\to x_H^-} \frac{\partial}{\partial x}\Big(\nu(x)^2 u(x, x_0; t)\Big) = 0. \tag{3.313}$$

Moreover, note that (regardless of whether $u$ is a barrier-free kernel or a kernel with absorption at a barrier for $t > 0$) any kernel $u$ integrates to unity in the limit $t \to 0$ because of the imposed delta function initial condition: $u(x, x_0; t) \to \delta(x - x_0)$ as $t \to 0$. The no-absorption conditions (3.312) and (3.313), if satisfied, ensure that $p(x_L, x_H|x_0, t)$ is constant as a function of $t$ and therefore that conservation of probability is satisfied, with kernel $u$ integrating to unity for *all* $t \geq 0$.

For the Bessel process $\nu(x) = 2\sqrt{x}$, hence, equation (3.311) simplifies to give the absorption rate $r(x_L, x_H|x_0, t)$ proportional to

$$\lim_{x\to x_H^-} x\frac{\partial u(x, x_0; t)}{\partial x} - \lim_{x\to x_L^+} x\frac{\partial u(x, x_0; t)}{\partial x}. \tag{3.314}$$

Using the barrier-free kernel given by equation (3.215) [i.e., $x_L = 0$, $x_H = \infty$, $x \in (0, \infty)$] while making use of the asymptotic properties of the $I_\mu(z)$ function for argument $z \to 0$ and $z \to \infty$, it is readily shown that these limits are both zero, hence giving no absorption. This barrier-free kernel therefore conserves probability. Alternatively, this is readily shown by direct integration; see Section 3.7.1. In contrast, the kernels given by equations (3.232), (3.245), and (3.253) for the double- and single-barrier Bessel process are all readily proven to lead to absorption. Considering the double-barrier solution equation (3.232), for example, the absorption rate due to either endpoint involves terms of the form $x^{\frac{\mu}{2}+1}\phi_n'(x)$ and $x^{\frac{\mu}{2}+1}\phi_n(x)$, with $x \to x_L$ and $x \to x_H$. The eigenfunctions evaluated at the endpoints obviously give zero, by design $\phi_n(x = x_L) = \phi_n(x = x_H) = 0$. However, the derivative of the eigenfunctions $\phi_n'(x = x_L)$, $\phi_n'(x = x_H)$ are nonzero. Similar arguments can be used to show that the other Bessel barrier solutions also give rise to absorption; i.e., probability is not conserved as paths attaining either finite barrier level, $x_L > 0$ or $x_H > 0$, are absorbed.

Now consider any $F$-space kernel $U$ that is generated from an underlying $x$-space kernel $u$ as given by equation (3.274) and a fixed interval $F \in (F_a, F_b)$, $F_a = F(a)$, $F_b = F(b)$, where $x \in (a, b)$ maps one to one onto $F \in (F_a, F_b)$. Given an initial point $F_0 \in (F_a, F_b)$ at time $t = 0$, the probability $P(F_a, F_b|F_0, t)$ that a sample path $F_\tau$, $0 \leq \tau \leq t$, will have terminal value $F_t = F \in (F_a, F_b)$ at time $t \geq 0$ is then, in analogy with equation (3.310),

$$P(F_a, F_b|F_0, t) = \int_{F_a}^{F_b} U(F, F_0; t)dF. \tag{3.315}$$

The rate of absorption into the interval, denoted by $R(F_a, F_b|F_0, t)$, is given by $\partial P/\partial t$. [Note: The absorption rate outside the interval is then just $-R$.] Again, taking the time derivative inside the integral and now using the forward equation (3.258) gives

$$R(F_a, F_b|F_0, t) = \frac{1}{2}\frac{\partial}{\partial F}\Big(\sigma(F)^2 U(F, F_0; t)\Big)\Big|_{F=F_a}^{F=F_b}. \tag{3.316}$$

In carrying out further analysis, it is convenient simply to transform to $x$-space variables. In particular, using equation (3.274), and the chain rule,

$$R(F_a, F_b|F_0, t) = \frac{e^{-\rho t}}{2\hat{u}(x_0, \rho)}\frac{\nu(x)}{\sigma(F(x))}\frac{\partial}{\partial x}\Big(\nu(x)\sigma(F(x))\hat{u}(x, \rho)u(x, x_0; t)\Big)\Big|_{x=a}^{x=b}, \tag{3.317}$$

$x_0 = X(F_0)$, $a = X(F_a)$, $b = X(F_b)$, where $|\partial X(F)/\partial F| = \nu(x)/\sigma(F(x))$ is used. By mapping $x \in [x_L, x_H]$ onto $F \in [L, H]$, with endpoints[9] $F(x_L) = L$ and $F(x_H) = H$ ($x_L = X(L)$, $x_H = X(H)$), and letting $a = x_L$, $b = x_H$, then from equation (3.317) the no-absorption condition for the interval $F \in (L, H)$ can be written generally as

$$\left[\frac{\nu(x)}{\sigma(F(x))}\frac{\partial}{\partial x}\left(\nu(x)\sigma(F(x))\hat{u}(x;\rho)u(x,x_0;t)\right)\right]_{x=x_L}^{x=x_H} = 0. \tag{3.318}$$

In analogy with the $x$-space kernel, this condition, if satisfied, therefore represents probability conservation with total unit probability on the entire interval of the $F$-space solution, since the kernel $U$ also integrates to unity in the limit $t \to 0$; i.e., $U(F, F_0; t) \to \delta(F - F_0)$ as $t \to 0$.

The general condition given by equation (3.318) can hence be used to determine whether absorption arises for any $F$-space kernel obtained via Theorem 3.1. We now apply this condition to the general Bessel family. In particular, using equations (3.277) and (3.285), the general Bessel family of pricing kernels given by equation (3.292) then admits a no-absorption condition in the form

$$\left[\left(\frac{\mu}{2}\hat{u}(x,\rho) + \sqrt{\frac{x\rho}{2}}x^{-\frac{\mu}{2}}[q_1 I'_\mu(\sqrt{2\rho x}) + q_2 K'_\mu(\sqrt{2\rho x})]\right)u(x,x_0;t) - x\hat{u}(x,\rho)\frac{\partial u(x,x_0;t)}{\partial x}\right]_{x=x_L}^{x=x_H} = 0. \tag{3.319}$$

From our analysis on the $x$-space kernels we readily observe that all single- and double-barrier solutions with $u(x_L, x_0; t) = u(x_H, x_0; t) = 0$ for *finite* $x_L, x_H > 0$ lead to absorption. This is the case since the $I_\mu$, $K_\mu$, $I'_\mu$, and $K'_\mu$ functions are finite at finite nonzero endpoints; hence for the barrier kernels the foregoing condition reduces to

$$x\hat{u}(x;\rho)\frac{\partial u(x,x_0;t)}{\partial x}\bigg|_{x=x_L}^{x=x_H} = 0. \tag{3.320}$$

However, as just seen, this condition cannot generally be satisfied for any of the barrier kernels given by equations (3.232), (3.245), or (3.253). We therefore conclude that the only *possible* families of $F$-space kernels that can lead to no absorption are those with $u(x, x_0; t)$ given by equation (3.215), i.e. the barrier-free solutions on $x \in (0, \infty)$ with $x_L = 0$, $x_H = \infty$.

We therefore further specialize our analysis exclusively to families of barrier-free solutions with underlying barrier-free $x$-space kernel chosen for $u$. Upon substituting equation (3.215) into equation (3.319), it readily follows that the first term in equation (3.319) is zero in the limits $x \to 0, \infty$. Indeed, for the lower limit $x \to 0$ this is a consequence of the asymptotic identities: $I_\mu(z) \to c_1 z^\mu$, $K_\mu(z) \to c_2 z^{-\mu}$, as $z \to 0$, where $c_1$, $c_2$ are positive constants dependent on the order $\mu > 0$. For the upper limit the asymptotic properties $I_\mu(z) \to e^z/\sqrt{2\pi z}$, $K_\mu(z) \to \sqrt{\pi/2z}e^{-z}$, as $z \to \infty$, are used. The exponential factor $e^{-x/2t}$ in $u$ is hence more rapidly decreasing, and the term $\hat{u}u$ vanishes in the limit $x \to \infty$. Using equation (3.277), the no-absorption condition is then reduced to

$$x\hat{u}(x,\rho)\frac{\partial u(x,x_0;t)}{\partial x}\bigg|_{x=0}^{x=\infty} = x^{1-\frac{\mu}{2}}\frac{e^{-x_0/2t}}{2x_0^{\frac{\mu}{2}}t}[q_1 I_\mu(\sqrt{2\rho x}) + q_2 K_\mu(\sqrt{2\rho x})] \times \frac{\partial}{\partial x}\left(x^{\frac{\mu}{2}}e^{-x/2t}I_\mu(\sqrt{xx_0}/t)\right)\bigg|_{x=0}^{x=\infty} = 0.$$

[9]Note: The arguments follow in exactly the same way whether a monotonically increasing or decreasing map $F = F(x)$ is assumed.

The foregoing asymptotic properties for the $I_\mu$, $K_\mu$ functions give zero for the upper limit $x \to \infty$, for all choices of parameters $q_1$,$q_2$. On the other hand, evaluating the lower limit while using the small-argument expressions for $I_\mu$ and $K_\mu$ gives, to leading order:

$$\frac{e^{-x_0/2t}}{t^{\mu+1}}\left(C_1 q_1 x + C_2 q_2 x^{1-\mu}\right)\frac{\partial}{\partial x}\left(x^{\mu} e^{-x/2t}\right) \to A q_2, \tag{3.321}$$

where $C_1$, $C_2$, $A$ are positive constants and $A$ depends on $t$, $\rho$, $x_0$, and $\mu$. Hence, we conclude that if $q_2 \neq 0$, then there is a nonzero finite rate of absorption (i.e., absorbed outside of the solution region) at the lower boundary; otherwise for $q_2 = 0$ there is no absorption, and probability is conserved for all time. The latter is the case of the barrier-free four-parameter subfamily kernel as given by equation (3.294). Notice that this conclusion is indeed consistent with equation (3.296).

## 3.8.5 *Barrier Pricing Formulas for Multiparameter Volatility Models*

In concluding this chapter we give a brief discussion of how pricing kernels and European option formulas can be obtained in analytically closed form for multiparameter state-dependent models and in particular for the Bessel family of models.

Let us assume we have solved for an underlying $x$-space barrier kernel in the form of an exact eigenfunction expansion given by equation (3.202) for a domain $x, x_0 \in (x_L, x_H)$ with zero-boundary conditions at the endpoints of the domain. Consider any $F_t$ process that is mapped onto an underlying $x_t$-process and thereby satisfying the general assumptions of Theorem 3.1. From the discussion in Section 3.8.1 it follows that reduction transformation formula (3.274) can be used together with equation (3.202) to obtain a general family of exact eigenfunction expansions for an $F$-space pricing kernel that takes the generic form

$$U(F, F_0; t) = \frac{\nu(x_0)}{\hat{u}(x_0, \rho)}\frac{\hat{u}(x, \rho)}{\sigma(F)} e^{\int_{x_0}^{x} \frac{\lambda(x')}{\nu(x')^2}dx'} \sum_{n=1}^{\infty} e^{-(\rho+|\epsilon_n|)t}\phi_n(x)\phi_n(x_0). \tag{3.322}$$

The generating function $\hat{u}$ solves equation (3.272) and is used to obtain $x_0 = X(F_0)$, $x = X(F)$ by inverting equation (3.273), where one uses either appropriate branch of the map $F = F(x)$ (e.g., monotonically increasing or decreasing) and the volatility function $\sigma(F)$ for the $F_t$ process is given by equation (3.271). The eigenfunctions $\phi_n$ solve equation (3.198). The $x$-space endpoints are mapped onto the corresponding barrier levels in $F$-space: $H = F(x_H)$ and $L = F(x_L)$.

By specializing to the four-parameter subfamily of Bessel models of Section 3.8.3, the pricing kernels are given by relation (3.293), where $u$ is taken to be the Bessel kernel as in either equation (3.232), (3.245), or (3.253), depending on whether we are seeking an $F$-space pricing kernel $U$ for a double barrier *or* a single barrier, respectively. For instance, in the case of a double barrier with absorption of paths $F_t$ at levels $L$ and $H$, we insert equation (3.232) into equation (3.293) to obtain the pricing kernel as a closed-form eigenfunction series solution:

$$U^{DB}(F, F_0, L, H; t) = \left(\frac{2}{a}\right) x \frac{I_\mu^3(\sqrt{2\rho x})}{I_\mu(\sqrt{2\rho x_0})} \sum_{n=1}^{\infty} e^{-(\rho+|\epsilon_n|)t}\phi_n(x)\phi_n(x_0), \tag{3.323}$$

with $x = X(F)$, $x_0 = X(F_0)$ given by inverting equation (3.290). Note that this barrier kernel is a special case of equation (3.322). The mapping in equation (3.290) provides us with the unique condition used to fix the two barrier levels $L$, $H$ by appropriate choice of $x$-space endpoints:

$$L = \bar{F} + a\frac{K_\mu(\sqrt{2\rho x_L})}{I_\mu(\sqrt{2\rho x_L})}, \qquad H = \bar{F} + a\frac{K_\mu(\sqrt{2\rho x_H})}{I_\mu(\sqrt{2\rho x_H})}. \tag{3.324}$$

Or alternatively, given $L$ and $H$ values, these equations are uniquely inverted to give the endpoint values $x_L = X(L)$ and $x_H = X(H)$. The eigenfunctions in equation (3.230) [and the eigenvalues satisfying equation (3.221)] are then given uniquely for all $n > 1$. Note that since the mapping $F(x)$ is decreasing, $F, F_0 \in [H, L]$, so the lower barrier is at $H$ and the upper barrier is at $L$ in our present notation. That is, the lower (upper) $x$-space endpoints are mapped to the upper (lower) barrier levels in $F$-space.

Similar formulas for the pricing kernel also follow for the single-barrier cases. Cumulative probability densities can also be computed in analytically closed form. These are in turn used to provide closed-form pricing formulas for European barrier calls and puts. For the case of the double barrier with $L, H > \bar{F}$, we define the cumulative density for $H \leq F \leq L$ as

$$\Phi_c(F, F_0, t) = \int_H^F U^{DB}(f, F_0, L, H; t)df. \tag{3.325}$$

Using equation (3.323) and changing integration variables from $f$ to $x = X(f)$ via the mapping (3.290) we obtain

$$\Phi_c(F, F_0, t) = \sum_{n=1}^{\infty} e^{-(\rho+|\epsilon_n|)t}\frac{\phi_n(X(F_0))}{I_\mu(\sqrt{2\rho X(F_0)})}\int_{X(F)}^{X(H)} I_\mu(\sqrt{2\rho x})\phi_n(x)dx. \tag{3.326}$$

Using equation (3.230) for $\phi_n(x)$ and making a simple change of variables, one can then use integral indentities (3.362) and (3.363) to evaluate the resulting integrals. After collecting terms and simplifying with the use of the Wronskian identity (3.384) we arrive at the closed-form series

$$\Phi_c(F, F_0, t) = \frac{1}{I_\mu(\sqrt{2\rho X(F_0)})}\sum_{n=1}^{\infty} e^{-(\rho+|\epsilon_n|)t}\phi_n(X(F_0))\Psi_{n,\rho}(X(F)), \tag{3.327}$$

where we have defined the functions

$$\Psi_{n,\rho}(x) \equiv \frac{1}{\rho+|\epsilon_n|}\Big[\sqrt{2|\epsilon_n|x}I_\mu(\sqrt{2\rho x})\tilde{\phi}_n(x) - \sqrt{2\rho x}I_{\mu+1}(\sqrt{2\rho x})\phi_n(x) - \frac{2}{\pi}\mathcal{N}_n I_\mu(\sqrt{2\rho X(H)})\Big] \tag{3.328}$$

and

$$\tilde{\phi}_n(x) \equiv \mathcal{N}_n\Big[Y_\mu(\sqrt{2|\epsilon_n|x_H})J_{\mu+1}(\sqrt{2|\epsilon_n|x}) - J_\mu(\sqrt{2|\epsilon_n|x_H})Y_{\mu+1}(\sqrt{2|\epsilon_n|x})\Big]. \tag{3.329}$$

The normalization factor $\mathcal{N}_n$ is given by equation (3.231). In a similar manner, the related cumulative density given by

$$\bar{\Phi}_c(F, F_0, t) = \int_H^F U^{DB}(f, F_0, L, H; t)f df, \tag{3.330}$$

for $H \leq F \leq L$, can also be evaluated analytically. Again using equation (3.323) and changing integration variables from $f$ to $x = X(f)$ via the mapping (3.290),

$$\bar{\Phi}_c(F, F_0, t) = \bar{F}\Phi_c(F, F_0, t) + a \sum_{n=1}^{\infty} e^{-(\rho+|\epsilon_n|)t} \frac{\phi_n(X(F_0))}{I_\mu(\sqrt{2\rho X(F_0)})} \times \int_{X(F)}^{X(H)} K_\mu(\sqrt{2\rho x})\phi_n(x)dx. \tag{3.331}$$

This last integral is evaluated using equation (3.230) for $\phi_n(x)$; and, after changing variables, we use the integral indentities equations (3.364) and (3.365). Collecting terms, using (3.384), and simplifying we obtain the closed-form series

$$\bar{\Phi}_c(F, F_0, t) = \bar{F}\Phi_c(F, F_0, t) + \frac{a}{I_\mu(\sqrt{2\rho X(F_0)})} \sum_{n=1}^{\infty} e^{-(\rho+|\epsilon_n|)t} \phi_n(X(F_0))\psi_{n,\rho}(X(F)), \tag{3.332}$$

where

$$\psi_{n,\rho}(x) \equiv \frac{1}{\rho + |\epsilon_n|} \Big[ \sqrt{2\rho x} K_{\mu+1}(\sqrt{2\rho x})\phi_n(x) + \sqrt{2|\epsilon_n| x} K_\mu(\sqrt{2\rho x})\tilde{\phi}_n(x) - \frac{2}{\pi} \mathcal{N}_n K_\mu(\sqrt{2\rho X(H)}) \Big]. \tag{3.333}$$

Under the four-parameter Bessel family of volatility models, a European double-knockout call maturing in time $t$ with payoff $(F - K)_+$ therefore has value given by (excluding discounting)

$$C^{DB}(F_0, K, t) = \bar{\Phi}_c(L, F_0, t) - \bar{\Phi}_c(K, F_0, t) - K\big[\Phi_c(L, F_0, t) - \Phi_c(K, F_0, t)\big] \tag{3.334}$$

for $H \leq K < L$ and

$$C^{DB}(F_0, K, t) = \bar{\Phi}_c(L, F_0, t) - K\Phi_c(L, F_0, t) \tag{3.335}$$

for strike values below the barriers, $K < H < L$. An analogous formula for the put option is also readily obtained. Analogous formulas for the option values for single barriers can also be derived in similar fashion. By applying a similar limiting procedure to the one discussed in Section 3.8.3.1, the foregoing families of formulas can also be used to recover closed-form formulas for barrier pricing kernels (as well as barrier call and put option values) for the CEV model with zero drift function. In particular, one can recover the double-barrier kernel for the CEV model (see Problem 1). Moreover, as a special case of the CEV solutions, even simpler closed-form expressions for the barrier kernels and barrier option values arise for the quadratic model of Section 3.8.3.2. As already discussed, in this case $\mu = \frac{1}{2}$ and the modified Bessel functions are just the elementary hyperbolic sine and exponential functions, while the ordinary Bessel functions are just the sine and cosine functions (see Problem 2).

## Problems

**Problem 1.** By using a similar limiting procedure to the one in Section 3.8.3.1 for the barrier-free case, with $\mu = 1/(2\theta)$, obtain an exact eigenfunction series expansion for the double-barrier pricing kernel for the CEV model with volatility given by equation (3.304) and zero drift function in the $F_t$ process. Express your answer explicitly in terms of $F$, $F_0$, $L$, $H$, $\sigma_0$, $\theta$, and time $t$. Also, provide the equation for the eigenvalues.

**Problem 2.** Obtain an exact eigenfunction series expansion for the double-barrier pricing kernel for the quadratic model of Section 3.8.3.2. This can be achieved by specializing the CEV formula from Problem 1 using $\theta = 1$. Another (simpler) way (which makes no use of the CEV result of Problem 1) is to set $\mu = \frac{1}{2}$ in the four-parameter Bessel family map (3.290) and in equation (3.323). Then by letting $a = C\sqrt{\rho}$ (for an appropriate choice of constant $C$), take the limit $\rho \to 0$ of equation (3.323). The series should in fact reduce to elementary functions, with a simple exact expression for the eigenvalues $\epsilon_n$. Express your answer explicitly in terms of $F$, $F_0$, $L$, $H$, $\sigma_0$, and $t$. Hint: For half-integer order the Bessel functions are $J_{\frac{1}{2}}(z) = \sqrt{2/\pi z}\sin z$, $Y_{\frac{1}{2}}(z) = -\sqrt{2/\pi z}\cos z$, $I_{\frac{1}{2}}(z) = \sqrt{2/\pi z}\sinh z$, $K_{\frac{1}{2}}(z) = \sqrt{\pi/2z}e^{-z}$.

**Problem 3.** Derive a closed-form series expression for a double-barrier call option price for the quadratic model of Section 3.8.3.2. You may use the result in Problem 2.

**Problem 4.** Following the same limiting procedure as in Problem 1, obtain an exact eigenfunction series expansion for the price of a double-barrier call option for the CEV model; i.e., obtain the analogues of equations (3.334) and (3.335) for the CEV model, with zero drift function in the $F_t$ process.

## 3.9 Appendix A: Proof of Lemma 3.1

Assume a relationship among the fundamental solutions in the form

$$U(F, F_0; t) = \frac{\nu(x)}{\sigma(F)} e^{\alpha t} \frac{\phi(x)}{\phi(x_0)} u(x, x_0; t), \tag{3.336}$$

with $\alpha$, $\phi(x)$ to be determined. By direct substitution of this Ansatz into equation (3.258), applying the chain rule of differentiation and collecting terms gives

$$\begin{aligned}\frac{\partial u}{\partial t} + \alpha u = {}& \frac{1}{2}\nu^2 \frac{\partial^2 u}{\partial x^2} + \frac{1}{2}\left[\frac{(\nu^2\phi)_x + \nu(\nu\phi)_x}{\phi} + \nu\sigma'(F)\right]\frac{\partial u}{\partial x} \\ & + \frac{1}{2}\left[\frac{\nu(\nu\phi)_{xx} + (\nu_x + \sigma'(F))(\nu\phi)_x}{\phi} + \sigma(F)\sigma''(F)\right]u,\end{aligned} \tag{3.337}$$

where primes and subscript variables denote derivatves with respect to the appropriate variable and function arguments and $u = u(x, x_0; t)$. Rewriting equation (3.256) by explicitly carrying out the derivatives gives

$$\frac{\partial u}{\partial t} = \frac{1}{2}\nu^2 \frac{\partial^2 u}{\partial x^2} + (2\nu\nu_x - \lambda)\frac{\partial u}{\partial x} + (\nu_x^2 + \nu\nu_{xx} - \lambda_x)u. \tag{3.338}$$

Combining the last two equations gives a linear equation in $u$ and $u_x$. Since this equation must be valid for arbitrary solution $u$, the coefficients in $u$ and $u_x$ must be zero identically. Setting the coefficient in $u_x$ to zero gives a first-order equation that can be cast in the form

$$\begin{aligned}\frac{d}{dx}\log\phi(x) &= \frac{1}{2}\frac{\nu'(x) - \sigma'(F)}{\nu(x)} - \frac{\lambda(x)}{\nu(x)^2} \\ &= \frac{1}{2}\frac{d}{dx}\log\frac{\nu(x)}{\sigma(F(x))} - \frac{\lambda(x)}{\nu(x)^2}.\end{aligned} \tag{3.339}$$

Here we used $F'(x) = \sigma/\nu$.

Integrating from an arbitrary point $x_0$ to $x$, we find

$$\frac{\phi(x)}{\phi(x_0)} = \exp\left(\frac{1}{2}\log\frac{\nu(x)/\sigma(F(x))}{\nu(x_0)/\sigma(F(x_0))} - \int_{x_0}^{x}\frac{\lambda(s)}{\nu(s)^2}ds\right). \tag{3.340}$$

Setting the coefficient in $u$ to zero and using equation (3.339) gives a second-order equation in $\phi$:

$$\nu^2\frac{\phi_{xx}}{\phi} + \frac{1}{2}(\nu_x^2 - \sigma'(F)^2) + \sigma(F)\sigma''(F) - \nu\nu_{xx} - (3\nu_x + \sigma'(F))\frac{\lambda}{\nu} + 2\lambda_x - 2\alpha = 0. \tag{3.341}$$

For a solution to exist, this equation must be consistent with equation (3.339). Hence, by differentiating equation (3.339) once with respect to $x$ while using equation (3.339) in the resulting expression, we obtain

$$\nu^2\frac{\phi_{xx}}{\phi} = \frac{1}{2}(\nu\nu_{xx} - \sigma(F)\sigma''(F)) + \frac{1}{4}(\sigma'(F)^2 - \nu_x^2)$$
$$+ (\nu_x + \sigma'(F))\frac{\lambda}{\nu} + \frac{\lambda^2}{\nu^2} - \lambda_x. \tag{3.342}$$

Inserting the value of $\nu^2\phi_{xx}/\phi$ in this equation into the previous one and simplifying finally leads to an expression for $\alpha$:

$$\alpha = \frac{1}{2}\left[\lambda_x + \frac{\lambda^2}{\nu^2} - 2\lambda\frac{\nu_x}{\nu} + \frac{1}{2}(\sigma(F)\sigma''(F) - \nu\nu_{xx}) + \frac{1}{4}(\nu_x^2 - \sigma'(F)^2)\right]$$

This is equation (3.257) and must be a constant, as assumed throughout the derivations. Hence, combining equations (3.336), (3.340), and (3.257), we conclude that $U$ given by equation (3.259) indeed solves equation (3.258). Moreover, the Dirac delta function initial condition in $F$-space is also satisfied, since

$$\lim_{t\to 0} U(F, F_0; t) = \frac{\nu(X(F))}{\sigma(F)}\delta(X(F) - X(F_0)) = \delta(F - F_0), \tag{3.343}$$

where $X'(F) = \nu(X(F))/\sigma(F)$.

## 3.10 Appendix B: Alternative "Proof" of Theorem 3.1

Here we show how Theorem 3.1 arises as an application of the (continuous-time) fundamental theorem of asset pricing presented in Chapter 1. The argument can be formulated by making reference to a financial model. Consider a multicurrency financial model where domestic interest rates are zero, the process $x_t$ is interpreted as a price process for an asset denominated in a foreign currency, and $F_t = F(x_t)$ is the price process for a contingent claim (a quanto option) in the domestic currency. Assume that under the pricing measure where $F_t$ has zero drift function, the underlying foreign price process $x_t$ obeys the equation

$$dx_t = \mu(x_t)dt + \nu(x_t)dW_t \tag{3.344}$$

for some drift function $\mu(x)$. Assume also that the volatility $\nu(x)$ is such that, for some choice of the drift function $\lambda(x)$, one can solve stochastic differential equation (3.254). By solving, we mean that it is possible to find the pricing kernel $u(x, t; x_0)$, which can be interpreted as the current time-zero price of an infinitesimally narrow butterfly spread option of maturity time $t$, i.e., with delta function payoff $\delta(x - x_0)$, where $x_0$ is the spot price of the underlying foreign asset.

Our objective is to show that if the volatility function for the quanto option $F_t$ is defined as [i.e., equation (3.271)]

$$\sigma(F) = \frac{\sigma_0 \nu(X(F)) e^{-2\int^{X(F)} \frac{\lambda(y)}{\nu(y)^2} dy}}{\hat{u}(X(F), \rho)^2}, \tag{3.345}$$

then it is possible to find the pricing kernel for the quanto option $F_t$ (which will be in analytically closed form assuming the kernel for the $x_t$-process is given analytically). Here, $\rho$ is a real valued parameter and the function $\hat{u} = \hat{u}(x, \rho)$ is defined as the solution of equation (3.272), i.e.,

$$\frac{\nu^2}{2} \hat{u}_{xx} = \rho \hat{u} - \lambda \hat{u}_x. \tag{3.346}$$

Finally, the function $X(F)$ in equation (3.345) and its inverse, $F(x)$, are defined as the solutions of the equation

$$\frac{dX(F)}{dF} = \frac{\nu(x)}{\sigma(F)}. \tag{3.347}$$

The key in this derivation involves a change of numeriare asset given by a process $g_t$, defined as

$$g_t = \frac{e^{\rho t}}{\hat{u}(x_t, \rho)}, \tag{3.348}$$

and by applying Itô's lemma to this function of $x_t$ and $t$ we have the SDE

$$dg = \left( \rho - \mu \frac{\hat{u}_x}{\hat{u}} + \nu^2 \left[ \left( \frac{\hat{u}_x}{\hat{u}} \right)^2 - \frac{1}{2} \frac{\hat{u}_{xx}}{\hat{u}} \right] \right) g\, dt + \sigma^g g\, dW_t, \tag{3.349}$$

where the lognormal volatility of $g_t$ (denoted by $\sigma^g$) is given by

$$\sigma^g = -\nu \frac{\hat{u}_x}{\hat{u}}. \tag{3.350}$$

Substituting equation (3.346), we find that

$$\frac{dg}{g} = \left( \frac{\mu - \lambda}{\nu} \sigma^g + (\sigma^g)^2 \right) dt + \sigma^g\, dW_t. \tag{3.351}$$

To demonstrate that $g_t$ defines a domestic asset price process, consider this equation in the original pricing measure, where the domestic quanto option price process $F_t$ has zero drift function. In this case, using Itô's lemma on the inverse mapping $x_t = X(F_t)$, we arrive at an SDE of the form of equation (3.344), with drift given by

$$\mu(x) = \frac{\sigma(F)^2}{2} \frac{d}{dF} \frac{dX(F)}{dF} = \frac{\sigma(F)^2}{2} \frac{d}{dF} \frac{\nu(x)}{\sigma(F)}, \tag{3.352}$$

where $F = F(x)$. Using the chain rule for differentiation and expressing all functions in terms of $x$, we then have

$$\mu(x) = \frac{\sigma\nu}{2}\frac{d}{dx}\left(\frac{\nu}{\sigma}\right) = \frac{\nu}{2}\left[\nu_x - \frac{\nu}{\sigma}\sigma_x\right], \tag{3.353}$$

where $\sigma \equiv \sigma(F(x))$ is the volatility function for the quanto option of price $F_t$. Hence, by substituting into the expression for the risk-neutral drift of $g_t$ in equation (3.351) we find

$$\frac{\mu - \lambda}{\nu}\sigma^g + (\sigma^g)^2 = \left[\lambda + \frac{\nu^2}{2}\frac{\sigma_x}{\sigma} - \frac{1}{2}\nu\nu_x\right]\frac{\hat{u}_x}{\hat{u}} + \nu^2\left(\frac{\hat{u}_x}{\hat{u}}\right)^2. \tag{3.354}$$

Using expression (3.345) for the volatility of the quanto option $F_t$, we find that

$$\frac{\sigma_x}{\sigma} = \frac{\nu_x}{\nu} - \frac{2\lambda}{\nu^2} - \frac{2\hat{u}_x}{\hat{u}}. \tag{3.355}$$

Substituting into equation (3.354), we find that the drift of $g_t$ under the pricing measure vanishes, as it ought to for a domestic asset. Hence, $g_t$ can be interpreted as the process for a numeraire asset.

Next, consider equation (3.351) again, but now under the measure having $g_t$ as numeraire. Under this pricing measure the price of risk is $\sigma^g$, hence the lognormal drift is just $(\sigma^g)^2$, and

$$dg_t = (\sigma^g)^2 g_t\, dt + \sigma^g g_t\, dW_t. \tag{3.356}$$

Comparison with equation (3.351) shows that under this measure the drift $\mu$ of the underlying process $x_t$ is $\lambda$, as stated. This implies that the pricing kernel for the quanto option, of volatility given by equation (3.345), is given by equation (3.274) with equation (3.276), as required.

## 3.11 Appendix C: Some Properties of Bessel Functions

Integral relations:[10]

$$\int_0^\infty x^{\nu/2} e^{-\alpha x} I_\nu(2\beta\sqrt{x})dx = \beta^\nu \frac{e^{\beta^2/\alpha}}{\alpha^{\nu+1}}, \tag{3.357}$$

$$\int_0^\infty e^{-\alpha x} I_\nu(2\beta\sqrt{x}) I_\nu(2\gamma\sqrt{x})dx = \frac{e^{(\beta^2+\gamma^2)/\alpha}}{\alpha} I_\nu\left(\frac{2\beta\gamma}{\alpha}\right), \tag{3.358}$$

$$\int_0^\infty e^{-\alpha x} J_\nu(2\beta\sqrt{x}) J_\nu(2\gamma\sqrt{x})dx = \frac{e^{-(\beta^2+\gamma^2)/\alpha}}{\alpha} I_\nu\left(\frac{2\beta\gamma}{\alpha}\right). \tag{3.359}$$

In these integrals, the order of the Bessel functions is such that Re $\nu > -1$.

$$\int xJ_\nu(ax)J_\nu(bx)dx = \frac{x}{a^2 - b^2}\Big[aJ_{\nu+1}(ax)J_\nu(bx) - bJ_\nu(ax)J_{\nu+1}(bx)\Big], \tag{3.360}$$

$$\int xJ_\nu(ax)Y_\nu(bx)dx = \frac{x}{b^2 - a^2}\Big[bJ_\nu(ax)Y_{\nu+1}(bx) - aJ_{\nu+1}(ax)Y_\nu(bx)\Big]. \tag{3.361}$$

[10] Indefinite integrals are given within an arbitrary constant.

$b \neq a$ in equations (3.360) and (3.361).

$$\int xJ_\nu(ax)I_\nu(bx)dx = \frac{x}{a^2+b^2}\Big[aJ_{\nu+1}(ax)I_\nu(bx) + bJ_\nu(ax)I_{\nu+1}(bx)\Big], \tag{3.362}$$

$$\int xI_\nu(ax)Y_\nu(bx)dx = \frac{x}{b^2+a^2}\Big[bI_\nu(ax)Y_{\nu+1}(bx) + aI_{\nu+1}(ax)Y_\nu(bx)\Big], \tag{3.363}$$

$$\int xY_\nu(ax)K_\nu(bx)dx = \frac{x}{a^2+b^2}\Big[aY_{\nu+1}(ax)K_\nu(bx) - bY_\nu(ax)K_{\nu+1}(bx)\Big], \tag{3.364}$$

$$\int xJ_\nu(ax)K_\nu(bx)dx = \frac{x}{a^2+b^2}\Big[aJ_{\nu+1}(ax)K_\nu(bx) - bJ_\nu(ax)K_{\nu+1}(bx)\Big], \tag{3.365}$$

$$\int xJ_\nu^2(ax)dx = \frac{x^2}{2}\Big[J_\nu^2(ax) - J_{\nu-1}(ax)J_{\nu+1}(ax)\Big], \tag{3.366}$$

$$\int xY_\nu^2(ax)dx = \frac{x^2}{2}\Big[Y_\nu^2(ax) - Y_{\nu-1}(ax)Y_{\nu+1}(ax)\Big], \tag{3.367}$$

$$\int xJ_\nu(ax)Y_\nu(ax)dx = \frac{x^2}{4}\Big[2J_\nu(ax)Y_\nu(ax) - J_{\nu+1}(ax)Y_{\nu-1}(ax) - J_{\nu-1}(ax)Y_{\nu+1}(ax)\Big], \tag{3.368}$$

$$\int x^{2\nu+1}Y_\nu^2(x)dx = \frac{x^{2(1+\nu)}}{2(1+2\nu)}\Big[Y_\nu^2(x) + Y_{\nu+1}^2(x)\Big], \tag{3.369}$$

$$\int x^{2\nu+1}J_\nu^2(x)dx = \frac{x^{2(1+\nu)}}{2(1+2\nu)}\Big[J_\nu^2(x) + J_{\nu+1}^2(x)\Big], \tag{3.370}$$

$$\int x^{2\nu+1}J_\nu(x)Y_\nu(x)dx = \frac{x^{2(1+\nu)}}{2(1+2\nu)}\Big[J_\nu(x)Y_\nu(x) + J_{\nu+1}(x)Y_{\nu+1}(x)\Big]. \tag{3.371}$$

The Wronskian $W\,[I_\nu(x), K_\nu(x)] = -1/x$ leads to other useful indefinite integrals:

$$\int \frac{dx}{x[aI_\nu(x)+bK_\nu(x)]^2} = \frac{(1/b)I_\nu(x)}{aI_\nu(x)+bK_\nu(x)}, \qquad b \neq 0 \tag{3.372}$$

or equivalently:

$$\int \frac{dx}{x[aI_\nu(x)+bK_\nu(x)]^2} = \frac{-(1/a)K_\nu(x)}{aI_\nu(x)+bK_\nu(x)}, \qquad a \neq 0. \tag{3.373}$$

Analogous integral identities involving the ordinary Bessel $\{J, Y\}$ pair also obtain from the Wronskian $W[J_\nu(x), Y_\nu(x)] = 2/\pi x$.

Differential equations:

$$Z_\nu''(x) + \frac{1}{x}Z_\nu'(x) - (1+\nu^2/x^2)Z_\nu(x) = 0; \qquad Z_\nu = I_\nu, K_\nu, \tag{3.374}$$

$$Z_\nu''(x) + \frac{1}{x}Z_\nu'(x) + (1-\nu^2/x^2)Z_\nu(x) = 0; \qquad Z_\nu = J_\nu, Y_\nu. \tag{3.375}$$

Recurrence relations:

$$2I'_\nu(x) = I_{\nu-1}(x) + I_{\nu+1}(x), \tag{3.376}$$

$$-2K'_\nu(x) = K_{\nu-1}(x) + K_{\nu+1}(x), \tag{3.377}$$

$$(2\nu/x)I_\nu(x) = I_{\nu-1}(x) - I_{\nu+1}(x), \tag{3.378}$$

$$-(2\nu/x)K_\nu(x) = K_{\nu-1}(x) - K_{\nu+1}(x), \tag{3.379}$$

$$xI'_\nu(x) = \pm\nu I_\nu(x) + xI_{\nu\pm1}(x), \tag{3.380}$$

$$xK'_\nu(x) = \pm\nu K_\nu(x) - xK_{\nu\pm1}(x), \tag{3.381}$$

$$xZ'_\nu(x) = \pm\nu Z_\nu(x) \mp xZ_{\nu\pm1}(x), \tag{3.382}$$

$$(2\nu/x)Z_\nu(x) = Z_{\nu+1}(x) + Z_{\nu-1}(x), \tag{3.383}$$

where $Z_\nu = J_\nu, Y_\nu$. Combining the Wronskian with recurrence relations gives

$$J_\nu(x)Y_{\nu+1}(x) - J_{\nu+1}(x)Y_\nu(x) = \frac{-2}{\pi x}. \tag{3.384}$$

Leading-order asymptotic expansions for $|z| \to \infty$:

$$I_\nu(z) \sim \frac{e^z}{\sqrt{2\pi z}}, \tag{3.385}$$

$$K_\nu(z) \sim \sqrt{\frac{\pi}{2z}}e^{-z}. \tag{3.386}$$

Jump discontinuities across the complex branch cut $z = e^{i\pi}x \to e^{-i\pi}x$, $x > 0$:

$$I_\nu(e^{i\pi}x) - I_\nu(e^{-i\pi}x) = 2i\sin\pi\nu I_\nu(x), \tag{3.387}$$

$$K_\nu(e^{i\pi}x) - K_\nu(e^{-i\pi}x) = -i\pi[I_\nu(x) + I_{-\nu}(x)], \tag{3.388}$$

$$I_\nu(e^{i\pi}x) + I_\nu(e^{-i\pi}x) = 2\cos\pi\nu K_\nu(x). \tag{3.389}$$

Leading order expansions for small argument $z \to 0$

$$I_\nu(z) \sim \frac{1}{\Gamma(\nu+1)}\left(\frac{z}{2}\right)^\nu + O(z^{\nu+2}), \text{ for complex } \nu \neq -1, -2, -3, \ldots \tag{3.390}$$

$$I_\nu(z) \sim \frac{1}{\Gamma(1-\nu)}\left(\frac{z}{2}\right)^{-\nu} + O(z^{2-\nu}), \text{ for } \nu = -1, -2, -3, \ldots \tag{3.391}$$

$$K_\nu(z) \sim \frac{1}{2}\Gamma(|\nu|)\left(\frac{z}{2}\right)^{-|\nu|} + O(z^{2-|\nu|}), \text{ for real } \nu \neq 0 \tag{3.392}$$

# CHAPTER • 4

# Numerical Methods for Value-at-Risk

Portfolios of financial assets are exposed to many types of risks, future events that if they occurred would result in financial losses. The purpose of risk management is to quantify and control these dangers. Value-at-risk (VaR) is a measure of the market risk, the chance of a loss in a company's portfolio caused by unfavorable changes in prices and rates. Minimum risk management standards for financial institutions are set and enforced by national regulators. The Basel Accord [Bas88], the market risk amendment [Bas96a, Bas96b], and the recent update [Bas88] contain the international guidelines implemented by the national agencies.[1] Value-at-risk has become the industry standard for quantifying market risk, partly because of its intuitive appeal and, more importantly, because it is endorsed in the Basel Accord.

For a given portfolio, value-at-risk is defined as the maximum loss forecast over a specified holding period and within a given confidence level (see Figure 4.1). In other words, it is a percentile of the distribution for changes in portfolio value. If $\Delta\Pi$ is the change in portfolio value during the holding period, then value-at-risk is the solution to a nonlinear equation:

$$P[\Delta\Pi \leq -\mathrm{VaR}] = 1 - \alpha, \tag{4.1}$$

where $\alpha$ is the confidence level. Another interpretation is that in the long term we expect losses exceeding value-at-risk with frequency $1-\alpha$. For $\alpha = 99\%$, we expect losses exceeding value-at-risk 1 out of every 100 days. Regulators require value-at-risk to be computed daily with a confidence level of 99% and for a holding period of 10 days. However, since the rules allow for value-at-risk for 1 day to be scaled to approximate the risk for 10 days, we choose to consider daily holding periods in our examples. There are many review papers about value-at-risk simulation: Stambaugh [Sta96] gives a high-level introduction; for more

[1] The Basel Accord and related documents are available from the Bank of International Settlements (www.bis.org).

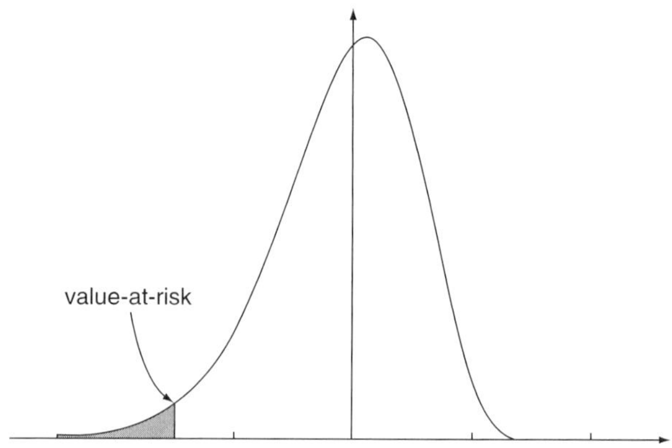


FIGURE 4.1 The probability that a loss is greater than value-at-risk, the density of the shaded region, is equal to $1-\alpha$.

in-depth, general, algorithmic, and mathematical discussions, we have a personal preference for [Mor96a, Hul00, DP97].[2]

In financial markets, risk is caused by uncertainty about the value of an investment in the future. The value of a portfolio is a function of a set of *risk factors*. *Risk factor* is the generic term for a financial variable related to market prices of selected reference securities, for example, equity indices, interest rates, foreign exchange rates, and commodity futures prices. *Market risk* is the risk that the value of a portfolio declines as a consequence of changes in the risk-factor values. Therefore, to model market risk we need to understand how risk factors evolve over time.

Consistently with the hypothesis of absence of arbitrage, we will assume that the changes in risk factors are random. Although historical data is of limited use to predict changes in risk factors, it can be used to estimate statistical models to model risk factors and their correlations. In our examples, we use stocks as elementary risk factors, although the methodology applies to a wide range of financial instruments.

A simple formula for value-at-risk can be obtained in the case where an $n \times 1$ vector of relative changes $\mathbf{R}$ in the market risk factors is a multivariate normal random variable with mean vector $\boldsymbol{\mu}$ and covariance matrix $\mathbf{C}$, and if one assumes that the change in portfolio value can be approximated by an affine function of the relative changes in the risk factors:

$$\Delta\Pi \approx \Xi + \boldsymbol{\Delta}^T\mathbf{R}. \tag{4.2}$$

Throughout this chapter we shall use superscript $T$ to denote the transpose. Note: We are using $\Delta\Pi$ to denote the change in portfolio, i.e., $\Delta\Pi = \Pi_t - \Pi_0$ for a time lapse $t$, whereas $\boldsymbol{\Delta}$ in the dot product is the vector of sensitivities w.r.t. the returns (i.e., the delta Greeks of the portfolio), as defined later.

[2] The Web site www.gloriamundi.org is an excellent source for information and links to papers on value-at-risk.

Since $\Delta\Pi - \Xi = \mathbf{\Delta}^T\mathbf{R}$ is a sum of normal random variables, then it is itself normal. The distribution is determined by its mean and variance,

$$\mu_{(\Delta\Pi-\Xi)} = E[\Delta\Pi - \Xi] = E[\mathbf{\Delta}^T\mathbf{R}] = \mathbf{\Delta}^T E[\mathbf{R}] = \mathbf{\Delta}^T \boldsymbol{\mu}, \tag{4.3}$$

$$\sigma^2_{(\Delta\Pi-\Xi)} = \sigma^2_{\Delta\Pi} = E[(\mathbf{\Delta}^T(\mathbf{R}-\boldsymbol{\mu}))^2] = \mathbf{\Delta}^T E[(\mathbf{R}-\boldsymbol{\mu})(\mathbf{R}-\boldsymbol{\mu})^T]\mathbf{\Delta} = \mathbf{\Delta}^T\mathbf{C}\mathbf{\Delta}. \tag{4.4}$$

So $\Delta\Pi$ is the random normal variable

$$\Delta\Pi = z\sqrt{\mathbf{\Delta}^T\mathbf{C}\mathbf{\Delta}} + \Xi + \mathbf{\Delta}^T\boldsymbol{\mu}, \tag{4.5}$$

where $z \sim N(0, 1)$. Hence, inverting equation (4.1) while using $N^{-1}(1-\alpha) = -N^{-1}(\alpha)$ gives the value-at-risk

$$\mathrm{VaR} = N^{-1}(\alpha)\sqrt{\mathbf{\Delta}^T\mathbf{C}\mathbf{\Delta}} - \Xi - \mathbf{\Delta}^T\boldsymbol{\mu}, \tag{4.6}$$

where $N^{-1}(\cdot)$ is the inverse of the standard normal cdf.

The linear model with normal relative changes has a closed-form solution, but it suffers from two serious problems. First, real-world returns have fatter tails than normal distributions. The model will therefore underestimate the likelihood of extreme returns, which as a consequence may lead to inaccurate estimates of value-at-risk. Second, for portfolios with derivatives, the change in value is a nonlinear function. The local error in the linear approximation will therefore often be unacceptable, a property that is exacerbated by dynamic hedging strategies that use the linearization to eliminate risk locally. To compute value-at-risk for models that take these difficulties into account is a substantially harder task.

Let $S_t$ be the process for a risk factor. Returns on $S_t$ over the time horizon $[0, t]$ can be defined either as arithmetic returns

$$R_t = \frac{S_t - S_0}{S_0} = \frac{\Delta S_t}{S_0}$$

or as the log-return,

$$\tilde{R}_t = \log S_t - \log S_0. \tag{4.7}$$

Log-returns have the advantage that one can aggregate returns over time by addition. In the multivariate case, $S_t$ is a vector of prices and returns are taken componentwise. Of course the two are closely related. The difference,

$$R_t - \tilde{R}_t = \frac{1}{2}\left(\frac{\Delta S_t}{S_0}\right)^2 + O\left(\left[\frac{\Delta S_t}{S_0}\right]^3\right),$$

is typically negligibly small for estimation purposes, and either type of return can safely be approximated by the other. In the examples that follow, we choose log-returns.

Because the return is dimensionless, i.e., the quantity does not have a unit, return models are preferred over models for prices. We consider a model in which the returns, sampled at equally spaced points in time, form a sequence $\{R_i\}_{i=1}^{\infty}$ of independent and identically distributed random variables. This means that stock prices are discrete time Markov chains with an infinite state space [Ros00]. Choosing different distributions gives different models in this family.

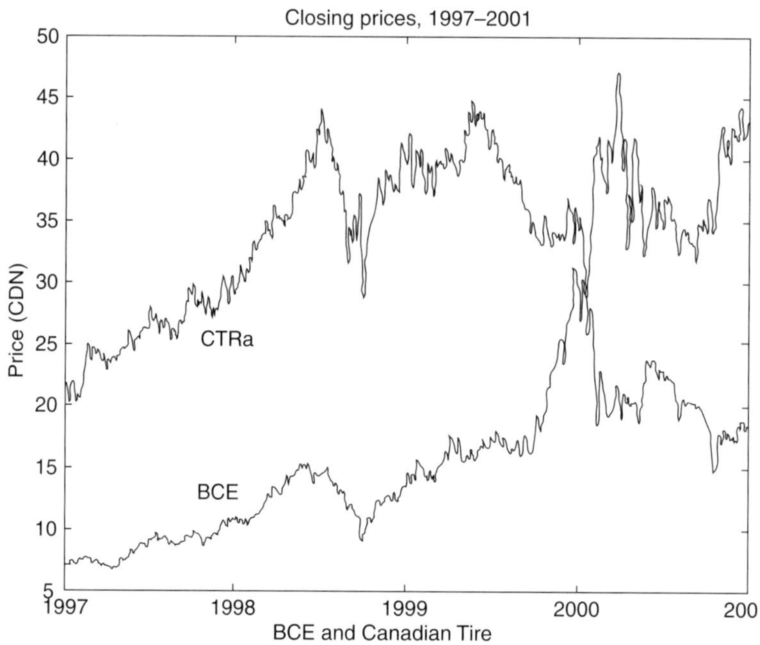


FIGURE 4.2 Daily closing prices for BCE and Canadian Tire from January 1997 to December 2001.

Visual inspection of historical time series gives clues on the key statistical properties. Figure 4.2 shows the daily closing prices over 4 years for two Canadian stocks traded on the Toronto Stock Exchange (TSX): Bell Canada Enterprises (BCE) and Canadian Tire (CTRa). The scatter plot in Figure 4.3 shows that the daily returns form a cloud of samples around the origin in what resembles a multivariate unimodal distribution. The time series can be divided into segments with the same time span as the returns in the model $\{R_i\}_{i=1}^{\infty}$. For each time interval, the relative return can be computed as

$$R_i = \frac{S_i - S_{i-1}}{S_{i-1}}, \qquad i = 1, \ldots, d, \tag{4.8}$$

where $S_{i-1}$ and $S_i$ are, respectively, the prices at the beginning and end of the time interval. Since the returns $\{R_i\}_{i=1}^{\infty}$ in the model are independent and identically distributed, the computed (observed) returns $r_i$ are viewed, rightly or wrongly, as independent samples from the same distribution. After settling on a family of distributions for the random-walk increments, the parameters of this distribution can be estimated from the time series of returns $\{r_i\}_{i=1}^{d}$.

Many generalizations of the random-walk model have been proposed to correct shortcomings revealed in empirical studies; see, for instance, [CLM97]. Over time periods of a few days one can make the simplifying assumption that the returns $\{R_i\}_{i=1}^{\infty}$ are independent and identically distributed. First, for time periods spanning more than a few years, the returns are not identically distributed. To obtain the current reading and forecast for the volatility, it is standard practice either to use only recent data or to use a weighting scheme to attribute a lesser weight to older data or to model the intertemporal dependencies by means of more elaborate statistical models, such as ARCH and GARCH [Eng82, Bol86, Nel91, Hul00].

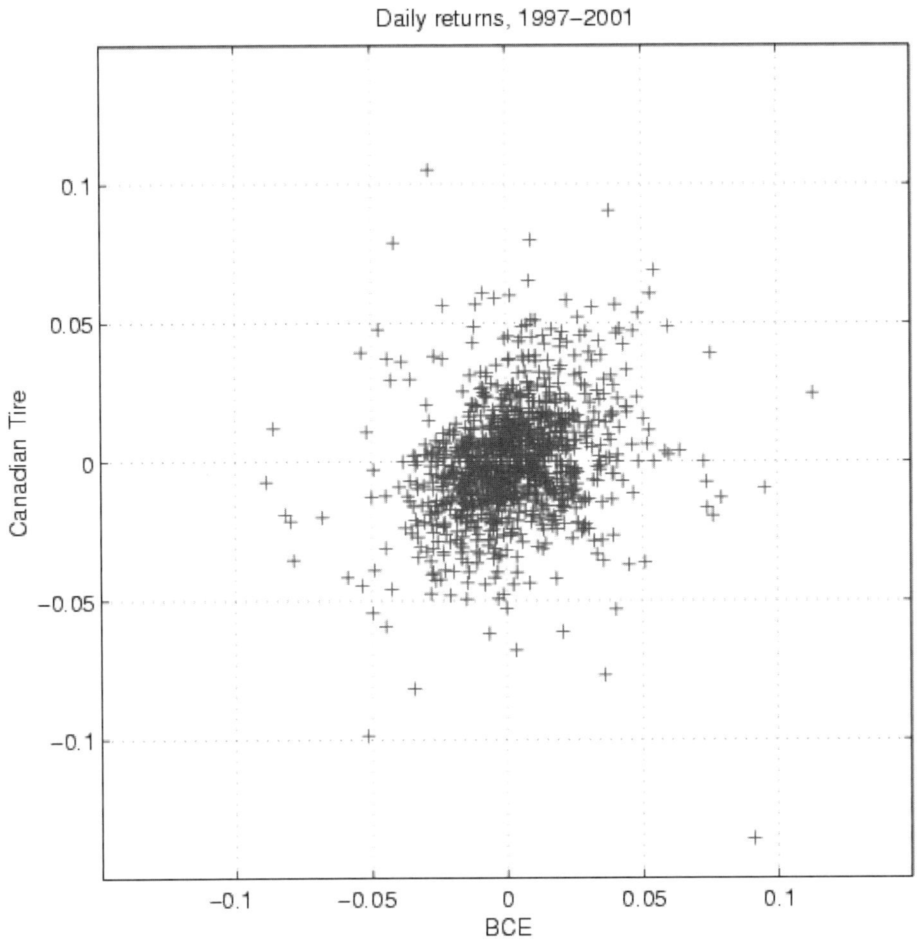


FIGURE 4.3 Scatter plot of relative returns for BCE and Canadian Tire.

## .1 Risk-Factor Models

Recall that, in the random-walk model, returns are modeled as a sequence $\{R_i\}_{i=1}^{\infty}$ of independent and identically distributed random variables. In this section, we discuss three different instances of this model, three different alternatives for the distribution of the random variables: the normal random walk, the asymmetric Student's t-distribution and the nonparametric density estimator due to Parzen [Par61]. The methods will be generalized to the multivariate case in the next section.

### .1.1 *The Lognormal Model*

In the lognormal model, the distribution of log-returns

$$R_i \sim N(\mu, \sigma^2), \qquad i = 1, 2, \ldots, \tag{4.9}$$

is normal with mean $\mu$ and volatility $\sigma$. The mean can be estimated using the sample returns

$$\widehat{\mu} = \frac{1}{d} \sum_{i=1}^{d} r_i \tag{4.10}$$

and the variance by

$$\widehat{\sigma}^2 = \frac{1}{d-1}\sum_{i=1}^{d}(r_i - \widehat{\mu})^2. \tag{4.11}$$

See, for instance, [LM86]. Some authors advocate using estimators that give more weight to recent returns than to old ones (see, for example, [Mor96a, Hul00]).

To illustrate the performance, we estimate the parameters $\widehat{\mu}$ and $\widehat{\sigma}^2$ for daily returns for the BCE time series. Figure 4.4 shows the quantile-quantile plot[3] for the fitted normal distribution. It is clear that the normal model is a good approximation for small returns, but

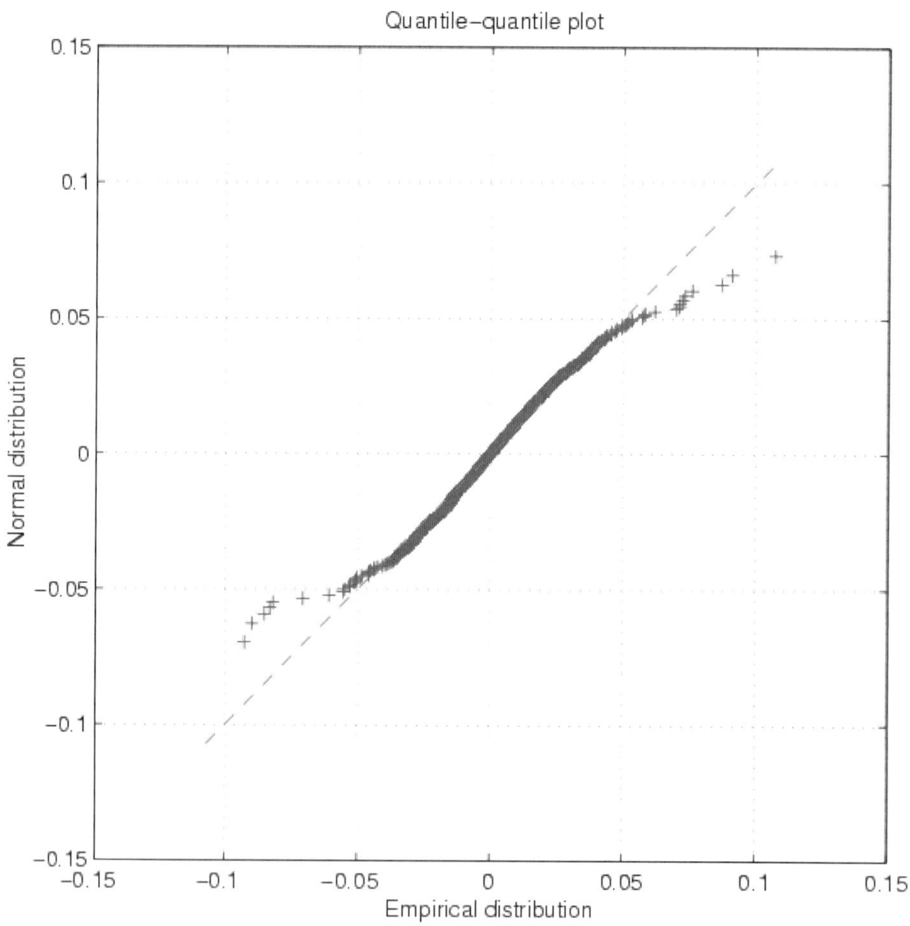


FIGURE 4.4 Quantile-quantile plot for the normal random walk with parameters estimated from 4 years of daily returns for BCE.

[3] A quantile-quantile plot is a method for comparing two distributions. Given a set of observations, we use it to compare the empirical distribution and a distribution fitted to this data. Sorting the observations gives the empirical cumulative distribution functions (cdfs). Each observation, which corresponds to a quantile, and the corresponding quantile for the fitted distribution are marked in the plot. If the two distributions are the same, the points fall on the diagonal reference line. Deviations from the diagonal line indicate that one distribution has fatter or thinner tails with respect to the other. To learn more about this, the reader is referred to the relevant numerical project in Part II.

for both the negative and positive tails the distribution does not fit the data. Fat tails are typical for stock returns; to estimate value-at-risk, where we need to compute tail quantiles, the normal model is less suitable. The next two subsections explore different approaches to construct random-walk models with more realistic tails.

## .1.2 *The Asymmetric Student's t Model*

Student's t-distributions have fat tails. The density for a t-distributed random variable is

$$p_T(x;\nu) = \frac{\Gamma\left(\frac{\nu+1}{2}\right)}{\Gamma\left(\frac{\nu}{2}\right)\sqrt{\nu\pi}}\left(1+\frac{x^2}{\nu}\right)^{-\frac{\nu+1}{2}}, \qquad x \in \mathbb{R}; \tag{4.12}$$

the mean is $\mu = 0$, and the variance for $\nu > 2$ is

$$\sigma^2 = \frac{\nu}{\nu-2}. \tag{4.13}$$

The normalization factor involves the gamma function $\Gamma(\cdot)$. The degrees of freedom $\nu$ control the fatness of the tails; as $\nu \to \infty$, the distribution converges to the normal distribution.

An alternative to the normal model is to define a random walk with t-distributed increments. Since the fatness of the tails can be different for negative and positive returns, we generalize this idea and let each random variable in the sequence $\{R_i\}_{i=1}^{\infty}$ be distributed as

$$A = m + \sigma\sqrt{(1-\rho)\left(\frac{\nu_+ - 2}{\nu_+}\right)}B|T_+| + \sigma\sqrt{\rho\left(\frac{\nu_- - 2}{\nu_-}\right)}(B-1)|T_-|. \tag{4.14}$$

The random variables $T_+$ and $T_-$ are t-distributed with degrees of freedom $\nu_+$ and $\nu_-$, respectively. The random variable $B$ is a Bernoulli random variable; $B$ takes the value 0 or 1 with probability .5. The random variables $T_-$, $T_+$, and $B$ are independent. We say that $A$ is an asymmetric Student's t-distributed random variable. The density, figuratively a density made up of a Student's t pdf cut in half, is

$$p(x) = \begin{cases} \dfrac{p_T\left(\sqrt{\frac{\nu_-}{\nu_- - 2}}\frac{(x-m)}{\sigma\sqrt{\rho}};\nu_-\right)\sqrt{\frac{\nu_-}{\nu_- - 2}}}{\sigma\sqrt{\rho}} & \text{if } x \le m, \\ \dfrac{p_T\left(\sqrt{\frac{\nu_+}{\nu_+ - 2}}\frac{(x-m)}{\sigma\sqrt{1-\rho}};\nu_+\right)\sqrt{\frac{\nu_+}{\nu_+ - 2}}}{\sigma\sqrt{1-\rho}} & \text{if } x > m. \end{cases} \tag{4.15}$$

Since the two regions each make up half of the density, $m$ is the median of the distribution, and, with a little algebra, it is easy to derive moment properties relative to the median. We then have the following result, whose proof is left as an exercise.

**Proposition 4.1.** *Suppose that $\nu_- > 4$ and $\nu_+ > 4$. Then an asymmetric t-distributed random variable, defined by equation (4.14), satisfies the following moment properties:*

*(i) The expectation is*

$$\eta^{(1)} = E[A-m] = \sigma\left[\frac{\Gamma\left(\frac{\nu_+ + 1}{2}\right)\sqrt{(1-\rho)(\nu_+ - 2)}}{\Gamma\left(\frac{\nu_+}{2}\right)\sqrt{\pi}(\nu_+ - 1)} - \frac{\Gamma\left(\frac{\nu_- + 1}{2}\right)\sqrt{\rho(\nu_- - 2)}}{\Gamma\left(\frac{\nu_-}{2}\right)\sqrt{\pi}(\nu_- - 1)}\right].$$

*(ii) The second moment is*

$$\eta^{(2)} = E[(A-m)^2] = \sigma^2.$$

*(iii) The second conditional moments are, for negative values,*

$$\eta_-^{(2)} = E[(A-m)^2 | A \le m] = 2\sigma^2 \rho$$

*and, for positive values,*

$$\eta_+^{(2)} = E[(A-m)^2 | A > m] = 2\sigma^2 (1-\rho).$$

*(iv) The fourth conditional moments are, for negative values,*

$$\eta_-^{(4)} = E[(A-m)^4 | A \le m] = 2\sigma^4 \rho^2 \left[3 + \frac{6}{\nu_- - 4}\right]$$

*and, for positive values,*

$$\eta_+^{(4)} = E[(A-m)^4 | A > m] = 2\sigma^4 (1-\rho)^2 \left[3 + \frac{6}{\nu_+ - 4}\right].$$

Once the moment properties are known, estimating the parameters in the model is straightforward. The first step is to compute the median $\widehat{m}$ of the observed returns $\{r_i\}_{i=1}^d$ by sorting the samples and taking $\widehat{m}$ to be the order-$k$ value if $d = 2k+1$ is odd, or the average of the order-$k$ and-$(k+1)$ values if $d = 2k$ is even. Then find the sample estimate for the second moment

$$\widehat{\sigma}^2 = \frac{1}{d-1} \sum_{i=1}^{d} (r_i - \widehat{m})^2.$$

We then estimate the contribution to the second moment $\widehat{\rho}$ from the negative and the positive halves. Let $d = d_- + d_+$, where $d_-$ and $d_+$ are the number of observations less than and greater than $\widehat{m}$, respectively. Then

$$\widehat{\rho} = \frac{1}{2\widehat{\sigma}^2 d_-} \sum_{r_i \le \widehat{m}} (r_i - \widehat{m})^2.$$

Finally, using the sample estimates for the fourth moments,

$$\widehat{\eta}_-^{(4)} = \frac{2}{d_-} \sum_{r_i \le \widehat{m}} (r_i - \widehat{m})^4 \quad \text{and} \quad \widehat{\eta}_+^{(4)} = \frac{2}{d_+} \sum_{r_i > \widehat{m}} (r_i - \widehat{m})^4,$$

we can solve for estimates of the degrees of freedom $\nu_+$ and $\nu_-$,

$$\widehat{\nu}_- = \frac{6}{\frac{\widehat{\eta}_-^{(4)}}{2\widehat{\sigma}^4 \widehat{\rho}^2} - 3} + 4 \quad \text{and} \quad \widehat{\nu}_+ = \frac{6}{\frac{\widehat{\eta}_+^{(4)}}{2\widehat{\sigma}^4 (1-\widehat{\rho})^2} - 3} + 4.$$

The advantage of the asymmetric t model over the normal model is that, as illustrated by the quantile-quantile plot in Figure 4.5, the tails of the empirical distribution can be reproduced more accurately. However, this improvement comes at a price, since the pdf has a discontinuity at the center. The jump is counterintuitive and the implementation of this model is more difficult, but in comparison to the advantage of increased accuracy these are minor concerns.

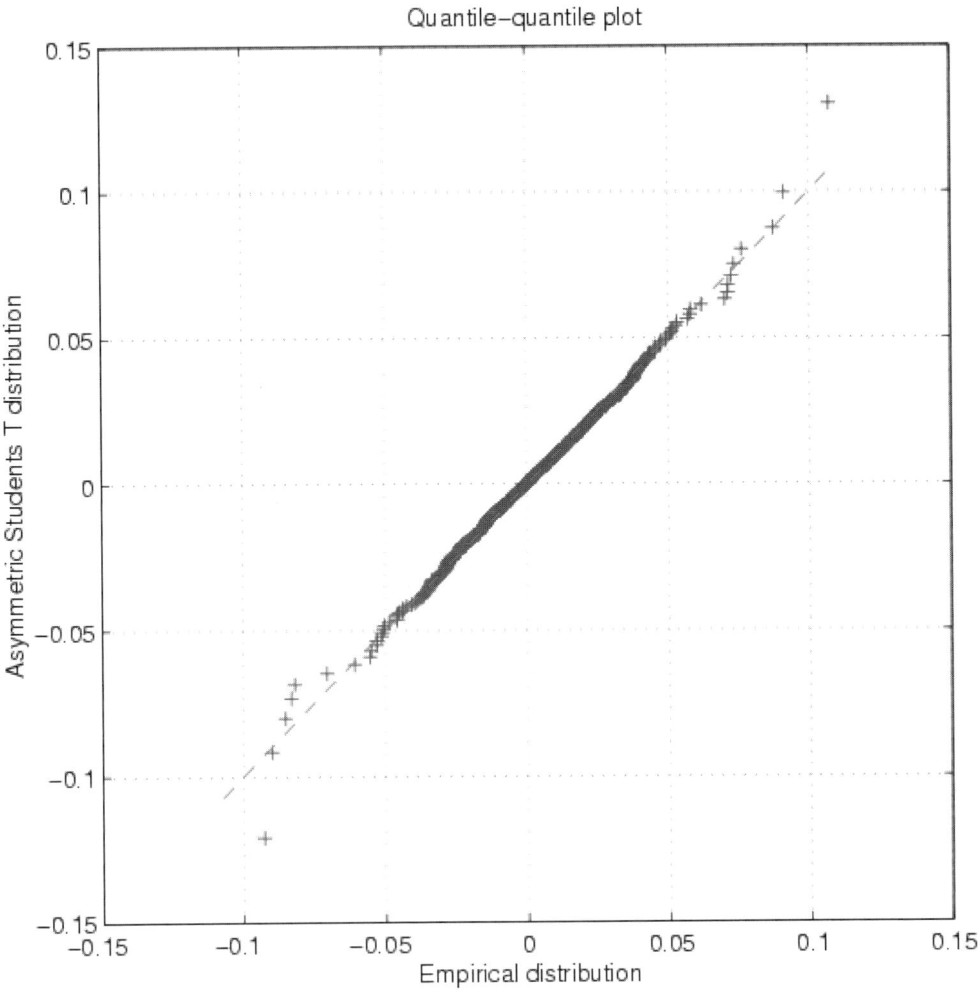


FIGURE 4.5 BCE quantile-quantile plot for the random walk model with the asymmetric t model.

### .1.3 *The Parzen Model*

A nonparametric density estimator is an alternative to using a parametric method, such as either of the first two examples. Let $\{r_i\}_{i=1}^{d}$ be samples from a distribution with an unknown pdf, $p(x)$. In [Par61] Parzen develops and analyzes a family of estimates of the form

$$\widehat{p}_d(x) = \frac{1}{dh}\sum_{i=1}^{d} K\left(\frac{x - r_i}{h}\right), \tag{4.16}$$

initially suggested by Rosenblatt in [Ros56]. In our examples, we use the weighting function [TT90]

$$K(x) = \frac{15}{16}(1 - x^2)^2 \qquad \text{for } |x| \le 1. \tag{4.17}$$

Note that $K(x) \geq 0$ is a kernel function that integrates to unity. Parzen shows that, if $p(x)$ is sufficiently smooth, $\widehat{p}_d(x)$ is asymptotically unbiased and, for an optimal sequence of $h$-values, the mean square error converges to zero as[4]

$$E[(\widehat{p}_d(x) - p(x))^2] = O(d^{-\frac{4}{5}}).$$

We refer to a random walk using the Parzen estimate (4.16) for the pdf as the Parzen model.

Similar to the asymmetric t model, the Parzen model can recreate the fat tails more accurately than the normal model, and it also seems to have a slight advantage over the asymmetric t model, as illustrated by the quantile-quantile plot in Figure 4.6. The advantage

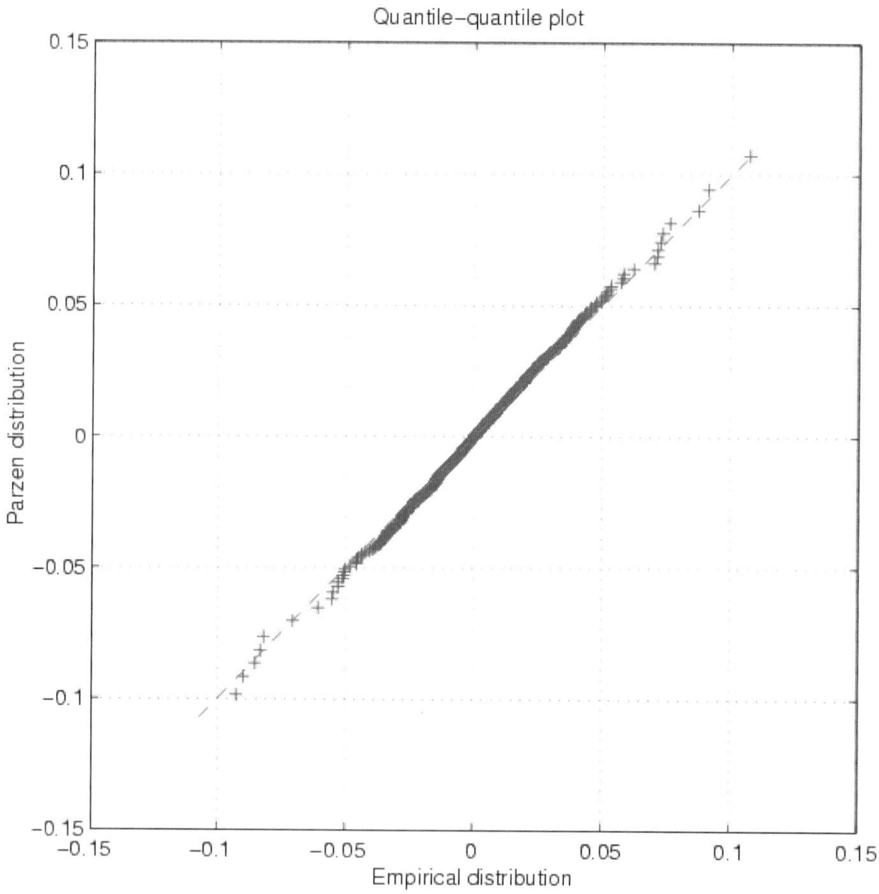


FIGURE 4.6 BCE quantile-quantile plot for the random-walk model with the Parzen density estimate.

[4]Parzen presents a theory for density estimates of the form of equation (4.16), with general weighting functions $K(x)$. Let $h_d \to 0$ as the number of samples $d \to \infty$. He shows that density estimates of the form of equation (4.16) converge (pointwise in a mean square sense) to a continuous pdf as $d \to \infty$, More precisely, given a sequence of smoothing parameters $\{h_d\}_{d=1}^{\infty}$ with $\lim_{d\to\infty} h_d = 0$ and $\lim_{d\to\infty} dh_d = \infty$,

$$E[(\widehat{p}_d(x) - p(x))]^2 \to 0 \qquad \text{as } d \to \infty.$$

The sequence of smoothing parameters giving optimal rate of convergence depends on both the point $x$ and the pdf $p(x)$ as well as the weighting function $K(x)$. See Parzen [Par61] for examples of and details about general weighting functions.

of using a nonparametric model is that it does not rely on specific assumptions about the shape of the density. There are three disadvantages to the Parzen model. First, the optimal smoothing parameter $h$ is unknown. While experimenting with different stocks, we have found that taking $h$ equal to the standard deviation works well.[5] Second, for our choice of weighting function, the density estimate has compact support. However, the support covers the region of interest for value-at-risk calculations, so it should have a minor influence on the result. Third, evaluating equation (4.16) or the corresponding cumulative distribution function (cdf) for different values of $x$ is expensive for large samples. In our implementation, we avoid summing over all sample points by using cubic splines to approximate the cdf and the pdf.

## 1.4 Multivariate Models

So far we have only considered models for the return on a single risk factor. In general, portfolios depend on many risk factors. Therefore we must extend the one-dimensional random-walk models, presented in the previous sections, to the multivariate case.

In the multivariate random walk, $\{\mathbf{R}_i\}_{i=0}^{\infty}$ is a sequence of $\mathbb{R}^n$-valued vectors of random variables. The random vectors are independent and identically distributed. The difficulty in constructing a realistic multivariate model is that returns on the risk factors are typically dependent, as exemplified by Figure 4.7. To approximate the dependence structure without introducing an overly complex model, we restrict our attention to multivariate models where the random vectors $\{\mathbf{R}_i\}_{i=1}^{\infty}$ satisfy

$$\mathbf{R}_i = \mathbf{A}^{-1}\mathbf{X}_i + \mathbf{b}. \tag{4.18}$$

Moreover, we assume that the random vector $X$ has independent components and the pdf is a product of one-dimensional density functions:

$$p(x) = p_1(x_1)\cdots p_n(x_n).$$

We postpone the discussion about how to choose the linear transformation, i.e., the matrix $\mathbf{A}$ and the vector $\mathbf{b}$, to Section 4.3, after discussing portfolios of derivatives.

To find a stochastic process to model stock prices in continuous time is a more difficult problem. Returns are often modeled by stochastic differential equations (SDEs). As discussed in Chapter 1, Brownian motion is the natural continuous-time generalization of a random walk with normal increments. In this model, the return process is a constant-coefficient SDE, $dR = \mu\ dt + \sigma\ dW_t$. Like the normal model for stock prices, geometric Brownian motion underestimates the likelihood of large returns: It does not have fat tails.

Many different types of continuous-time models have been proposed and studied in the literature, in particular for pricing derivatives. If the returns are a stationary Markov process, then, for example, the sequence $\{r_i\}_{i=1}^{d}$ of historical returns can be used to find an estimate for the transition density — the time-dependent probability density $p(r, t)$ representing the density for the return $r$ at time $t$. Figure 4.8 shows the Parzen estimate for the transition density for the stock BCE. A good model has a transition density that is close to this estimate.

[5] This choice of $h$ may work well in our examples, but it is not a satisfactory solution in general since a fixed smoothing parameter does not give convergence as the number of samples $d \to \infty$. The estimate converges for a sequence of smoothing parameters that decrease to zero as the number of samples increases (see [Par61] for details).

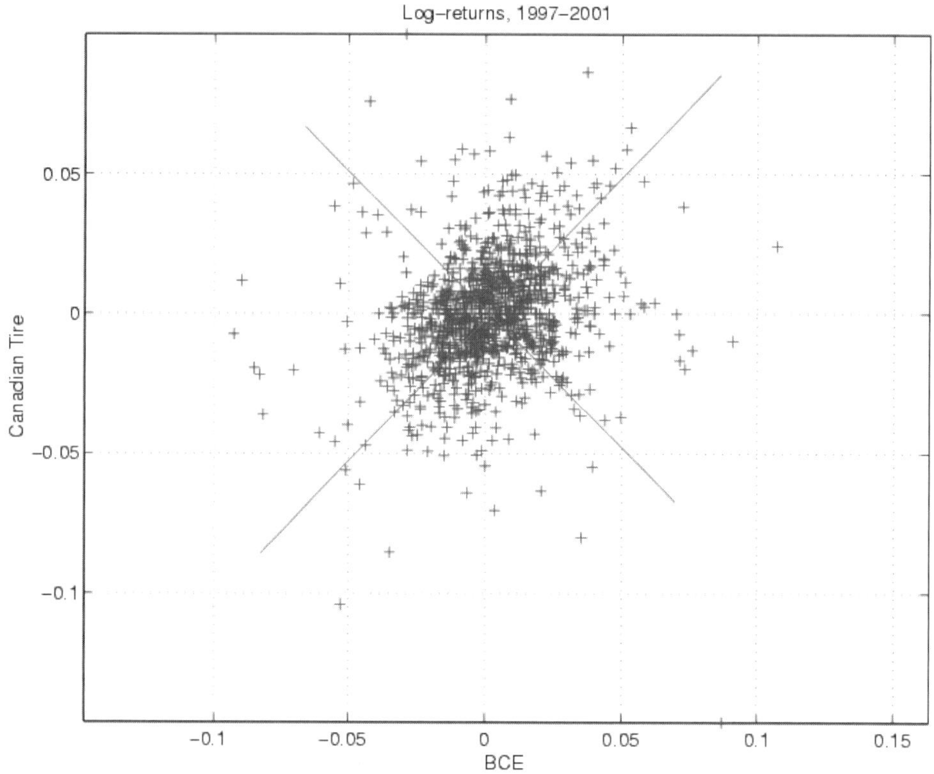


FIGURE 4.7 Principal components superimposed on the scatter plot for the returns on BCE and CTRa.

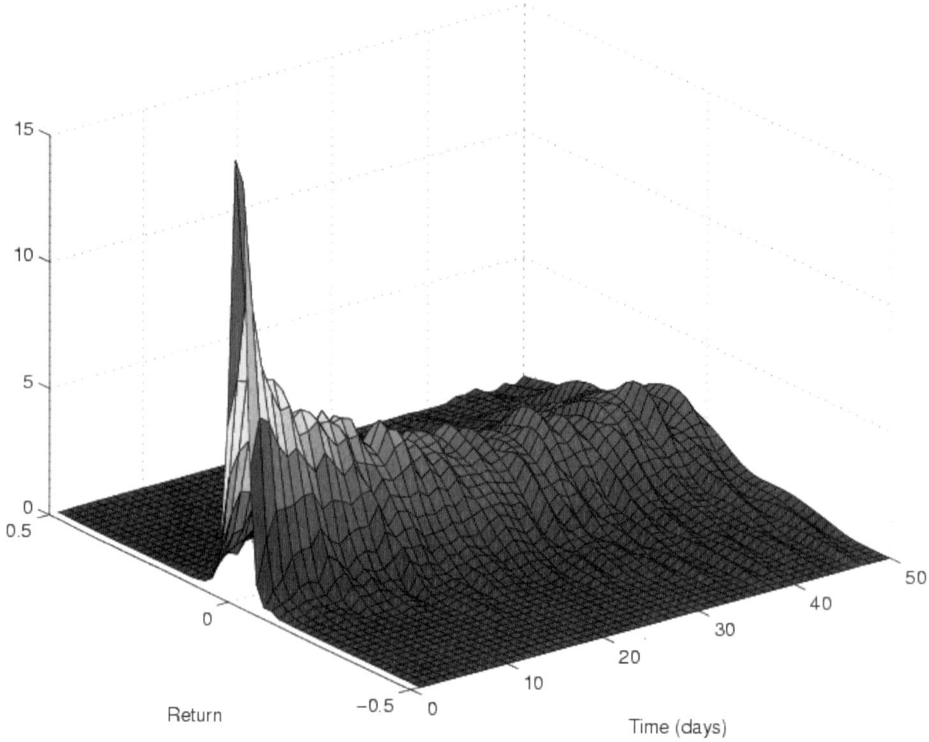


FIGURE 4.8 Parzen estimate for BCE to the transition density.

## 2 Portfolio Models

In this section we discuss portfolios and introduce a new method for portfolio-dependent parameter estimation. The purpose of this section is first to connect the idea of a portfolio and the changes in its value to the risk-factor models considered in the previous section. The second purpose is to give a detailed presentation of a portfolio-dependent estimation procedure and to discuss the related computational issues.

A portfolio is represented as a $k$-vector $\boldsymbol{\theta}$, where $\theta_i$ is the position in the $i$th security. When considering a market without derivatives one assumes the number of securities $k$ and the number of risk factors $n$ are the same, $k = n$. Generally, in a market with derivatives, the number of securities is greater than the number of risk factors, $k > n$. In practice, if $\theta_i$ represents shares, then it is an integer; but in modeling portfolios it is convenient to let $\theta_i$ be a real number. Furthermore, in a market where short selling is allowed, $\theta_i$ can be either positive or negative.

Let $V_t^i$ be the price of the $i$th security at time $t$. If the security is a direct investment in the risk factor, then $V_t^i$ satisfies the identity

$$V_t^i = S_t^i = S_0^i(1 + R_t^i). \tag{4.19}$$

If, on the other hand, it is a derivative security, then $V_t^i$ is a function of the risk factors. Assuming that the portfolio remains unchanged, the dollar value of the portfolio at time $t$ is the sum

$$\Pi_t = \sum_{i=1}^{k} \theta_i V_t^i. \tag{4.20}$$

The change in the value from time 0 to time $t$ is

$$\Delta\Pi_t = \Pi_t - \Pi_0 = \sum_{i=1}^{k} \theta_i (V_t^i - V_0^i). \tag{4.21}$$

If the Black–Scholes model were correct, the drift would be the only difference between the stock price processes for the probability spaces with the risk-neutral versus the real-world measures. However, this approach does not reproduce observable prices for traded options. Therefore, instead of using a volatility estimated from stock price data, pricing models typically use parameters implied by option prices; i.e., the pricing model is used as a form of interpolation scheme. In value-at-risk simulation, we are interested in changes in value over a short time period, and what is needed is a model that captures the local dynamics of the value. Hence, compared to the models used for derivatives trading, where the whole lifespan of the contract must be considered, the quality of the pricing model is less critical. In our examples we use the Black–Scholes model to construct such local approximations, but the ideas could in principle be extended to more complex pricing models.

Two problems we touch lightly upon in this chapter are volatility risk and mapping of risk factors. Both topics are important in the implementation of market-risk models. As mentioned, parameters of option-pricing models must be chosen to reproduce prices in the market. Unfortunately, parameters such as the volatility $\sigma$ in the Black–Scholes model tend to change over time. Therefore, a natural extension is to make volatility stochastic (see, for example, [Wil00]). This makes potential changes in volatility a source of risk, and it can be introduced as a risk factor in a value-at-risk model. In Sections 4.5 and 4.7, we study a very simple version of such a model, and we see that it leads to some interesting qualitative changes to the problem. Mapping of risk factors is the process where some risk factors are replaced by a few general factors. An example is replacing a continuum of interest rates with different maturities by a few representative rates. The dimension-reduction problem in Section 4.7 can be viewed as an automatic mapping method.

As we have seen, the price for a derivative may be a complicated function. Often there is no explicit formula and it must be priced using a separate simulation. Furthermore, the number of different types of securities in a derivatives portfolio will typically be much larger than the number of basic risk factors. In such cases, Taylor's theorem provides a tool to approximate the value of a portfolio by a function with a simple mathematical form.

## 4.2.1 Δ-Approximation

Taylor approximations are accurate close to the point of expansion if the function is sufficiently smooth. Under this assumption and using a first-order approximation, we obtain[6]

$$\Pi_t = \Pi_0 + \left( \sum_{i=1}^{k} \theta_i \frac{\partial V_t^i}{\partial t} \right) t + \sum_{j=1}^{n} \left( \sum_{i=1}^{k} \theta_i \frac{\partial V_t^i}{\partial s_j} \right) S_0^j r_j + O(t^2 + r_1^2 + \cdots + r_n^2).$$

The $\mathbb{R}^n$-valued vector $\mathbf{R}_t$ of returns has components $R_t^j = r_j = (s_j - S_0^j)/S_0^j$, where $S_0^j$ are initial prices and $S_t^j = s_j$, $j = 1, \ldots, n$, are time-$t$ prices of the underlying assets. Collecting the coefficients, we obtain

$$\Pi_t \approx \Pi_0 + \Theta t + \mathbf{\Delta}^T \mathbf{R}_t \tag{4.22}$$

where $\mathbf{\Delta}^T = (\Delta_1, \ldots, \Delta_n)$,

$$\Theta = \left( \sum_{i=1}^{k} \theta_i \frac{\partial V_t^i}{\partial t} \right) \quad \text{and} \quad \Delta_j = \left( \sum_{i=1}^{k} \theta_i \frac{\partial V_t^i}{\partial s_j} \right) S_0^j. \tag{4.23}$$

In finance, such a linear approximation is called a Δ-approximation. As seen in Chapter 1, the $\Delta_j$ are often used to hedge a portfolio. By taking a position, for example, by buying the risk factor or future contracts, that offsets the derivative, a portfolio's sensitivity to changes in the underlying can be reduced. As a consequence, in risk models for derivative portfolios a significant component of the risk will be made up of higher-order effects.

Consider a call option on a single stock, and suppose the current price of the underlying $S_0$ is equal to the strike price, $K = 100$, i.e., the option is at-the-money. The derivatives of the call option value give[7]

$$\frac{\partial V_t}{\partial s} = N(d_+), \tag{4.24}$$

$$\frac{\partial V_t}{\partial t} = -\frac{s}{2} N'(d_+) \frac{\sigma}{\sqrt{T-t}} - rKe^{-r(T-t)} N(d_-), \tag{4.25}$$

[6] Taylor's theorem gives an explicit formula for the error. For the first-order approximation the error is

$$\text{error} = \frac{1}{2!} \left( \sum_i \theta_i \frac{\partial^2 V_t^i}{\partial t^2} \right) (t')^2 + \frac{1}{2!} \left( \sum_{i,j} \theta_i \frac{\partial^2 V_t^i}{\partial t \partial s_j} S_0^j r_j' \right) t' + \frac{1}{2!} \sum_{i,j,k} \theta_i \frac{\partial^2 V_t^i}{\partial s_j \partial s_k} S_0^j S_0^k r_j' r_k',$$

where the derivatives are evaluated at some $t' \in (0, t)$ and $r_j' \in (0, r_j)$ for $j = 1, \ldots, n$.

[7] The Δ and Θ are derived by differentiating the Black–Scholes equation, using $d_- = d_+ - \sigma\sqrt{T-t}$, as discussed in Chapter 1.

where $d_{\pm} = \frac{\log(s/S_0)+(r\pm\sigma^2/2)(T-t)}{\sigma\sqrt{T-t}}$, and we set the risk-free rate equal to the return $r = (s-S_0)/S_0$, with $s$ as the spot. Choosing appropriate parameters,

$$\Delta = \frac{\partial V_t}{\partial r} = \frac{\partial V_t}{\partial s} S_0 \approx 59.77 \qquad \text{and} \qquad \Theta \approx -8.12,$$

and the $\Delta$-approximation of the gain of the call option is

$$\Pi_t \approx \Pi_0 + \Theta t + \Delta r \approx 6.89 - 8.12t + 59.77r. \tag{4.26}$$

Figure 4.9 shows the $\Delta$-approximation (as a function of the stock price $s$ for one day, $t = 1/250$) compared to the Black–Scholes price. It is accurate for small returns but quickly deteriorates as $|r|$ increases. Finally, when the return $|r|$ is large enough, the approximation is negative or less than $(s - Ke^{-r(T-t)})_+$, and therefore it violates the basic principle of no arbitrage.

### .2.2 ΔΓ-Approximation

For nonlinear portfolios, the $\Delta$-approximation is useful when the time period considered is relatively small. But as Figure 4.9 illustrates, ignoring the curvature for an option portfolio may lead to a large truncation error. This feature becomes particularly important if $\Delta$-hedging is used, since the linear term is hedged out, leaving a higher-order residual for the portfolio value.

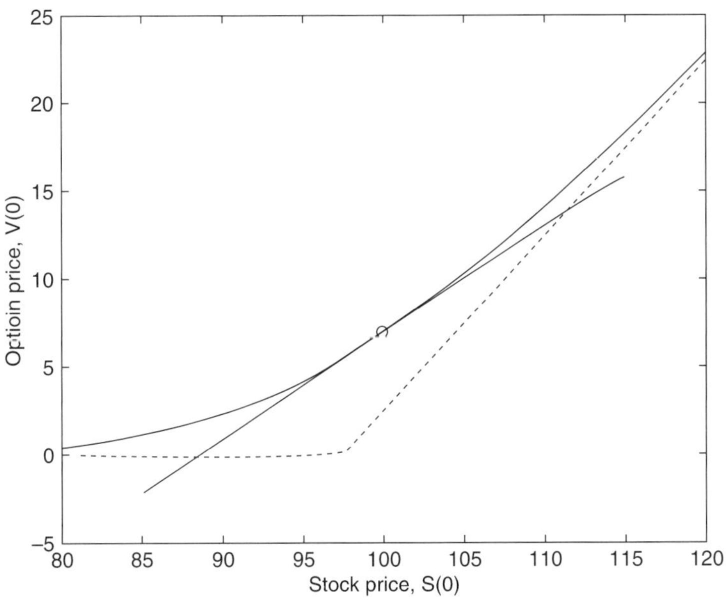


FIGURE 4.9 The price of a European call option in the Black–Scholes model compared to the at-the-money $\Delta$-approximation for the value.

The nonlinearity can be better approximated by including more terms from the Taylor series. Keeping terms in second order in the returns gives the quadratic approximation[8]

$$\Pi_t = \Pi_0 + \left(\sum_{i=1}^{k} \theta_i \frac{\partial V_t^i}{\partial t}\right) t + \sum_{j=1}^{n} \left(\sum_{i=1}^{k} \theta_i \frac{\partial V_t^i}{\partial s_j}\right) S_0^j r_j$$
$$+ \frac{1}{2!} \sum_{l=1}^{n} \sum_{j=1}^{n} \left(\sum_{i=1}^{k} \theta_i \frac{\partial^2 V_t^i}{\partial s_j \partial s_l}\right) S_0^j S_0^l r_j r_l + O(t^2 + t\sum_i r_i + \sum_i r_i^3)$$

or

$$\Pi_t \approx \Pi_0 + \Theta t + \mathbf{\Delta}^T \mathbf{R}_t + \frac{1}{2} \mathbf{R}_t^T \mathbf{\Gamma} \mathbf{R}_t. \tag{4.27}$$

The vector $\mathbf{\Delta}$ and the scalar $\Theta$ are as in equation (4.22), and

$$\Gamma_{jl} = \left(\sum_{i=1}^{k} \theta_i \frac{\partial^2 V_t^i}{\partial s_j \partial s_l}\right) S_0^j S_0^l, \qquad j, l = 1, 2, \ldots, n.$$

In finance, quadratic approximation (4.27) is called a $\Delta\Gamma$-approximation. Because it approximates the curvature of the value function, it is a more accurate local approximation to the portfolio value than the linear $\Delta$-approximation.

We return to the (single risk factor) example with $\Delta$-approximation (4.26) for a call option. In this case, the second derivative for the value is

$$\frac{\partial^2 V_t}{\partial s^2} = \frac{N'(d_+)}{s\sigma\sqrt{T-t}}.$$

Then

$$\Gamma = \frac{\partial^2 V_t}{\partial s^2} S_0^2 \approx 273.6,$$

which gives the $\Delta\Gamma$-approximation

$$\Pi_t \approx \Pi_0 + \Theta t + \Delta r + \Gamma \frac{r^2}{2} \approx 6.89 - 8.12t + 59.77r + 273.6\frac{r^2}{2}, \tag{4.28}$$

where $r = (s - S_0)/S_0$. The approximation compared to the exact values is shown in Figure 4.10. In comparison to the $\Delta$-approximation, the $\Delta\Gamma$-approximation is, not surprisingly, much closer to the price in the Black–Scholes model.

[8] The truncation error is

$$\text{error} = \frac{1}{2!}\left(\sum_i \theta_i \frac{\partial^2 V_t^i}{\partial t^2}\right) t^2 + \frac{1}{2!}\left(\sum_{i,j} \theta_i \frac{\partial^2 V_t^i}{\partial t \partial s_j} S_0^j r_j'\right) t'$$
$$+ \frac{1}{3!} \sum_{i,j,k} \left(\theta_i \frac{\partial^3 V_t^i}{\partial t \partial s_j \partial s_k} S_0^j S_0^k r_j' r_k'\right) t' + \frac{1}{3!} \sum_{i,j,k,l} \theta_i \frac{\partial^3 V_t^i}{\partial s_j \partial s_k \partial s_l} S_0^j S_0^k S_0^l r_j' r_k' r_l'$$

for some $t' \in (0, t)$ and $r_j' \in (0, r_j)$ for $j = 1, \ldots, n$.

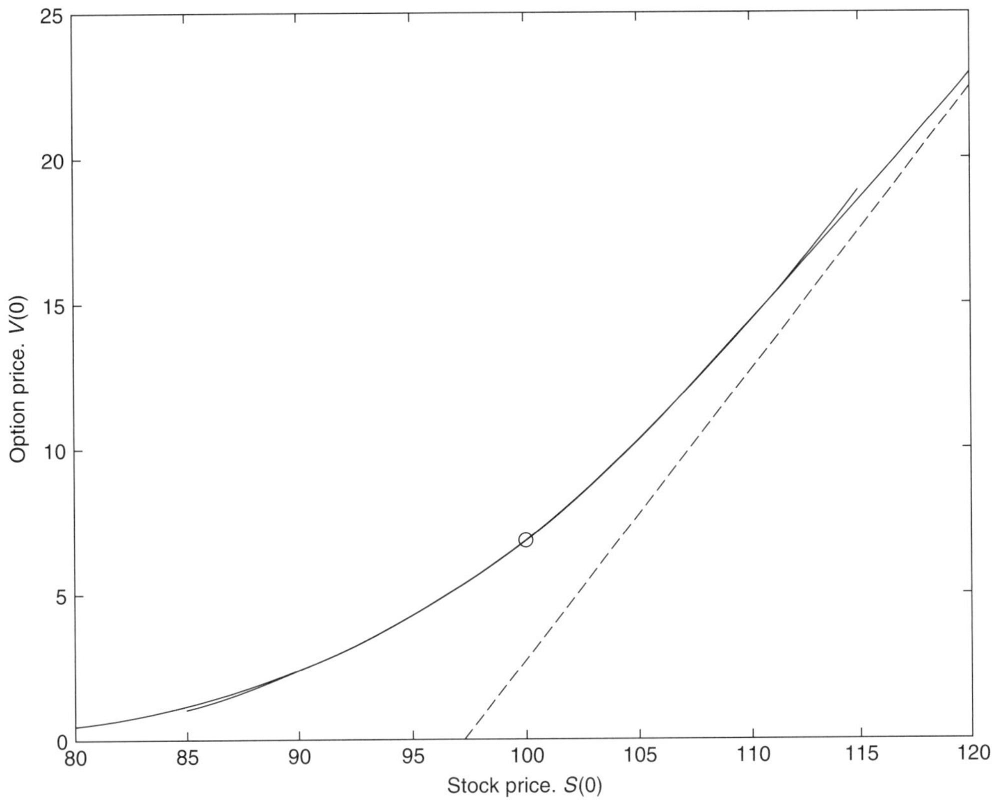


FIGURE 4.10 The price of a European call option compared to the at-the-money ΔΓ-approximation.

## 4.3 Statistical Estimations for ΔΓ-Portfolios

In the algorithm developed later in Section 4.5, we assume that derivative portfolios are assumed to be represented as ΔΓ-approximations (4.27). The change in value, over the time period from time 0 to time $t$, is then

$$\Delta\Pi_t = \Pi_t - \Pi_0 \quad \approx \quad \Delta\tilde{\Pi}_t = \Theta t + \mathbf{\Delta}^T \mathbf{R}_t + \frac{1}{2}\mathbf{R}_t^T \mathbf{\Gamma} \mathbf{R}_t. \tag{4.29}$$

The return vector $\mathbf{R}_t$ on the risk factors is a vector of random variables, and, when $\Delta\Pi_t$ is viewed as a function of $\mathbf{R}_t$, it is a random variable modeling the change in the portfolio's value over the time period $t$. We assume that the stochastic model for returns is of the type presented in Section 4.1.

The parameters of the risk-factor model are estimated, independent from the pricing model, from a time series of historical returns. This choice comes at the cost of ignoring any connections, suggested by the theory of arbitrage-free pricing, about the relation of the real-world and risk-neutral processes. However, it has the important advantage of more flexibility in choosing the underlying model for price changes. Furthermore, as illustrated by the Black–Scholes model, the conclusion that risk-neutral and real-world processes have the same variance, properties implied by the pricing model, may not hold up to scrutiny. For these reasons, we believe that this pragmatic approach is justified.

### 4.3.1 Portfolio Decomposition and Portfolio-Dependent Estimation

In this section, we present a new method for portfolio-dependent estimation of parameters for the risk-factor models. The strategy for parameter estimation builds on the observation that a $\Delta\Gamma$-approximation can be decomposed as a sum of one-dimensional quadratic functions. This decomposition has a long history in applied mathematics and statistics, and it has been used by other authors in the computation of value-at-risk. Still, the multivariate risk-factor model we present is fundamentally different, in that the resulting risk-factor models are portfolio dependent.

Suppose that $\{\mathbf{r}_i\}_{i=1}^d$ is a time series of returns, where $\mathbf{r}_i$ is an $n$-vector with component observations $r_{ji}$, $j = 1, \ldots, n$. We know, as discussed in Section 4.1, that the mean $\widehat{\boldsymbol{\mu}}$ can be estimated with standard sample estimators (in the case of the asymmetric t model, the mean is replaced by the median $\widehat{m}$). Similarly, the standard sample statistics

$$\widehat{C}_{ij} = \frac{1}{d-1}\sum_{k=1}^{d}(r_{ik} - \widehat{\mu}_i)(r_{jk} - \widehat{\mu}_j), \tag{4.30}$$

can be used to estimate the matrix elements $C_{ij} = \mathrm{Cov}(R_t^i, R_t^j)$ of the covariance matrix. For the normal model, $\widehat{\boldsymbol{\mu}}$ and $\widehat{\mathbf{C}}$ characterize the model completely. For the other two models, we explain how to estimate the remaining parameters.

Recall that, as discussed in Section 4.1.4, we want to approximate the dependence structure of the risk-factor returns with a product pdf. The first step to construct such a model is to observe that a $\Delta\Gamma$-approximation can be factored by solving the generalized eigenvalue problem

$$\begin{cases} \boldsymbol{\Lambda} = \mathbf{A}^T\boldsymbol{\Gamma}\mathbf{A}, \\ \widehat{\mathbf{C}} = \mathbf{A}\mathbf{A}^T. \end{cases} \tag{4.31}$$

The matrix $\mathbf{A}$ is nonsingular and

$$\boldsymbol{\Lambda} = \mathrm{diag}(\lambda_1, \ldots, \lambda_n).$$

Let

$$\mathbf{X} = \mathbf{A}^{-1}(\mathbf{R}_t - \widehat{\boldsymbol{\mu}})$$

define a new vector of random variables. Then, since $E[\mathbf{R}_t - \widehat{\boldsymbol{\mu}}] = \mathbf{0}$ and $\widehat{\mathbf{C}} = E[(\mathbf{R}_t - \widehat{\boldsymbol{\mu}})(\mathbf{R}_t - \widehat{\boldsymbol{\mu}})^T]$, we have

$$E[\mathbf{X}] = \mathbf{0} \qquad \text{and} \qquad E[\mathbf{X}\mathbf{X}^T] = \mathbf{I}. \tag{4.32}$$

Also, by substituting $\mathbf{R}_t = \mathbf{A}\mathbf{X} + \widehat{\boldsymbol{\mu}}$ we can express $\Delta\widetilde{\Pi}_t$ in terms of our new variables:

$$\Delta\widetilde{\Pi}_t = \left(\Theta t + \widehat{\boldsymbol{\mu}}^T\boldsymbol{\Delta} + \frac{1}{2}\widehat{\boldsymbol{\mu}}^T\boldsymbol{\Gamma}\widehat{\boldsymbol{\mu}}\right) + \mathbf{X}^T\mathbf{A}^T(\boldsymbol{\Delta} + \boldsymbol{\Gamma}\widehat{\boldsymbol{\mu}}) + \frac{1}{2}\mathbf{X}^T\boldsymbol{\Lambda}\mathbf{X}. \tag{4.33}$$

Defining

$$\Xi \equiv \Theta t + \widehat{\boldsymbol{\mu}}^T\boldsymbol{\Delta} + \frac{1}{2}\widehat{\boldsymbol{\mu}}^T\boldsymbol{\Gamma}\widehat{\boldsymbol{\mu}}, \tag{4.34}$$

$$\boldsymbol{\Delta}' \equiv \mathbf{A}^T(\boldsymbol{\Delta} + \boldsymbol{\Gamma}\widehat{\boldsymbol{\mu}}) \tag{4.35}$$

gives the ΔΓ-approximation to the change in portfolio value:

$$\Delta\widetilde{\Pi}_t = \Xi + \sum_{i=1}^{n} \left(\Delta_i' X_i + \frac{\lambda_i}{2} X_i^2\right). \tag{4.36}$$

This is simply a sum of one-dimensional quadratic functions

$$\pi_i = \Delta_i' x_i + \frac{\lambda_i}{2} x_i^2. \tag{4.37}$$

We postulate that the joint probability density for $\mathbf{X} = \mathbf{x}$ is a product pdf,

$$p(\mathbf{x}) = p_1(x_1) \cdots p_n(x_n), \tag{4.38}$$

or equivalently that the components of $\mathbf{X}$ are independent. As equation (4.32) shows, the components of $\mathbf{X}$ are uncorrelated by construction, so the approximation with a product pdf extrapolates from this property to a more general assumption. Finally, the remaining parameters for the asymmetric t and Parzen models can be estimated for each component $X_i$ individually. The complete parameter estimation and factorization procedure is summarized in Algorithm 1.

**Algorithm 1**
**Parameter Estimation and Factorization**

**Input:** *A ΔΓ-approximation. A time series* $\{\mathbf{r}_i\}_{i=1}^{d}$ *of daily returns.*
**Output:** *Parameter estimates for the market model. A factorization of the ΔΓ-approximation.*

- *Compute estimates for the mean* $\widehat{\boldsymbol{\mu}}$ *and the covariance matrix* $\widehat{\mathbf{C}}$.
- *Solve the eigenvalue problem*

$$\begin{cases} \boldsymbol{\Lambda} = \mathbf{A}^T \boldsymbol{\Gamma} \mathbf{A}, \\ \widehat{\mathbf{C}} = \mathbf{A}\mathbf{A}^T. \end{cases}$$

- *Compute*

$$\Xi = \Theta t + \widehat{\boldsymbol{\mu}}^T \boldsymbol{\Delta} + \frac{1}{2} \widehat{\boldsymbol{\mu}}^T \boldsymbol{\Gamma} \widehat{\boldsymbol{\mu}},$$

$$\boldsymbol{\Delta}' = \mathbf{A}^T (\boldsymbol{\Delta} + \boldsymbol{\Gamma} \widehat{\boldsymbol{\mu}}).$$

- *For each variable* $X_i$, *estimate the remaining parameters from the transformed returns* $\{\mathbf{A}^{-1}(\mathbf{r}_i - \widehat{\boldsymbol{\mu}})\}_{i=1}^{d}$ *using the methods in Section 4.1.*

## 3.2 *Testing Independence*

The assumption that the risk-factor returns can be modeled by a product pdf is central to what follows in Section 4.5. It is therefore natural to question whether it is a valid assumption.

We note that returns tend to be scattered around a central point, as in Figure 4.3. Also, the parameter estimation procedure produces uncorrelated random variables for which the sample correlation is zero by construction. These two points form the basis for our belief that a product pdf is a reasonable model.

For our two-dimensional example, with returns on BCE and CTRa, it is possible to construct a statistical test.[9] To test the assumption, we use a binomial test, as discussed, for example, in [LM86]. As we will show, the experiment suggests that in this case, the independence assumption is valid.[10]

Consider a portfolio with at-the-money call options on BCE and CTRa. Using the Black–Scholes model to price this portfolio, we get a $\Delta\Gamma$-approximation with

$$\boldsymbol{\Delta} = \begin{bmatrix} 25.4932 \\ 10.9513 \end{bmatrix} \quad \text{and} \quad \boldsymbol{\Gamma} = \begin{bmatrix} 69.4300 & 0 \\ 0 & 29.7313 \end{bmatrix}.$$

The portfolio-dependent estimation procedure, with the standard sample estimates and the data in Figure 4.3, gives us

$$\lambda_1 = 0.0064, \qquad \Delta_1' = 0.1371,$$
$$\lambda_2 = 0.0169, \qquad \Delta_2' = 0.6382.$$

We want to test if the two portfolio components $\pi_1$ and $\pi_2$ are independent. To formulate this question as a binomial test, viewing $\pi_1$ and $\pi_2$ as random variables, define the events

$$A_1 = \{\pi_1 : \pi_1 > \widehat{\mu}_{\pi_1}\},$$
$$A_2 = \{\pi_2 : \pi_2 > \widehat{\mu}_{\pi_2}\},$$

where $\widehat{\mu}_{\pi_i}$ is the sample mean for $\pi_i$. From the time series we estimate the probabilities of the events:

$$\widehat{p}_1 = \frac{\#\text{ samples in } A_1}{\#\text{ samples}} = \frac{476}{1006},$$
$$\widehat{p}_2 = \frac{\#\text{ samples in } A_2}{\#\text{ samples}} = \frac{490}{1006}.$$

[9] In fact, the possibilities for statistical tests are infinite; see, for example, [Feu93] and the references therein. If $X$ and $Y$ are independent, then for any functions $h$ and $g$

$$E[h(X)g(Y)] = E[h(X)]E[g(y)],$$

provided the integrals exist. This equality is taken as the null hypothesis for a statistical test. Given a set of samples, it is possible to compare sample estimates for the left- and right-hand sides. By the law of large numbers, the two converge for independent random variables. If it is possible to find confidence intervals for the sample estimates, then the estimates can be tested to accept or reject the null hypothesis.

Statistical tests cannot prove independence; they can only *reject* independence through rejection of the null hypothesis. Furthermore, although the same idea could in principle be used in high dimensions, formulating tests with sufficient power and obtaining a set of samples large enough makes such tests practically unfeasible.

[10] Whether the independence assumption holds is not the central question. The important question is whether the approximation leads to good simulation results. Our experience using different stocks and portfolios is that for four years of data, rejections become more common as the scales of the portfolio components become more different. However, a difference in scales indicates that the influence of one component dominates the dynamics, making the independence a secondary issue. For shorter time series, the rejection rate decreases, and rejections with two years of data appear to be rare.

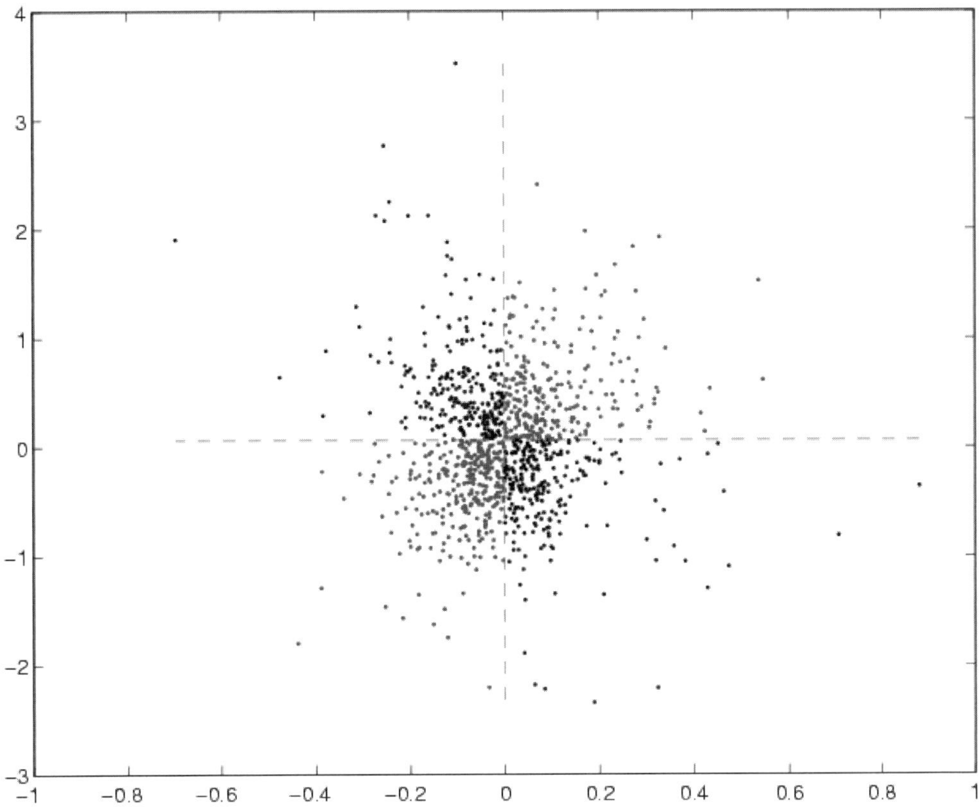


FIGURE 4.11 Scatter plot for the empirical returns on the two portfolio components $\pi_1$ and $\pi_2$. The event $B$ corresponds to the bottom-left and top-right quarters of the plane.

Consider the event

$$B = (A_1 \cap A_2) \cup (\overline{A_1} \cap \overline{A_2}),$$

i.e., the event that the pair of returns are both either larger or smaller than their estimates for the mean (see Figure 4.11). We then formulate our null hypothesis: The probability $P(B)$ is equal to

$$q = \widehat{p}_1 \widehat{p}_2 + (1 - \widehat{p}_1)(1 - \widehat{p}_2).$$

Estimate the probability of $B$:

$$\widehat{q} = \frac{\#\text{ samples in } B}{\#\text{ samples}} = \frac{522}{1006}.$$

Treating each sample as an independent Bernoulli trial and normalizing the random variable of the number of successes leads to

$$\frac{522 - 1006q}{\sqrt{1006q(1-q)}} = 1.15$$

which is within both the confidence intervals for 95%, $[-1.96, 1.96]$, and 90%, $[-1.64, 1.64]$. Therefore, the null hypothesis should be accepted (i.e., not rejected) and the portfolio components have passed the independence test.

### 4.3.3 *A Few Implementation Issues*

In the generalized eigenvalue problem (4.31), the Hessian $\mathbf{\Gamma}$ is symmetric and the covariance matrix $\widehat{\mathbf{C}}$ is nonnegative definite. Symmetric-definite eigenvalue problems arise in other applications. The standard method [GL89] is to compute the Cholesky factorization

$$\widehat{\mathbf{C}} = \mathbf{U}^T\mathbf{U}, \tag{4.39}$$

where $\mathbf{U}$ is upper triangular. We know that a matrix $\mathbf{A}$ can be factored in the form $\mathbf{QU}$, where $\mathbf{Q}$ is orthogonal (see, for example, the so-called QR algorithm: [PTVF92]). Then, since $\mathbf{\Lambda} = \mathbf{A}^T\mathbf{\Gamma}\mathbf{A}$,

$$\mathbf{\Lambda} = \mathbf{U}^T(\mathbf{Q}^T\mathbf{\Gamma}\mathbf{Q})\mathbf{U}.$$

To solve this eigenvalue problem requires $O(n^3)$ floating-point operations. Combined with the estimation of the covariance matrix, this gives a total of $O(n^3+n^2d)$ floating-point operations.

However, if we take advantage of the structure of problem (4.31), it is possible to improve slightly on this procedure. Let $\{\mathbf{r}_i\}_{i=1}^d$ be the time series of returns and define the $d \times n$ matrix

$$\mathbf{W} = \begin{bmatrix} \mathbf{r}_1^T - \boldsymbol{\mu}^T \\ \vdots \\ \mathbf{r}_d^T - \boldsymbol{\mu}^T \end{bmatrix}.$$

The estimate of covariance matrix (4.30) can be written as

$$\widehat{\mathbf{C}} = \frac{1}{d-1}\mathbf{W}^T\mathbf{W}.$$

A variety of estimators for the covariance matrix have been proposed in the finance literature (see, for example, [CLM97, Hul00, Wil00]). Many of the estimators are of the form

$$\widehat{\mathbf{C}} = \mathbf{W}^T\mathbf{D}\mathbf{W}, \tag{4.40}$$

where $\mathbf{D}$ is a weight matrix.[11]

Depending on whether the number of dates $d$ is greater or smaller than the number of risk factors $n$, we get two cases. Suppose $d > n$. Rather than explicitly forming the covariance matrix $\widehat{\mathbf{C}}$, it is preferable to factor it directly by computing the (QR) factorization

$$\mathbf{D}^{\frac{1}{2}}\mathbf{W} = \mathbf{QU}, \tag{4.41}$$

where $\mathbf{Q}$ is a $d \times n$ matrix with orthonormal columns and $\mathbf{U}$ is a $n \times n$ upper triangular matrix. The *QR* algorithm (4.41) is then applied to the matrix $\mathbf{U}\mathbf{\Gamma}\mathbf{U}^T$. The QR factorization takes about

[11] Two examples are the exponentially weighted moving average with

$$\mathbf{D} = \frac{1-\lambda}{1-\lambda^d}\operatorname{diag}(1, \lambda, \ldots, \lambda^{d-1})$$

and the multivariate GARCH(1,1) model [Bol86, CLM97, Hul00] $\Sigma_n = (1-\alpha-\beta)V + \alpha\mathbf{r}_i\mathbf{r}_i^T + \beta\,\Sigma_{n-1}$, where $V$ is the standard estimate for the long-run average volatility, which has

$$\mathbf{D} = \operatorname{diag}(\alpha, \alpha\beta, \alpha\beta^{d-1}) + \frac{1-\beta^d}{1-\beta}(1-\alpha-\beta)\mathbf{I}.$$

$2n^2d$ floating-point operations, compared to approximately $n^2d + n^3/3$ for equation (4.39). The advantage of the QR algorithm is that it can be shown that the forward error is smaller than for the Cholesky factorization.[12]

On the other hand, if $d < n$, then the matrix $\widehat{\mathbf{C}}$ will be singular. The best approach when $d \leq n$ is to use factorization (4.40) and compute the Schur decomposition

$$\mathbf{\Lambda} = \mathbf{Q}^T(\mathbf{D}^{\frac{1}{2}}\mathbf{W})\mathbf{\Gamma}(\mathbf{D}^{\frac{1}{2}}\mathbf{W})^T\mathbf{Q}.$$

This effectively reduces the size of the eigenvalue problem from order $n$ to order $d$, and therefore the whole step requires only $O(dn^2 + d^2n + d^3)$ floating-point operations. We conclude that performing computations directly on the time series matrix, rather than forming the covariance matrix explicitly, is convenient, effective, and numerically sound.

## 4 Numerical Methods for ΔΓ-Portfolios

Singular portfolios have to be estimated by means of straightforward Monte Carlo simulations; for ΔΓ-portfolios, more methods have been proposed.

### 4.1 *Monte Carlo Methods and Variance Reduction*

Assume a multivariate distribution for the returns $\mathbf{r} \in \mathbb{R}^n$:

$$p^G(\mathbf{r}) = \frac{1}{\sqrt{(2\pi)^n|\mathbf{C}|}} \exp\left(-\tfrac{1}{2}\mathbf{r}^T\mathbf{C}^{-1}\mathbf{r}\right), \tag{4.42}$$

where $\mathbf{C}$ is the $n \times n$ covariance matrix and $|\mathbf{C}|$ is its determinant. The plain Monte Carlo (MC) method for computing VaR for ΔΓ-portfolios in this case is based on sampling the returns $\mathbf{r}$ from $p^G \sim N_n(\mathbf{0}, \mathbf{C})$. The precise steps are described in the MC VaR numerical project in Part II. The basic steps are summarized as follows.

- Cholesky factorize the covariance matrix, $\mathbf{C} = \mathbf{U}^T\mathbf{U}$.
- For each scenario, generate an $n \times 1$ vector $\mathbf{y}$ of identically and independently distributed normal variates. For each scenario vector compute $\mathbf{r} = \mathbf{U}^T\mathbf{y}$, and evaluate the portfolio variation $\Delta V(\mathbf{r})$, e.g., within the ΔΓ-approximation, then $\Delta V(\mathbf{r}) = \Delta\tilde{\Pi}(\mathbf{r})$.
- Sort the returns from the complete simulation in increasing order and evaluate the VaR as a percentile.

Other popular methods are based on *importance sampling*. The idea is to improve accuracy, i.e., reduce the variance of a simulation within the same number of scenarios. For VaR calculations the idea is to generate *weighted* scenario sets that populate the tails of the distribution more accurately than the body. In the general theory of the evaluation of integrals, importance sampling increases sampling efficiency within certain regions of the integration space. There are several ways of implementing importance sampling, with various degrees of sophistication, depending on the integral dimensionality, the integrand variability, and so

[12] Sun [Sun92, Hig96] proves an upper bound for the forward error for Cholesky factorization. Sun's result can be adapted for our purposes to the positive-definite matrix $\mathbf{W}^T\mathbf{W}$ by considering a small perturbation $\mathbf{W} \to \mathbf{W} + \delta\mathbf{W}$.

on. What essentially underlies the approach is the technique of changing probability measure from which the scenarios are sampled. [We have seen various examples of the use of changing measure for pricing options in previous chapters.]

A brief overview of importance sampling as applied to the evaluation of an integral is as follows. Suppose we wish to evaluate an integral or expectation of some function $h : \mathbb{R}^n \to \mathbb{R}$,

$$I = E^{(f)}[h(\mathbf{X})] = \int_{\mathbb{R}^n} h(\mathbf{x})f(\mathbf{x})d\mathbf{x}, \tag{4.43}$$

where $\mathbf{X}$ is a random vector in $\mathbb{R}^n$. The superscript $f$ denotes, as usual, the fact that the expectation is taken w.r.t. a given probability measure or distribution $f$. If we use plain MC, then scenarios $\mathbf{X}_j$, $j = 1, \ldots, N$, are sampled with $f$ as density; i.e., the MC estimator (w.r.t. $f$ as density) of $I$ is

$$\hat{I}_f = \frac{1}{N}\sum_{j=1}^{N} h(\mathbf{X}_j). \tag{4.44}$$

Alternatively, the integral $I$ can be equivalently recast as an expectation w.r.t. any other density $g$, as long as this density satisfies $f(\mathbf{x}) > 0 \implies g(\mathbf{x}) > 0$, $\mathbf{x} \in \mathbb{R}^n$:

$$I = E^{(g)}\left[h(\mathbf{X})\frac{f(\mathbf{X})}{g(\mathbf{X})}\right] = \int_{\mathbb{R}^n} (h(\mathbf{x})w(\mathbf{x}))g(\mathbf{X})d\mathbf{x}, \tag{4.45}$$

where $w(\mathbf{x}) = \frac{f(\mathbf{x})}{g(\mathbf{x})}$ is a weight function (also called *the Radon–Nikodym derivative* or *likelihood ratio*). This factor is introduced by taking $g$ as density in place of the original density $f$. Applying standard MC, with samples $\mathbf{X}_j$, $j = 1, \ldots, N$, now drawn from $g$, gives an estimator w.r.t. $g$:

$$\hat{I}_g = \frac{1}{N}\sum_{j=1}^{N} h(\mathbf{X}_j)\frac{f(\mathbf{X}_j)}{g(\mathbf{X}_j)}. \tag{4.46}$$

By taking expectations in this expression and treating the $\mathbf{X}_j$ as identically distributed random vectors with $g$ as density, the reader can readily show from equation (4.45) that $\hat{I}_g$ is an unbiased estimator of $I$; i.e., $E^{(g)}[\hat{I}_g] = I$. Likewise, $E^{(f)}[\hat{I}_f] = I$. In practice, it is of interest to compare the difference in the variances $\mathrm{Var}_g(\hat{I}_g)$ and $\mathrm{Var}_f(\hat{I}_f)$ of the two estimators $\hat{I}_f$ and $\hat{I}_g$, respectively. Since both estimators have the same mean (equal to $I$), it suffices to consider the second moments. In particular,

$$\mathrm{Var}_g(\hat{I}_g) = E^{(g)}[h^2(\mathbf{X})w^2(\mathbf{X})] \qquad \text{and} \qquad \mathrm{Var}_f(\hat{I}_f) = E^{(f)}[h^2(\mathbf{X})]. \tag{4.47}$$

For arbitrary choices of $g$, the variance $\mathrm{Var}_g$ (with importance sampling) may be either larger or smaller than $\mathrm{Var}_f$ (without importance sampling). The goal of a successful implementation of importance sampling is to choose a density $g$ that is effective in sampling and thereby reducing the variance. In order to implement an effective importance-sampling algorithm one should at best attempt to sample in proportion to the integrand $h \cdot f$. Recall the typical situation in option pricing, where the price is given by an expectation integral over a product of the risk-neutral transition density and the discounted payoff function. Hence, we can view the transition density as playing the role of $f$ and $h$ as the discounted payoff function. A plain MC calculation for the option price would proceed by sampling asset price paths with transition density as the sampling distribution. This is the basis of the MC basket option pricer numerical project in Part II. In contrast, a more effective importance-sampling MC algorithm for option pricing would be to consider a different sampling distribution — one that gives more "importance" to the payoff function as well as the risk-neutral density. A good choice of density $g$ should be such that a greater percentage of sample paths lies within the more significant contributions of the integrand.

For the purposes of computing VaR, we observe from upcoming equation (4.72) that a more efficient importance-sampling procedure should be one in which the chosen sampling density generates a substantial number of return scenarios that are in the tails of the distribution for the portfolio variation $\Delta V$. This can be seen from the fact that the cdf of $\Delta V$ is an integral over the product of the return distribution and the step function. For VaR calculations, the step function is significant only in the left tail of the distribution of $\Delta V$. One possibile importance-sampling MC strategy therefore consists of generating *scaled* returns by introducing a scale factor $f_s$ so as to transform the return scenarios $\mathbf{r}^{(i)}$ into $(\sqrt{1-f_s})^{-1}\mathbf{r}^{(i)}$. In some sense one can also think of $f_s$ as a stress-testing factor. The factor $f_s$ is strictly between 0 and 1. Another possibility is to shift the returns by a common vector, i.e., to transform $\mathbf{r}$ into $\mathbf{r}+\mathbf{r}_0$. A more general approach is to make an affine transformation, i.e., to transform $\mathbf{r}$ into $\mathbf{A}\mathbf{r}+\mathbf{r}_0$ for some matrix $\mathbf{A}$.

To show how to compute the weights in an importance-sampling implementation, let's consider the simple case of scaling for a univariate, standard normal distribution $\phi(x)$. The particular technique we now present is readily generalizable into the multivariate case. In the one-dimensional case, we are assuming that our original sampling density is the standard normal, i.e., $f(x) = \phi(x)$, and that we wish to evaluate an integral of the form

$$I = \int_{-\infty}^{\infty} \phi(x)h(x)dx, \tag{4.48}$$

where $h$ has significant contributions in the tails of $\phi$. Sampling from the pdf $\phi$ itself would probably not constitute an optimal importance-sampling strategy. The trick we employ is to rewrite the original density as follows:

$$\phi(x) = \frac{1}{\sqrt{2\pi}}e^{-\frac{x^2}{2}} = \frac{1}{\sqrt{2\pi q}}e^{-\frac{x^2}{2q}} \cdot \sqrt{q}e^{-\frac{x^2}{2p}}, \tag{4.49}$$

where $p$ and $q$ are positive numbers chosen such that

$$\frac{1}{p} + \frac{1}{q} = 1. \tag{4.50}$$

By defining the factor $f_s$ so that $p = \frac{1}{f_s}$ and $q = \frac{1}{1-f_s}$, we have

$$\phi(x) = w(x)\tilde{\phi}(x), \tag{4.51}$$

where $w$ is the weight function

$$w(x) = (1-f_s)^{-\frac{1}{2}} \exp(-f_s x^2/2) \tag{4.52}$$

and $\tilde{\phi}$ is the new sampling density

$$\tilde{\phi}(x) = \frac{1}{\sqrt{2\pi(1-f_s)^{-1}}}e^{-\frac{x^2}{2(1-f_s)^{-1}}}. \tag{4.53}$$

The original integral is therefore transformed into

$$I = \int_{-\infty}^{\infty} \tilde{\phi}(x)(h(x)w(x))dx, \tag{4.54}$$

which can now be evaluated by sampling from the pdf $\tilde{\phi}$ — a normal density with rescaled variance $(1-f_s)^{-1}$. The parameter $f_s$ can hence be chosen so as to reduce variance in the corresponding MC estimate of $I$.

## 4.4.2 Moment Methods

Moment methods are approximate analytical methods that propose to estimate portfolio VaR by evaluating the first few moments of the distribution of the portfolio variation analytically and then matching these moments with a model distribution. The first four moments are of particular interest, for they provide us with measures of the mean, variance, skewness, and kurtosis. We discuss two methods in this class, one named Cornish–Fisher and the other the Johnson method.

In what follows we shall consider portfolio variations within the delta-gamma approximation in the form

$$\Delta V \equiv \Delta V(\mathbf{r}) = \mathbf{\Delta}^T \mathbf{r} + \frac{1}{2}\mathbf{r}^T \mathbf{\Gamma} \mathbf{r}. \tag{4.55}$$

[Note: This is $\Delta\tilde{\Pi}_t$ as defined previously but without the theta factor $\Theta t$, which is trivial to include if desired.] We denote $\mu_1 = E[\Delta V]$ as the first moment of the distribution of $\Delta V$ and $\mu_m = E[(\Delta V - \mu_1)^m]$ as $m$th central moment for $m \geq 2$. Assuming the return density is given by equation (4.42), a straightforward, though lengthy, calculation gives

$$\mu_1 = E[\Delta V] = \frac{1}{2}\text{tr}\{\mathbf{\Gamma C}\}, \tag{4.56}$$

$$\mu_2 = E[(\Delta V - \mu_1)^2] = \mathbf{\Delta}^T \mathbf{C} \mathbf{\Delta} + \frac{1}{2}\text{tr}\{(\mathbf{\Gamma C})^2\}, \tag{4.57}$$

$$\mu_3 = E[(\Delta V - \mu_1)^3] = 3(\mathbf{C\Delta})^T \mathbf{\Gamma}(\mathbf{C\Delta}) + \text{tr}\{(\mathbf{\Gamma C})^3\}, \tag{4.58}$$

$$\mu_4 = E[(\Delta V - \mu_1)^4] = 12(\mathbf{C\Delta})^T (\mathbf{\Gamma C})^2 \mathbf{\Delta} + 3\text{tr}\{(\mathbf{\Gamma C})^4\} + 3\mu_2^2, \tag{4.59}$$

where tr denotes the matrix trace. For example, the first moment is simple to derive:

$$\mu_1 = E[\Delta V] = \mathbf{\Delta}^T E[\mathbf{r}] + \frac{1}{2}E[\mathbf{r}^T \mathbf{\Gamma r}] = \frac{1}{2}\sum_{i,j=1}^{n} \Gamma_{ij} E[r_i r_j]$$
$$= \frac{1}{2}\sum_{i,j=1}^{n} \Gamma_{ij} C_{ji} = \frac{1}{2}\text{tr}\{\mathbf{\Gamma C}\},$$

where we used $E[\mathbf{r}] = \mathbf{0}$ and $\text{Cov}(r_i, r_j) = C_{ij}$, since the density $p^G \sim N_n(\mathbf{0}, \mathbf{C})$. The higher moments can be derived by a similar procedure and using known identities for integrals of products such as $p^G r_i^k r_j^l$ and higher products, with $k, l = 1, 2, 3, 4$. For moments $\mu_3, \mu_4$ this is rather tedious. Alternatively, we can obtain the (noncentral) $m$th moments (and thereby the central moments) by evaluating the $m$th derivative (at the origin) of the moment-generating function (mgf) for the random variable $\Delta V$. That is, $E[(\Delta V)^m] = M^m(0)$, where the mgf $M(u) = E[e^{u\Delta V}]$ is given analytically as derived in the next section.

The Cornish–Fisher (CF) method stems from the fact that it is possible to derive explicit polynomial asymptotic expansions for standardized quantiles (or percentiles) of a general distribution in terms of its (standardized) moments and the quantiles (or percentiles) of the standard normal distribution. For a detailed mathematical discussion of the general technique, see, for example, [HD68]. For our purposes it suffices to point out the main result of the CF expansion. Generally, given a probability distribution $g(x)$ having cumulants[13] $\kappa_j, j = 0, \ldots,$ the distribution $f(x)$ generated by the expansion

$$f(x) = \sum_{i=0}^{\infty} \frac{\left[ \sum_{j=1}^{\infty} \epsilon_j \frac{(-D)^j}{j!} \right]^i}{i!} g(x) \tag{4.60}$$

has cumulants $\kappa_j + \epsilon_j, j = 0, \ldots,$ where $D^j \equiv d^j/dx^j$ defines the $j$th-order differential operator. By truncating this expansion, one can hence obtain approximate analytical formulas for the density of a distribution $f$ using only its first few known central moments and the first few derivatives of an analytically known distribution function $g$ and its central moments. Similarly, we also obtain analytical formulas that relate the quantiles of $f$ with those of $g$. When $g$ is chosen to be the standard normal distribution, then what arises is the known Cornish–Fisher formula.

Given the first four central moments in equations (4.56)–(4.59), computing VaR with the CF expansion (to fourth order) is particularly simple, for there are explicit formulas for it. In particular, the random variable $(\Delta V - \mu_1)/\sqrt{\mu_2}$ has $\alpha$-quantile given by

$$\tilde{z}_\alpha = z_\alpha + \frac{1}{6}\rho_3(z_\alpha^2 - 1) + \frac{1}{24}\rho_4 z_\alpha (z_\alpha^2 - 3) - \frac{1}{36}\rho_3^2 z_\alpha (2z_\alpha^2 - 5),$$

where $\rho_3 = \mu_3/\mu_2^{3/2}$, $\rho_4 = \mu_4/\mu_2^2 - 3$, and $z_\alpha$ is defined as the $\alpha$-quantile of the standard normal distribution $z_\alpha = N^{-1}(\alpha)$.

Within the CF approximation, VaR with confidence level $\alpha\%$ (as defined by equation (4.1)) is given by

$$\text{VaR} = \tilde{z}_\alpha \sqrt{\mu_2} - \mu_1. \tag{4.61}$$

Note that within the simpler centered normal distribution approximation for $\Delta V$ (with assumed zero gamma matrix), $\rho_3 = \rho_4 = 0$, so $\tilde{z}_\alpha = z_\alpha$ and this equation is consistent with equation (4.6).

In the Johnson method, one seeks to match the first four moments of the pdf of $\Delta V$ to the cumulative distribution of the random variable

$$f(X) = \lambda \sinh((X - \gamma)/\delta) + \xi, \tag{4.62}$$

where $X$ is a standard normal and $\xi, \delta, \gamma, \lambda$ are model parameters. The pdf of the random variable $Y = f(X)$ can be found by means of changing coordinates and is given by

$$p(y) = \frac{\delta}{\sqrt{2\pi}\lambda} \frac{e^{-[\gamma + \delta \sinh^{-1}((y-\xi)/\lambda)]^2/2}}{\cosh[\sinh^{-1}((y-\xi)/\lambda)]}. \tag{4.63}$$

[13] The cumulants of a distribution are defined by coefficients in a power series expansion of the logarithm of the characteristic function or the mgf. Cumulants are related to the central moments: $\kappa_1 = \mu_1$, $\kappa_2 = \mu_2$, $\kappa_3 = \mu_3$, $\kappa_4 = \mu_4 - 3\mu_2^2$, etc.

The expectation integrals $E[y^n]$ for the Johnson distribution, for $n = 0, 1, \ldots,$ can be expressed as follows:

$$E[y^n] = \frac{\delta}{\sqrt{2\pi}} \int_{-\infty}^{+\infty} (\xi + \lambda \sinh x)^n e^{-(\gamma+\delta x)^2/2} dx. \tag{4.64}$$

The integrals

$$I_n(\delta, \gamma) \equiv \frac{\delta}{\sqrt{2\pi}} \int_{-\infty}^{+\infty} \sinh^n x e^{-(\gamma+\delta x)^2/2} dx \tag{4.65}$$

are obtained recursively using the recursion relation

$$\begin{aligned} I_{n+1}(\delta, \gamma) = \tfrac{1}{2}[&e^{-(\gamma^2-(\gamma-1/\delta)^2)/2} I_n(\delta, \gamma - 1/\delta) \\ &- e^{-(\gamma^2-(\gamma+1/\delta)^2)/2} I_n(\delta, \gamma + 1/\delta)] \end{aligned} \tag{4.66}$$

and the formula for $n = 1$:

$$I_1(\delta, \gamma) = -e^{1/2\delta^2} \sinh(\gamma/\delta). \tag{4.67}$$

From equations (4.64) and (4.65) we find that

$$\begin{aligned} E[y] &= \xi + \lambda I_1, \\ E[y^2] &= \xi^2 + 2\xi\lambda I_1 + \lambda^2 I_2, \\ E[y^3] &= \xi E[y^2] + \lambda\xi^2 I_1 + 2\xi\lambda^2 I_2 + \lambda^3 I_3, \\ E[y^4] &= \xi E[y^3] + \lambda\xi^3 I_1 + \lambda^4 I_4 + 3\xi^2\lambda^2 I_2 + 3\xi\lambda^3 I_3, \end{aligned}$$

where $I_n = I_n(\delta, \gamma)$ for all $n$. The moments $\mu_i^J = \mu_i^J(\xi, \delta, \gamma, \lambda)$ for the Johnson distribution are given by

$$\begin{aligned} \mu_1^J &= E[y], \\ \mu_2^J &= E[(y - \mu_1^J)^2] = E[y^2] - \mu_1^{J2}, \\ \mu_3^J &= E[(y - \mu_1^J)^3] = E[y^3] - 2\mu_1^J E[y^2] + \mu_1^{J3} - \mu_1^J \mu_2^J, \\ \mu_4^J &= E[(y - \mu_1^J)^4] = E[y^4] - 3\mu_1^J E[y^3] + 3\mu_1^{J2} E[y^2] - \mu_1^{J4} - \mu_1^J \mu_3^J. \end{aligned}$$

These four moments are explicitly functions of the four parameters $\xi, \delta, \gamma, \lambda$, which are then fitted by matching the $\mu_i^J$ with the $\mu_i$ in equations (4.56)–(4.59). This results in a nonlinear system of four equations,

$$\mu_1 = \mu_1^J(\xi, \delta, \gamma, \lambda), \tag{4.68}$$

$$\mu_2 = \mu_2^J(\xi, \delta, \gamma, \lambda), \tag{4.69}$$

$$\mu_3 = \mu_3^J(\xi, \delta, \gamma, \lambda), \tag{4.70}$$

$$\mu_4 = \mu_4^J(\xi, \delta, \gamma, \lambda), \tag{4.71}$$

which is solved for $\xi, \delta, \gamma, \lambda$.

### 4.4.3 Fourier Transform of the Moment-Generating Function

In the Fourier transform method proposed by Milne–Ulma, the idea is to compute the moment-generating function for the portfolio return distribution and then to invert a Fourier transform to obtain the density $f(V)$ of the portfolio P&L. From this density one then computes VaR from the area under the left tail of the density $f$ corresponding to a given percentile.

Consider the cumulative distribution function for the portfolio variation $\Delta V$

$$\Phi(V) = P(\Delta V \le V) = \int_{\mathbb{R}^n} p^G(\mathbf{r})\Theta(V - \Delta V(\mathbf{r}))d\mathbf{r}, \tag{4.72}$$

where $p^G$ is given by equation (4.42) and the integration is over the complete space of all risk-factor returns and $\Theta(\cdot)$ is the unit step function.

Differentiating this gives $f(V)$:

$$f(V) = \frac{d\Phi(V)}{dV}. \tag{4.73}$$

Taking the derivative w.r.t. $V$ inside the integral gives

$$f(V) = \frac{1}{\sqrt{\det(2\pi\mathbf{C})}}\int_{\mathbb{R}^n} \exp(-\tfrac{1}{2}\mathbf{r}^T\mathbf{C}^{-1}\mathbf{r})\delta(V - \Delta V(\mathbf{r}))d\mathbf{r}, \tag{4.74}$$

where we have used the property $d\Theta(x)/dx = \delta(x)$. The Dirac delta function $\delta(x)$ is then written in terms of its integral representation, and, assuming a delta-gamma approximation, $\Delta V(\mathbf{r}) \approx \mathbf{\Delta}^T\mathbf{r} + \frac{1}{2}\mathbf{r}^T\mathbf{\Gamma}\mathbf{r}$, we find that

$$f(V) = \frac{1}{2\pi}\int_{-\infty}^{\infty} e^{-iuV}M(iu)du, \tag{4.75}$$

$i \equiv \sqrt{-1}$, where $M$ is the moment-generating function (mgf)

$$M(u) = \frac{1}{\sqrt{\det(2\pi\mathbf{C})}}\int_{\mathbb{R}^n} \exp(-\tfrac{1}{2}\mathbf{x}^T[\mathbf{C}^{-1} - u\mathbf{\Gamma}]\mathbf{x} + u\mathbf{\Delta}^T\mathbf{x})d\mathbf{x}. \tag{4.76}$$

The mgf is given by a Gaussian integral and can be explicitly computed by using the integral identity

$$\int_{\mathbb{R}^n} \exp(-\mathbf{x}^T[\mathbf{A} + i\mathbf{B}]\mathbf{x} + \mathbf{v}^T\mathbf{x})d\mathbf{x} = \frac{\pi^{n/2}}{\sqrt{\det(\mathbf{A} + i\mathbf{B})}}\exp(\tfrac{1}{4}\mathbf{v}^T(\mathbf{A} + i\mathbf{B})^{-1}\mathbf{v}) \tag{4.77}$$

for any $n \times 1$ vector $\mathbf{v}$ and $n \times n$ (complex) matrices $\mathbf{A}$, $\mathbf{B}$. Setting $\mathbf{A} = \frac{1}{2}\mathbf{C}^{-1}$, $\mathbf{B} = i\frac{u}{2}\mathbf{\Gamma}$, and $\mathbf{v} = u\mathbf{\Delta}$ we find

$$M(u) = \frac{1}{\sqrt{\det(\mathbf{I} - u\mathbf{\Gamma}\mathbf{C})}}\exp\left(\frac{u^2}{2}(\mathbf{C}\mathbf{\Delta})^T(\mathbf{I} - u\mathbf{\Gamma}\mathbf{C})^{-1}\mathbf{\Delta}\right), \tag{4.78}$$

where $\mathbf{I}$ is the $n \times n$ identity matrix. Note that an equivalent expression also follows in terms of the transpose matrix $(\mathbf{I} - u\mathbf{\Gamma}\mathbf{C})^T = \mathbf{I} - u\mathbf{C}\mathbf{\Gamma}$.

The last step is to cast the given mgf into the following computationally tractable form:

$$M(u) = \prod_{j=1}^{n}(1 - u\lambda_j)^{-1/2}\exp\left[\frac{u^2}{2}\frac{b_j^2}{(1 - u\lambda_j)}\right]. \tag{4.79}$$

Here the $b_j$ are the components of the vector $\mathbf{b}$ given by

$$\mathbf{b} = \mathbf{O}^T \mathbf{U} \boldsymbol{\Delta}, \tag{4.80}$$

where $\mathbf{O}$ is the matrix of eigenvectors of the symmetric matrix $\mathbf{U}\boldsymbol{\Gamma}\mathbf{U}^T$,

$$\mathbf{O}^T(\mathbf{U}\boldsymbol{\Gamma}\mathbf{U}^T)\mathbf{O} = \boldsymbol{\lambda}. \tag{4.81}$$

The latter equation gives the diagonal matrix $\boldsymbol{\lambda}$, whose diagonal elements define the given $\lambda_j$ components. The matrix $\mathbf{U}$ is defined (as usual) in the Cholesky factorization of the covariance matrix $\mathbf{C} = \mathbf{U}^T\mathbf{U}$.

We finally obtain the real part of the mgf,

$$\mathrm{Re}\{M(iu)\} = \prod_{j=1}^{n}(1 + u^2\lambda_j^2)^{-\frac{1}{4}} \exp[-(u^2 b_j^2/2)/(1 + u^2\lambda_j^2)] \cos(\phi), \tag{4.82}$$

and an identical expression for the imaginary part $\mathrm{Im}\{M(iu)\}$, with the cosine function replaced by the sine function. The phase function is given by

$$\phi = \frac{1}{2}\sum_{j=1}^{n} \arctan(u\lambda_j) - \frac{u^3}{2}\sum_{j=1}^{n} \lambda_j/(1 + u^2\lambda_j^2). \tag{4.83}$$

The final form of the Fourier transform involves only real quantities:

$$f(V) = \frac{1}{\pi}\int_0^{\infty} [\cos(uV)\mathrm{Re}\{M(iu)\} + \sin(uV)\mathrm{Im}\{M(iu)\}]du. \tag{4.84}$$

This is a sum of cosine and sine transforms, which can be evaluated using a number of appropriate numerical routines for integrating one-dimensional oscillatory functions. It is particularly important to implement an algorithm that gives an accurate representation of the pdf within the left tail. From equation (4.73), then, VaR for a chosen percentile $1 - \alpha$ is obtained by evaluating the area under the left tail of $f$; i.e., the cumulative density gives $\Phi(V = -\mathrm{VaR}) = 1 - \alpha$.

## 4.5 The Fast Convolution Method

The risk-factor model and the pricing model provide the necessary ingredients to make equation (4.1), defining value-at-risk, meaningful. In this section, we present a new Fourier transform algorithm for computing value-at-risk. The method is different from existing Fourier methods (see [MU99, DP01, GHS02]) in that it does not assume that the characteristic function (or mgf) of the density is explicitly known. The method therefore has the advantage of greater freedom in choosing the risk-factor model. We present an extended example illustrating the performance of the algorithm and the importance of risk-factor models with fat tails. In later sections, we also extend the method to compute the gradient of value-at-risk. The section concludes with two computational examples. The first is a simple linear approximation to the change in value-at-risk with changes in portfolio composition. In the second example we hedge a derivatives portfolio by solving an optimization problem to minimize the value-at-risk.

The local dynamics of changes in portfolio value is again approximated by

$$\Delta\widetilde{\Pi} = \Theta t + \mathbf{\Delta}^T\mathbf{r} + \frac{1}{2}\mathbf{r}^T\mathbf{\Gamma}\mathbf{r},$$

and we approximate the value-at-risk by the solution to

$$P(\Delta\widetilde{\Pi} \le -\mathrm{VaR}) = 1 - \alpha. \tag{4.85}$$

Because $\Delta\widetilde{\Pi}$ is a quadratic function, it is generally easier to solve equation (4.85) than equation (4.1). At the same time, since $\Delta\widetilde{\Pi}$ is locally accurate, the solution VaR to equation (4.85) is a good approximation to value-at-risk, provided the probability of large changes in the risk factors is relatively small. This is exemplified by the relative closeness of the $\Delta$ and $\Delta\Gamma$ distributions (see Figure 4.12). We return to study accuracy in Section 4.8.

Equation (4.85) is solved in two steps. We find the pdf of $\Delta\widetilde{\Pi}$ and compute the value-at-risk from this distribution. Consider a risk-factor model and a factorization of the portfolio of the type produced by Algorithm 1. That is, the joint distribution of the transformed risk factors has a product pdf

$$p(x) = p_1(x_1) \cdots p_n(x_n),$$

and the $\Delta\Gamma$-approximation is a sum of independent quadratic functions

$$\Delta\widetilde{\Pi} = \Xi + \sum_{i=1}^{n} \pi_i(x_i) \qquad \text{where } \pi_i = \Delta_i' x_i + \frac{\lambda_i}{2} x_i^2.$$

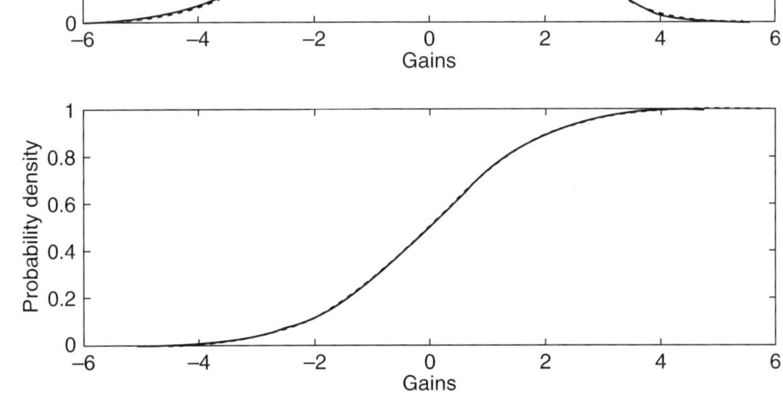


FIGURE 4.12 $\Delta\Gamma$ distribution (solid line) compared to the $\Delta$ distribution (dashed line) for relatively small changes in risk factors.

Let $p_{\pi_i}(x)$ be the pdf of $\pi_i$ for $i = 1, \ldots, n$. Then the pdf of $\Delta\widetilde{\Pi}$ has the form of a multiple convolution,[14]

$$p_{\Delta\widetilde{\Pi}} = \tau_{\Xi}(p_{\pi_1} * \cdots * p_{\pi_n}). \tag{4.86}$$

## 4.5.1 The Probability Density Function of a Quadratic Random Variable

Given a single risk factor $x_i$ with pdf $p_i$ and a quadratic portfolio component

$$\pi_i = \Delta_i' x_i + \frac{\lambda_i}{2} x_i^2,$$

it is easy to derive the pdf $p_{\pi_i}$ for $\pi_i$. Let $x_+(u)$ and $x_-(u)$ be the two roots of $\pi_i(x) - u = 0$,

$$x_{\pm}(u) = \frac{-\Delta_i' \pm \sqrt{(\Delta_i')^2 + 2\lambda_i u}}{\lambda_i}, \qquad (-(\Delta_i')^2/2\lambda_i \le u). \tag{4.87}$$

The pdf $p_{\pi_i}$ is the derivative of the probability

$$\mathrm{P}(\pi_i \le u) = \mathrm{P}(x \in x_+([-\infty, u])) + \mathrm{P}(x \in x_-([-\infty, u])).$$

The sets $x_{\pm}([-\infty, u])$ are empty for $u < -(\Delta_i')^2/2\lambda_i$, and otherwise

$$p_{\pi_i}(u) = p_i(x_+(u))x_+'(u) - p_i(x_-(u))x_-'(u).$$

It follows that the pdf is

$$p_{\pi_i}(u) = \begin{cases} 0, & \text{if } u < -(\Delta_i')^2/2\lambda_i, \\ \frac{p_i(x_+(u)) + p_i(x_-(u))}{\sqrt{(\Delta_i')^2 + 2\lambda_i u}}, & \text{if } u \ge -(\Delta_i')^2/2\lambda_i. \end{cases}$$

The pdf has a singularity at $u = -(\Delta_i')^2/2\lambda_i$, i.e., at the critical point $x = -\Delta_i'/\lambda_i$ of $\pi_i$.

## 4.5.2 Discretization

Let $[-a, a]$, $a > 0$, be a closed interval. Consider a regular grid $\{\xi_j = -a + jh\}_{j=0}^{N-1}$, where $h = 2a/N$. Because $p_{\pi_i}$ has a singularity, it is necessary to use a discretization scheme that conserves probability (see Figure 4.13). Therefore, we take

$$p_{\pi_i}(x) \approx p_i^D(x) = h \sum_{j=1}^{N-1} p_i^j \delta_j(x) \qquad \text{where} \quad p_i^j = \frac{1}{h} \int_{\xi_j - h/2}^{\xi_j + h/2} p_{\pi_i}(u)\, du \tag{4.88}$$

where $\delta$ is the delta function and $\delta_j(x) \equiv \delta(x - \xi_j)$. For convenience, let $x_0 = -a$ and define $p_i^0 = 0$.

[14] The function $f * g$ defines the convolution product

$$f * g(x) = \int_{-\infty}^{\infty} f(y) g(x - y)\, dy.$$

The functional $\tau_a$ is the shift operator defined by

$$\tau_a f(x) = f(x + a).$$

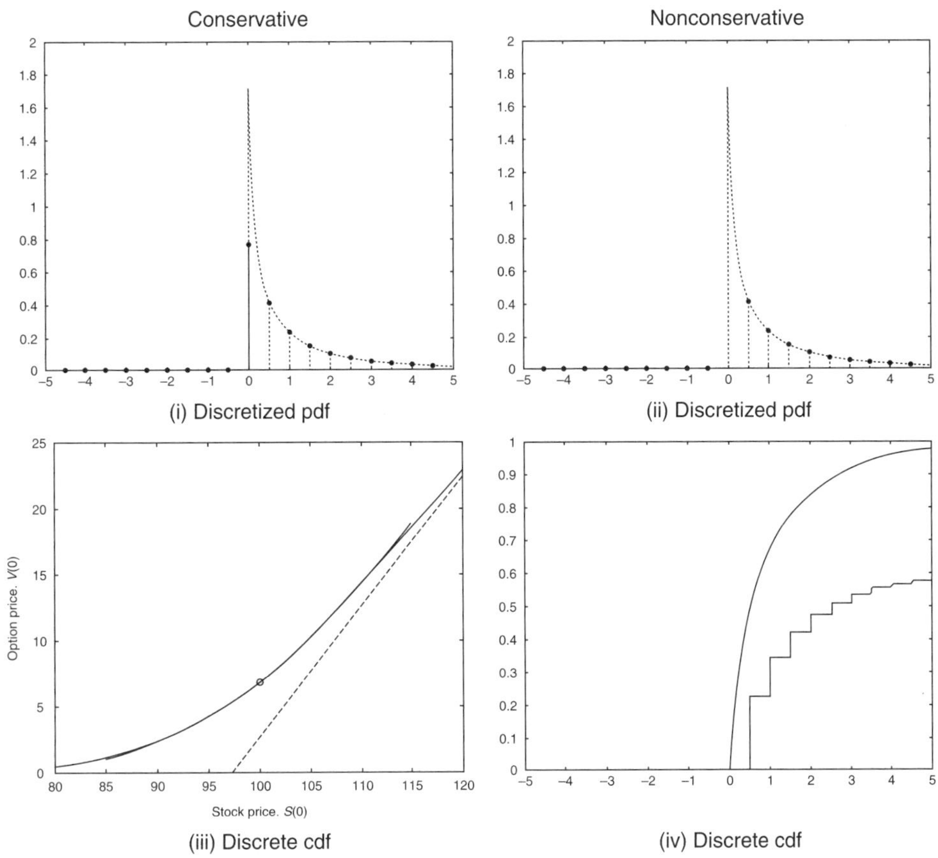


FIGURE 4.13 Example of the importance of a discretization method that conserves probability density. Consider $X^2$, where $X$ is normal — a $\chi^2$ random variable with one degree of freedom. For the conservative method, the discrete pdf and cdf are close to the pdf and cdf of the $\chi^2$ random variable. In the nonconservative method, the pdf is sampled at the grid points to get a discrete approximation. As the graph for the cdf shows, the distribution function for the discrete approximation is not close to the original cdf.

### .5.3 *Accuracy and Convergence*

The accuracy of the discretized function depends on the density outside the interval $[-a, a]$ and the number of grid points $N$. Suppose that

$$\mathrm{P}(\pi_i \leq -a + h/2) < \epsilon/2 \quad \text{and} \quad \mathrm{P}(\pi_i \geq a - h/2) < \epsilon/2,$$

for $i = 1, \ldots, n$, and some small $\epsilon > 0$. On $[-a, a]$, the discretization converges to the probability density in the weak sense.

As a special case of convergence, the approximate cdf converges linearly to the exact cumulative density: If $\xi_k - h/2 \leq y < \xi_k + h/2$, then

$$\left| \int_{\xi_1 - h/2}^{y} [p_{\pi_j}(x) - h \sum_{j=0}^{N-1} p_i^j \delta_j(x)] dx \right| \leq p_i^k h,$$

and they agree exactly for $y = \xi_k + h/2$. For the cdf, we have

$$\left| \int_{-\infty}^{y} [p_{\pi_j}(x) - p_j^D(x)]dx \right| \leq \epsilon + h \max_k p_i^k.$$

### 4.5.4 The Computational Details

To compute the coefficients in equation (4.88) is a bit messy. We consider the case $\lambda_i \neq 0$ (the case $\lambda_i = 0$ is simple). Considering an interval $[\xi_j - h/2, \xi_j + h/2]$, the computation of $p_i^j$ falls into one of three categories:

(i) The polynomial $\pi_i(x) - u$ does not have any zeros for any $u \in [\xi_j - h/2, \xi_j + h/2]$. This gives

$$p_i^j = 0. \tag{4.89}$$

(ii) The polynomial $\pi_i(x) - u$ has a double zero for some $u \in [\xi_j - h/2, \xi_j + h/2]$. It follows that

$$p_i^j = \begin{cases} \frac{1}{h} \int_{x_-(\xi_j+h/2)}^{x_+(\xi_j+h/2)} p_i(x)dx, & \text{if } \lambda_i > 0, \\ \frac{1}{h} \int_{x_+(\xi_j-h/2)}^{x_-(\xi_j-h/2)} p_i(x)dx, & \text{if } \lambda_i < 0. \end{cases} \tag{4.90}$$

(iii) The polynomial $\pi_i(x) - u$ has two distinct zeros for each $u \in [\xi_j - h/2, \xi_j + h/2]$. This yields

$$p_i^j = \frac{1}{h} \int_{x_+(\xi_j-h/2)}^{x_+(\xi_j+h/2)} p_i(x)dx + \frac{1}{h} \int_{x_-(\xi_j+h/2)}^{x_-(\xi_j-h/2)} p_i(x)dx. \tag{4.91}$$

### 4.5.5 Convolution with the Fast Fourier Transform

Because the pdf for the $\Delta\Gamma$-approximation $\Delta\widetilde{\Pi}$ is a convolution product, it can be computed using ideas from Fourier analysis [GW98]. The convolution product and its Fourier transform[15] satisfies

$$p_{\pi_1} * \cdots * p_{\pi_n}(x) = \left( \prod_{k=1}^{n} \widehat{p_{\pi_k}} \right)^{\vee} (x).$$

[15] The continuous Fourier transform of a function $f(x)$ is

$$\widehat{f}(\omega) = \int_{-\infty}^{\infty} e^{i\omega x} f(x)dx.$$

The inverse Fourier transform is

$$(\widehat{f})^{\vee}(x) = \frac{1}{2\pi} \int_{-\infty}^{\infty} e^{-i\omega x} \widehat{f}(\omega)d\omega.$$

It is possible to compute an approximate pdf for the $\Delta\Gamma$-approximation $\Delta\widetilde{\Pi}$ by multiplying and inverting the discrete Fourier transform of the coefficients of the discretized densities:[16]

$$p_{\pi_1} * \cdots * p_{\pi_n}(x) \approx p^D_{\Delta\widetilde{\Pi}}(x) = h\sum_{j=0}^{N-1} p_j\delta_j(x) \qquad \text{where} \quad P^j = (-1)^{j(n-1)}h^{n-1}\prod_{k=1}^{n} P^j_k. \tag{4.92}$$

The sequences $\{P^j\}_{j=0}^{N-1}$ and $\{P^j_k\}_{j=0}^{N-1}$ are defined as the DFT of the sequences $\{p_j\}_{j=0}^{N-1}$ and $\{p^j_k\}_{k=0}^{N-1}$ in $p^D_{\Delta\widetilde{\Pi}}$ and $p^D_j(x)$, respectively. The DFT and the inverse DFT of a sequence with $N$ points can be computed with the fast Fourier transform (FFT) using $O(N\log N)$ floating-point operations [BH95, GW98]. To compute the discrete approximation $p^D_{\Delta\widetilde{\Pi}}(\xi_i)$ therefore requires a total of $O(nN\log N)$ floating-point operations.

To prove this method works and that the computed distribution converges linearly as $h$ and $\epsilon$ decrease requires a bit of work. Essentially, the proof is an exercise in Fourier analysis and it proceeds in two steps. In the first step, we prove that, for a fixed interval $[-a, a]$, equation (4.92) converges to the cyclic convolution. In the second step, we show that as $a$ grows the cyclic convolution approximates the standard convolution.

For the set of integrable functions with compact support in $[-a, a]$, we define the *cyclic convolution* by the integral

$$f \circledast g(x) \equiv \int_{-a}^{a} f^p(x-y)g(y)dy \qquad x \in [-a, a] \tag{4.93}$$

and to be zero elsewhere. The function $f^p(x)$ is the periodic extension of $f(x)$; in other words $f^p(x) = \sum_k f(x-2ak)$. It is easy to show that the cyclic convolution is commutative and linear. Furthermore, if the two functions are in $L^1$, we have

$$|f \circledast g|_1 \le |f|_1|g|_1,$$

mimicking the similar property for the standard convolution.

**Theorem 4.1.** *Assume $f$ and $g$ are Riemann integrable and have a finite number of discontinuities. Let*

$$f^D = h\sum_{k=0}^{N-1} f^k\delta_k \qquad \textit{and} \qquad g^D = h\sum_{k=0}^{N-1} g^k\delta_k$$

[16] Given a sequence $\{y_k\}_{k=0}^{N-1}$, the discrete Fourier transform (DFT) $\{Y^k\}_{k=0}^{N-1}$ and its inverse are defined by

$$y_k = \frac{1}{N}\sum_{j=0}^{N-1} Y^j\omega^{jk}, \qquad k = 0, 1, \ldots, N-1,$$

$$Y^k = \sum_{j=0}^{N-1} y_j\omega^{-jk}, \qquad k = 0, 1, \ldots, N-1,$$

where $\omega = e^{-i\frac{2\pi}{N}}$.

*be discretizations (in the manner of Section 4.5.2). Then*

$$f^D \circledast g^D = h \sum_{k=0}^{N-1} a^k \delta_k \qquad where \qquad A^j = (-1)^j h F^j G^j,$$

*and* $f^D \circledast g^D \to f \circledast g$ *and* $f^D \circledast g^D \to (f \circledast g)^D$ *in the weak sense.*

**Proof.** We prove the two statements separately, starting with the first one.

1. The cyclic convolution $f^D \circledast g^D$ is

$$f^D \circledast g^D(x) = h^2 \sum_{k=0}^{N-1} \left( \sum_{l=0}^{N-1} f^{k-l+N/2} g^l \right) \delta_k(x),$$

and

$$a^k = h \sum_{l=0}^{N-1} f^{k-l+N/2} g^l, \qquad k = 0, 1, \ldots, N-1.$$

By applying the DFT with $\omega = e^{-i\frac{2\pi}{N}}$, it follows that

$$\begin{aligned} A^k &= h \sum_{j=0}^{N-1} a^j \omega^{-jk} \\ &= h\omega^{-kN/2} \sum_{l=0}^{N-1} g^l \omega^{-lk} \left( \sum_{j=0}^{N-1} f^{(j-l)-N/2} \omega^{-k(j-l)+kN/2} \right) \\ &= (-1)^k h F^k G^k. \end{aligned}$$

2. It remains to show that $f^D \circledast g^D \to f \circledast g$ and $f^D \circledast g^D \to (f \circledast g)^D$ in the weak sense. Since $f$ and $g$ are piecewise continuous, we have, for all points of continuity,

$$\begin{aligned} \frac{1}{h} \int_{x-h/2}^{x+h/2} f(z-y)dy &= f(z-x) + q(z-x, h)h, \\ \frac{1}{h} \int_{x-h/2}^{x+h/2} g(y)dy &= g(x) + r(x, h)h, \end{aligned}$$

where $q(\cdot, h)h \to 0$ and $r(\cdot, h)h \to 0$ as $h \to 0$. Let $\phi(x)$ be a smooth test function. Then

$$\begin{aligned} \int_{-a}^{a} \phi(x) f^D \circledast g^D(x)dx &= \sum_{k=0}^{N-1} \phi(\xi_k) \sum_{l=0}^{N-1} \left( \int_{\xi_l-h/2}^{\xi_l+h/2} f^p(\xi_k - y)dy \right) \times \left( \int_{\xi_l-h/2}^{\xi_l+h/2} g^p(y)dy \right) \\ &= \sum_{k=0}^{N-1} h\phi(\xi_k) \sum_{l=0}^{N-1} h\left(f(\xi_k - \xi_l) + hq(\xi_k - \xi_l, h)\right) \\ &\quad \times \left(g(\xi_l) + hr(\xi_l, h)\right). \end{aligned}$$

We note that

$$\sum_{l=0}^{N-1} hf(\xi_k - \xi_l)g(\xi_l) \to f \circledast g(\xi_k)$$

and

$$\sum_{k=0}^{N-1} h\phi(\xi_k) f \circledast g(\xi_k) \to \int_{-a}^{a} \phi(x) f \circledast g(x) dx.$$

If necessary, the finite number of points with discontinuities may be excluded without changing the limits. The remaining integrals involving $r$ and $q$ vanish, proving that $f^D \circledast g^D \to f \circledast g$.
Since $f \circledast g$ is Riemann integrable and piecewise continuous, it follows that

$$\frac{1}{h}\int_{x-h/2}^{x+h/2} f \circledast g(y)dy = f \circledast g(x) + hs(x,h)$$

at all points of continuity. Hence,

$$\begin{aligned}\int_{-a}^{a} \phi(x)(f \circledast g)^D(x)dx &= \sum_{k=0}^{N-1} h\phi(\xi_k)\frac{1}{h}\int_{\xi_k-h/2}^{\xi_k+h/2} f \circledast g(y)dy \\ &\to \int_{-a}^{a} \phi(x) f \circledast g(x)dx\end{aligned}$$

as $h \to 0$. □

By repeated applications of Theorem 4.1, we conclude that the right-hand side of equation (4.92) converges weakly to the cyclic convolution of the truncated densities. Our remaining obligation is to show that the standard convolution can be approximated by the cyclic convolution. The intuition being that since a pdf decays in the tails, for a large enough interval the density of the overlapping regions in the cyclic convolution decreases.

In the first lemma, we show that the error from restricting the pdfs to an interval containing the majority of the density is small.

**Lemma 4.1.** *Given the probability density functions $p_{\pi_i}$ where $i = 1, \ldots, n$, let*

$$\widetilde{p}_{\pi_i}(x) = \begin{cases} p_{\pi_i}(x), & \text{if } -a \le x \le a, \\ 0, & \text{otherwise,} \end{cases}$$

*where $|p_{\pi_i} - \widetilde{p}_{\pi_i}|_1 \le \epsilon$ for some $\epsilon > 0$ and for all $i = 1, \ldots, n$. Then,*

$$|p_{\pi_1} * \cdots * p_{\pi_n} - \widetilde{p}_{\pi_1} * \cdots * \widetilde{p}_{\pi_n}|_1 \le \epsilon n.$$

**Proof.** The statement follows immediately from the inequality $|f * g|_1 \le |f|_1 |g|_1$:

$$\begin{aligned}|p_{\pi_1} * \cdots * p_{\pi_n} - \widetilde{p}_{\pi_1} * \cdots * \widetilde{p}_{\pi_n}|_1 \le& |(p_{\pi_1} - \widetilde{p}_{\pi_1}) * \cdots * p_{\pi_n}|_1 \\ &+ \cdots + |\widetilde{p}_{\pi_1} * \cdots * (p_{\pi_n} - \widetilde{p}_{\pi_n})|_1 \\ \le& \epsilon n.\end{aligned}$$

□

In the next lemma, we show that if two pdfs are small outside an interval around the origin and if the cyclic convolution is taken over a large enough interval, then the error is small.

**Lemma 4.2.** *Assume that the pdfs $f$ and $g$ satisfy*

$$f(x) \le C_1 \frac{1}{|x|^{\alpha+1}} \quad \text{and} \quad g(x) \le C_2 \frac{1}{|x|^{\alpha+1}}, \tag{4.94}$$

*for some $\alpha > 0$ for all $x$ outside some interval $[-b, b]$. Suppose that $a > 2b$. Let $\widetilde{f}$ and $\widetilde{g}$ be the restrictions of $f$ and $g$, respectively, to the interval $[-a, a]$. Then*

$$0 \le \widetilde{f} \circledast \widetilde{g}(x) - \widetilde{f} * \widetilde{g}(x) \le \frac{1}{(a-|x|/2)^{\alpha+1}} \big(C_1 D_2(x) + C_2 D_1(x)\big)$$

*where*

$$D_1(x) = \begin{cases} \int_{-a+x/2}^{-a+x} f(y)dy, & \text{if } x > 0, \\ \int_{a+x}^{a+x/2} f(y)dy, & \text{if } x \le 0, \end{cases}$$

*and*

$$D_2(x) = \begin{cases} \int_{-a+x/2}^{-a+x} g(y)dy, & \text{if } x > 0, \\ \int_{a+x}^{a+x/2} g(y)dy, & \text{if } x \le 0. \end{cases}$$

**Proof.** We prove the bound in two steps. Consider a point $x$ in the interval $[-a, a]$. Suppose that $0 < x \le a$.

1. Then

$$\widetilde{f} * \widetilde{g}(x) = \int_{-a+x}^{a} f(x-y)g(y)dy$$

and

$$\widetilde{f} \circledast \widetilde{g}(x) = \int_{-a}^{-a+x} f(x-y-2a)g(y)dy + \int_{-a+x}^{a} f(y-x)g(y)dy.$$

Since $f$ and $g$ are positive, the error satisfies

$$0 \le \widetilde{f} \circledast \widetilde{g}(x) - \widetilde{f} * \widetilde{g}(x) = \int_{-a}^{-a+x} f(x-y-2a)g(y)dy.$$

2. Consider the interval $[-a, -a+x]$. The functions $g(y)$ and $f(y)$ satisfy equation (4.94) in $[-a, -a+x/2]$ and $[-a+x/2, -a+x]$, respectively. Hence,

$$\int_{-a}^{-a+x/2} f(x-y-2a)g(y)dy \le C_2 \int_{-a}^{-a+x/2} \frac{1}{|y|^{\alpha+1}} f(x-y-2a)dy,$$

$$\int_{-a+x/2}^{-a+x} f(x-y-2a)g(y)dy \le C_1 \int_{-a}^{-a+x/2} \frac{1}{|y|^{\alpha+1}} g(x-y-2a)dy.$$

Since $|y|^{\alpha+1} < (a - |x|/2)^{\alpha+1}$, by combining the two equalities we find

$$0 \leq \int_{-a}^{-a+x} f(x-y-2a)g(y)dy \leq \frac{1}{(a-|x|/2)^{\alpha+1}}(C_2 D_1(x) + C_1 D_2(x)),$$

as required.

The proof for negative $x$ is similar. The case $x \notin [-a, a]$ is trivial since both functions are zero outside the interval. □

With the help of the two lemmas, we are in a position to relate the cyclic approximation to the standard convolution.

**Theorem 4.2.** *Let $p_{\pi_i}$ for $i = 1, \ldots, n$ be probability density functions that satisfy the assumptions in Lemma 4.1 and Lemma 4.2. Then*

$$|p_{\pi_1} * \cdots * p_{\pi_n} - \widetilde{p}_{\pi_1} \circledast \cdots \circledast \widetilde{p}_{\pi_n}|_1 = O(\epsilon + a^{-\alpha}).$$

**Proof.** Use Lemma 4.2 to bound the error from the cyclic convolution. Since the result is asymptotic, we may assume that $a > 2b$, so the the lemma can be applied. Note that

$$\begin{aligned}\widetilde{p}_{\pi_1} * \cdots * \widetilde{p}_{\pi_n} - \widetilde{p}_{\pi_1} \circledast \cdots \circledast \widetilde{p}_{\pi_n} &= (\widetilde{p}_{\pi_1} * \widetilde{p}_{\pi_2} - \widetilde{p}_{\pi_1} \circledast \widetilde{p}_{\pi_2}) * \cdots * \widetilde{p}_{\pi_n} \\ &+ \cdots + \widetilde{p}_{\pi_1} \circledast \cdots \circledast (\widetilde{p}_{\pi_{n-1}} * \widetilde{p}_{\pi_n} - \widetilde{p}_{\pi_{n-1}} \circledast \widetilde{p}_{\pi_n}).\end{aligned}$$

By the lemma and because $|D(x)| \leq 1$, it follows that

$$\begin{aligned}|\widetilde{p}_{\pi_k} * \widetilde{p}_{\pi_{k+1}} - \widetilde{p}_{\pi_k} \circledast \widetilde{p}_{\pi_{k+1}}|_1 &\leq 2(C_k + C_{k+1}) \int_0^a \frac{1}{(a-x/2)^{\alpha+1}} dx \\ &= (C_k + C_{k+1}) \frac{4}{\alpha} \left( \left( \frac{2}{a} \right)^{\alpha} - \frac{1}{a^{\alpha}} \right).\end{aligned}$$

Hence,

$$\left|\widetilde{p}_{\pi_1} * \cdots * \widetilde{p}_{\pi_n} - \widetilde{p}_{\pi_1} \circledast \cdots \circledast \widetilde{p}_{\pi_n}\right| = O(a^{-\alpha}).$$

By Lemma 4.1 and the triangle inequality, it follows that

$$|p_{\pi_1} * \cdots * p_{\pi_n} - \widetilde{p}_{\pi_1} \circledast \cdots \circledast \widetilde{p}_{\pi_n}|_1 = O(\epsilon + a^{-\alpha}).$$

□

Theorems 4.1 and 4.2 show that equation (4.92) works. As later examples demonstrate, the rate of convergence for numerical experiments appears to be linear (see upcoming Figure 4.14). It is therefore reasonable to guess that a stronger version of Theorem 4.1 may be true.

### 4.5.6 *Computing Value-at-Risk*

In this section, we discuss how to compute value-at-risk from discrete approximation (4.92) to the pdf of the $\Delta\Gamma$-approximation. Recall that the value-at-risk is defined by nonlinear equation (4.1), and we approximate it by the solution to equation (4.85). Given the pdf $p^D_{\Delta\tilde{\Pi}}(x) = h\sum_{j=0}^{N-1} p_j\delta_j(x)$, we simply add the coefficients to get the cdf at the grid points

$$P_{\Delta\tilde{\Pi}}(\xi_k) = h\frac{p_0}{2} + h\sum_{j\le k} p_j \qquad \text{and} \qquad P_{\Delta\tilde{\Pi}}(\xi_N) = P_{\Delta\tilde{\Pi}}(a) = hp_0 + h\sum_{j=1}^{N-1} p_j. \tag{4.95}$$

Since the first point of the grid corresponds to the interval $[-a, -a+h/2]\cup[a-h/2, a]$, we choose to assign half of the density to the right-end grid point, $a$, and half to the left-end grid point, $-a$. Value-at-risk can then be computed by interpolating the cdf. Since the cdf is an increasing function, we search for the index $k$ such that $P_{\Delta\tilde{\Pi}}(\xi_k) \le 1-\alpha \le P_{\Delta\tilde{\Pi}}(\xi_{k+1})$, and therefore $-\text{VaR} - \Xi$ is in the interval $[\xi_k, \xi_{k+1}]$. The linear interpolant to the inverse cdf is

$$L(p) = \xi_k + \left(\frac{p - P_{\Delta\tilde{\Pi}}(\xi_k)}{P_{\Delta\tilde{\Pi}}(\xi_{k+1}) - P_{\Delta\tilde{\Pi}}(\xi_k)}\right)h, \qquad p \in [P_{\Delta\tilde{\Pi}}(\xi_k), P_{\Delta\tilde{\Pi}}(\xi_{k+1})].$$

The desired approximation to value-at-risk is $\text{VaR} = -L(1-\alpha) - \Xi$.

### 4.5.7 *Richardson's Extrapolation Improves Accuracy*

In practice, we have found that the observed rate of convergence can be improved with a step of Richardson's extrapolation. In our implementation, we compute two solutions $\text{VaR}_f$ and $\text{VaR}_c$ on a fine grid with step size $h$ and a coarse grid with step size $2h$, respectively. Richardson's extrapolation then gives the solution[17]

$$\text{VaR} = 2\text{VaR}_f - \text{VaR}_c.$$

Although Richardson's extrapolation can be extended to eliminate higher-order errors, we have found that additional levels of extrapolation do not lead to further improvements. The steps of the fast convolution method for value-at-risk are summarized in Algorithm 2.

[17] Suppose we want to compute $y$. If we have two approximations satisfying

$$y_1 = y + ch + o(h),$$
$$y_2 = y + c(2h) + o(h),$$

then the linear error term can be eliminated with Richardson's extrapolation

$$2y_1 - y_2 = y + o(h).$$

It is easy to derive similar formulas for higher-order terms.

**Algorithm 2**
**Fast Convolution Method for Value-at-Risk**

**Input:** *A $\Delta\Gamma$-approximation. A time series of daily returns. A confidence level $0 < \alpha < 1$. A number of grid points $N$ and a bound for the grid interval $a$.*
**Output:** *The value-at-risk for one day with confidence level $\alpha$.*

- *Estimate the parameters and factorize the $\Delta\Gamma$-approximation with Algorithm 1.*
- *Create the grid. Define $\{\xi_k = -a + hk\}_{k=0}^{N-1}$, where $h = 2a/N$.*
- *Discretize the densities,*

$$p_i^D = h \sum_{j=1}^{N-1} p_i^j \delta_j \qquad \textit{where} \quad p_i^j = \frac{1}{h} \int_{\xi_j - h/2}^{\xi_j + h/2} p_\pi(u) du.$$

- *Compute the density for the $\Delta\Gamma$-approximation with the FFT:*

$$p_{\Delta\tilde{\Pi}}^D = h \sum_{j=0}^{N-1} p_j \delta_j \qquad \textit{where} \quad P^j = (-1)^{j(n-1)} h^{n-1} \prod_{k=1}^{n} P_k^j.$$

- *Compute the discrete cdf $P_{\Delta\tilde{\Pi}}$ and find $k$ such that*

$$P_{\Delta\tilde{\Pi}}(\xi_k) \leq 1 - \alpha \leq P_{\Delta\tilde{\Pi}}(\xi_{k+1}).$$

- *Compute the linear interpolant $L(\cdot)$ for the inverse over $[\xi_k, \xi_{k+1}]$ and let*

$$\mathrm{VaR} = -L(1-\alpha) - \Xi.$$

- *Repeat from step 2 for a grid with step size $2h$. Extrapolate to get a more accurate approximation.*

The advantage of Richardson's extrapolation is clearly illustrated by the computational example in Figure 4.14. The four graphs show the error as a function of the step size $h$. The functions are

$$(i) \ \Delta\tilde{\Pi} = X_1 + X_2,$$
$$(ii) \ \Delta\tilde{\Pi} = X_1^2 + X_2^2,$$
$$(iii) \ \Delta\tilde{\Pi} = -(X_1^2 + X_2^2),$$
$$(iv) \ \Delta\tilde{\Pi} = X_1^2 + X_2^2 - X_3^2 - X_4^2 + X_5,$$

and the random variables $X_i$ are normal. The graphs show the error for confidence level $\alpha$ equal to 1%, 5%, 10%, and 20% and for value-at-risk computed with and without Richardson's extrapolation. The error is estimated as the difference between the value-at-risk for consecutive grid sizes. The rate of convergence was computed using linear regression.

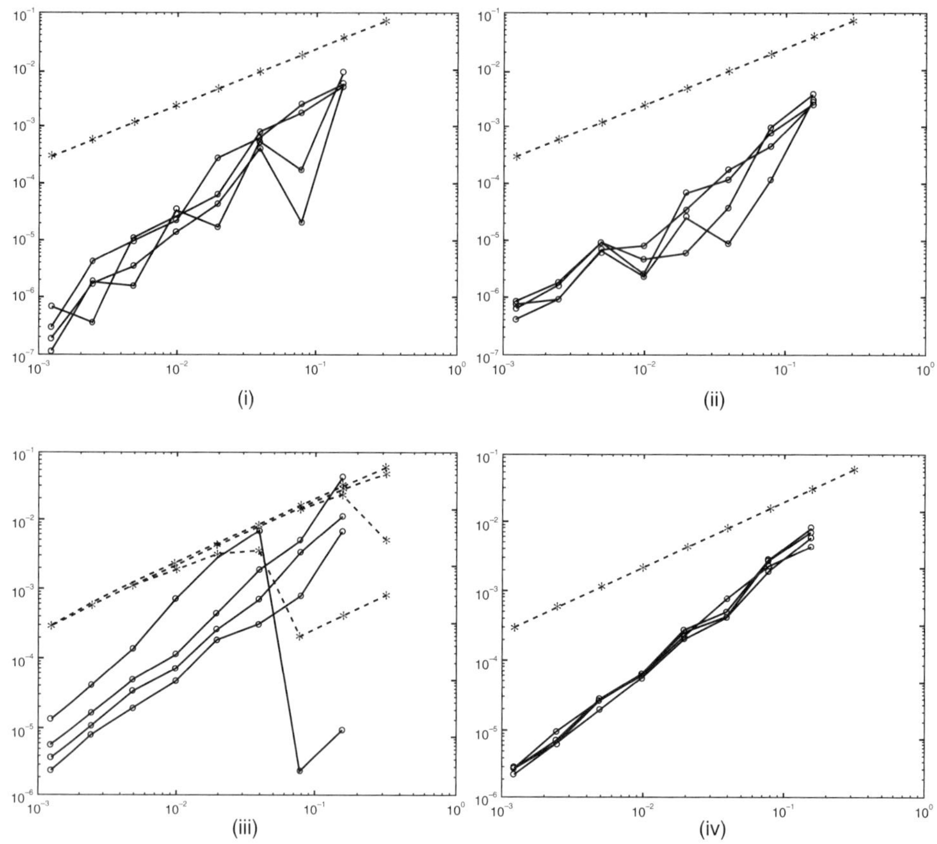


FIGURE 4.14 Error in value-at-risk as a function of step size for four abstract problems. Each graph shows the error for the confidence levels 1%, 5%, 10%, and 20%, and for value-at-risk computed with and without Richardson's extrapolation. Extrapolation is superior since the errors are smaller, and the observed rate of convergence improves from 1 to between 1.5 and 2, depending on the problem.

Without extrapolation, the observed rate of convergence is very close to 1, for all four problems and all confidence levels. When extrapolation is used, all problems show a faster rate of convergence, but the systematic relationship is less clear. The estimated rates of convergence range from approximately 1.5 for problems (ii) and (iii) to approximately 2 for problem (i).

## 4.5.8 *Computational Complexity*

The number of floating point operations for Algorithm 2 is $O(n^2 \min(d, n) + nN \log N)$, where $n$ is the number of risk factors, $d$ is the number of dates in the time series, and $N$ is the grid size. Figure 4.15 shows the computation time for portfolios for increasing $n$; the remaining parameters are fixed, with $d = 1000$ and $N = 4096$. The figure shows that the time is essentially linear in $n$. In addition, we note that a large portion of the computation time is spent in the parameter estimation step. As expected, the parameter estimation for the Parzen model is much slower than for the asymmetric t model, which in turn is slower than the normal model.

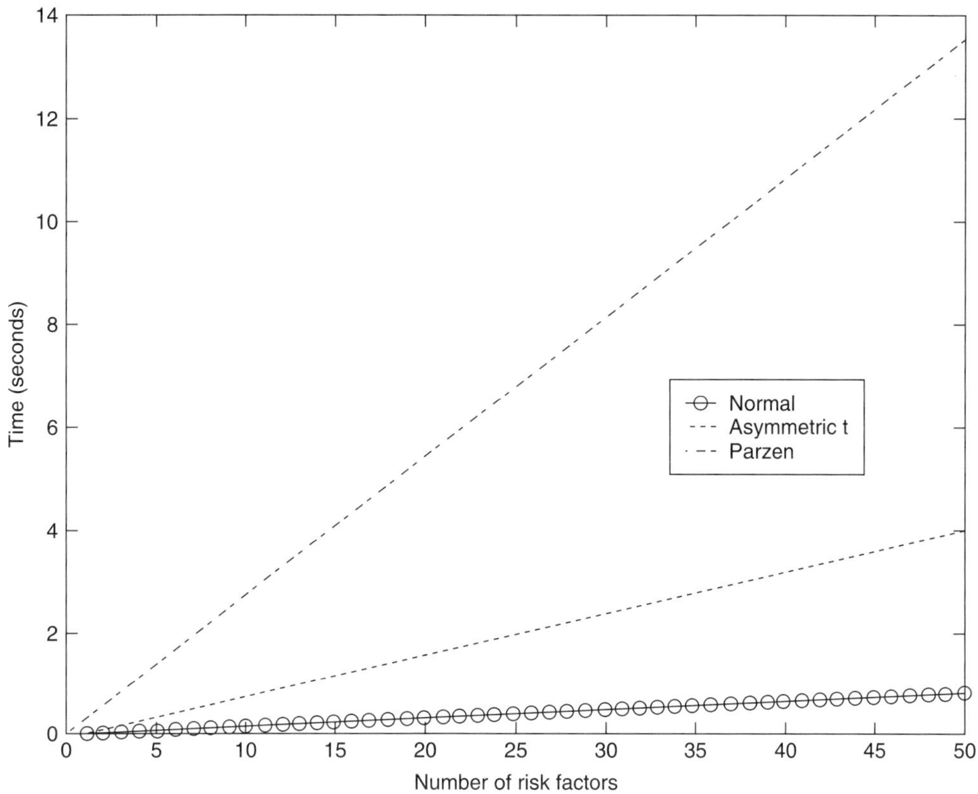


FIGURE 4.15 Computation time versus the number risk factors for the three return models. The parameter estimation takes a large portion of the total time.

# .6 Examples

## .6.1 *Fat Tails and Value-at-Risk*

We illustrate the performance of the algorithm with an example. The example also demonstrates the importance of a return model that incorporates fat tails and that includes information about the curvature of the function for the portfolio value. Consider a portfolio containing one European call option on each of BCE and Canadian Tire. The options are at-the-money and have 3 months to maturity. The Black–Scholes price of the options are \$3.23 and \$1.39. The Hessian in the $\Delta\Gamma$-approximation is

$$\boldsymbol{\Gamma} = \begin{bmatrix} 99.0967 & 0 \\ 0 & 42.6138 \end{bmatrix}.$$

The portfolio is similar to the example in Section 4.3.2, but it has shorter time to maturity, which increases the curvature.

An investor who has sold this portfolio will see her holdings decrease in value if the stock prices increase. To hedge the portfolio, she might take a linear position that offsets the $\boldsymbol{\Delta}$ of the portfolio. To examine how the value-at-risk changes with the $\boldsymbol{\Delta}$ vector, we computed the 95% and 99% value-at-risk on a grid with $-10 \leq \Delta_i \leq 10$, where $i = 1, 2$. Figures 4.16 and 4.17 show the level sets for value-at-risk. Each graph is computed with a $30 \times 30$ grid for $\boldsymbol{\Delta}$.

The dynamics of portfolio value are approximated by a linear function, and the returns on the risk factors are modeled by a multivariate normal with zero mean. The remaining six graphs in Figures 4.16 and 4.17 were computed with the fast convolution method with $N = 4096$ grid points. The dynamics of portfolio value are approximated by the same $\Delta\Gamma$-approximation in all six simulations, and the risk factors are modeled using the three models introduced in Section 4.1. We see that the linear model oversimplifies the problem. In particular, it suggests that the risk is eliminated completely by the hedging strategy, which is not true. From the graph it is clear that it also underestimates the risk away from the origin.

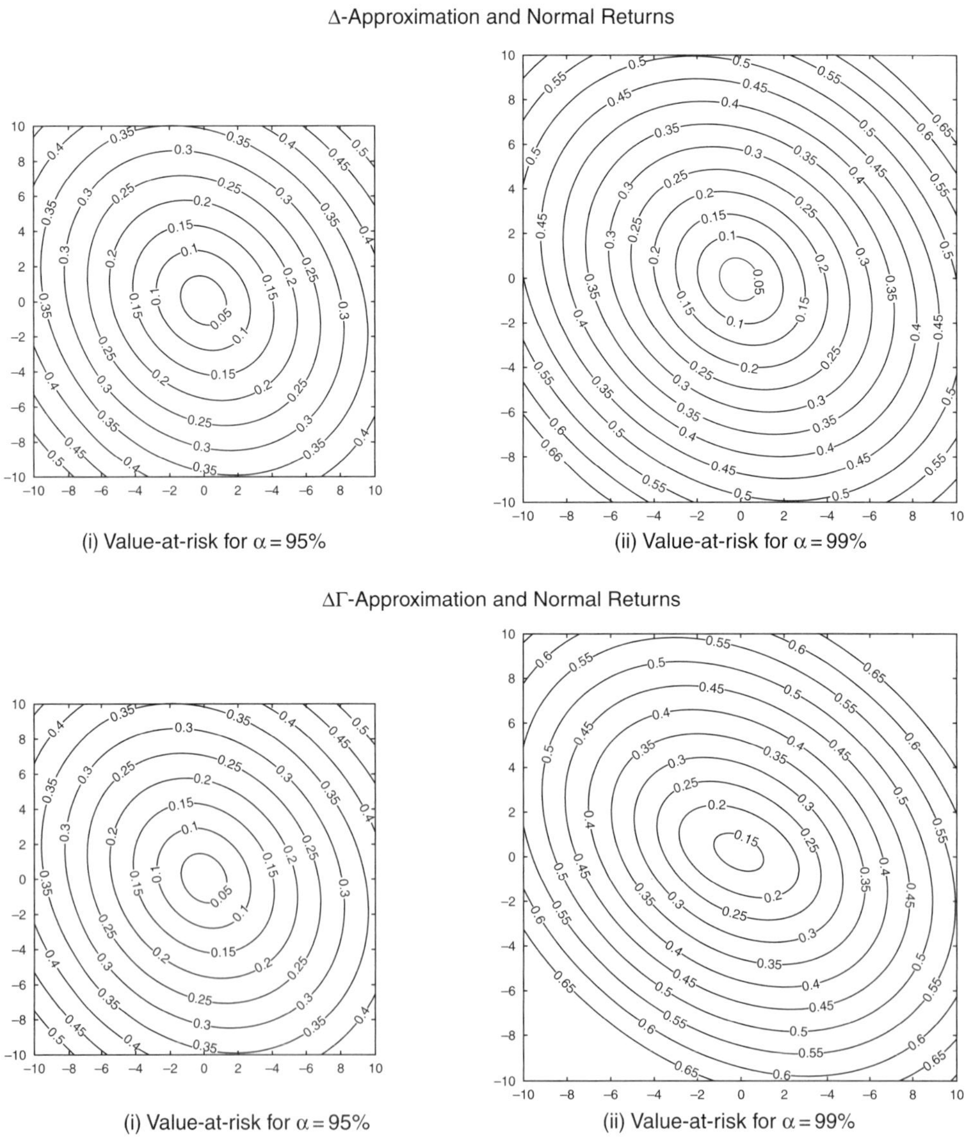


FIGURE 4.16 Part I. Value-at-risk for a short position in an option portfolio as a function of $\Delta_1$ and $\Delta_2$. The horizontal axis is $\Delta_1$, the linear position in BCE; the vertical axis is $\Delta_2$, the linear position in CTRa.

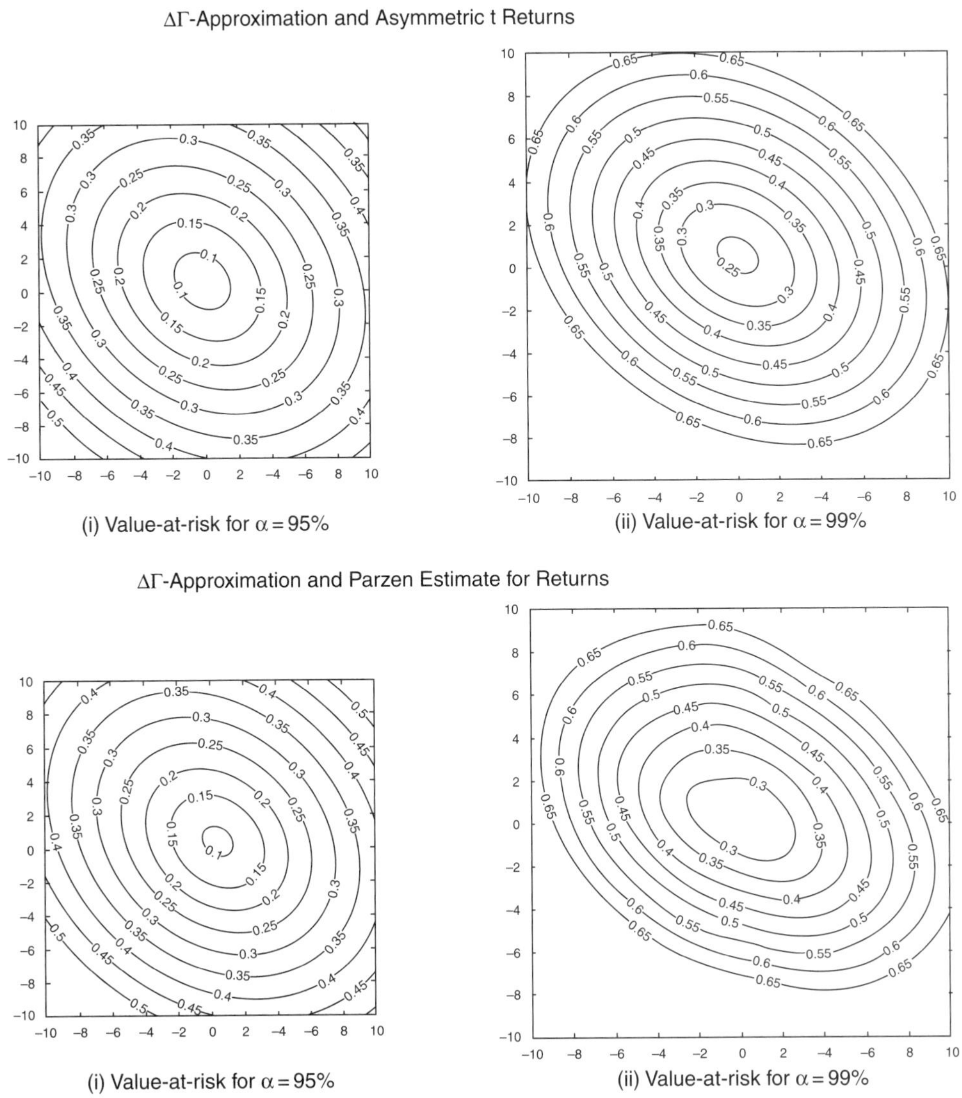


FIGURE 4.17 Part II. Value-at-risk for a short position in an option portfolio as a function of $\Delta_1$ and $\Delta_2$. The horizontal axis is $\Delta_1$, the linear position in BCE; the vertical axis is $\Delta_2$, the linear position in CTRa.

Although the remaining six graphs are more in agreement with each other, we can see some interesting differences. Again the estimate for the value-at-risk is smaller for the normal model as compared to the asymmetric t and Parzen models. For $\mathbf{\Delta} = \mathbf{0}$, the relative differences are approximately 16% (38%) and 80% (92%) for the asymmetric t model (Parzen model) and for the 95% and 99% value-at-risk, respectively. The differences between the asymmetric t and Parzen models are about 19% and 7% for the 95% and 99% value-at-risk, respectively. The Parzen model gives a larger value-at-risk as compared to both of the other models. The level sets of the normal and asymmetric t models are elliptical, whereas the level sets for the Parzen model display less symmetry. The fat tails are primarily a concern for portfolios with negative curvature. Figure 4.18 shows the 95% value-at-risk for a long position in the

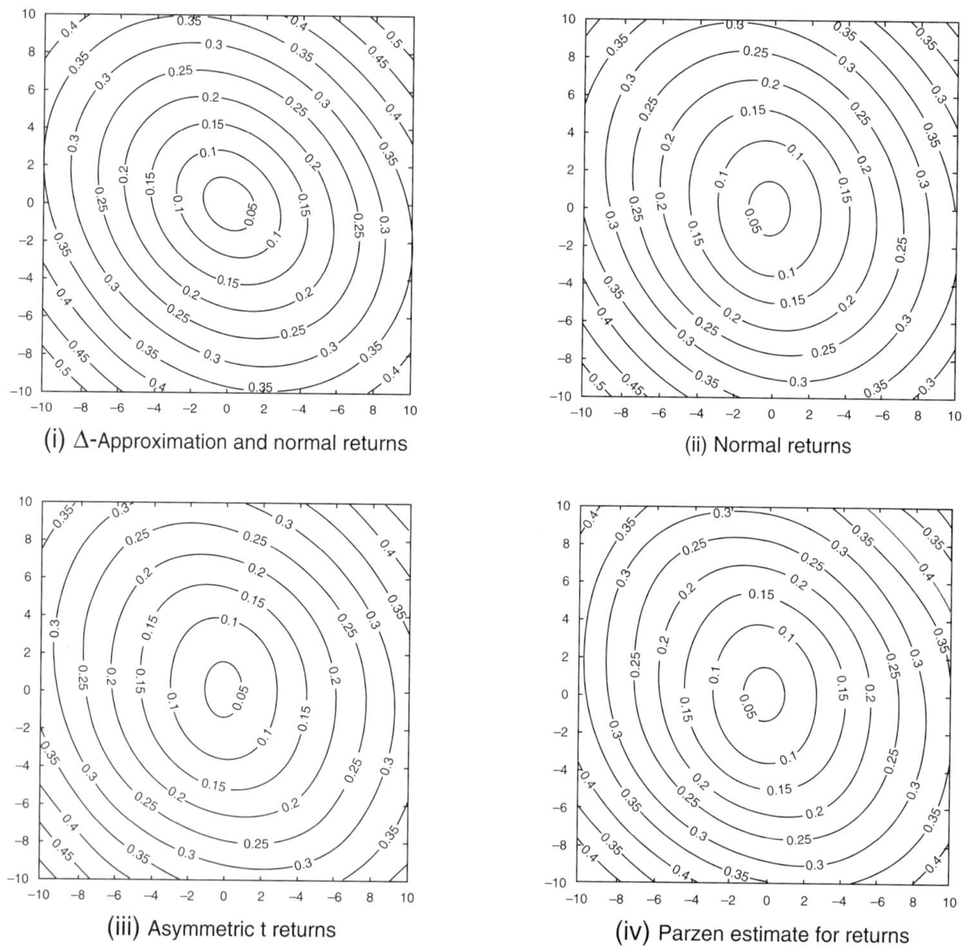


FIGURE 4.18 Part III. The 95% value-at-risk for a long position in an option portfolio as a function of $\Delta_1$ and $\Delta_2$. The horizontal axis is $\Delta_1$, the linear position in BCE; the vertical axis is $\Delta_2$, the linear position in CTRa.

example portfolio. The differences between the four graphs are much smaller, and the last three are almost identically close to the origin. Of course, this just confirms that buying the call options is much less risky than selling them.

## 4.6.2 *So Which Result Can We Trust?*

To better understand the simulations results, we can take a closer look at a long position in the delta-hedged portfolio, the portfolio with $\mathbf{\Delta} = \mathbf{0}$. Recall that if the value-at-risk is correct, we expect to have approximately 5 (25) losses exceeding the 99% (95%) value-at-risk for a sample of 500 returns. We repeated the value-at-risk simulation for 500 consecutive days and computed the number of losses greater than value-at-risk over the 500 returns used in the calculation. Figure 4.19 shows histograms for the normal, asymmetric t, and Parzen models.

From the examples in Section 4.1, we know that the normal model does not produce an accurate model for the tails. This is confirmed by the simulation, since the number

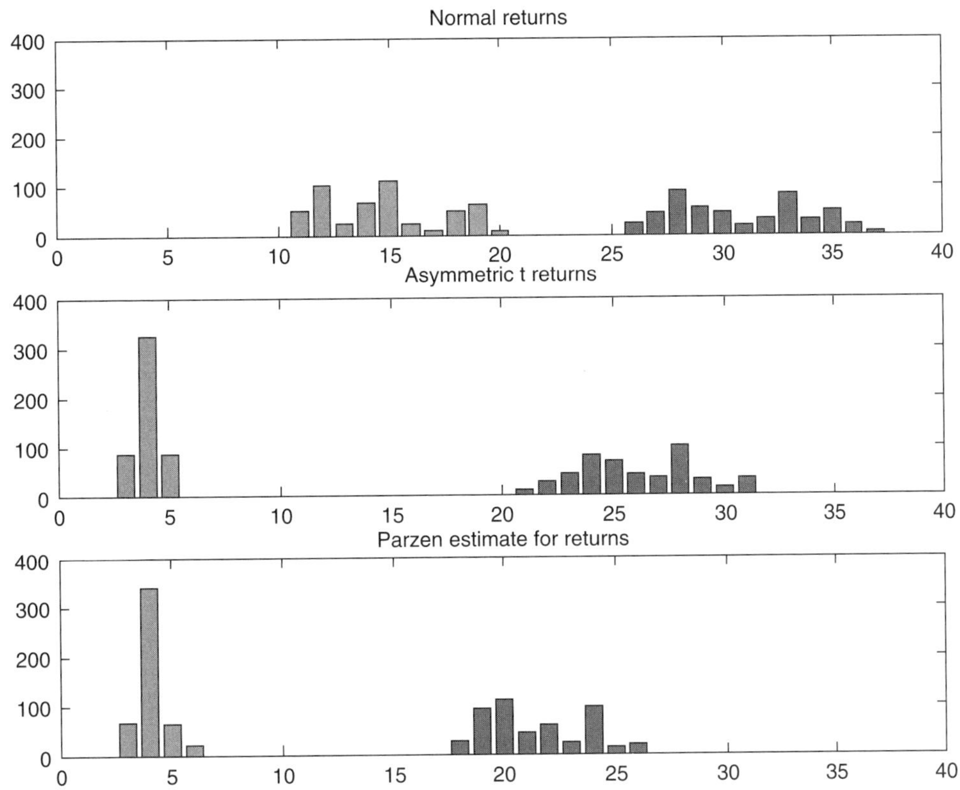


FIGURE 4.19 Histograms of the number of losses larger than value-at-risk for $\alpha = 95\%$ and $\alpha = 99\%$. Each graph shows two superimposed histograms, where the left "bump" is the result for 99% value-at-risk and the right bump is the result for the 95% value-at-risk.

of losses greater than value-at-risk deviates from the expected values. The histograms for the asymmetric t model are centered close to the expected values. The Parzen model produces a good estimate for the 99% value-at-risk, but it seems to overestimate the smaller confidence level. We conclude that the asymmetric t and Parzen models are preferable to the normal model. Furthermore, for this example, both models give acceptable results for the 99% value-at-risk, and the asymmetric t model produces a better estimate for the 95% value-at-risk.

### 4.6.3 *Computing the Gradient of Value-at-Risk*

For small portfolios, such as the one in our example, it is possible to understand how value-at-risk changes with portfolio composition by computing it many times, varying the parameters, and visualizing the result. For large portfolios and for increasing number of parameters, this is a time-consuming strategy, and the result becomes harder to interpret. In this section, we extend the fast convolution method to compute the gradient of value-at-risk. It is an interesting problem because the gradient gives local information about how the value-at-risk changes with portfolio composition. Gradient information is important to understand and evaluate decisions about changes in the portfolio. In [MR98] Mausser and Rosen review applications and methods for value-at-risk gradients. Monte Carlo methods to compute value-at-risk have been proposed independently by several authors; see, for example, the methods developed by Păun [Pău99] and Mausser and Rosen [MR98]. For the linear model with normal returns, the gradient may be interpreted in terms of "risk contributions" from instruments or risk factors; see, for example, the paper by Hallerbach [Hal99].

Recall that value-at-risk is defined by nonlinear equation (4.1). In the fast convolution method, it is approximated by the solution to equation (4.85). We can reformulate equation (4.85) as a one-dimensional integral over the pdf of $\Delta\widetilde{\Pi}$,

$$\int_{-\infty}^{-\mathrm{VaR}} p_{\Delta\tilde{\Pi}}(x)dx = 1-\alpha.$$

Then the gradient (i.e., gradients w.r.t., the "Greeks") can be computed by implicit differentiation:

$$\frac{\partial \mathrm{VaR}}{\partial \Theta} = \frac{1}{p_{\Delta\tilde{\Pi}}(-\mathrm{VaR})}\int_{-\infty}^{-\mathrm{VaR}} \frac{\partial p_{\Delta\tilde{\Pi}}}{\partial \Theta}dx, \tag{4.96}$$

$$\frac{\partial \mathrm{VaR}}{\partial \Delta_i'} = \frac{1}{p_{\Delta\tilde{\Pi}}(-\mathrm{VaR})}\int_{-\infty}^{-\mathrm{VaR}} \frac{\partial p_{\Delta\tilde{\Pi}}}{\partial \Delta_i'}dx, \tag{4.97}$$

$$\frac{\partial \mathrm{VaR}}{\partial \Lambda_{ij}} = \frac{1}{p_{\Delta\tilde{\Pi}}(-\mathrm{VaR})}\int_{-\infty}^{-\mathrm{VaR}} \frac{\partial p_{\Delta\tilde{\Pi}}}{\partial \Lambda_{ij}}dx. \tag{4.98}$$

Of course, rather than the derivatives for the parameters in the portfolio factorization, we want the derivatives for the parameters in the original $\Delta\Gamma$-approximation or the portfolio positions $\theta_i$. Fortunately, these can be computed from equations (4.96)–(4.98).[18]

## 4.6.4 The Value-at-Risk Gradient and Portfolio Composition

Suppose that the VaR gradient with respect to the parameters in the $\Delta\Gamma$-approximation are known. The gradient with respect to $\theta_i$, the quantity invested in the $i$th security, follows from the chain rule:

$$\frac{\partial \mathrm{VaR}}{\partial \theta_i} = \frac{\partial \mathrm{VaR}}{\partial \Theta}\frac{\partial \Theta}{\partial \theta_i} + \sum_{k=1}^{n}\frac{\partial \mathrm{VaR}}{\partial \Delta_k}\frac{\partial \Delta_k}{\partial \theta_i} + \sum_{k=1}^{n}\sum_{j=1}^{n}\frac{\partial \mathrm{VaR}}{\partial \Gamma_{kj}}\frac{\partial \Gamma_{kj}}{\partial \theta_i}. \tag{4.99}$$

The derivatives for the parameters of the $\Delta\Gamma$-approximation can in turn be computed from the derivatives w.r.t. parameters in the portfolio factorization, equations (4.31), (4.34), and (4.35). The $\Theta$ derivative, in equation (4.96), immediately gives

$$\frac{\partial \mathrm{VaR}}{\partial \Theta} = -t.$$

For the remaining two derivatives, it follows after some calculation that the gradient vector for $\Delta_i$ is

$$\left[\frac{\partial \mathrm{VaR}}{\partial \Delta_i}\right]_{i=1}^{n} = \mathbf{A}\left(\left[\frac{\partial \mathrm{VaR}}{\partial \Delta_i'}\right]_{i=1}^{n} - \hat{\boldsymbol{\mu}}\right) \tag{4.100}$$

[18] In the gradient computation, we have left out the contribution to the gradient from the parameter estimates.

and the Jacobian matrix for $\Gamma_{ij}$ is

$$\left[\frac{\partial \mathrm{VaR}}{\partial \Gamma_{ij}}\right]_{i,j=1}^{n} = \frac{1}{2}\mathbf{A}\left(\left[\frac{\partial \mathrm{VaR}}{\partial \Lambda_{ij}}\right]_{i,j=1}^{n} + \left[\widehat{\mu}_i \frac{\partial \mathrm{VaR}}{\partial \Delta'_j}\right]_{i,j=1}^{n} + \left[\frac{\partial \mathrm{VaR}}{\partial \Delta'_i}\widehat{\mu}_j\right]_{i,j=1}^{n} - \widehat{\boldsymbol{\mu}}\widehat{\boldsymbol{\mu}}^T\right)\mathbf{A}^T. \tag{4.101}$$

Here we have used the shorthand notation $[v_j]_{j=1}^n \equiv (v_1, \ldots, v_n)^T$ and $[M_{ij}]_{i,j=1}^n = \mathbf{M}$ for any $n \times n$ matrix $\mathbf{M}$ with elements $M_{ij}$.

## 6.5 *Computing the Gradient*

Consider integrals (4.96)–(4.98). We note that the density $p_{\Delta\tilde{\Pi}}(-\mathrm{VaR})$ is directly available in Algorithm 2; it can be computed by interpolation. Therefore, the task that remains is to approximate the two integrands in equations (4.97) and (4.98).

The derivative with respect to $\Delta'_i$ is

$$\frac{\partial p_{\Delta\tilde{\Pi}}}{\partial \Delta'_i} = \tau_{\Xi}\left(p_{\pi_1} * \cdots * \frac{\partial p_{\pi_i}}{\partial \Delta'_i} * \cdots * p_{\pi_n}\right).$$

Similarly, the derivative with respect to the diagonal element $\Lambda_{ii}$ is

$$\frac{\partial p_{\Delta\tilde{\Pi}}}{\partial \Lambda_{ii}} = \tau_{\Xi}\left(p_{\pi_1} * \cdots * \frac{\partial p_{\pi_i}}{\partial \Lambda_{ii}} * \cdots * p_{\pi_n}\right).$$

To find the derivatives for $\Lambda_{ij}$, where $i \neq j$, we have to resort to a slightly different technique.[19] The matrix $\boldsymbol{\Lambda}$ is a Hessian and hence is symmetric. Therefore, we only have to consider derivatives in directions that preserve symmetry. Consider a perturbation in the direction $E_{ij} + E_{ji}$ $(i \neq j)$. Differentiating

$$\begin{pmatrix} 1 & -t \\ t & 1 \end{pmatrix}\begin{pmatrix} \Lambda_{ii} & 0 \\ 0 & \Lambda_{jj} \end{pmatrix}\begin{pmatrix} 1 & t \\ -t & 1 \end{pmatrix}$$

we get

$$\left.\frac{d}{dt}\right|_{t=0}\begin{pmatrix} 1 & -t \\ t & 1 \end{pmatrix}\begin{pmatrix} \Lambda_{ii} & 0 \\ 0 & \Lambda_{jj} \end{pmatrix}\begin{pmatrix} 1 & t \\ -t & 1 \end{pmatrix} = (\Lambda_{ii} - \Lambda_{jj})\begin{pmatrix} 0 & 1 \\ 1 & 0 \end{pmatrix}.$$

Since the curve is tangent to $E_{ij} + E_{ji}$, we consider $p_{\Delta\tilde{\Pi}}$ along this curve as a function of $t$:

$$p_{\Delta\tilde{\Pi}} = p_{\pi_1} * \cdots * p_{\pi_i}(\Lambda_{ii}(t), \Delta'_i(t)) * \cdots * p_{\pi_j}(\Lambda_{jj}(t), \Delta'_j(t)) * \cdots * p_{\pi_n}.$$

For this curve, we get

$$\left.\frac{d}{dt}\right|_{t=0}\boldsymbol{\Delta}'(t) = \left.\frac{d}{dt}\right|_{t=0}\begin{pmatrix} 1 & t \\ -t & 1 \end{pmatrix}^{-1}\boldsymbol{\Delta}' = \begin{pmatrix} 0 & -1 \\ 1 & 0 \end{pmatrix}\boldsymbol{\Delta}'$$

[19] Differentiating a function $f: \mathbb{R}^n \to R$ is often simplified if the derivation is carried out in a convenient basis. The derivative in the direction of $\boldsymbol{v}$ at $\boldsymbol{x}$ satisfies

$$\nabla_{\boldsymbol{v}} f = \boldsymbol{v}^T \nabla f(\boldsymbol{x})$$

where $\nabla f(\boldsymbol{x})$ is the gradient of $f$ at $\boldsymbol{x}$. Hence, if $\boldsymbol{b}_1, \ldots, \boldsymbol{b}_n$ is a basis for $\mathbb{R}^n$, then the gradient can be computed from the derivatives $\nabla_{\boldsymbol{b}_i} f$ by

$$\nabla f(\boldsymbol{x}) = B^{-1}\left[\nabla_{\boldsymbol{b}_i} f\right],$$

where the rows of the matrix $B$ are the basis vectors.

The derivative of $p_{\Delta\tilde{\Pi}}$ in the direction of $T = \frac{1}{2}(E_{ij} + E_{ji})$ is the same as the derivative of $p_{\Delta\tilde{\Pi}}$ along the curve renormalized to compensate for the curve's not having unit speed:

$$\nabla_T p_{\Delta\tilde{\Pi}} = \frac{d}{dt}\bigg|_{t=0} \frac{\tau_{\Xi}\left(p_{\pi_1} * \cdots * p_{\pi_i}(\Lambda_{ii}(t), \Delta_i' + t\Delta_j') * \cdots * p_{\pi_j}(\Lambda_{jj}(t), \Delta_j' - t\Delta_i') * \cdots * p_{\pi_n}\right)}{2(\Lambda_{ii} - \Lambda_{jj})}.$$

Hence, it follows that

$$\frac{\partial p_{\Delta\tilde{\Pi}}}{\partial \Lambda_{ij}} = \frac{1}{2(\Lambda_{ii} - \Lambda_{jj})}\left(\Delta_j' \frac{\partial p_{\Delta\tilde{\Pi}}}{\partial \Delta_i'} - \Delta_i' \frac{\partial p_{\Delta\tilde{\Pi}}}{\partial \Delta_j'}\right).$$

**Algorithm 3**
**Fast Convolution Method for Value-at-Risk Gradient**

**Input:** *A $\Delta\Gamma$-approximation. A time series of daily returns. A confidence level $0 < \alpha < 1$. A number of grid points $N$ and a bound for the grid interval $a$.*
**Output:** *The value-at-risk for one day with confidence level $\alpha$. The gradient of value-at-risk for the parameters of the $\Delta\Gamma$-approximation.*

- *Compute value-at-risk with Algorithm 2.*
- *Compute the discretized partial derivative for $p_{\pi_i}$:*

$$\frac{\partial p^D_{\pi_i}}{\partial \Delta_i'} = h \sum_{j=0}^{N-1} q_i^j \delta_j.$$

- *Convolve the functions with the FFT:*

$$\left(\frac{\partial p_{\Delta\tilde{\Pi}}}{\partial \Delta_i'}\right)^D = h \sum_{j=0}^{N-1} r_j \delta_j \qquad \textit{where} \qquad R^j = (-1)^{j(n-1)} h^{n-1} Q_i^j \prod_{k \neq i} P_k^j$$

- *Integrate over $[-a, -\mathrm{VaR}]$ by adding the coefficients $\{r_j\}_{j=0}^{N-1}$ and linear interpolation. Let $I$ be this approximation.*
- *Compute $p_{\Delta\tilde{\Pi}}(-\mathrm{VaR})$ with linear interpolation for $p^D_{\Delta\tilde{\Pi}}$.*
- *Set the vector of components:*

$$\left[\frac{\partial \mathrm{VaR}}{\partial \Delta_i'}\right]_i = \frac{1}{p_{\Delta\tilde{\Pi}}(-\mathrm{VaR})} I.$$

- *Repeat from step 2 for each $\Delta_i'$ and for $\Lambda_{ii}$.*
- *Repeat for a grid with step size $2h$. Extrapolate both value-at-risk and the gradients to get more accurate answers.*
- *Perform the change of coordinates using equations (4.100) and (4.101).*

The computation boils down to finding derivatives of the pdf $p_{\pi_i}$ with respect to $\Delta'_i$ and $\Lambda_{ii}$. In our implementation, we approximate the derivatives by differentiating the coefficients of the discretized pdf $p^D_{\Pi_i}$, i.e., by differentiating equations (4.89)–(4.91) with respect to $\Delta'_i$ and $\Lambda_{ii}$. This procedure gives two discretized functions:

$$\frac{\partial p^D_{\pi_i}}{\partial \Delta'_i} = h \sum_{j=0}^{N-1} q_i^j \delta_j \quad \text{and} \quad \frac{\partial p^D_{\pi_i}}{\partial \Lambda_{ii}} = h \sum_{j=0}^{N-1} r_i^j \delta_j.$$

Hence, we can compute discrete approximations to the partial derivatives with the DFT. The algorithm is identical to the corresponding step, in equation (4.92), in the value-at-risk algorithm. Finally, integrals (4.97) and (4.98) are approximated by summing the coefficients and with linear interpolation over the final interval. The integration step is similar to equation (4.95). The complete computational procedure is summarized in Algorithm 3.

## 4.6.6 *Sensitivity Analysis and the Linear Approximation*

Consider a portfolio with two European call options on each of BCE and Canadian Tire. Both options are at-the-money and mature in 3 months. Let $\theta_1$ and $\theta_2$ be the number of contracts held for each of the two options. When priced using the Black–Scholes model, the value of the portfolio is

$$\Pi = 3.23 \cdot \theta_1 + 1.39 \cdot \theta_2.$$

The $\Delta\Gamma$-approximation is given by

$$\Theta = -0.0277 \cdot \theta_1 - 0.0119 \cdot \theta_2,$$

$$\boldsymbol{\Delta} = \begin{bmatrix} 24.3797 \cdot \theta_1 \\ 10.4725 \cdot \theta_2 \end{bmatrix},$$

$$\boldsymbol{\Gamma} = \begin{bmatrix} 99.0967 \cdot \theta_1 & 0 \\ 0 & 42.6138 \cdot \theta_2 \end{bmatrix}.$$

As an example of a gradient calculation, we computed value-at-risk for a $25 \times 25$ grid of $(\theta_1, \theta_2)$ portfolios. To understand the sensitivities of value-at-risk to the portfolio parameters, we also computed the gradient at each grid point. Figure 4.20 shows the computed gradient field superimposed over the level curves of value-at-risk. The vectors show the direction of largest sensitivity to changes in the portfolio. The six graphs show the results for the 95% and 99% value-at-risk and for the three return models introduced in Section 4.1. From the figure we make three observations. First, for all plots the computed vector field agrees with the level sets and shows that changes in $\theta_1$, the position in the BCE option, has the most impact on the risk. Second, similar to the previous example, the asymmetric t and Parzen models for return give a larger estimate of the risk as compared to the normal model. Third, the level sets for the Parzen model are less smooth, but the computed gradients still seem to agree quite well with the macro scale of change for the function. For value-at-risk simulations the lack of smoothness of computed value-at-risk is a minor concern, but when applied to optimization problems this is a serious shortcoming.

Consider portfolios where $\theta_2 = 1$ is fixed and $\theta_1$ varies, i.e., the portfolios on a horizontal line in each of the graphs in Figure 4.20. For value-at-risk as a function of $\theta_1$, the derivative of value-at-risk gives a linear approximation. Figure 4.21 shows the value-at-risk and the linear approximation computed for $\theta_1 = -0.2$. In all six cases, the linear approximations accurately describe the local dynamics of the value-at-risk. Neglecting to differentiate the market model does not lead to an inaccurate derivative.

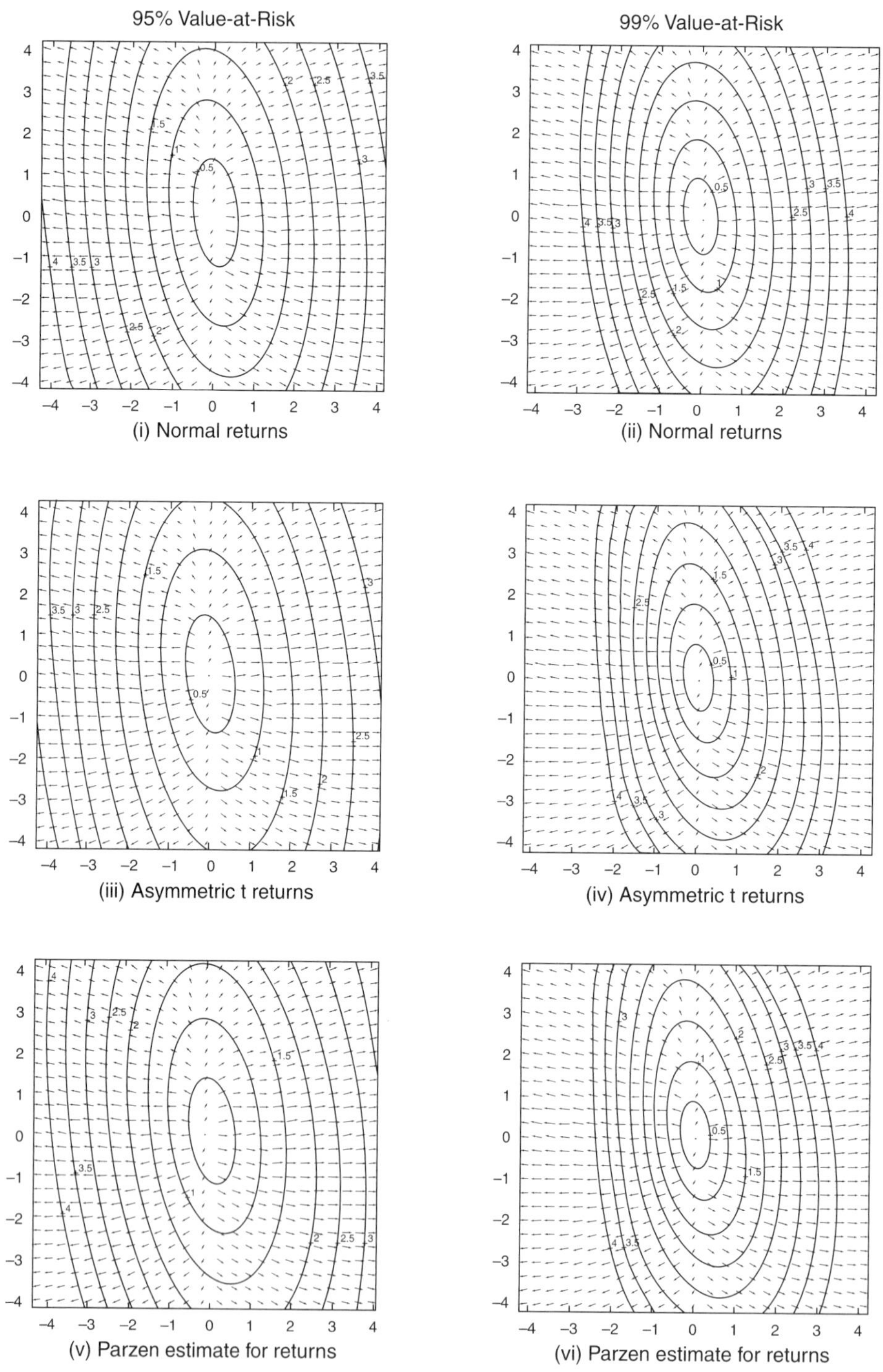


FIGURE 4.20 Level sets for value-at-risk and the gradient field for value-at-risk. The horizontal axis is $\theta_1$, the position in the BCE call option, and the vertical axis is $\theta_2$, the position in the Canadian Tire option.

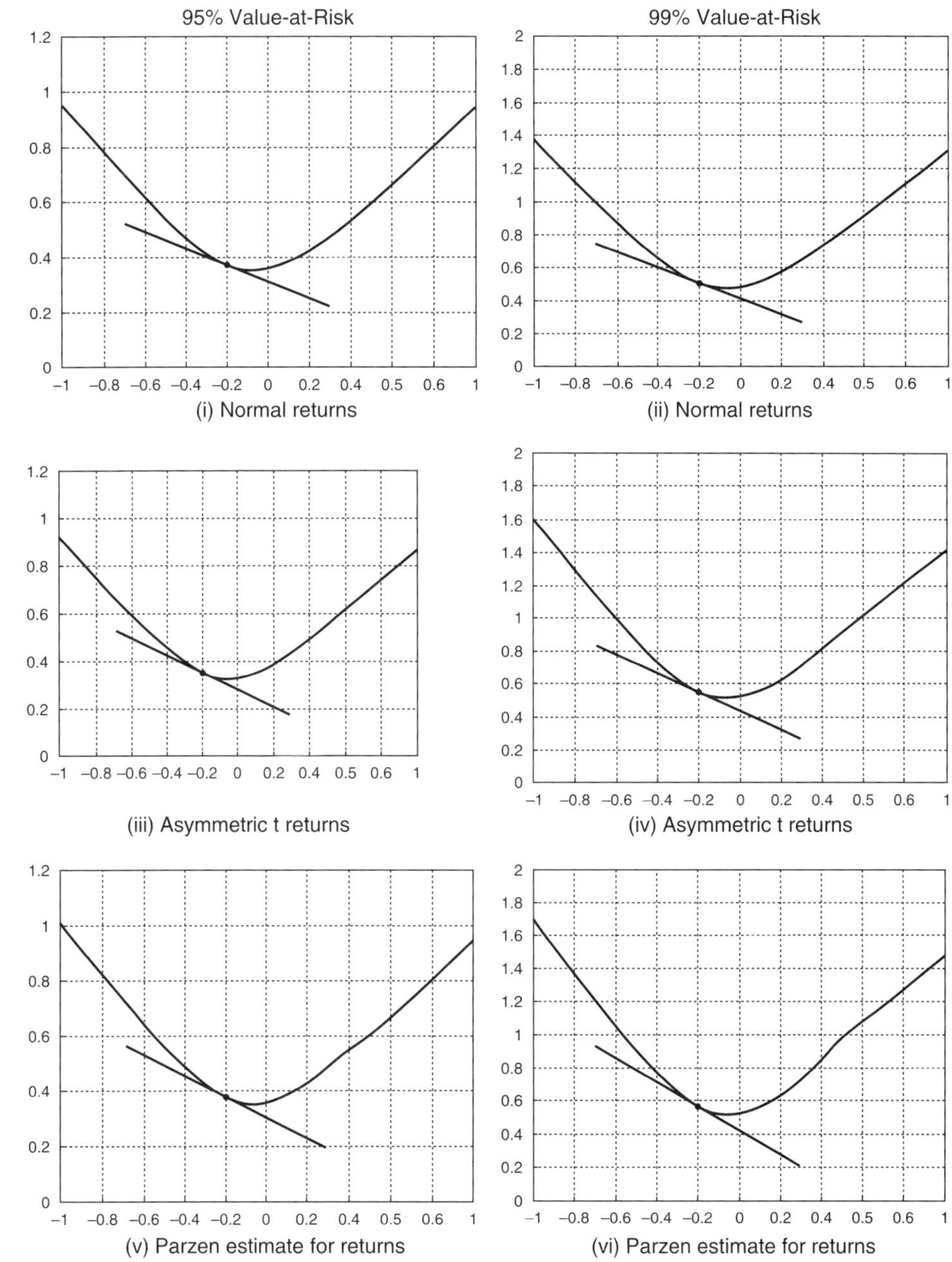


FIGURE 4.21 Linear approximations to the 95% and 99% value-at-risk as a function of $\theta_1$, the position in the call option on BCE. The linear approximation is computed for $\theta_1 = -0.2$.

## 6.7 *Hedging with Value-at-Risk*

We conclude this section with an optimization example. Similar to the previous examples, we consider a portfolio with a long position in one European call option each on BCE and Canadian Tire. The two options both mature in 3 months and are at-the-money. The TSE35 is an index with 35 stocks trading on the Toronto Stock Exchange. Extend the model with two more securities: the index itself and an at-the-money call option on the index. Since the TSE35 includes both BCE and Canadian Tire and there is significant positive correlation of the returns, it is possible to find a position that reduces the value-at-risk of the original portfolio. Let $\theta_1$ and $\theta_2$ be the number of index and call options on the index in the portfolio. The $\Delta\Gamma$-approximation for this portfolio is given by

$$\Theta = 0.0396 - 0.2354 \cdot \theta_2,$$

$$\boldsymbol{\Delta} = \begin{bmatrix} -24.3797 \\ -10.4725 \\ 564.75 \cdot \theta_1 + 321.71 \cdot \theta_2 \end{bmatrix},$$

$$\boldsymbol{\Gamma} = \begin{bmatrix} -99.0967 & 0 & 0 \\ 0 & -42.6138 & 0 \\ 0 & 0 & 2233.2 \cdot \theta_2 \end{bmatrix}.$$

This leads to the following optimization problem:

$$\min_{\theta_1, \theta_2} \mathrm{VaR}(\theta_1, \theta_2).$$

The value-at-risk surface has a flat fold; see Figure 4.22. The portfolios along the fold all have approximately the same value-at-risk, which corresponds to portfolios where $\Delta_3$ is constant. To study the performance of the fast convolution method when used in combination with optimization software, we computed the minimum for each of the three return models.[20] The computed solutions are marked in the plots in Figure 4.22. From studying the surface and inspecting the iterations, we note that the computed solutions are different and that the solutions for all three problems are very close to degenerate. Hence, hedging the portfolio using both index and index options does not lead to any significant reduction in the value-at-risk as compared to using the index alone.

### 4.6.8 *Adding Stochastic Volatility*

The picture changes when the model is extended by making volatility a risk factor. We chose to consider a simple extension. Suppose that the volatility for all three risk factors is the same; i.e., the changes in volatilities satisfy

$$\frac{\Delta\sigma^1}{\sigma_0^1} = \frac{\Delta\sigma^2}{\sigma_0^2} = \frac{\Delta\sigma^3}{\sigma_0^3},$$

and we use the Black–Scholes equation to extract a $\Delta\Gamma$-approximation. Although this model is too simple to be of use in practice, it captures the main property of a stochastic volatility model and introduces a risk factor that can only be hedged using the option. We therefore believe that the qualitative properties of the example are correct, and for the purpose of exploring value-at-risk optimization it is therefore an appropriate model problem.

[20] We used the quasi-Newton method.

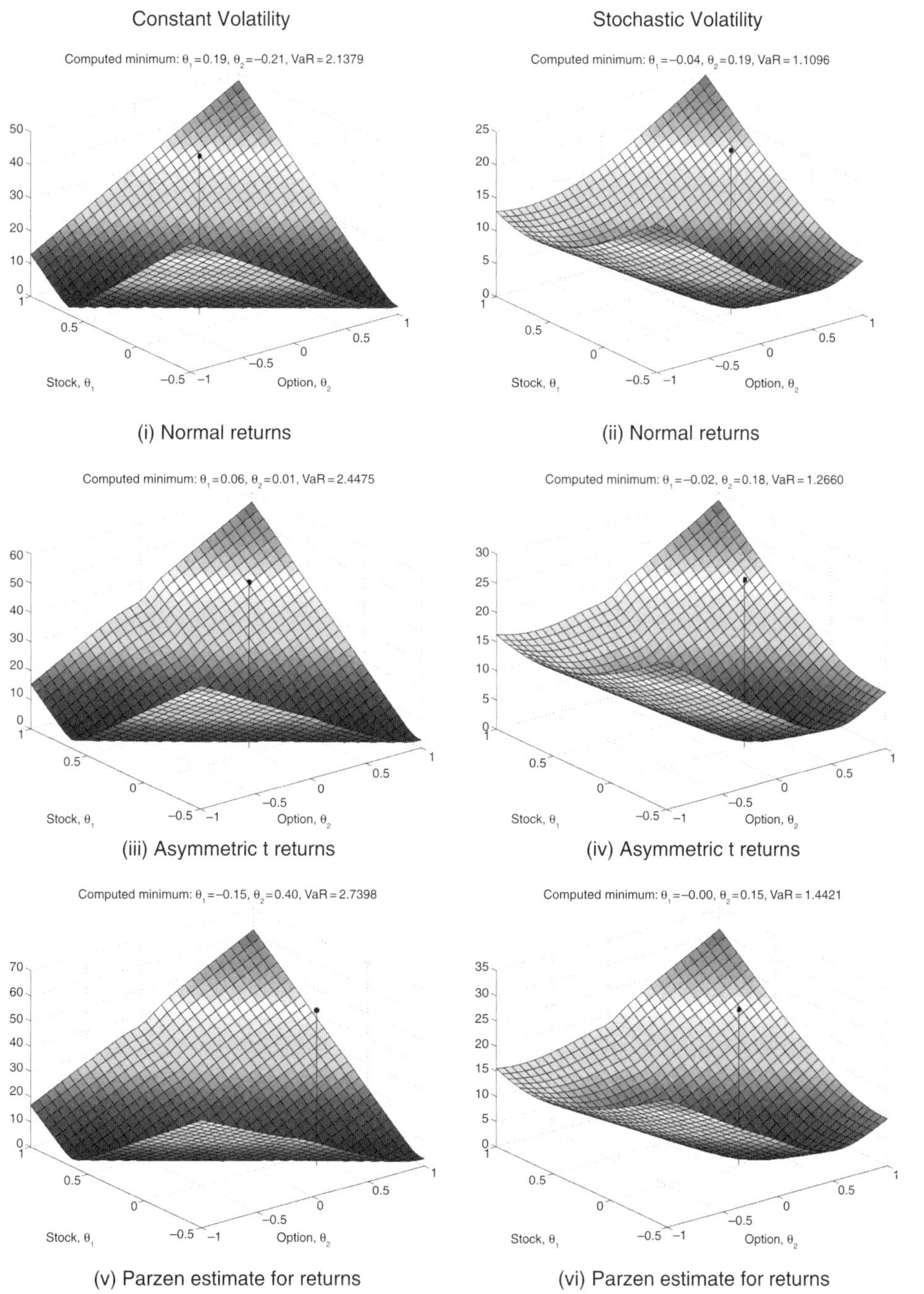


FIGURE 4.22 The 99% value-at-risk surface as a function of $\theta_1$, the position in TSE35, and $\theta_2$, the position in call options on TSE35. The point marked is the computed solution to the optimization problem. Volatility as a risk factor changes the shape of the value-at-risk surface and makes the minimum well defined.

If we include first-order volatility risk, this gives a new $\Delta\Gamma$-approximation for the portfolio:

$$\Theta = 0.0396 - 0.2354 \cdot \theta_2,$$

$$\boldsymbol{\Delta} = \begin{bmatrix} -24.3797 \\ -10.4725 \\ 564.75 \cdot \theta_1 + 321.71 \cdot \theta_2 \\ -4.1963 + 22.0409 \cdot \theta_2 \end{bmatrix},$$

$$\boldsymbol{\Gamma} = \begin{bmatrix} -99.0967 & 0 & 0 & 0 \\ 0 & -42.6138 & 0 & 0 \\ 0 & 0 & 2233.2 \cdot \theta_2 & 0 \\ 0 & 0 & 0 & 0 \end{bmatrix}.$$

In Figure 4.22, we see that the value-at-risk surfaces for the stochastic volatility model are different from the model without volatility risk. In the stochastic volatility model, the degenerate minimum is replaced by a well-defined minimum. Furthermore, the portfolio with optimal value-at-risk combines a position in the index and the index option to reduce the risk, and using only the index would give a less efficient hedge. Finally, the minima computed using optimization software are close to each other for all three return models.

From the example we can draw several interesting conclusions. First, the example shows that value-at-risk leads to nontrivial optimization problems. The fast convolution method provides an efficient basis for solving hedging problems with value-at-risk as the risk measure. Second, the shape of the value-at-risk surface changes when volatility risk is included in the model. In the stochastic volatility model, options are important to reduce the value-at-risk. Third, our experience is that optimization with the Parzen model is less reliable than the other two return models. The reason is that the surface for the Parzen model has small-scale variations caused by the nonparametric density estimator. When it is used in an optimization algorithm, the lack of smoothness makes the use of finite-difference approximations to the derivatives less stable. To some extent this can be handled by varying the tolerances used in these computations, but it does not change the fact that the steps and stopping criteria are less reliable, and special care must be taken to check the accuracy of the computed solution.

## 4.7 Risk-Factor Aggregation and Dimension Reduction

For many portfolios, a few main directions determine the dominant behavior of the dynamics of the portfolio value. This is a combined effect of correlation of risk-factor returns and the quantities of each security held in the portfolio. Therefore, it is natural to search for a simpler approximation depending on fewer factors that is close to the original model. In this section, we develop two portfolio-dependent methods for dimension reduction (see also [AJW02]).

This problem has been examined by other authors; e.g., Hull [Hul00] shows how to use principal component analysis in an interest rate model; Kreinin et al. [KMRZ98] present a principal-component–based method for linear portfolios with normally distributed risk factors; and Reimers and Zerbs [RZ98] develop a reduction method where asset blocks are represented by their dominant principal components and cross-block covariances by the covariance for the largest principal component of each block. The method presented here take a $\Delta\Gamma$-approximation and compute a new approximation of lower dimensionality that is close to the original function. The section concludes with two examples. In the first example, we compare the performance of the methods for a sample portfolio. We conclude that the method based on mean square error is more accurate, easier to implement, and slightly more efficient.

In the second example, we apply the dimension reduction method to an optimization problem. This experiment shows that dimension reduction can be effective in reducing computation time while maintaining good accuracy. Parts of these results have appeared in Albanese, Jackson, and Wiberg [AJW02], but the numerical experiments presented here are more extensive and the conclusions about nonnormal models are new.

A $\Delta\Gamma$-approximation can be written as

$$\Delta\tilde{\Pi} = \Xi + \mathbf{X}^T\boldsymbol{\Delta}' + \frac{1}{2}\mathbf{X}^T\boldsymbol{\Lambda}\mathbf{X} = \Xi + \sum_{i=1}^{n}\left(\Delta_i' X_i + \frac{\lambda_i}{2}X_i^2\right). \tag{4.102}$$

The $n \times 1$ vector $\mathbf{X}$ is an affine transformation of the original risk-factor returns (see Section 4.3). The vector of risk-factor returns is a random variable, and the the transformed random vector $\mathbf{X}$ satisfies

$$E[\mathbf{X}] = \mathbf{0} \qquad \text{and} \qquad E[\mathbf{X}\mathbf{X}^T] = \mathbf{I}.$$

If the number of risk factors $n$ is large, the time to compute value-at-risk is large. The objective in dimension reduction is to find a $\Delta\Gamma$-approximation $\Delta\tilde{\Pi}_1$ that captures the main properties of $\Delta\tilde{\Pi}$, but $\Delta\tilde{\Pi}_1$ depends on $k \ll n$ dimensions. Successful dimension reduction reduces the time to compute value-at-risk. This is particularly important for problems where value-at-risk must be computed repeatedly, as, for example, in solving optimization problems.

The strategy proposed here is to restrict $\Delta\tilde{\Pi}$ to a subspace such that $\Delta\tilde{\Pi}$ and the restriction $\Delta\tilde{\Pi}_1$ are close. Let $\mathcal{P}$ be a projection[21] onto the subspace spanned by the orthonormal columns of an $n \times k$ matrix $Q_1$, and let $\mathcal{P}^\perp$ be the projection onto the complementary subspace spanned by the orthonormal columns of $Q_2$. Let $\mathbf{z} = (\mathbf{z}_1, \mathbf{z}_2)$ be defined by

$$\mathbf{X} = \mathcal{P}\mathbf{X} + \mathcal{P}^\perp\mathbf{X} = Q_1\mathbf{z}_1 + Q_2\mathbf{z}_2. \tag{4.103}$$

Based on this factorization of the risk-factor space, we conclude that the reduced approximation has the form

$$\Delta\tilde{\Pi}(\mathbf{X}) \approx \Delta\tilde{\Pi}(\mathcal{P}\mathbf{X}) = \Delta\tilde{\Pi}_1(\mathbf{z}_1). \tag{4.104}$$

In the sections that follow, we present two methods for finding a projection so that $\Delta\tilde{\Pi}$ and $\Delta\tilde{\Pi}_1$ are close. The methods are motivated by two different views of the meaning of closeness. The first method solves the problem by finding a lower-dimension approximation with a small mean square error. The second method uses the identification of quadratic forms with matrices and solves the problem, after rescaling the variables, by finding a lower-rank matrix close to the original $\Delta\Gamma$-approximation.

### 7.1 *Method 1: Reduction with Small Mean Square Error*

In Method 1, we find $\Delta\tilde{\Pi}_1$ in equation (4.104) by insisting that the mean square error $E[(\Delta\tilde{\Pi} - \Delta\tilde{\Pi}_1)^2]$ be small. To motivate the algorithm for dimension reduction, we need the following lemma.

[21] A *projection* is a Hermitian matrix $\mathcal{P}$ such that $\mathcal{P}^2 = \mathcal{P}$.

**Lemma 4.3.** *Let $A$ be an $n \times n$ matrix with $x$, $a$, $b$ as $n \times 1$ vectors. Suppose that*

$$\max_{|a|_2=1} \max_{|b|_2=1} E[(a^T x)^2 (b^T x)^2] \leq \beta.$$

*Then*

$$E[(x^T Ax)^2] \leq n\beta |A|_F^2.$$

$||_F$ denotes the Frobenius norm of a matrix; i.e., $|A|_F = (\sum_{i,j=1}^n |A_{ij}|^2)^{1/2}$.

**Proof.** From the Cauchy–Schwartz inequality, it follows that

$$E[(x^T Ax)^2] \leq E[(x^T x)(x^T A^T Ax)].$$

Since $A^T A$ is symmetric, there is an orthogonal matrix $Q$ such that $A^T A = Q^T \Sigma^2 Q$ where $\Sigma$ is diagonal. Hence,

$$E[(x^T x)(x^T A^T Ax)] = E[(x^T Q^T Qx)(x^T Q^T \Sigma^2 Qx)].$$

Define $y_i = q_i^T x$, where $q_i^T$ is the $i$th row vector of $Q$. By assumption, we have $E[y_i^2 y_j^2] \leq \beta$ for all $i$ and $j$, and it follows that

$$\begin{aligned} E[(y^T y)(y^T \Sigma^2 y)] &= E\left[ (\sum_k y_k^2)(\sum_l \sigma_l^2 y_l^2) \right] \\ &= \sum_l \sigma_l^2 \sum_k E[y_k^2 y_l^2] \\ &\leq n\beta \sum \sigma_j^2. \end{aligned}$$

Since $\sum_j \sigma_j^2 = |A|_F^2$, the lemma follows. □

Suppose that the 4th-order moments in the lemma are bounded and that we have an estimate $\beta$ for this bound. Consider a tolerance $\epsilon > 0$. By reordering the components of $\mathbf{X}$, we can order the eigenvalues, the diagonal elements of $\mathbf{\Lambda}$, such that $|\lambda_i| \geq |\lambda_{i+1}|$ for all $i$. Let $k$ be the smallest index for which

$$\sum_{i=k+1}^{n} \lambda_i^2 \leq \epsilon. \tag{4.105}$$

Partition $\mathbf{X} = (\mathbf{X}_1, \mathbf{X}_2)$, with $\mathbf{X}_1$ containing the first $k$ components. We can then write equation (4.102) as

$$\begin{aligned} \Delta\widetilde{\Pi} = \Xi &+ [\mathbf{X}_1^T, 0] \begin{bmatrix} \mathbf{\Delta}_1' \\ 0 \end{bmatrix} + \frac{1}{2} [\mathbf{X}_1^T, 0] \begin{bmatrix} \mathbf{\Lambda}_1 & 0 \\ 0 & 0 \end{bmatrix} \begin{bmatrix} \mathbf{X}_1 \\ 0 \end{bmatrix} \\ &+ [0, \mathbf{X}_2^T] \begin{bmatrix} 0 \\ \mathbf{\Delta}_2' \end{bmatrix} + \frac{1}{2} [0, \mathbf{X}_2^T] \begin{bmatrix} 0 & 0 \\ 0 & \mathbf{\Lambda}_2 \end{bmatrix} \begin{bmatrix} 0 \\ \mathbf{X}_2 \end{bmatrix}. \end{aligned}$$

The contribution to the mean square error from $\mathbf{\Lambda}_2$ is small relative to $\epsilon$, but the effect on the gain from $\mathbf{\Delta}_2'$ may still be large. With a simple trick we can keep all the information in $\mathbf{\Delta}_2'$. Let $\mathbf{V} = [\mathbf{v}_1, \mathbf{V}_2]$ be an orthogonal matrix with the first column $\mathbf{v}_1 = \mathbf{\Delta}_2' / |\mathbf{\Delta}_2'|_2$. Since $\mathbf{\Delta}_2'$ is orthogonal to the column vectors in $\mathbf{V}_2$, we define the $(k+1)$-vector $\mathbf{z}_1$ and the $(n-k-1)$-vector $\mathbf{z}_2$ in equation (4.103) by

$$\mathbf{z}_1 = \begin{bmatrix} \mathbf{x}_1 \\ \mathbf{v}_1^T \mathbf{x}_2 \end{bmatrix} \quad \text{and} \quad \mathbf{z}_2 = \mathbf{V}_2^T \mathbf{x}_2.$$

Hence, the reduced $\Delta\Gamma$-approximation is

$$\Delta\widetilde{\Pi}_1(\mathbf{z}_1) = \Xi + \mathbf{z}_1^T \begin{bmatrix} \mathbf{\Delta}_1' \\ |\mathbf{\Delta}_2'|_2 \end{bmatrix} + \frac{1}{2} \mathbf{z}_1^T \begin{bmatrix} \mathbf{\Lambda}_1 & 0 \\ 0 & \mathbf{v}_1^T \mathbf{\Lambda}_2 \mathbf{v}_1 \end{bmatrix} \mathbf{z}_1.$$

This dimension reduction method is summarized in Algorithm 4.

To relate the reduced $\Delta\Gamma$-approximation to the mean square error, we observe that the residual is a pure quadratic term:

$$\Delta\widetilde{\Pi} - \Delta\widetilde{\Pi}_1 = \frac{1}{2}[\mathbf{z}_1^T, \mathbf{z}_2^T] \left[\begin{array}{cc|c} 0 & 0 & 0 \\ 0 & 0 & \mathbf{v}_1^T \mathbf{\Lambda}_2 \mathbf{V}_2 \\ \hline 0 & \mathbf{V}_2^T \mathbf{\Lambda}_2 \mathbf{v}_1 & \mathbf{V}_2^T \mathbf{\Lambda}_2 \mathbf{V}_2 \end{array}\right] \begin{bmatrix} \mathbf{z}_1 \\ \mathbf{z}_2 \end{bmatrix}.$$

It is easy to show that

$$\left\| \begin{bmatrix} 0 & \mathbf{v}_1^T \mathbf{\Lambda}_2 \mathbf{V}_2 \\ \mathbf{V}_2^T \mathbf{\Lambda}_2 \mathbf{v}_1 & \mathbf{V}_2^T \mathbf{\Lambda}_2 \mathbf{V}_2 \end{bmatrix} \right\|_F \leq |\mathbf{V}^T \mathbf{\Lambda}_2 \mathbf{V}|_F.$$

By the unitary invariance of the Frobenius norm and criterion (4.105), we can apply the lemma to prove the bound for the mean square error summarized in the following theorem.

**Algorithm 4**
**Dimension Reduction, Method 1**

**Input:** *A $\Delta\Gamma$-approximation. A time series $\{\mathbf{r}_i\}_{i=1}^d$ of daily returns. A tolerance $\epsilon > 0$.*

**Output:** *Parameter estimates for the market model. A factorization of the reduced $\Delta\Gamma$-approximation.*

- *Compute estimates for the mean $\widehat{\boldsymbol{\mu}}$ and the covariance matrix $\widehat{\mathbf{C}}$.*
- *Solve the eigenvalue problem*

$$\begin{cases} \mathbf{\Lambda} = \mathbf{A}^T \mathbf{\Gamma} \mathbf{A}, \\ \widehat{\mathbf{C}} = \mathbf{A}\mathbf{A}^T. \end{cases}$$

*Order the eigenvalues so that $|\lambda_i| \geq |\lambda_{i+1}|$.*

- *Compute*

$$\Xi = \Theta t + \widehat{\boldsymbol{\mu}}^T \mathbf{\Delta} + \frac{1}{2} \widehat{\boldsymbol{\mu}}^T \Gamma \widehat{\boldsymbol{\mu}},$$

$$\mathbf{\Delta}' = \mathbf{A}^T (\mathbf{\Delta} + \mathbf{\Gamma} \widehat{\boldsymbol{\mu}}).$$

- *Find the smallest integer k such that*

$$\sum_{i=k+1}^{n} \lambda_i^2 \leq \epsilon.$$

- *Compute*

$$\begin{aligned}
\boldsymbol{\Delta}_1 &= [\Delta(1:k), |\Delta(k+1:n)|_2],\\
\mathbf{v}_1 &= \Delta(k+1:n)/|\Delta(k+1:n)|_2,\\
\boldsymbol{\Lambda}_1 &= \begin{bmatrix} \boldsymbol{\Lambda}(1:k,1:k) & \mathbf{0} \\ \mathbf{0} & \mathbf{v}_1^T \boldsymbol{\Lambda}(k+1:n,k+1:n)\mathbf{v}_1 \end{bmatrix}.
\end{aligned}$$

- *For each $Z_i$, estimate the remaining parameters from the transformed returns $\{\mathbf{A}^{-1}(\mathbf{r}_i - \widehat{\boldsymbol{\mu}})\}_{i=1}^d$ and $\{\mathbf{v}_1^T \mathbf{A}^{-1}(\mathbf{r}_i - \widehat{\boldsymbol{\mu}})\}_{i=1}^d$, using the methods in Section 4.1.*

**Theorem 4.3.** *Let $\epsilon > 0$ and $\beta > 0$, with x, a, b as $n \times 1$ vectors such that*

$$\max_{|a|_2=1} \max_{|b|_2=1} E[(a^T x)^2 (b^T x)^2] \leq \beta.$$

*Then the $\Delta\Gamma$-approximation $\Delta\widetilde{\Pi}_1$ produced by Algorithm 4 satisfies*

$$E[(\Delta\widetilde{\Pi} - \Delta\widetilde{\Pi}_1)^2] \leq \frac{\beta\epsilon n}{4}.$$

## 4.7.2 Method 2: Reduction by Low-Rank Approximation

The random variables $X_i$ in equation (4.102) have zero mean, are uncorrelated, and have variance 1. Therefore, the impact of each random variable is approximately the same. The $\Delta\Gamma$-approximation can be written as a quadratic form:[22]

$$\Delta\widetilde{\Pi} = \Xi + \frac{1}{2}[\mathbf{X}^T, 1] \begin{bmatrix} \boldsymbol{\Lambda} & \boldsymbol{\Delta}' \\ (\boldsymbol{\Delta}')^T & 0 \end{bmatrix} \begin{bmatrix} \mathbf{X} \\ 1 \end{bmatrix}.$$

Therefore, we can define the distance between two $\Delta\Gamma$-approximations, with the same constant term $\Xi$, as

$$|\Delta\widetilde{\Pi} - \Delta\widetilde{\Pi}_1| = \left\| \begin{bmatrix} \boldsymbol{\Lambda} & \boldsymbol{\Delta}' \\ (\boldsymbol{\Delta}')^T & 0 \end{bmatrix} - \begin{bmatrix} \boldsymbol{\Lambda}_1 & \boldsymbol{\Delta}_1' \\ (\boldsymbol{\Delta}_1')^T & 0 \end{bmatrix} \right\|_2. \tag{4.106}$$

In Method 2, this is the definition of closeness we use for dimension reduction.

[22] In practice, it is not necessary to transform the problem all the way to form (4.102). It is sufficient to compute the Cholesky factor of the covariance matrix and transform the $\Delta\Gamma$-approximation accordingly. This way the number of eigenvalue problems solved in upcoming Algorithm 5 can be reduced from three to two.

Consider the Schur decomposition for $\Delta\widetilde{\Pi}$,

$$\begin{bmatrix} \mathbf{\Lambda} & \mathbf{\Delta}' \\ (\mathbf{\Delta}')^T & 0 \end{bmatrix} = \mathbf{O}\mathbf{\Sigma}\mathbf{O}^T,$$

where $\mathbf{O}$ is an orthogonal matrix and $\mathbf{\Sigma}$ is a diagonal matrix. The diagonal elements of $\mathbf{\Sigma}$ are the eigenvalues, and we may assume that they are ordered in decreasing absolute value $|\sigma_1| \geq \ldots \geq |\sigma_n|$. Let the $n \times n$ diagonal matrix $\mathbf{\Sigma}_k$ be defined by

$$\mathbf{\Sigma}_k = \text{diag}(\sigma_1, \ldots, \sigma_k, 0, \ldots, 0).$$

The Schmidt–Mirsky theorem says that $\mathbf{O}\Sigma_k\mathbf{O}^T$ solves the minimization problem

$$\min_{\text{rank}(\mathbf{B})=k} |\Delta\widetilde{\Pi} - \mathbf{B}|_2 = |\Delta\widetilde{\Pi} - \mathbf{O}\mathbf{\Sigma}_k\mathbf{O}^T|_2 = |\sigma_{k+1}|$$

(for a proof, see, for example, [SS90 p. 208 or GL89 p. 71]).

Given a tolerance $\epsilon > 0$, the Schmidt–Mirsky theorem gives a tool to find the optimal function $\Delta\widetilde{\Pi}_1$. Let $k$ be the smallest $k$ such that

$$|\sigma_{k+1}| \leq \epsilon.$$

Then

$$|\Delta\widetilde{\Pi} - \mathbf{O}\mathbf{\Sigma}_k\mathbf{O}^T|_2 \leq \epsilon.$$

The terms of the reduced function $\Delta\widetilde{\Pi}_1$ can be computed from the matrix $\mathbf{O}\mathbf{\Sigma}_k\mathbf{O}^T$. Partition the orthogonal matrix $U$ as

$$\mathbf{O} = \begin{bmatrix} \mathbf{U}_{11} & \mathbf{U}_{12} \\ \mathbf{u}_{21}^T & \mathbf{u}_{22}^T \end{bmatrix} = \left[\begin{array}{ccc|ccc} u_{11} & \cdots & u_{1k} & u_{1,k+1} & \cdots & u_{1,n+1} \\ \vdots & \ddots & \vdots & \vdots & \ddots & \vdots \\ u_{n1} & \cdots & u_{nk} & u_{n,k+1} & \cdots & u_{n,n+1} \\ \hline u_{n+1,1} & \cdots & u_{n+1,k} & u_{n+1,k+1} & \cdots & u_{n+1,n+1} \end{array}\right].$$

The matrix $Q_1$ in equation (4.103) can be taken as the $n \times k$ matrix with orthonormal columns in the QR factorization $Q_1\mathbf{R} = \mathbf{U}_{11}$. For this choice, we get the reduced $\Delta\Gamma$-approximation

$$\Delta\widetilde{\Pi}_1 = \Xi + \mathbf{u}_{21}^T\mathbf{\Sigma}_k\mathbf{u}_{21} + \mathbf{z}_1^T\mathbf{R}\mathbf{\Sigma}_k\mathbf{u}_{21} + \frac{1}{2}\mathbf{z}_1^T\mathbf{R}\mathbf{\Sigma}_k\mathbf{R}^T\mathbf{z}_1,$$

which can be factorized again. The steps of Method 2 are summarized in Algorithm 5. In some special cases, the matrix $\mathbf{U}_{11}$ may be rank deficient. This is not a serious problem since it is easy to show that this leads to an approximation of size $(k-1) \times (k-1)$ — it is a lucky break that earns us the additional reduction of one dimension.

**Algorithm 5**
**Dimension Reduction, Method 2**

**Input:** *A $\Delta\Gamma$-approximation. A time series $\{r_i\}_{i=1}^d$ of daily returns. A tolerance* $\epsilon > 0$.

**Output:** *Parameter estimates for the market model. A factorization of the reduced $\Delta\Gamma$-approximation.*

- *Compute estimates for the mean* $\widehat{\mu}$ *and the covariance matrix* $\widehat{\mathbf{C}}$.
- *Solve the eigenvalue problem*

$$\begin{cases} \mathbf{\Lambda} = \mathbf{A}^T\mathbf{\Gamma}\mathbf{A}, \\ \widehat{\mathbf{C}} = \mathbf{A}\mathbf{A}^T. \end{cases}$$

- *Compute*

$$\Xi = \Theta t + \widehat{\boldsymbol{\mu}}^T\mathbf{\Delta} + \frac{1}{2}\widehat{\boldsymbol{\mu}}^T\mathbf{\Gamma}\widehat{\boldsymbol{\mu}},$$

$$\mathbf{\Delta}' = A^T(\mathbf{\Delta} + \mathbf{\Gamma}\widehat{\boldsymbol{\mu}}).$$

- *Compute the Schur decomposition*

$$\begin{bmatrix} \mathbf{\Lambda} & \mathbf{\Delta}' \\ (\mathbf{\Delta}')^T & 0 \end{bmatrix} = \mathbf{O}\mathbf{\Sigma}\mathbf{O}^T.$$

  *Order the eigenvalues so that* $|\sigma_i| \geq |\sigma_{i+1}|$.
- *Find the smallest k such that*

$$|\sigma_{k+1}| \leq \epsilon.$$

- *Compute the QR factorization*

$$Q_1\mathbf{R} = \mathbf{U}_{11}.$$

- *Compute*

$$\mathbf{\Gamma}_1 = \mathbf{R}\mathbf{\Sigma}_k\mathbf{R}^T,$$

$$\Delta_1 = \mathbf{R}\mathbf{\Sigma}_k\mathbf{U}(n, 1:k),$$

$$\Xi_1 = \Xi + \mathbf{u}_{21}^T\mathbf{\Sigma}_k\mathbf{u}_{21}.$$

- *Factorize* $\Delta\widetilde{\Pi}_1$ *and estimate the remaining parameters from the transformed returns, using the methods in Section 4.1.*

### 4.7.3 Absolute versus Relative Value-at-Risk

Instead of computing the value-at-risk directly from the value-at-risk equation (4.1) with $\Delta\Pi$ replaced by $\Delta\widetilde{\Pi}_1$, we have found it is more accurate to compute the *relative value-at-risk*, defined by

$$\mathrm{VaR}_{\mathrm{rel}}(\Delta\widetilde{\Pi}) = \mathrm{VaR}(\Delta\widetilde{\Pi}) - E[\Delta\widetilde{\Pi}].$$

Instead of computing the value-at-risk from the reduced approximation as

$$\mathrm{VaR}(\Delta\widetilde{\Pi}) \approx \mathrm{VaR}(\Delta\widetilde{\Pi}_1), \tag{4.107}$$

we use the formula

$$\mathrm{VaR}(\Delta\widetilde{\Pi}) \approx \mathrm{VaR}_{\mathrm{rel}}(\Delta\widetilde{\Pi}_1) + E[\Delta\widetilde{\Pi}]. \tag{4.108}$$

The motivation for this choice becomes more clear when we consider the errors in the two formulas. The error for equation (4.107) is

$$\mathrm{VaR}(\Delta\widetilde{\Pi}) - \mathrm{VaR}(\Delta\widetilde{\Pi}_1) = \mathrm{VaR}_{\mathrm{rel}}(\Delta\widetilde{\Pi}) - \mathrm{VaR}_{\mathrm{rel}}(\Delta\widetilde{\Pi}_1) + E[\Delta\widetilde{\Pi} - \Delta\widetilde{\Pi}_1], \tag{4.109}$$

and for equation (4.108) it is

$$\mathrm{VaR}(\Delta\widetilde{\Pi}) - \mathrm{VaR}(\Delta\widetilde{\Pi}_1) - E[\Delta\widetilde{\Pi}] = \mathrm{VaR}_{\mathrm{rel}}(\Delta\widetilde{\Pi}) - \mathrm{VaR}_{\mathrm{rel}}(\Delta\widetilde{\Pi}_1).$$

The extra term $E[\Delta\widetilde{\Pi} - \Delta\widetilde{\Pi}_1]$ in equation (4.109) can be large, while the difference between the relative value-at-risk term is still small. Since computing the expectation for the full portfolio is easy, separating the terms and using the reduction $\Delta\widetilde{\Pi}_1$ only to compute the relative value-at-risk is both easy to do and leads to better accuracy.

The ideas used to develop the two methods for dimension reduction are very different. Method 1 has a direct connection to probability theory; it finds a reduced model with small mean square error. Method 2 is based on a linear algebra argument, computing reduced model using a low-rank approximation of the matrix for the quadratic form. Method 1 has several advantages over Method 2. First, Method 1 is easier to implement and slightly more efficient. Second, in Method 1 the structure of the $\Delta\Gamma$-approximation is preserved in the new quadratic form, except for the final risk factor, which captures the residual linear term. Finally, the numerical experiment in the next section shows that Method 1 gives a more accurate reduced model in practice.

### 7.4 *Example: A Comparative Experiment*

In this section, we present the first of two computational examples. The purpose of the first example is to study the performance of dimension reduction and to compare the results from the two methods.

The portfolios in our previous examples have few dimensions. The advantage of dimension reduction is to reduce computation time for portfolios with many risk factors. So we consider a portfolio with options on each of the stocks in the TSE35 index. The returns for the 35 stocks have significant correlation, and we expect that dimension reduction will produce accurate simulation results for relatively few dimensions. The portfolio consists of short positions with one call option and one put option on each of the stocks. As before, the options are European and at-the-money and have 3 months to maturity. Furthermore, we include one independent risk factor for changes in volatility, which is shared by all options; see Section 4.6.8 for details. Experiments with a similar problem, without volatility risk factor and with normal returns, are reported.

To study the effect of dimension reduction, the daily value-at-risk is computed with dimensions $k = 1, \ldots, 36$ (see Figure 4.23). Each graph shows the 99% value-at-risk and 95% value-at-risk. There is one graph for each of Method 1 and Method 2 applied to the three risk factor models. Our expectation, that only a few dimensions essentially characterize the risk of such a portfolio, is confirmed by the simulations.

Figure 4.24 shows the relative errors for the simulations in Figure 4.23. The figures lead to some interesting observations. Dimension reduction produces results that resemble the result for the full model, but the quality of the result differs for Method 1 and Method 2. In general, the result for the 95% value-at-risk seems to be better than for the 99% value-at-risk. In the case of normal returns, the error is very small for both methods [see graphs (i) and (ii)]. The error for Method 1 is small for both the asymmetric t and the Parzen models [see graphs

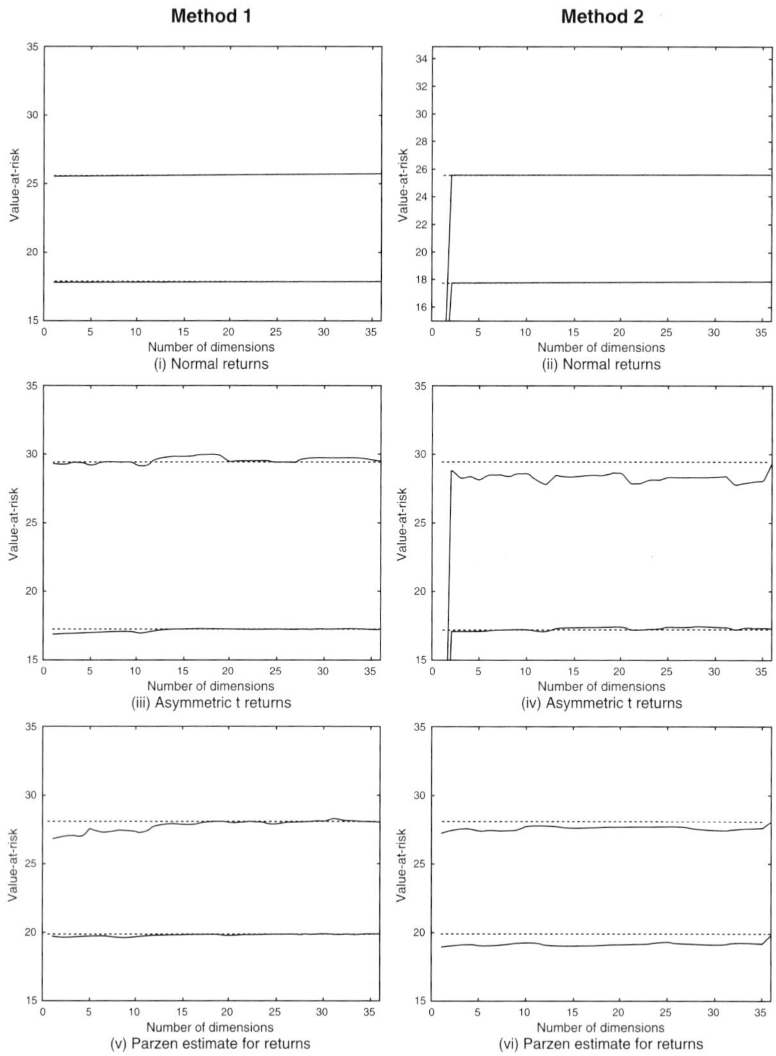


FIGURE 4.23 Value-at-risk with dimension reduction. The graphs show the 95% and 99% value-at-risk as a function of the number of dimensions used in the simulation. The $x$-axis spacings are marked for every five dimensions.

(iii) and (v)]. Method 2 does not produce accurate results for the last two return models [see graphs (iv) and (vi)], and, more seriously, it is not clear that the results improve when more dimensions are included.

To try to understand the failure of Method 2, we examined the intermediate results produced during the execution. We believe that the problem arises from a large $\Delta$ component that is split over two dimensions in the reduced model, one with positive and one with negative curvature. In the normal model, where the model is characterized by the mean and variance of the returns, this does not lead to a deterioration of the simulation result. In the other two models, the estimation error destroys the balance between the two components, and this corrupts the result. This is a serious drawback for Method 2 and one that is not shared by Method 1.

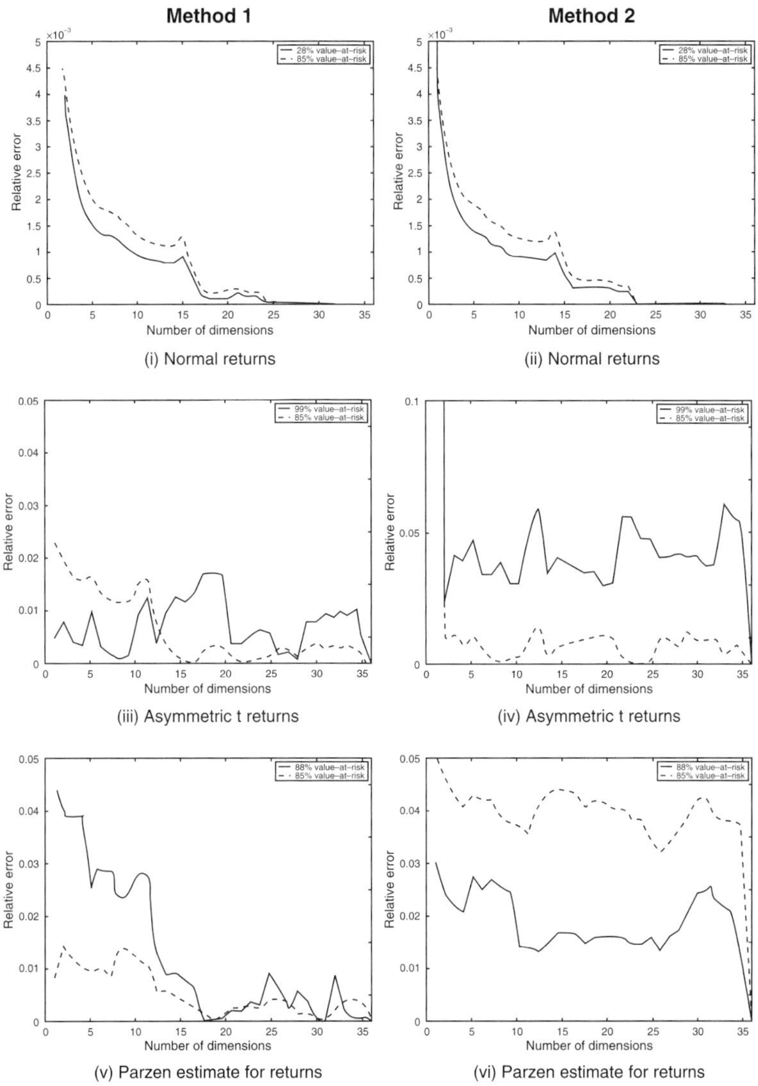


FIGURE 4.24 Relative errors for the simulations in Figure 4.23. The solid lines correspond to 99% VaR. Observe that the scales for graphs (iii) and (iv) are different.

To conclude, the value-at-risk computed with Method 1 has a small relative error. Although the error does not decrease monotonically, the trend is clear — more dimensions give more accurate results. As demonstrated by this example, Method 1 produces better results than Method 2.

### 7.5 *Example: Dimension Reduction and Optimization*

We conclude with an optimization example similar to the one in Section 4.6.7. The example shows that dimension reduction leads to significant savings in computation time and that the accuracy is preserved despite the reduction. Since Method 1 has clear advantages over Method 2, we restrict our attention to the first method.

Consider a portfolio with a short position in one call option on each of the stocks in the TSE35 index. All options are European and at-the-money and have 3 months to maturity.

Suppose we want to minimize the value-at-risk by buying or selling a combination of a linear position in the index, such as the index itself or a future, and a position in a call option on the index. Let $\theta_1$ and $\theta_2$ be, respectively, the number of index units and call options in the portfolio. To see how dimension reduction affects the shape of the value-at-risk surface, we computed the 99% value-at-risk for the full portfolio and for a reduced model with five dimensions. As Figure 4.25 shows, the differences between the surfaces are small. Of course, we are therefore led to believe that the same applies to the optimization problem

$$\min_{\theta_1,\theta_2} \mathrm{VaR}(\theta_1, \theta_2).$$

Table 4.1 show the computed solutions to the optimization problem and statistics about time and number of value-at-risk computations required by the numerical procedure.[23]

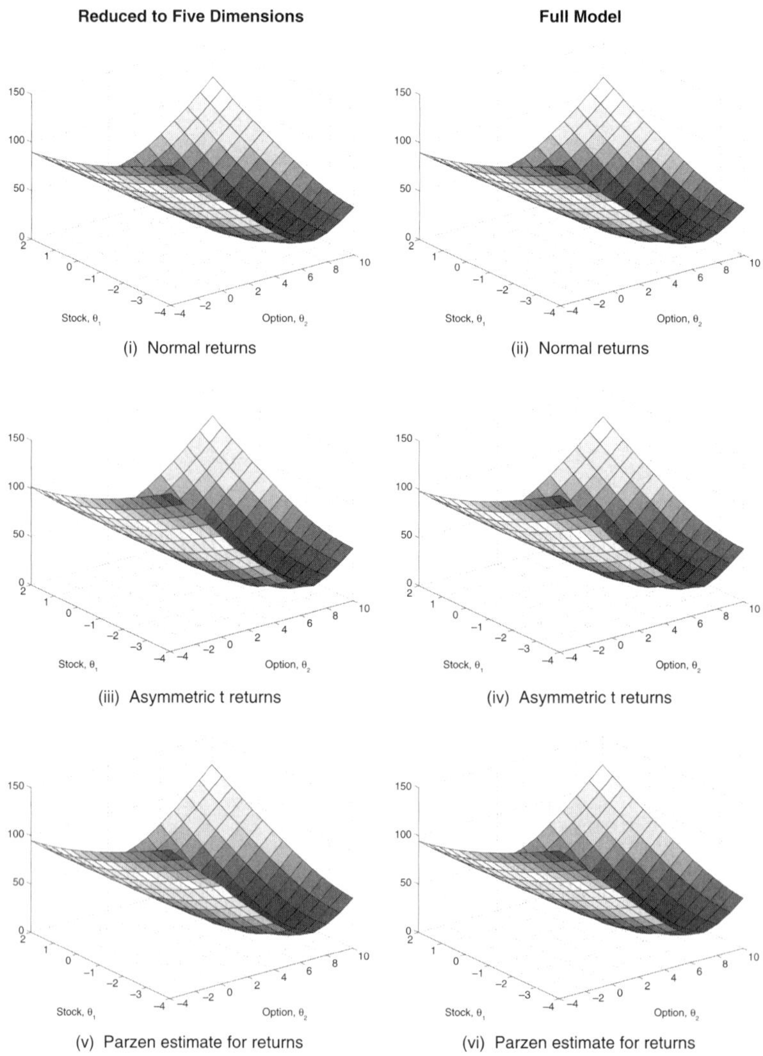


FIGURE 4.25 The 99% value-at-risk surfaces, as functions of $\theta_1$ and $\theta_2$, computed with the number of dimensions reduced to five and for the full model.

[23] The results were created using the quasi-Newton method fminunc in Matlab's optimization toolbox.

TABLE 4.1 Computed Solutions to the Optimization Problem (Quoted Times Are in Seconds of CPU Time)

| # Dimensions | $\theta_1$ | $\theta_2$ | VaR | Time | Function calls | Time/function call |
|---|---|---|---|---|---|---|
| 1 | −1.89 | 5.31 | 11.93 | 8.70 | 101 | 0.0861 |
| 2 | −1.87 | 5.30 | 12.15 | 12.87 | 129 | 0.0998 |
| 3 | −1.87 | 5.31 | 12.16 | 14.79 | 129 | 0.1147 |
| 4 | −1.88 | 5.32 | 12.16 | 15.53 | 122 | 0.1273 |
| 5 | −1.87 | 5.31 | 12.16 | 18.54 | 130 | 0.1426 |
| 6 | −1.87 | 5.31 | 12.16 | 13.41 | 84 | 0.1596 |
| 8 | −1.87 | 5.29 | 12.16 | 24.50 | 127 | 0.1929 |
| 10 | −1.86 | 5.27 | 12.17 | 29.60 | 131 | 0.2260 |
| 12 | −1.89 | 5.34 | 12.16 | 30.40 | 118 | 0.2576 |
| 17 | −1.86 | 5.27 | 12.16 | 33.06 | 102 | 0.3241 |
| 22 | −1.86 | 5.29 | 12.17 | 31.18 | 79 | 0.3947 |
| 27 | −1.92 | 5.28 | 12.16 | 38.70 | 83 | 0.4663 |
| 32 | −1.82 | 5.19 | 12.21 | 50.85 | 95 | 0.5353 |
| 37 | −1.86 | 5.27 | 12.17 | 77.33 | 127 | 0.6089 |

(i) Normal returns

| # Dimensions | $\theta_1$ | $\theta_2$ | VaR | Time | Function calls | Time/function call |
|---|---|---|---|---|---|---|
| 1 | −1.91 | 5.46 | 13.15 | 18.83 | 85 | 0.2215 |
| 2 | −1.91 | 5.43 | 13.30 | 29.10 | 102 | 0.2853 |
| 3 | −1.91 | 5.40 | 13.55 | 36.36 | 100 | 0.3636 |
| 4 | −1.82 | 5.26 | 13.61 | 55.51 | 128 | 0.4337 |
| 5 | −1.80 | 5.23 | 13.62 | 41.45 | 80 | 0.5181 |
| 6 | −1.87 | 5.30 | 13.56 | 65.75 | 111 | 0.5923 |
| 8 | −1.81 | 5.28 | 13.59 | 29.70 | 39 | 0.7615 |
| 10 | −1.81 | 5.23 | 13.58 | 114.8 | 127 | 0.9038 |
| 12 | −1.83 | 5.31 | 13.57 | 146.5 | 142 | 1.0316 |
| 17 | −1.86 | 5.28 | 13.57 | 115.3 | 83 | 1.3889 |
| 22 | −1.80 | 5.22 | 13.60 | 151.0 | 87 | 1.7356 |
| 27 | −1.81 | 5.23 | 13.62 | 167.2 | 80 | 2.0905 |
| 32 | −1.80 | 5.22 | 13.60 | 316.5 | 129 | 2.4537 |
| 37 | −1.84 | 5.28 | 13.60 | 252.9 | 89 | 2.8412 |

(ii) Asymmetric t returns

| # Dimensions | $\theta_1$ | $\theta_2$ | VaR | Time | Function calls | Time/function call |
|---|---|---|---|---|---|---|
| 1 | −2.05 | 5.26 | 14.21 | 33.50 | 128 | 0.2617 |
| 2 | −1.85 | 5.27 | 14.51 | 50.69 | 110 | 0.4608 |
| 3 | −1.86 | 5.32 | 14.70 | 90.81 | 140 | 0.6486 |

continued

TABLE 4.1 continued

| # Dimensions | $\theta_1$ | $\theta_2$ | VaR | Time | Function calls | Time/function call |
|---|---|---|---|---|---|---|
| 4 | −1.82 | 5.18 | 13.01 | 69.43 | 82 | 0.8467 |
| 5 | −1.91 | 5.27 | 13.03 | 73.63 | 70 | 1.0519 |
| 6 | −2.00 | 5.00 | 13.32 | 153.0 | 122 | 1.2543 |
| 8 | −1.88 | 4.96 | 13.28 | 153.4 | 93 | 1.6495 |
| 10 | −1.86 | 5.16 | 12.98 | 108.9 | 53 | 2.0555 |
| 12 | −1.86 | 5.15 | 13.00 | 365.4 | 150 | 2.4357 |
| 17 | −1.83 | 5.16 | 13.11 | 210.1 | 61 | 3.4434 |
| 22 | −1.64 | 4.98 | 13.34 | 725.2 | 165 | 4.3950 |
| 27 | −1.77 | 5.10 | 13.22 | 749.9 | 139 | 5.3947 |
| 32 | −1.85 | 5.18 | 12.89 | 562.6 | 88 | 6.3928 |
| 37 | −1.89 | 5.27 | 12.90 | 601.8 | 81 | 7.4291 |

(iii) Parzen estimate for returns

The variations in value-at-risk at the computed minima are small for the normal and asymmetric t models, and the changes in the location of the minima are relatively small. Similar to the observations in Section 4.6.7, we note that the performance of the minimization procedure is less reliable for the Parzen model; the density estimator introduces small fluctuations in the value-at-risk. In all three cases, we see that dimension reduction is very effective in reducing the computation time per function evaluation. The total time for the optimization is also reduced. Although the total time is important, it is not a good indicator, since it mostly depends on the success of the stopping criteria used by the optimization algorithm.

## 4.8 Perturbation Theory

The experiments in previous sections show that the different models for risk-factor returns can lead to large differences in the estimate for value-at-risk. Similar observations have been made by several other authors; see Beder [Bed95] and Jorion [Jor96]. In this section, we derive a perturbation result that describes how value-at-risk changes with perturbations to the model for risk-factor returns. It shows that value-at-risk becomes increasingly sensitive as the confidence level increases and that the tail is more sensitive than the center of the distribution. We present a computable-condition number for value-at-risk and illustrate the theory with a numerical example.

### 4.8.1 *When Is Value-at-Risk Well Posed?*

Value-at-risk is defined as the solution to nonlinear equation (4.1). If $p(\mathbf{r})$ is the pdf for the risk-factor returns, then equation (4.1) is equivalent to the integral equation (i.e., equation (4.72):

$$\int_{\{\mathbf{r}\in\mathbb{R}^n:\Delta\Pi(\mathbf{r})\leq-\mathrm{VaR}\}} p(\mathbf{r})d\mathbf{r} = 1-\alpha. \tag{4.110}$$

The integration domain is a subset of risk-factor returns in $\mathbb{R}^n$. The coarea formula[24] gives

$$\int_{-\infty}^{-\mathrm{VaR}} p_{\Delta\Pi}(\rho)d\rho = 1-\alpha, \tag{4.111}$$

where

$$p_{\Delta\Pi}(\rho) = \int_{\{\mathbf{r}\in\mathbb{R}^n:\Delta\Pi(r)=\rho\}} \frac{p(\mathbf{r})}{|D\Delta\Pi(\mathbf{r})|^{1/2}}dA \tag{4.112}$$

and

$$|D\Delta\Pi(\mathbf{r})| = \sum_{i=1}^{n}\left(\frac{\partial\Delta\Pi}{\partial r_i}\right)^2.$$

So the coarea formula transforms integral (4.110) into one-dimensional integral (4.111) over a new pdf (4.112). In other words, equation (4.111) is an integral over the pdf for $\Delta\Pi$, which is defined by equation (4.112) as the $(n-1)$-dimensional surface integral (i.e., $dA$ is the surface differential) over the level sets of $\Delta\Pi$. Equations (4.110) and (4.111) are equivalent, but we find the second form more convenient in our perturbation analysis.

Hadamard's classic definition says that a problem is well posed if it has a unique solution that depends continuously on the initial data. The properties of pdf (4.112) determine if value-at-risk is a well-posed problem. The first condition, existence of a solution, holds without additional assumptions. The cdf

$$\Phi(x) = \mathrm{P}(\Delta\Pi \le x) = \int_{-\infty}^{x} p_{\Delta\Pi}(\rho)d\rho \tag{4.113}$$

has range $(0,1)$. Hence, since the cdf $\Phi$ is a continuous function, the intermediate value theorem implies that equation (4.111) has a solution for all $\alpha \in (0,1)$.

For uniqueness and continuity to hold, equation (4.113) must be strictly increasing in a neighborhood around the solution $x=-\mathrm{VaR}$. Equivalently, it is unique if equation (4.112) is positive almost everywhere in a neighborhood of $-\mathrm{VaR}$. Uniqueness follows by observing that, if there were to be two solutions with $\mathrm{VaR}_1 > \mathrm{VaR}_2$, then since equation (4.113) is an increasing function, any $x$ with $-\mathrm{VaR}_1 < x < -\mathrm{VaR}_2$ must be a solution too. Hence, equation (4.113) is not strictly increasing, or, equivalently, the pdf is not positive almost everywhere, in any neighborhood of a solution.

Continuity, the third condition for equation (4.111) to be well posed, requires that the solution depend continuously on the data. We show that value-at-risk is continuous for changes in the pdf $p(\mathbf{r})$. Suppose that the density in equation (4.112) is positive almost everywhere in some interval $(-\mathrm{VaR}-\epsilon, -\mathrm{VaR}+\epsilon)$. Let $\{p_i\}_{i=1}^{\infty}$ be a sequence of pdfs that converges to $p$; that is, $|p-p_i|_1 \to 0$ as $i\to\infty$. Moreover, let $\mathrm{VaR}_i$ be the solutions to equation (4.110) corresponding to $p_i$ and some fixed $\alpha$. Combining equations (4.111) and (4.112) gives

$$\int_{-\infty}^{-\mathrm{VaR}} d\rho \int_{\{\Delta\Pi(\mathbf{r})=\rho\}} \frac{p(\mathbf{r})}{|D\Delta\Pi(\mathbf{r})|^{1/2}}dA = \int_{-\infty}^{-\mathrm{VaR}_i} d\rho \int_{\{\Delta\Pi(\mathbf{r})=\rho\}} \frac{p_i(\mathbf{r})}{|D\Delta\Pi(\mathbf{r})|^{1/2}}dA. \tag{4.114}$$

[24] We refer to Evans and Gariepy [EG92] for a proof and discussion of the coarea formula. The formula can be applied assuming that $\Delta\Pi$ is Lipschitz differentiable and ess inf $|Df| > 0$.

We then obtain

$$\left| \int_{-\mathrm{VaR}_i}^{-\mathrm{VaR}} p_{\Delta\Pi}(\rho)d\rho \right| \le \int_{\Delta\Pi(\mathbf{r})\le -\mathrm{VaR}_i} |p(\mathbf{r}) - p_i(\mathbf{r})| d\mathbf{r} \le |p - p_i|_1. \tag{4.115}$$

Furthermore, since $p_{\Delta\Pi}(\rho)$ is positive almost everywhere in a neighborhood of $-\mathrm{VaR}$ and the left-hand side of the inequality goes to zero, we must have $\mathrm{VaR}_i \to \mathrm{VaR}$, as $i \to \infty$. So value-at-risk is continuous with respect to the return distribution model $p$.

Hence, value-at-risk is a well-posed problem, given that cdf (4.113) is increasing close to the solution. This condition holds for the market-risk models we consider in this chapter. However, for credit-risk models, this condition is often violated and value-at-risk is a questionable measure of risk. For a detailed discussion and an axiomatic system of desirable properties of general risk measures, see Artzner et al. [ADEH99]. In the preceding analysis we examined perturbations of the model for risk-factor returns. In the future, it would be interesting to extend the analysis to perturbations of $\Delta\Pi$.

## 4.8.2 *Perturbations of the Return Model*

The foregoing analysis shows that value-at-risk, defined by equation (4.111), is well posed if the cdf $\Phi(x)$ is strictly increasing for values of $x$ in a neighborhood of the solution. We argued that value-at-risk is continuous for changes in $p(\mathbf{r})$, but the analysis does not indicate the size of the resulting perturbations. In this section, we quantify the change in value-at-risk for a perturbation of the pdf $p(\mathbf{r})$.

### 4.8.2.1 Proof of a First-Order Perturbation Property

We now derive a variational, first-order perturbation property for value-at-risk. Consider a differentiable pdf $p_{\Delta\Pi}(\rho)$ as given by equation (4.112). The set of probability density functions is the subset of functions in $L^1$ that integrate to 1.[25] Furthermore, if $p$ and $q$ are pdfs, then $hp + (1-h)q$ is a pdf for all $h \in [0, 1]$; i.e., the pdfs are a convex set. Consider a variation $v$ where

$$v_{\Delta\Pi}(\rho) = \int_{\{\Delta\Pi(\mathbf{r})=\rho\}} \frac{v(\mathbf{r})}{|D\,\Delta\Pi(\mathbf{r})|^{1/2}} dA \tag{4.116}$$

is continuous. In addition, for $v$ to be an admissible variation, the function $p + hv$ must be a pdf for all $h$ in some interval $[0, \epsilon)$, $\epsilon > 0$.

For an admissible variation, the value-at-risk is a function of $h$ that satisfies

$$\int_{-\infty}^{-\mathrm{VaR}(h)} d\rho \int_{\{\Delta\Pi(\mathbf{r})=\rho\}} \frac{p(\mathbf{r}) + hv(\mathbf{r})}{|D\,\Delta\Pi(\mathbf{r})|^{1/2}} dA = 1 - \alpha.$$

The function on the left side has the form

$$F(x, h) = \int_{-\infty}^{x} d\rho \int_{\{\Delta\Pi(\mathbf{r})=\rho\}} \frac{p(\mathbf{r}) + hv(\mathbf{r})}{|D\,\Delta\Pi(\mathbf{r})|^{1/2}} dA,$$

[25] A function $f$ is in the function space $L^1$ if it is measurable and has finite $L^1$ norm, i.e., if

$$|f|_1 = \int_{\mathbb{R}^n} |f(\mathbf{x})| d\mathbf{x} < \infty.$$

with first partial derivatives

$$\frac{\partial F}{\partial x} = \int_{\{\Delta\Pi(r)=x\}} \frac{p(\mathbf{r}) + hv(\mathbf{r})}{|D\ \Delta\Pi(\mathbf{r})|^{1/2}} dA \tag{4.117}$$

and

$$\frac{\partial F}{\partial h} = \int_{-\infty}^{x} d\rho \int_{\{\Delta\Pi(\mathbf{r})=\rho\}} \frac{v(\mathbf{r})}{|D\ \Delta\Pi(\mathbf{r})|^{1/2}} dA.$$

Since the density in equation (4.112) is continuous, value-at-risk is well posed for $h = 0$ if and only if equation (4.117) is positive for $h = 0$,

$$\frac{\partial F}{\partial x}\Big|_{h=0} = p_{\Delta\Pi}(x) > 0.$$

Assuming that value-at-risk is well posed, the implicit function theorem guarantees that the solution $\mathrm{VaR}(h)$ to

$$F(-\mathrm{VaR}(h), h) = 1 - \alpha$$

is continuously differentiable for $h$ in some interval $[0, \delta)$, where $\delta \leq \epsilon$ (see Rudin [Rud76]). The derivative of $\mathrm{VaR}(h)$ at $h = 0$ is

$$\mathrm{VaR}'(0) = \frac{1}{p_{\Delta\Pi}(-\mathrm{VaR})} \int_{\Delta\Pi(\mathbf{r}) \leq -\mathrm{VaR}} v(\mathbf{r}) d\mathbf{r}. \tag{4.118}$$

In the terminology of variational calculus, $\mathrm{VaR}'(0)$ is the Gateaux variation in the direction of $v$. Taylor's theorem gives the linear approximation

$$\mathrm{VaR}(h) = \mathrm{VaR}(0) + \mathrm{VaR}'(0)h + O(h^2). \tag{4.119}$$

#### 4.8.2.2 Error Bounds and the Condition Number

Taking the absolute value of linear approximation (4.119), we get an estimate of the absolute error,

$$|\mathrm{VaR}(h) - \mathrm{VaR}(0)| \leq |\mathrm{VaR}'(0)| \cdot |h| + O(|h|^2).$$

Note that derivative $\mathrm{VaR}'(0)$ depends on the variation $v$.[26] So in general, different variations give different derivatives, and we write $\mathrm{VaR}'_v(0)$ for the Gateaux variation in the direction $v$ to emphasize this dependency. Also, since $\mathrm{VaR}_v(0)$ is independent of $v$, we write VaR instead.

There are many possible metrics that could be used to measure the distance between two pdfs. We choose to consider the metric induced by the $L^1$-norm,

$$d(p, q) = |p - q|_1,$$

[26] Similarly, the constant in the asymptotic term $O(h^2)$ depends on $v$. Recall that $O(h^2)$ denotes a function $f(h)$ such that

$$\lim_{h \to 0} \left|\frac{f(h)}{h^2}\right| = C < \infty.$$

For our error bound, $C$ depends on the variation $v$.

mostly because it leads to an elegant result. We may without loss of generality assume that the variations have unit length, since we may rescale $v$ and $h$ simultaneously to achieve this. Remember that the admissible variations satisfy the constraint that $p+hv$ is a pdf for all $h$ in some interval $[0, \epsilon)$. Since

$$\int_{\mathbb{R}^n} [p(\mathbf{r}) + hv(\mathbf{r})]d\mathbf{r} = 1, \qquad h \in [0, \epsilon),$$

it follows that

$$\int_{\mathbb{R}^n} v(\mathbf{r})d\mathbf{r} = 0. \tag{4.120}$$

The function $v = v_+ + v_-$ can be separated into its positive $v_+$ and negative $v_-$ parts. Since $v$ has unit length and satisfies equation (4.120), we obtain

$$\int_{\mathbb{R}^n} v_+(\mathbf{r})d\mathbf{r} = -\int_{\mathbb{R}^n} v_-(\mathbf{r})d\mathbf{r} = \frac{1}{2}. \tag{4.121}$$

Property (4.121) can be used to bound the integral term in equation (4.118),

$$|\int_{\Delta\Pi(\mathbf{r}) \le -\mathrm{VaR}} v(\mathbf{r})d\mathbf{r}| \le \frac{1}{2}.$$

Therefore, the derivative is bounded by

$$|\mathrm{VaR}_v'(0)| \le \frac{1}{2p_{\Delta\Pi}(-\mathrm{VaR})}$$

for all admissible variations $v$ with unit length. The absolute error is bounded by

$$|\mathrm{VaR}_v(h) - \mathrm{VaR}| \le \frac{|h|}{2p_{\Delta\Pi}(-\mathrm{VaR})} + O(|h|^2), \tag{4.122}$$

and the relative error is bounded by

$$\frac{|\mathrm{VaR}_v(h) - \mathrm{VaR}|}{\mathrm{VaR}} \le \frac{|h|}{2\mathrm{VaR}\, p_{\Delta\Pi}(-\mathrm{VaR})} + O(|h|^2). \tag{4.123}$$

The condition number of value-at-risk is[27]

$$\kappa = \frac{1}{2\mathrm{VaR}\, p_{\Delta\Pi}(-\mathrm{VaR})}. \tag{4.124}$$

Relative error bound (4.123) provides the first-order error estimate

$$\frac{|\mathrm{VaR}_v(h) - \mathrm{VaR}|}{\mathrm{VaR}} \lesssim \kappa |p - \tilde{p}|_1, \tag{4.125}$$

where $p$ is the original pdf, $\tilde{p} = p + hv$ is a perturbed pdf, and $v = \frac{p-\tilde{p}}{h}$ is the direction of the perturbation.

[27] Note that this is a variational condition number. The standard condition number (see Rice [Ric66]) for the problem $y = f(x)$ is defined as

$$\kappa = \frac{xf'(x)}{f(x)}.$$

This can be interpreted in a directional sense. Consider $f: \mathbb{R}^n \to \mathbb{R}$, then the absolute error is

$$f(x+\Delta x) - f(x) \approx \nabla f(x) \cdot \Delta x.$$

#### 4.8.2.3 Example: Mixture Model

We consider two portfolios that depend on a single risk factor, the stock price for BCE. The first portfolio consists of a single stock. The second portfolio is a short position in a European call option. The option is delta-hedged with a position in the stock; i.e., a stock position has been chosen so that the $\Delta$ is zero. The option is at-the-money and has 3 months to maturity. The returns are assumed to be normal,

$$p(r) = \frac{1}{\sqrt{2\pi\sigma^2}} e^{-\frac{(r-\mu)^2}{2\sigma^2}}.$$

Although the theory is valid for a large class of return models, we chose this example for its simplicity and for accurate computations.

Figure 4.26 shows plots with some results of our experiment. The plots are for the stock portfolio (for the option portfolio, similar plots were obtained). The plots are for value-at-risk with $\alpha = 95\%$ and $\alpha = 99\%$. The continuous line shows equation (4.125) as a function of the size of the perturbation $|p - \tilde{p}|_1$. We see that the relative error in value-at-risk grows rapidly for small errors in $p$. For the stock portfolio, $\kappa$ can be computed directly. For the option portfolio, we used a Monte Carlo method to compute value-at-risk and a Parzen estimator for $p_{\Delta\Pi}(-\text{VaR})$. The Monte Carlo method used 20,000 random normal samples and variance reduction with antithetic variables. We computed the option price for each sample via the Black–Scholes formula.

To compare equation (4.125) and actual relative errors for perturbations of the model, we generated random densities of the form

$$\tilde{p} = H_1 p_1 + H_2 p_2 + (1 - H_1 - H_2)p,$$

where $p_1$ and $p_2$ are normal pdfs. The random variables $H_1$ and $H_2$ are uniform with ranges $[0, 0.1]$ and $[0, 0.01]$, respectively. The parameters of $p_1$ and $p_2$ were also generated at random. The parameters were generated as

$$\mu_1 = \mu(1 + 2M_1), \qquad \sigma_1 = \sigma|0.5 + 2U_1|,$$
$$\mu_2 = \mu(1 + 10M_2), \qquad \sigma_2 = \sigma|0.5 + U_2|,$$

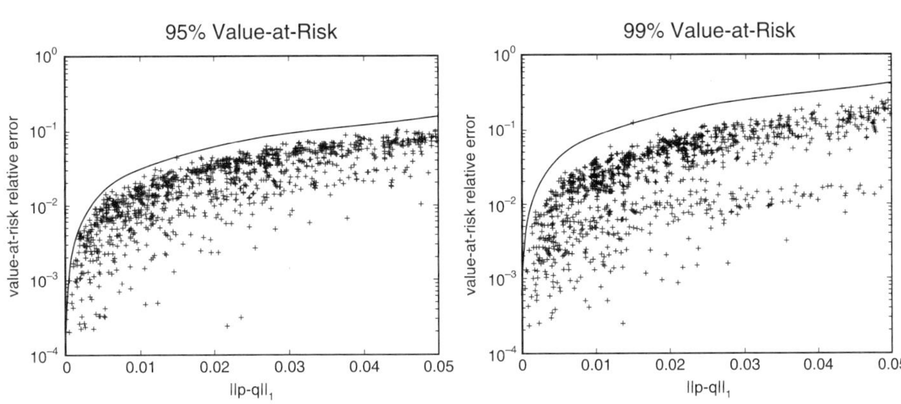


FIGURE 4.26 The plots show the relative error in value-at-risk for a stock portfolio versus the size of the perturbation $|p - \tilde{p}|_1$. The continuous line is the error bound, in equation (4.125). The plus signs corresponed to random perturbations.

where $M_1$ and $M_2$ are standard normal random variables and $U_1$ and $U_2$ are standard uniform random variables.

For each random mixture, we computed the relative error in value-at-risk and the size of the perturbation, $|p-\tilde{p}|_1$. Each plus sign in the plots in Figure 4.26 marks the result for a randomly perturbed problem. For the stock portfolio, we used 1000 randomly perturbed densities. The relative error in value-at-risk is indeed smaller than the first-order, worst-case estimate, equation (4.125). For the option portfolio, we computed the value-at-risk and the norm $|p-\tilde{p}|_1$ with the Monte Carlo method. Since this procedure is much more time consuming than for the stock portfolio, we had to limit the experiment to 300 random densities. Although some samples are larger than the approximate bound, we conclude that equation (4.125) is a good estimate of the relative error. The accuracy of the Monte Carlo method is limited, and we see that these plots contain some simulation noise.

In this section, we have discussed the properties of value-at-risk equation (4.111). In our analysis, we argued that in most cases value-at-risk is a well-posed problem. The requirement for being well posed is that cdf (4.113) be strictly increasing close to $-\text{VaR}$. An equivalent condition is that equation (4.112) be positive almost everywhere in a neighborhood of the solution. Credit risk is one important exception where these assumptions will typically not hold, but it is a reasonable assumption for standard value-at-risk models.

Nevertheless, being well posed alone does not guarantee that a small error in the model for returns translates into a small relative error in value-at-risk. To understand how such errors affect the simulation, a variational perturbation theory was developed. The theory applies to problems that are sufficiently smooth and for smooth variations in the model density of the returns. The advantage of the variational approach is a theory that is model independent; it can therefore be used to quantify model risk. The theory provides estimate (4.125) for the relative error where the condition number can be computed. The stumbling block is to find $p_{\Delta\Pi}(-\text{VaR})$. In some methods, such as the fast convolution method, $p_{\Delta\Pi}(-\text{VaR})$ is computed. In other methods, for example, a Monte Carlo method, $p_{\Delta\Pi}(-\text{VaR})$ must be computed with a density estimator; see, for example, [TT90].

The condition number, equation (4.124), controls the size of the relative error. The problem becomes increasingly ill conditioned as $\text{VaR}\,p_{\Delta\Pi}(-\text{VaR})$ decreases to zero. This confirms that the empirical observations of small perturbations to the return model — caused by changes either in the model, in the data, or in the estimation procedure — that cause large changes in the simulation result is an intrinsic property for large $\alpha$. Value-at-risk is ill conditioned for extreme levels of confidence.

ART • II

# Numerical Projects in Pricing and Risk Management

# CHAPTER • 5

# Project: Arbitrage Theory

The purpose of this exercise is to detect arbitrage opportunities given a payoff matrix and a set of asset prices. The arbitrage theorem is analyzed within the simple context of a single-period financial model. Some of the basic finance concepts, terminology, and formalism are reintroduced in a more practical form. Following this, an example that illustrates the logic behind derivative asset pricing within the single-period model is presented. The arbitrage theorem is discussed in explicit matrix format for a single-period model with a finite number of assets and states. This provides all of the background needed to automate arbitrage with a chosen number of states. As is shown, the problems underlying the single-period model are simply related to finding solutions to a linear system of equations.

*Worksheet:* **arb**

*Required Libraries:* **MFioxl, MFBlas, MFRangen, MFLapack**

## 5.1 Basic Terminology and Concepts: Asset Prices, States, Returns, and Pay-Offs

We let the index $t$ represent time. The first object we introduce is a price vector. That is, all securities (options, futures, forwards, bonds, stocks, etc.) are represented by a vector of $N$ asset prices, which we can denote simply by $\mathbf{p}(t)$:

$$\mathbf{p}(t) = \begin{pmatrix} p_1(t) \\ p_2(t) \\ \vdots \\ p_N(t) \end{pmatrix} \tag{5.1}$$

The asset price $p_1(t)$ can typically represent riskless borrowing or lending, such as a U.S. Treasury bill, $p_2(t)$ can denote a stock price $S_t$, $p_3(t)$ a call or put option on the same stock $S_t$, etc. In a discrete time series the prices are given by a series of vectors $\mathbf{p}(0)$, $\mathbf{p}(1)$, . . . , $\mathbf{p}(t)$, $\mathbf{p}(t+1)$, . . . . Note that in a single period model $t =$ present time and $T = t+1$ is the terminal time of any trading period $[t, t+1]$. In terms of the base assets then $A_t^i = p_i(t), i = 1, \ldots, N$.

Next we introduce the concept of states of the world. That is, we assume that each possible outcome or scenario corresponds to an elementary event, or state of the world, $\omega_i$, where there is only a finite number $M$ of them: $i = 1, \ldots, M$. These states are mutually exclusive, with at least one of them occurring with nonzero probability. All possible states are represented by the set $\Omega = \{\omega_1, \ldots, \omega_M\}$.

Financial assets will attain different values and give rise to differing payouts corresponding to the different states $\omega_i$. Shortly we discuss in detail an instructive example. Before that, however, we recall a couple of other concepts. One is that of *payoffs* $D_{i,j}$, which represent the number of units of account paid out per unit of security $i$ in the state $j$. Generally, for an $N$-asset and $M$-state system we can represent all single-period pay-offs by an $N \times M$ dividend matrix for an interval $[t, t+1]$:

$$\mathbf{D} = \begin{pmatrix} D_{11} \cdots D_{1M} \\ \vdots \quad \vdots \\ D_{N1} \cdots D_{NM} \end{pmatrix}. \tag{5.2}$$

This payoff matrix can be interpreted in two different ways. The first is that each $i$th row of the matrix corresponds to pay-offs for one unit of a given $i$th security in all the different states of the world. In the second interpretation, each $j$th column represents pay-offs for all the different assets within a given $j$th state of the world.

The other concept of importance is that of a *portfolio*. Recall from Chapter 1 that a portfolio is defined as a linear combination of assets or securities. That is, one can generally have positions given by $\theta_i$ in the $i$th asset and by specifying all such $N$ positions $\theta_i$, $i = 1, \ldots, N$, we have uniquely specified a portfolio as a vector,

$$\theta = \begin{pmatrix} \theta_1 \\ \theta_2 \\ \vdots \\ \theta_N \end{pmatrix}. \tag{5.3}$$

Positive $\theta_i$ correspond to long positions and negative values correspond to short positions. A zero position $\theta_i = 0$ implies that the $i$th asset is not included in the portfolio. A portfolio that delivers the same pay-off regardless of any possible state of the world is defined as *riskless*. By taking the dot product of $\theta$ with the asset price vector $\mathbf{p}_t \equiv \mathbf{p}(\mathrm{t})$ we obtain the value of the portfolio at time $t$:

$$V_t^\theta = \theta \cdot \mathbf{p}_t = \sum_{i=1}^{N} \theta_i p_i(t) = \sum_{i=1}^{N} \theta_i A_t^i. \tag{5.4}$$

The payoff $V_{t+1}^\theta(\omega_j)$, denoted here by $\Lambda_j$, for the portfolio given by $\theta$ in a given $j$th state is then expressible as a sum over all asset pay-offs weighted by their respective positions, where $D_{ij} = A_{t+1}^i(\omega_j)$,

$$\Lambda_j = \sum_{i=1}^{N} D_{ij}\theta_i = \sum_{i=1}^{N} (\mathbf{D}')_{ji}\theta_i. \tag{5.5}$$

The superscript $'$ stands for matrix transpose. We can therefore express the payoff vector with components $\Lambda_j$, $j = 1, \ldots, M$, in matrix form, $\Lambda \equiv \mathbf{D}'\theta$,

$$\begin{pmatrix} \Lambda_1 \\ \Lambda_2 \\ \vdots \\ \Lambda_M \end{pmatrix} = \begin{pmatrix} D_{11} \cdots D_{N1} \\ \vdots \quad \vdots \\ D_{1M} \cdots D_{NM} \end{pmatrix} \begin{pmatrix} \theta_1 \\ \theta_2 \\ \vdots \\ \theta_N \end{pmatrix}. \tag{5.6}$$

## 5.2 Arbitrage Portfolios and the Arbitrage Theorem

As in Chapter 1, we define $\theta$ to be an arbitrage portfolio, or sometimes simply called an arbitrage, if either one of the following conditions applies:

(i) $\mathbf{p}_t \cdot \theta = 0$ and $\mathbf{D}' \cdot \theta \geq 0$, where $(\mathbf{D}'\theta)_j > 0$ for some $j$.
(ii) $\mathbf{p}_t \cdot \theta < 0$ and $\mathbf{D}' \cdot \theta \geq 0$.

Note that these vector inequalities are meant to be applicable component by component. In case (i) the portfolio guarantees a positive return in some states with no possible loss, yet costs nothing to purchase. In case (ii) the portfolio will guarantee a nonnegative return and has a negative cost to purchase.

Finally, we can state the arbitrage theorem as follows:

1. If there are no arbitrage opportunities then there exist positive constants $\psi_i > 0$, $i = 1, \ldots, M$ (in vector notation we write simply $\psi > 0$, where $\psi$ is the vector of $\psi_i$ components), such that

$$\mathbf{p}_t = \mathbf{D}\psi. \tag{5.7}$$

2. If condition 1 is true, then there is no arbitrage.

One notes that, apart from a positive constant [i.e., the inverse of the discount factor as shown in upcoming equation (5.10)], the $\psi_i$ correspond to certain nonzero probabilities of occurence for all the states $i = 1, \ldots, M$. In fact, these coefficients give the risk-neutral probabilities for the correct pricing of financial securities, as explained in the following section and as was observed in the discrete case of the fundamental theorem of asset pricing given in Chapter 1. In matrix form, equation (5.7) reads

$$\begin{pmatrix} p_1 \\ \vdots \\ p_N \end{pmatrix} = \begin{pmatrix} D_{11} \cdots D_{1M} \\ \vdots \quad \vdots \\ D_{N1} \cdots D_{NM} \end{pmatrix} \begin{pmatrix} \psi_1 \\ \vdots \\ \psi_M \end{pmatrix}. \tag{5.8}$$

In the **arb** spreadsheet assignment we consider the case $M = N$ and the special type of payoff matrix

$$\mathbf{D} = \begin{pmatrix} (1+R) \cdots (1+R) \\ D_{21} \cdots D_{2M} \\ \vdots \quad \vdots \\ D_{N1} \cdots D_{NM} \end{pmatrix}. \tag{5.9}$$

The first row has all equal payoff values and corresponds to the riskless return on a money-market or bond; i.e., $p_1 = p_1(t) = 1$, with single-period rate of return $R$. Without loss of generality, here we have simply set the bond's present value to one unit of worth. The first row in equation (5.8) of the arbitrage theorem then gives

$$\sum_{i=1}^{M}(1+R)\psi_i \equiv \sum_{i=1}^{M}\tilde{\psi}_i = 1. \tag{5.10}$$

The coefficients $\tilde{\psi}_i$ defined here correspond to the risk-neutral probabilities for all possible states. In fact, $\tilde{\psi}_i$ are recognized as being the $q_i$ probabilities used to define the pricing measure in the fundamental theorem of asset pricing discussed in Chapter 1. They sum up to unity as required and also satisfy the condition $0 < \tilde{\psi}_i < 1$. As noted earlier, these probabilities are very different from the real-world probabilities, which provide no information on the risk-neutral probabilities used for pricing. The risk-neutral probabilities therefore exist with the correct properties mentioned if, and only if, there is no arbitrage.

## 5.3 An Example of Single-Period Asset Pricing: Risk-Neutral Probabilities and Arbitrage

The single-period setting assumes that time consists of the present time $t$ and a later time $T = t+1$ and that there is a finite time separation. We consider here a portfolio consisting of just one bond with present value of unity, $B(t) = 1$, one asset (or stock) $S$, and a call option $C$ on the underlying stock $S$. Moreover, we assume only two possible states of the world. In this situation the stock, which has present value $S(t)$, can attain either of two values: $S_1(t+1)$ or $S_2(t+1)$ at time $t+1$. Accordingly, the option with present value $C(t)$ can take on the values given by $C_1(t+1)$ or $C_2(t+1)$ in state $\omega_1$ and $\omega_2$, respectively. No matter what the outcome, however, the bond has a fixed (riskless) return of $1+R$, with $R$ being the single-period rate of return. In this situation we have a $3 \times 2$ payoff matrix and the foregoing arbitrage theorem gives

$$\begin{pmatrix} 1 \\ S(t) \\ C(t) \end{pmatrix} = \begin{pmatrix} (1+R) & (1+R) \\ S_1(t+1) & S_2(t+1) \\ C_1(t+1) & C_2(t+1) \end{pmatrix} \begin{pmatrix} \psi_1 \\ \psi_2 \end{pmatrix}, \tag{5.11}$$

which implies a linear system of three equations:

$$\tilde{\psi}_1 + \tilde{\psi}_2 = 1, \tag{5.12}$$

$$\tilde{\psi}_1 S_1(t+1) + \tilde{\psi}_2 S_2(t+1) = (1+R)S(t), \tag{5.13}$$

$$\tilde{\psi}_1 C_1(t+1) + \tilde{\psi}_2 C_2(t+1) = (1+R)C(t). \tag{5.14}$$

Here we have used the same definition as before for the risk-neutral probabilities $\tilde{\psi}_i \equiv (1+R)\psi_i$. These equations have the familiar form of the binomial pricing equations for options, as discussed in the project that deals with binomial lattice pricing. That is, the price today of a security is given as the discounted sum of the risk-neutral expected payoff values for all possible future values of the security. We also note that if we allow for three states of the world, we then obtain pricing equations that resemble the trinomial pricing equations.

To demonstrate an example of arbitrage, let us consider $R = 7\%$ and the two possible values at time $t+1$: $S_1(t+1) = 50$ dollars , $S_2(t+1) = 150$ dollars, where $S(t) = 100$ dollars.

Say the call option $C$ has strike price of 100 dollars and expires exactly at time $t+1$. This option then has pay-off of zero and 50 dollars, respectively. If we denote the price of the call option today as $C$, then equations (5.12) to (5.14) give

$$\tilde{\psi}_1 + \tilde{\psi}_2 = 1, \tag{5.15}$$

$$0.5\tilde{\psi}_1 + 1.5\tilde{\psi}_2 = 1.07, \tag{5.16}$$

$$50\tilde{\psi}_2 = 1.07C. \tag{5.17}$$

By satisfying the first two equations we actually obtain the arbitrage-free price for $C$ by substituing the resulting risk-neutral values $\tilde{\psi}_1 = 0.43$, $\tilde{\psi}_2 = 0.57$ into the third equation. The correct (no-arbitrage) price is therefore $C = 26.6355$ dollars. If, however, we are given a market price for $C = 25$ dollars and we wish to answer the question of whether there is arbitrage or not in this case, then we solve equation (5.17), giving $\tilde{\psi}_2 = 0.535$, and then equation (5.16) gives $\tilde{\psi}_1 = 0.535$. These values, however, do not satisfy probability conservation equation (5.15), therefore, one concludes that there is indeed arbitrage at that market price.

# 4 Arbitrage Detection and the Formation of Arbitrage Portfolios in the N-Dimensional Case

The preceding example involves an overdetermined system of linear equations. Now, however, we shall consider a uniquely specified system where the number of unknowns is equal to the number of equations. Hence, we consider the case of $N$ states and $N$ assets, i.e., $M = N$. We shall assume that one of the assets always corresponds to a bond with fixed rate of return $R$. The payoff matrix has the form given in equation (5.9), where the first row has all equal elements of value $(1+R)$. The corresponding system of $N$ equations is given in equation (5.18). The problem is then the following. Generate an arbitrary price vector in one of two fashions: Set $p_1(t) = 1$ and then generate independent components $p_i(t)$ $(i \geq 2)$ distributed either (i) uniformly as integers lying within some given minimum and maximum integer values or (ii) continuously using some standard normal distribtuion, say, $p_i(t) \epsilon N(0,1)$ $(i \geq 2)$. Similarly, generate $N(N-1)$ arbitrary payoff matrix elements $D_{ij}$ $(i \geq 2)$ in the discrete or continuous cases, respectively. The numerical library called MFRangen is useful for random-number and random-matrix generation. For a given generated pair of price vector $\mathbf{p}(t)$ and payoff matrix $\mathbf{D}$ one obtains the vector of risk-neutral probabilities $\tilde{\psi} = (1+R)\psi$ by solving the linear system

$$\begin{pmatrix} 1 \\ p_2 \\ \cdot \\ \cdot \\ p_N \end{pmatrix} = \begin{pmatrix} (1+R) \cdots (1+R) \\ D_{21} \cdots D_{2N} \\ \cdot \quad \cdot \\ \cdot \quad \cdot \\ D_{N1} \cdots D_{NN} \end{pmatrix} \begin{pmatrix} \psi_1 \\ \cdot \\ \cdot \\ \cdot \\ \psi_N \end{pmatrix}. \tag{5.18}$$

In practice, one can solve this system numerically by inverting the payoff matrix using a routine based on the singular value decomposition. Note that the first equation in the system is that of probability conservation [this is equation (5.10) with $M = N$]. Arbitrage then exists if the solution gives at least one nonpositive component, that is, if for any given $i$, $\psi_i \leq 0$ (since we are enforcing probability conservation). For every such $i$ we then have a corresponding $i$th state, which we can use to form an arbitrage portfolio that we denote by $\theta^{(i)}$ with components

$\theta_1^{(i)}, \ldots, \theta_N^{(i)}$. According to the discussion on equations (5.5) and (5.6), then, the payoff vector corresponding to the $i$th state alone can be obtained by setting only the $i$th component to a nonzero positive value, $\Lambda_j^{(i)} = 1$ for $j = i$ (i.e., picking a number greater than zero) and setting all other $j$ components to zero. Note that this corresponds to the pay-off of an Arrow–Debreu security, yet with nonpositive initial value. The transpose of this $N$-dimensional pay-off column vector, denoted by $\Lambda^{(i)}$, has a row representation of $(0, \ldots, 0, 1, 0, \ldots, 0)$, where unity occurs in the $i$th position only. The arbitrage portfolio is then obtained by solving the linear system of $N$ equations in the $N$ unknowns $\theta_j^{(i)}$, $j = 1, \ldots, N$, as in equation (5.5) or, in matrix form:

$$\mathbf{D}' \cdot \theta^{(i)} = \Lambda^{(i)}. \tag{5.19}$$

To obtain more arbitrage portfolios, one can repeat the preceding steps for the other state components that led to arbitrage, i.e., for the other nonpositive $\psi_i$ components. To see why $\theta^{(i)}$ is an arbitrage portfolio note that $\mathbf{p}_t = \mathbf{D}\psi$. So the portfolio has present value $V_t^{\theta^{(i)}} = \theta^{(i)}.\mathbf{p}_t = (\mathbf{D}'\theta^{(i)}).\psi = \Lambda^{(i)}.\psi = \psi_i$. Since $\psi_i \leq 0$, then $V_t^{\theta^{(i)}} \leq 0$ and by construction the terminal value or pay-off of this portfolio is given by $V_{t+1}^{\theta^{(i)}}(\omega_i) = \Lambda_i^{(i)} = 1$ (hence greater than 0) when $\omega_j = \omega_i$, yet $V_{t+1}^{\theta^i}(\omega_j) = 0$ (hence $\geq 0$) for all other states. From the definition of single-period arbitrage we conclude that the portfolio $\theta^{(i)}$ is indeed an arbitrage.

CHAPTER · 6

# Project: The Black–Scholes (Lognormal) Model

The purpose of this project is to develop pricing routines for plotting and analyzing the Black–Scholes price for European calls, puts, and butterfly spreads as well as for the corresponding sensitivities — delta, gamma, rho, vega, and theta — as a function of the five basic parameters that make up the plain-vanilla Black–Scholes pricing formula.

*Worksheet:* **bs**

*Required Libraries:* **MFioxl, MFFuncs, MFStat**

## 6.1 Black–Scholes Pricing Formula

The celebrated Black–Scholes pricing formula is quite straightforward since it makes use of the standard normal distribution. Building the necessary Visual Basic code for this spreadsheet will, however, quickly familiarize the user with the use of ActiveX numerical library methods for input/output to Excel. One of the features of the spreadsheet is to allow the user the flexibility of inputting any values for the fixed parameters while also allowing a choice for the range of plotting.

Although symmetries of the Black–Scholes formula can be used to reduce the number of dependent functional parameters, the price of a call option can be most explicitly written (as seen in Chapter 1) as a function of five variables (or parameters): the interest rate $r$ (assumed constant), the stock price $S$, the time to maturity $\tau \equiv T - t$ ($t =$ current calendar time and $T =$ maturity calendar time), the volatility $\sigma$ (assumed constant), and the strike price $K$. The Black–Scholes formula for the value of a plain-vanilla European call option is

$$C(S, K, r, \sigma, \tau) = SN(d_+) - Ke^{-r\tau}N(d_-), \tag{6.1}$$

where

$$d_\pm = (\log(S/K) + (r \pm \tfrac{1}{2}\sigma^2)\tau)/(\sigma\sqrt{\tau}), \tag{6.2}$$

where $d_- = d_+ - \sigma\sqrt{\tau}$. The function $N(x)$ is the cumulative standard normal distribution at $x$.

As an example of the functionality built into the **bs** spreadsheet, a plot of the value of a call option $C$ as a function of $S$ in the range $S_{min}$ (the minimum spot price) to $S_{max}$ (the maximum spot price) is generated via equation (6.1) while holding $r$, $K$, $\tau$, and $\sigma$ fixed. A plot of the option price as a function of varying the interest rate while holding the other four variables constant is generated in a similar manner. The same plotting functionality is also generated for varying volatility, time-to-maturity, and strike price while simultaneously making use of the Black–Scholes formula at appropriate interval points. The interface for the **bs** spreadsheet also allows for the choice of plotting a variable input number of points for each graph.

Put-call parity

$$P = C - S + Ke^{-r\tau} \tag{6.3}$$

can also be used to study the corresponding prices and sensitivities of puts. The dimensionality of the variables is worth emphasizing and is as follows. Volatility refers to a per annum (i.e., yearly) time scale and has units of year$^{-1/2}$. Maturity is in years, so $\sigma\sqrt{\tau}$ is dimensionless. The interest rate is per annum and has units of year$^{-1}$, making $r\tau$ dimensionless. Both strike and spot are in units of currency (e.g., dollars). One noteworthy property of the Black–Scholes formula is its so-called numeraire invariance. This essentially implies that prices can be made dimensionless so that the formula is invariant with respect to the underlying currency. This is easily seen by dividing equation (6.1) throughout by the strike, giving

$$C/K = (S/K)N(d_+) - e^{-r\tau}N(d_+ - \sigma\sqrt{\tau}), \tag{6.4}$$

where $d_+$ is also a function of the dimensionless quantity $S/K$.

From the vanilla call or put options one can construct many other options with various payoff structures, as was discussed with the theory of static hedging in Chapter 1. One important pay-off that was discussed explicitly is the butterfly spread, as given by equation (1.228). Here we reconsider this option, with pay-off defined in a similar manner except for a trivial normalization constant. Namely, the pay-off is peaked at strike $K$ and has a nonzero width of $2\delta K$. This pay-off is statically replicated by taking a long position in a vanilla call struck at $K + \delta K$, another long position in a vanilla call struck at $K - \delta K$, and two short positions in a vanilla call struck at $K$:

$$\begin{aligned}\Lambda(S) &= (S - (K + \delta K))_+ + (S - (K - \delta K))_+ - 2(S - K)_+ \\ &= \begin{cases} (S - (K - \delta K))_+, & S \le K, \\ ((K + \delta K) - S)_+, & S > K. \end{cases}\end{aligned} \tag{6.5}$$

Note that from put-call parity one can also construct such a pay-off with a combination of puts. The exact analytical expression for the Black–Scholes price of such a butterfly contract maturing in time $\tau$ is hence

$$\begin{aligned}B_{\delta K}(S, K, r, \sigma, \tau) = {} & C(S, K + \delta K, r, \sigma, \tau) + C(S, K - \delta K, r, \sigma, \tau) \\ & -2C(S, K, r, \sigma, \tau),\end{aligned} \tag{6.6}$$

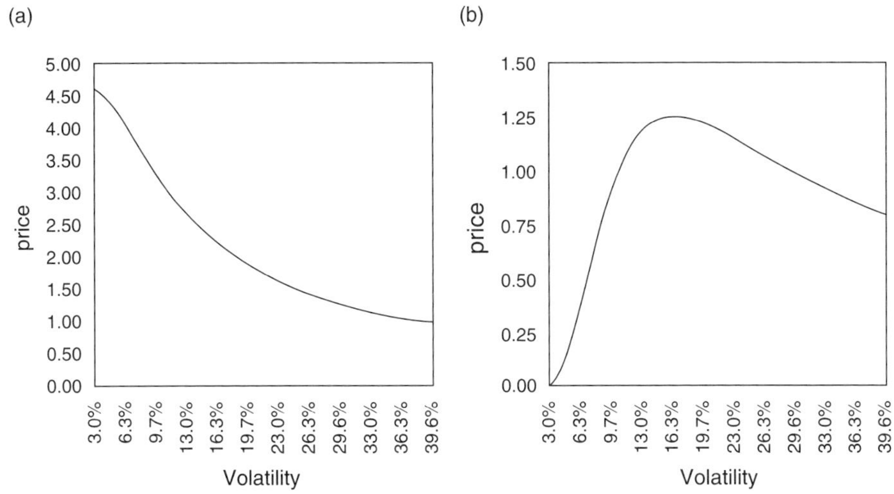


FIGURE 6.1 Price variations as volatility changes for an (a) in-the-money (spot $= 100$) versus an (b) out-of-the-money (spot $= 80$) European butterfly option with fixed spread $\Delta K = 10$, strike $K = 100$, $r = 5\%$ per annum, $\tau = 1$ year. Plot (a) is monotonically decreasing, whereas plot (b) displays a pronounced maximum, as is expected within a lognormal density model for the stock movements.

where the call formula is given by equation (6.1). Figure 6.1 gives an example of the results of the **bs** spreadsheet application obtained for two cases of chosen spot. As observed, the plots illustrate the differing effects of volatility on the price of a relatively narrow butterfly spread option for in-the-money versus out-of-the-money (below strike) options.

The observed changes in the option prices as one changes a parameter, such as volatility, time to maturity, spot, interest rate, or strike and shape of the payoff function, can be qualitatively understood by means of the risk-neutral pricing formula. Let us generally denote by $V(S, K, r, \sigma, \tau)$ the option price for a payoff function $\Lambda(K, S)$. This pay-off can, for instance, represent either a call, put, or butterfly spread struck at $K$. Note that for the case of the butterfly the pay-off is, of course, also a function of the spread $\delta K$. As stated in Chapter 1, the risk-neutral pricing formula gives

$$V(S, K, r, \sigma, \tau) = e^{-r\tau} \int_0^\infty p(S_\tau, S; \tau)\Lambda(K, S_\tau)dS_\tau, \tag{6.7}$$

where $p(S_\tau, S; \tau)$ is the lognormal transition probability density [i.e., equation (1.165)]

$$p(S_\tau, S; \tau) = \frac{1}{\sigma S_\tau \sqrt{2\pi\tau}} e^{-[\log(S/S_\tau)+(r-\frac{1}{2}\sigma^2)\tau]^2/2\sigma^2\tau}. \tag{6.8}$$

Observe that the density $p$ is actually a function of $\sigma\sqrt{\tau}$ as well as $r\tau$. The interest rate gives rise to part of the drift of the center of $p$. The quantity $\sigma\sqrt{\tau}$ gives a negative contribution to the drift. More importantly, however, $\sigma\sqrt{\tau}$ determines the width (or standard deviation) of the density. A direct interpretation of equation (6.7) shows that higher option prices correspond to situations for which there is maximal overlap between the density $p$ and the

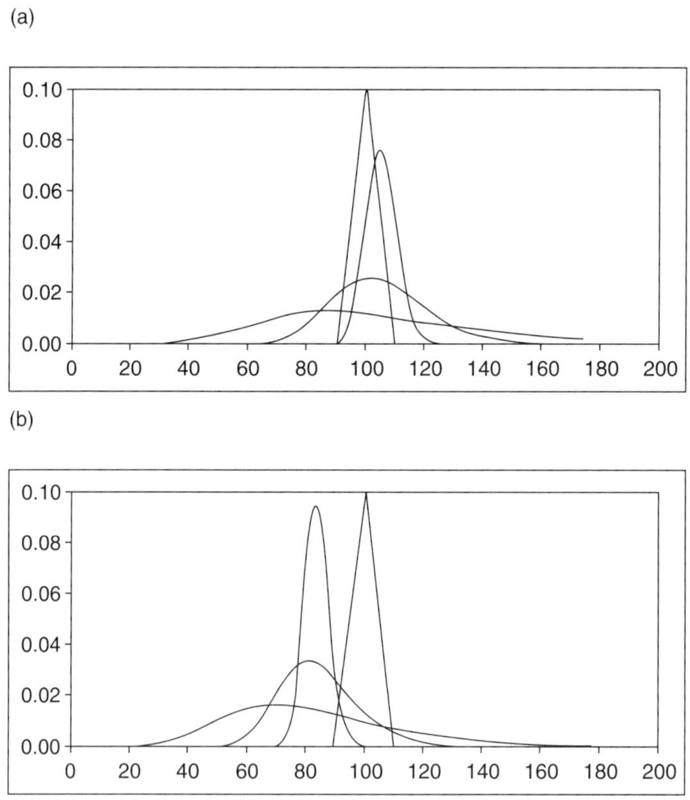


FIGURE 6.2 Variations in overlap between the risk-neutral pricing density and the pay-off, as functions of $S_T$, for an (a) in-the-money (spot $S = 100$) versus an (b) out-of-the-money (spot $S = 80$) European butterfly option with spread $\Delta K = 10$ and strike $K = 100$. The interest rate $r = 5\%$ per annum, and time to maturity $\tau = 1$ year. In both cases, the three lognormal density curves correspond to $\sigma = 5\%, 15\%, 35\%$ with horizontal axis as final stock level $S_T$.

payoff function $\Lambda$, and vice versa. For smaller values of $\sigma\sqrt{\tau}$ (i.e., smaller volatility values for fixed time-to-maturity or smaller time-to-maturity values for fixed volatility), the density is more highly concentrated and is centered about the spot $S$. Figure 6.2 shows the changes in overlap between the lognormal transition probability density and the butterfly pay-off struck at $K = 100$ ($\delta K = 10$). Note that in order to keep the two functions on the same scale, the pay-off has been multiplied by a normalization $1/(\delta K)^2$, giving unit payoff area with height $1/\delta K$. Increases in the volatility parameter $\sigma$ correspond to more dispersion in the density, hence giving less and less overlap with the butterfly pay-off in the (in-the-money) case where the spot is at strike, $S = K$. For moderately out-of-the-money cases, increases in volatility lead to a greater overlap for lower values of $\sigma$ (i.e., from $\sigma = 5\%$ to $15\%$), approaching a maximum at an intermediate value, followed by a decrease in overlap at relatively higher values (i.e., from $\sigma = 15\%$ to $35\%$). This argument is consistent with the price variations observed in Figure 6.1. One can use the same reasoning to obtain the qualitative picture of price variations one would expect in other circumstances. Another example, for instance, is the case of a deeply out-of-the money butterfly option whereby one expects a monotonically increasing price as function of $\sigma$ over a wider range of $\sigma$ values, with sharper increases at lower $\sigma$ values. Note that our overlap analysis can be applied to other pay-offs, such as calls and puts.

## 6.2 Black–Scholes Sensitivity Analysis

Sensitivities of option prices with respect to changes in the underlying parameters $r$, $\tau$, $S$, $\sigma$ were also discussed in Chapter 1. As noted, these are of importance to hedging and computing risk for nonlinear portfolios. Within the Black–Scholes formulation, these sensitivities are obtained analytically by taking the respective partial derivatives of the option-pricing formula. The $\Delta$, $\rho$, $\Theta$, and the vega, $\partial V/\partial\sigma$, of an option give the change in the option's price $V$ with respect to changes in spot $S$, $r$, $\tau$, and $\sigma$, respectively. The other sensitivity of interest is $\Gamma$, which gives the change in $\Delta$ with respect to a change in $S$.

For a vanilla call with price $C$ one can readily derive the following sensitivities:

$$\Delta_c \equiv \frac{\partial C}{\partial S} = N(d_+), \tag{6.9}$$

$$\Gamma_c \equiv \frac{\partial^2 C}{\partial S^2} = \frac{e^{-d_+^2/2}}{\sigma S\sqrt{2\pi\tau}}, \tag{6.10}$$

$$\rho_c \equiv \frac{\partial C}{\partial r} = K\tau e^{-r\tau}N(d_-), \tag{6.11}$$

$$\frac{\partial C}{\partial \sigma} = S\sqrt{\tau/2\pi}e^{-d_+^2/2}. \tag{6.12}$$

The Black–Scholes PDE can be used to give

$$\Theta = \frac{\partial V}{\partial \tau} = (\sigma^2 S^2/2)\Gamma + r\,(S\Delta - V) \tag{6.13}$$

for any European-style option with value $V$. Hence,

$$\Theta_c \equiv \frac{\partial C}{\partial \tau} = (\sigma^2 S^2/2)\Gamma_c + r\,(S\Delta_c - C), \tag{6.14}$$

with $\Delta_c$ and $\Gamma_c$ given by equations (6.9) and (6.10), respectively.

The sensitivities for a vanilla put follow from put-call parity:

$$\Delta_p = \Delta_c - 1, \tag{6.15}$$

$$\Gamma_p = \Gamma_c, \tag{6.16}$$

$$\rho_p = \rho_c - K\tau e^{-r\tau}, \tag{6.17}$$

$$\frac{\partial P}{\partial \sigma} = \frac{\partial C}{\partial \sigma}, \tag{6.18}$$

$$\Theta_p = (\sigma^2 S^2/2)\Gamma_p + r\,(S\Delta_p - P), \tag{6.19}$$

with $\Delta_p$ and $\Gamma_p$ given by equations (6.15) and (6.16), respectively.

The sensitivities for the butterfly spread option follow trivially by differentiation of equation (6.6) and the use of equations (6.9)–(6.14), giving

$$\Delta_B(K) = \Delta_c(K+\delta K) + \Delta_c(K-\delta K) - 2\Delta_c(K), \tag{6.20}$$

$$\Gamma_B(K) = \Gamma_c(K+\delta K) + \Gamma_c(K-\delta K) - 2\Gamma_c(K), \tag{6.21}$$

$$\rho_B(K) = \rho_c(K+\delta K) + \rho_c(K-\delta K) - 2\rho_c(K), \tag{6.22}$$

$$\frac{\partial B(K)}{\partial \sigma} = \frac{\partial C(K+\delta K)}{\partial \sigma} + \frac{\partial C(K-\delta K)}{\partial \sigma} - 2\frac{\partial C(K)}{\partial \sigma}, \tag{6.23}$$

$$\Theta_B(K) = \Theta_c(K+\delta K) + \Theta_c(K-\delta K) - 2\Theta_c(K). \tag{6.24}$$

Note that in these formulas, the explicit dependence of the option sensitivities as functions of the strike is depicted. Hence, equations (6.1), (6.3), (6.6), and (6.9)–(6.24) are used to generate all option prices and sensitivities required within the Black–Scholes spreadsheet **bs**. The numerical library MFStat is useful for computing the cumulative normal distribution function.

CHAPTER • 7

# Project: Quantile-Quantile Plots

The purpose of this project is to visualize kurtosis in risk-factor return distributions by means of quantile-quantile plots. The test cases include equity indices in 40 different currencies.

*Worksheet:* **qq**

*Required Libraries:* **MFioxl, MFFuncs, MFStat, MFSort**

## 1 Log-Returns and Standardization

Historical data series are provided in table format for the weekly returns on 40 different indices (e.g., the TSE100COMPX denotes the TSE100 composite index, SP500C denotes the Standard & Poor 500). The objective here is to create histograms for the P&L on the log-return time series for a given choice of index as well as plot the q-q (i.e., quantile-quantile) plot for the estimated cumulative distribution against the standardized log-returns on the same index. This allows one to display and study the deviations of the actual distributions from the standard normal distribution for the log-returns.

The log-returns over a time period $dt$ at time $t$ are defined by

$$r_{i,t}^{dt} = \log(S_{t,i}/S_{t-dt,i}) \tag{7.1}$$

for each index $i$. The value $S_{t,i}$ corresponds to a price for index $i$ at time $t$. To standardize the returns, we first estimate the mean using

$$E[r_{i,t}^{dt}] = \frac{1}{N_{\text{ret}}} \sum_{k=1}^{N_{\text{ret}}} \log\left(\frac{S_{t,i}}{S_{t-dt,i}}\right)|_{t=t_k} \tag{7.2}$$

over all $N_{\text{ret}}$ return dates $t_k$, and secondly estimate the standard deviation $\sigma_{i,dt}$ using

$$(\sigma_{i,dt})^2 = E[(r_{i,t}^{dt})^2] - (E[r_{i,t}^{dt}])^2, \tag{7.3}$$

where

$$E[(r_{i,t}^{dt})^2] = \frac{1}{N_{\text{ret}}} \sum_{k=1}^{N_{\text{ret}}} \log^2\left(\frac{S_{t,i}}{S_{t-dt,i}}\right)|_{t=t_k}. \tag{7.4}$$

Note: More precisely the usual factor of $1/(N_{\text{ret}} - 1)$ is used when computing sample standard deviations, however the return series are large (on the order of $100 \to 1000$) and using this factor instead of $1/N_{\text{ret}}$ is immaterial for the present calculations.

Next, we make use of the known result: If a random variable $x$ is distributed as $N(\mu, \sigma)$, then the variable $y = (x - \mu)/\sigma$ has standard normal distribution $N(0, 1)$. In order to compare the actual return series on an equal footing with the corresponding standard normal distribution, we *standardize* the return variables by considering the random variable defined by:

$$\tilde{y}_i \equiv \frac{r_{i,t}^{dt} - E[r_{i,t}^{dt}]}{\sigma_{i,dt}}. \tag{7.5}$$

Note that if the return series were normally distributed, then $\tilde{y}_i \sim N(0, 1)$. As observed next using a quantile-quantile analysis, however, actual return series are generally not normally distributed.

## 7.2 Quantile-Quantile Plots

To obtain the P&L and the quantile-quantile (q-q) plot for a given index we proceed by sampling the weekly log-return data for that index. Note that the data on the **qq** spreadsheet is given in terms of the standardized log-returns; i.e., the data corresponds to the foregoing $\tilde{y}_i$. Having sampled the data, the cumulative distribution in the variable $\tilde{y}_i$ is then estimated by sorting and counting occurrences, $N_k$, within subintervals $(x_{k-1}, x_k)$, where we divide up the P&L range of values into $n$ regions: $x_0 = \tilde{y}_i^{\min}$, $x_1 = x_0 + dx$, ..., $x_n = x_0 + n(dx) = \tilde{y}_i^{\max}$. The quantity $dx = (\tilde{y}_i^{\max} - \tilde{y}_i^{\min})/n$ is the spacing over the $n$ subintervals. One can then plot a histogram of the P&L by plotting $N_k$ against the midpoints $(x_k + x_{k-1})/2$ for all points given by $k = 1, \ldots, n$. Figure 7.1 shows an example of two such histograms. Note that the histograms have been created by eliminating extreme outliers. The actual (i.e., the realized) cumulative distribution $F$ of the standardized return $\tilde{y}_i$ at points $x_k$ is then estimated by

$$\beta_k \equiv F(x_k) \approx \frac{1}{N_{\text{ret}}} \sum_{j=1}^{k} N_j, \tag{7.6}$$

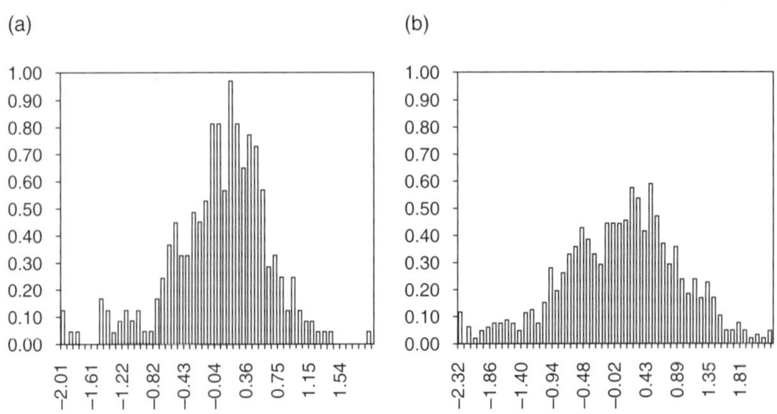


FIGURE 7.1 A comparison of return histograms: (a) SPINDMV and (b) SP500C, for time series during the period Jan. 1980 to Feb. 1999. The number of bins is set to 50. Note that the histogram densities are normalized to give an area of 1, with the returns in percentage units.

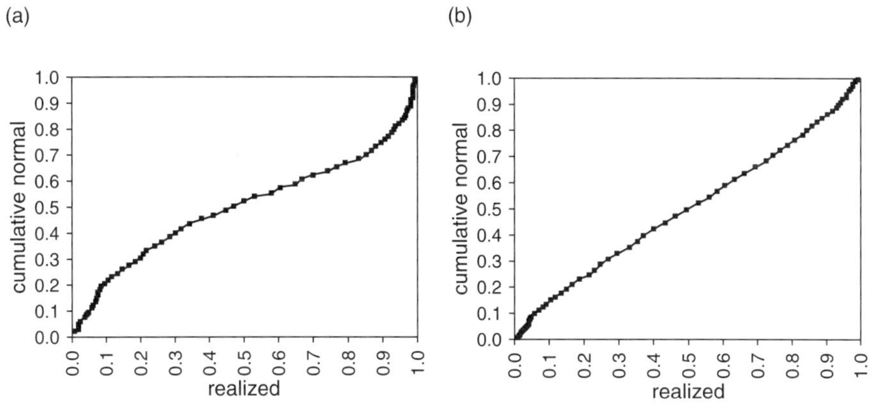


FIGURE 7.2 A comparison of quantile-quantile plots computed using the weekly log-returns of two indices: (a) SPINDMV and (b) SP500C, for time series during the period Jan. 1980 to Feb. 1999. The return distribution for series (a) shows a greater deviation from normality with a thinner tail to the left and a fatter tail to the right of the P&L.

where $N_{\text{ret}}$ is the total number of dates for which data is available on a given index $i$ (i.e., the length of the return time series for index $i$). The percentiles $\beta_k$ are then plotted against the percentiles $\alpha_k$, for all parameter values $k$. The latter percentiles $\alpha_k$ correspond to those of the standard cumulative normal distribution at $x_k$, i.e., $\alpha_k = N(x = x_k)$.

The results for the quantile-quantile (q-q) plots are used to demonstrate the deviations from normality for the log-returns of the realized distribution. The distribution for a given particular time series may show a more pronounced deviation when one compares the corresponding q-q plot with that of another time series, as shown in Figure 7.2. Fatter or thinner tails will skew the otherwise-straight-line q-q plot. The MFStat numerical library is useful for computing cumulative and inverse cumulative normal functions.

HAPTER • 8

# 'roject: Monte Carlo Pricer

This purpose of this project is to implement calibration and pricing of basket equity options within a Monte Carlo simulation. The calibration combines implied volatilities with historical correlations. A multidimensional correlated lognormal distribution is used as the model for the equity returns.

*Worksheet:* **mc** (uses parts of **qq** as input)

*Required Libraries:* **MFioxl, MFBlas, MFLapack, MFFuncs, MFRangen, MFZero**

## 1 Scenario Generation

Let us consider a group (i.e., basket) of $n$ stocks (or indices) with prices (or levels) denoted by $S_{T,i}$, $i = 1, 2, \ldots, n$, at maturity time $T$. Given an initial price vector $\mathbf{S}_0 = (S_{0,1}, \ldots, S_{0,n})$, a standard method of generating correlated Brownian motion for the stock prices then follows from (see Section 1.6):

$$S_{T,i} = S_{0,i} \exp\{(r - \tfrac{1}{2}\sigma_i^2)T + \sqrt{T} \sum_{k=1}^{n} U_{ki} x_k\}. \tag{8.1}$$

Here $r$ is the risk-free interest rate and $\sigma_i$ is the volatility with respect to the $i$th stock price. These quantities are assumed constant in equation (8.1). The set of variables $x_k$, $k = 1, 2, \ldots, n$, is made up of i.i.d. random variables drawn from the standard normal distribution $N(0, 1)$. Matrix $U$ is used to introduce correlations among the stock prices, as shown in detail later. Note, however, that equation (8.1) assumes time-independent volatilities. For the time-dependent case, the foregoing can be extended by considering small time increments $dt$ and writing

$$S_{t+dt,i} = S_{t,i} \exp\{(r - \tfrac{1}{2}\sigma_i(t)^2)dt + \sqrt{dt} \sum_{k=1}^{n} U_{ki} x_k\}. \tag{8.2}$$

To generate the stock prices, this equation is then applied $M$ times from any initial time, say, $t = 0$, to final time $t = T = M\,dt$ (over $M$ steps) while using the time-varying volatilities. Note that equation (8.2) is, of course, also valid for time-independent volatilities. Throughout this project, however, we shall assume time-independent volatilities for implementation. Using equation (8.2), with time-independent $\sigma_i$, we can relate the correlations of the standardized log-returns to the $U$ matrix:

$$\frac{\log(S_{t+dt,i}/S_{t,i}) - (r - \frac{1}{2}\sigma_i^2)dt}{\sqrt{dt}} = (\mathbf{U}'\mathbf{x})_i \equiv y_i, \tag{8.3}$$

where superscript $'$ is the matrix transpose and the vector $\mathbf{x}$ has components $x_k$. The $y_i$ components are closely related to the standardized log-returns $\tilde{y}_i$, as defined within the quantile-quantile project. Time series for these quantities can therefore be obtained from the **qq** spreadsheet. The $y_i$ variables have correlation matrix elements

$$\rho_{ij} \equiv \text{Corr}\,(y_i, y_j) = \frac{C_{ij}}{\sqrt{C_{ii}C_{jj}}}, \tag{8.4}$$

in terms of the covariance matrix elements $C_{ij} = E[y_i y_j]$, with $E[]$ being an expectation over the underlying probability distribution. The covariance matrix of standardized log-returns is then:

$$\begin{aligned}\text{Cov}(y_i, y_j) = E[y_i y_j] &= \sum_{k,l=1}^{n} U_{ki} U_{lj} E[x_k x_l] \\ &= \sum_{k=1}^{n} U_{ki} U_{kj} = (\mathbf{U}'\mathbf{U})_{ij} \equiv C_{ij},\end{aligned} \tag{8.5}$$

since $E[x_k x_l] = \delta_{kl}$ [i.e., the $x_k$ are independent standard normals, $x_k \sim N(0, 1)$]. This shows that the $U$ matrix used to generate correlated stock price movements is obtained from the Cholesky factorization of the covariance matrix. One also observes that uncorrelated stock price motion follows readily in the case of $C_{ij} = \delta_{ij}\sigma_i^2 = \delta_{ij}C_{ii}$, i.e., $U_{ij} = \delta_{ij}\sigma_i = \delta_{ij}\sqrt{C_{ii}}$. We also have the useful result that

$$\text{Cov}(\tilde{y}_i, \tilde{y}_j) = \frac{\text{Cov}(y_i, y_j)}{\sqrt{C_{ii}C_{jj}}} = \rho_{ij}. \tag{8.6}$$

## 8.2 Calibration

In the **mc** spreadsheet application, the first phase is to calibrate the scenario-generation engine to be used later for Monte Carlo pricing. This is accomplished by considering a basket of options with known market prices on plain-vanilla calls. The second phase, discussed in the next section, is to price the basket option of choice by running a Monte Carlo simulation based on the calibrated volatilities as input. The spreadsheet table for the calibration basket, duplicated in Figure 8.1, shows that for each $i$th stock (or index), we have a market plain-vanilla call option price $C_i$ on a single underlying equity $i$ with fixed spot $S_{0,i} = \$100$ (for example), present calendar time $t$ (e.g., today's date), given maturity $T_i$, and strike $K_i$. From this we extract an implied volatility $\sigma_i^I$ for each underlying index $i$ independently. This is done by inverting the Black–Scholes formula for a call with $\sigma_i = \sigma_i^I$,

$$\text{Market Call Price}_i = C(S_{0,i}, K_i, r, \sigma_i^I, T_i - t), \tag{8.7}$$

| | strike | maturity | price | implied vol |
|---|---|---|---|---|
| TSE100COMPX | 99.22 | 22-Jan-2000 | 5.18 | 12.850% |
| MEXISEX | 99.59 | 12-Feb-2000 | 5.44 | 13.003% |
| SP500C | 105.67 | 5-Feb-2000 | 2.85 | 15.075% |
| SPFINL | 106.64 | 14-Jan-2000 | 1.74 | 13.508% |
| SPINDMV | 101.54 | 9-May-2000 | 6.28 | 13.783% |
| BRINDEX | 95.48 | 12-Feb-2000 | 7.84 | 9.793% |
| FRCAC40X | 104.72 | 15-Jan-2000 | 2.03 | 11.927% |
| FT30X | 96.04 | 31-Jan-2000 | 7.38 | 11.443% |
| IRDAVYX | 100.14 | 14-May-2000 | 7.98 | 16.362% |
| NLCBSX | 101.15 | 6-May-2000 | 5.70 | 15.578% |

FIGURE 8.1 Calibration basket for 10 indices. All implied volatilities are computed with interest rate $r = 7\%$, spot 100, and present date 1-Sept-1999.

for each call contract in the calibration basket. Once the $\sigma_i^I$ are obtained, the covariance matrix of log-returns for the total number $n$ of underlyings is estimated using the historical returns in the **qq** spreadsheet. That is, we estimate the correlation from equation (8.6) using the average

$$\rho_{ij} \approx \frac{1}{N_{\text{ret}}} \sum_{k=1}^{N_{\text{ret}}} \tilde{y}_i^{(k)} \tilde{y}_j^{(k)} \tag{8.8}$$

over the total number of historical returns $N_{\text{ret}}$ contained in the time series table of the **qq** spreadsheet. Note that superscript $(k)$ denotes the standardized return at time $t_k$, and the $\tilde{y}_i^{(k)}$ are given by equation (7.5), where $t = t_k$. Equation (8.8) gives the correlation matrix. Note that volatility varies as $1/\sqrt{\text{time}}$, whereas covariance matrix elements vary as the square of volatility (i.e., as 1/time). Equation (8.8) is very useful as it stands, since the matrix elements are *dimensionless* and hence do not depend on the time scale of the returns (i.e., these can be daily, weekly, yearly, etc.).

The calibrated covariance matrix is then obtained by using the correlation matrix in conjunction with the yearly implied volatilities in equation (8.7). The covariance matrix that is actually used for the Monte Carlo sampling, and hence used for pricing as discussed in the next section, is given by

$$C_{ij} = \rho_{ij} \sigma_i^I \sigma_j^I. \tag{8.9}$$

Note that from use of equation (8.1) the time scale of the covariance matrix is automatically set by the unit used for the implied volatilities, i.e., yearly.

## 3 Pricing Equity Basket Options

The price $V(\mathbf{S}_0, t = 0)$ of a basket option at present time $t = 0$ and present stock price vector $\mathbf{S}_0 = (S_{0,1}, \ldots, S_{0,n})$ with maturity $t = T$ can be expressed as a closed-form $n$-dimensional integral. In particular, the transition probability density function for an initial stock vector $\mathbf{S}_0$ to attain value $\mathbf{S}_T = (S_{T,1}, \ldots, S_{T,n})$, in time $T$, is given by an $n$-dimensional correlated lognormal distribution [i.e., equation (1.198)]:

$$p(\mathbf{S}_T, \mathbf{S}_0; T) = (2\pi T)^{-\frac{n}{2}} (\det \mathbf{C})^{-\frac{1}{2}} \exp\left(-\tfrac{1}{2}\mathbf{z} \cdot \mathbf{C}^{-1} \cdot \mathbf{z}\right), \tag{8.10}$$

where the $n$-dimensional vector $\mathbf{z}$ is defined by the components

$$z_i \equiv \frac{\log(S_{T,i}/S_{0,i}) - (r - \frac{1}{2}\sigma_i^2)T}{\sqrt{T}} \tag{8.11}$$

and $\sigma_i = \sigma_i^I$ (the implied volatilities). Note that these components are essentially the $y_i$ defined in equation (8.3), with $dt$ replaced by $T$. The covariance matrix $\mathbf{C}$ is given in terms of the correlation matrix and the implied volatilities via equation (8.9). Risk-neutral pricing then gives [i.e., equation (1.187)]

$$V(\mathbf{S}_0, 0) = e^{-rT} \int_0^\infty p(\mathbf{S}_T, \mathbf{S}_0; T)\Pi(\mathbf{S}_T)d\mathbf{S}_\mathbf{T}, \tag{8.12}$$

where $\Pi$ is the payoff function.

For a Monte Carlo implementation it is useful to rewrite the $n$-dimensional integral in equation (8.12) using a change of variables defined by $\mathbf{z} = \mathbf{U}'\mathbf{x}$, i.e,

$$z_i = \sum_{k=1}^{n} U_{ki} x_k, \tag{8.13}$$

where matrix $\mathbf{U}$ is obtained from the (upper) Cholesky factorization of the covariance matrix with elements given in equation (8.9): $\mathbf{C} = \mathbf{U}'\mathbf{U}$. The Jacobian of the transformation $\mathbf{S}_T \to \mathbf{x}$ is $T^{\frac{n}{2}}\sqrt{\det \mathbf{C}}$, while for the inner product we have $\mathbf{z} \cdot \mathbf{C}^{-1} \cdot \mathbf{z} = \mathbf{x} \cdot \mathbf{x}$. Note that the inverse transformation $\mathbf{S}_T = \mathbf{S}_T(\mathbf{x})$ is given by equation (8.1). Combining these results with the integrand in equation (8.12) gives the pricing formula as a discounted expectation over the uncorrelated $n$-dimensional standard normal distribution:

$$\begin{aligned} V(\mathbf{S}_0, 0) &= \frac{e^{-rT}}{(2\pi)^{n/2}} \int_{-\infty}^{\infty} e^{-\frac{1}{2}|\mathbf{x}|^2} \Pi(\mathbf{S}_T(\mathbf{x}))d\mathbf{x} \\ &\approx e^{-rT} \frac{1}{N_s} \sum_{i=1}^{N_s} \Pi(\mathbf{S}_T(\mathbf{x}^{(i)})). \end{aligned} \tag{8.14}$$

This sum gives the Monte Carlo average of the pay-off evaluated at each $i$th scenario vector $\mathbf{S}_T(\mathbf{x}^{(i)})$, i.e., the stock price vector with components given by equation (8.1), where the $x_k^{(i)}$ are $n$ i.i.d. standard normal deviates for all $N_s$ scenarios. The MFRangen numerical library is useful for generating the standard normal deviates, while MFBlas can be used for matrix-vector multiplication in the scenario generation.

Within the **mc** spreadsheet we consider the pricing of three types of basket options, as entered within the user interface. These have the respective pay-offs

(i) Simple chooser: $\Pi(\mathbf{S}_T) = \max\{S_{T,i}; i = 1, \ldots, n\}$
(ii) Chooser call: $\Pi(\mathbf{S}_T) = \max\{C_i = \max(S_{T,i} - K, 0) : i = 1, \ldots, n\}$, corresponding to the choice of one underlying that gives the maximum call pay-off.
(iii) Chooser put: $\Pi(\mathbf{S}_T) = \max\{P_i = \max(K - S_{T,i}, 0) : i = 1, \ldots, n\}$, corresponding to the choice of maximum put pay-off. Note: The strike $K$ is also a user input.

Figure 8.2 shows the results of a Monte Carlo simulation for pricing a simple chooser option on a basket of 10 stocks. Fairly good convergence is obtained in the range of 5000 to

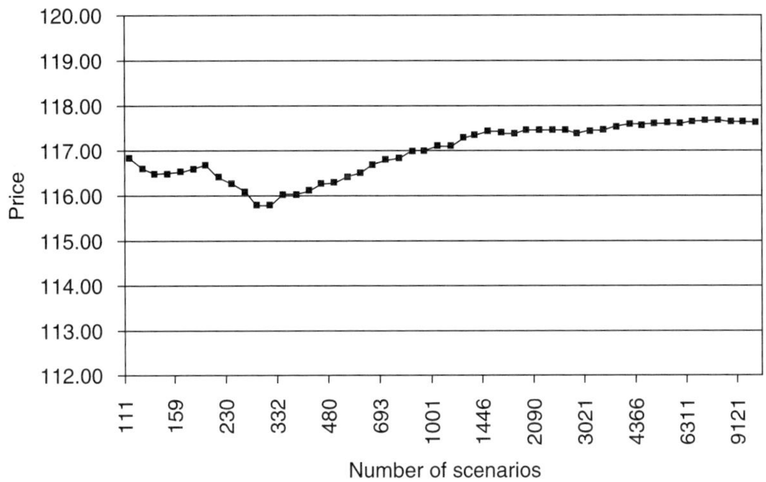


FIGURE 8.2 An example of the convergence pattern of an actual Monte Carlo simulation for the price of a simple chooser option on a basket of 10 correlated stocks.

10,000 scenarios. Note that the spacing in the $x$-axis scale is not constant since the increments were chosen using an exponentially increasing number of points. The user is encouraged to experiment with pricing various contracts that are in-the-money, at-the-money, and out-of-the-money for a varying number of total stocks in the basket. Whenever possible, compare the results of your Monte Carlo simulations with exact results, as in the special case of two correlated underlyings.

CHAPTER • 9

# Project: The Binomial Lattice Model

The purpose of this project is to build a binomial lattice model to price both European and American puts and calls. We demonstrate how to parameterize the lattice in terms of a drift and a volatility parameter, adjust the drift to match forward prices, and adjust the lattice volatility in such a way as to match the price of an at-the-money European call option. Once calibrated, the binomial lattice is used to price European and American options. Extensions to Derman–Kani trees are left to the interested reader.

*Worksheet:* **bin**

*Required Libraries:* **MFioxl, MFBlas, MFFuncs, MFZero, MFStat**

## 9.1 Building the Lattice

A binomial lattice is a recombining two-dimensional tree with a total number of time steps $M \geq 1$ over the time interval $[0, T]$. Lattice nodes parameterize stock prices and calendar time. Dates are denoted by $t_m$, $m = 0, 1, \ldots, M$, where $t_0$ is the date at which we seek the price and $t_m = t_0 + m\ \Delta t$, where $\Delta t = (T - t_0)/M$ is the elementary time step. At the $m$th time step of size $\Delta t$, there are $(m+1)$ nodes labeled by an index $n = 0, 1, \ldots, m$. The stock price at node $(m, n)$ is given by

$$S_n^m = d^{m-n} u^n S_0, \tag{9.1}$$

where $u > 1$ and $d < 1$. The value $S_0^0 = S_0$ is the spot price at the current time $t = t_0$ when the option is valued. Figure 9.1 depicts the binomial lattice geometry. The model is characterized by the parameters $d, u, \Delta t$ and by the risk-neutral probability $p$ of an upward jump. An upward move corresponds to a multiplication by $u$, whereas a downward move corresponds to a multiplication by $d$. The parameter $p$ is strictly between 0 and 1.

According to pricing theory covered in Chapter 1, arbitrage-free prices are achieved if the discrete stochastic process defined by the binomial lattice is risk neutral. One-period returns

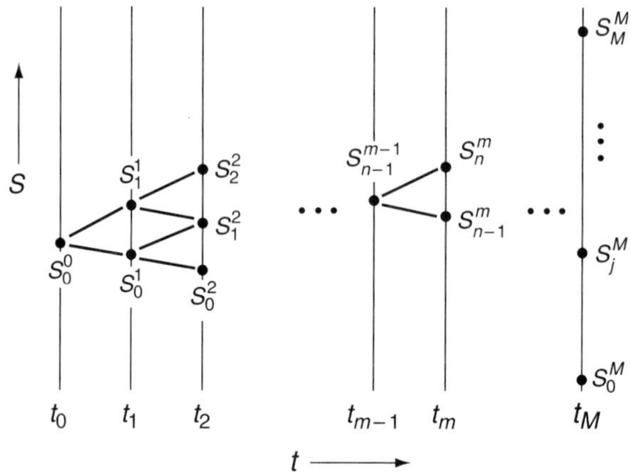


FIGURE 9.1 A binomial lattice originating at the current time $t = t_0$ with stock level $S_0^0$ to final time $t_M = T$. At every time slice $t_{m-1}$ a grid point $S_{n-1}^{m-1}$ gives rise to two points, $S_n^m = uS_{n-1}^{m-1}$ and $S_{n-1}^m = dS_{n-1}^{m-1}$, at a later time $t_m = t_{m-1} + \Delta t$.

on the stock must equal the return on the prevailing risk-free rate $r$. Assuming $r$ constant, we find that the condition

$$puS + (1-p)dS = e^{r\Delta t} S \tag{9.2}$$

must be satisfied at all nodes $S = S_n^m$. Hence,

$$pu + (1-p)d = e^{r\Delta t}. \tag{9.3}$$

Let us introduce a lattice volatility parameter $\sigma$ by means of the following equation:

$$pu^2 + (1-p)d^2 = e^{(2r+\sigma^2)\Delta t}. \tag{9.4}$$

**Proposition 9.1.** *In the limit as $\Delta t \to 0$, the lattice volatility converges to the continuous-time lognormal volatility in the Black–Scholes model.*

**Proof.** For a lognormal distribution, we have

$$S^{i+1} = S^i \exp\{(r - \tfrac{1}{2}\sigma^2)\Delta t + \sigma\sqrt{\Delta t}x\}, \qquad x \sim N(0, 1), \tag{9.5}$$

where $S^i$ denotes a stock price at time $t_i$ and $\Delta t = t_{i+1} - t_i$. Conditional on the stock price being $S^i$ at time $t_i$, the following expected values at a later time $t_{i+1} = t_i + \Delta t$ obtain using equation (9.5):

$$E[S^{i+1}] = S^i e^{r\Delta t}, \tag{9.6}$$

$$E[(S^{i+1})^2] = (S^i)^2 e^{(2r+\sigma^2)\Delta t}. \tag{9.7}$$

Within the binomial lattice we have instead:

$$E_b[S^{i+1}] = (pu + (1-p)d)S^i, \tag{9.8}$$

$$E_b[(S^{i+1})^2] = (pu^2 + (1-p)d^2)(S^i)^2. \tag{9.9}$$

Equating the variances $E[(S^{i+1})^2] - (E[S^{i+1}])^2$ and $E_b[(S^{i+1})^2] - (E_b[S^{i+1}])^2$, and using equation (9.3), gives equation (9.4). □

Equations (9.3) and (9.4) allow one to parameterize the three unknowns $d, u, p$ in the binomial lattice by means of the risk-free rate $r$, the lattice volatility $\sigma$, and a third degree of freedom. To resolve the indeterminacy we are at liberty to choose another constraining equation. Two choices are popular:

$$p = \frac{1}{2} \tag{9.10}$$

and

$$u = \frac{1}{d}. \tag{9.11}$$

For the case $p = \frac{1}{2}$, the lattice parameters can be expressed as follows in terms of a lattice volatility $\sigma$ and drift $r$:

$$d = e^{r\Delta t}\left(1 - \sqrt{e^{\sigma^2 \Delta t} - 1}\right), \tag{9.12}$$

$$u = e^{r\Delta t}\left(1 + \sqrt{e^{\sigma^2 \Delta t} - 1}\right), \tag{9.13}$$

$$p = \tfrac{1}{2}. \tag{9.14}$$

This is a recombining binomial tree that drifts upward in the stock price direction.

If $u = \frac{1}{d}$, the tree is symmetric about the line $S = S_0$ with zero drift and the lattice parameters are given as follows:

$$d = a - \sqrt{a^2 - 1}, \tag{9.15}$$

$$u = 1/d, \tag{9.16}$$

$$p = (e^{r\Delta t} - d)/(u - d), \tag{9.17}$$

where

$$a = (e^{-r\Delta t} + e^{(r+\sigma^2)\Delta t})/2. \tag{9.18}$$

## .2 Lattice Calibration and Pricing

Prices of European-style options are computed iteratively, starting from the maturity date $T$, where the payoff function $\phi(S)$ is ascribed to the terminal nodes $S_n^M$, $n = 0, 1, \ldots, M$. Let $f_n^m = V(S_n^m, t_m)$ be the option price at the node $S_n^m$. For a call option with strike $K$, the final time condition is

$$f_n^M = \phi(S_n^M) = \max(S_n^M - K, 0). \tag{9.19}$$

For the put struck at the same level, the condition is instead

$$f_n^M = \phi(S_n^M) = \max(K - S_n^M, 0). \tag{9.20}$$

The risk-neutral condition on the option price process applied to each node yields the following recurrence relation (i.e., valuation formula):

$$f_n^m = e^{-r\Delta t}(p f_{n+1}^{m+1} + (1-p) f_n^{m+1}). \tag{9.21}$$

The price of the option at current time $t_0$ and spot $S_0$ is given by the last iterate, $V(S_0, t_0) = f_0^0$.

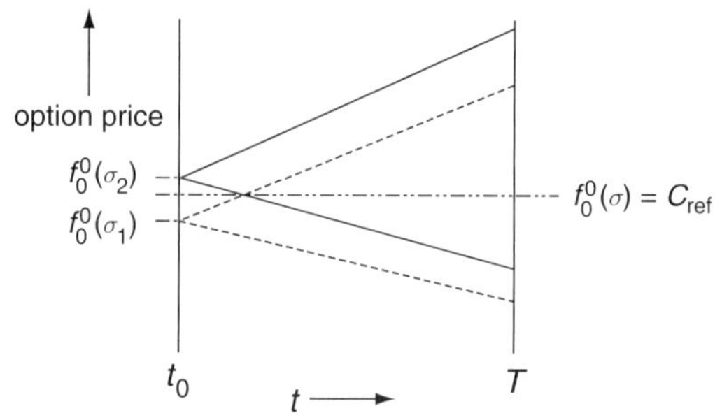


FIGURE 9.2 A schematic of upper and lower bands of option prices (i.e., the outer node values $f_0^m, f_m^m, m = 0, \ldots, M$) for two different lattice geometries corresponding to lattice volatilities $\sigma_1$ (dashed lines) and $\sigma_2$ (solid lines). The lower lattice volatility value $\sigma_1$ gives a lower estimate of the reference market option price, while the higher value $\sigma_2$ gives an upper estimate of the market option price. The lattice volatility $\sigma$ (for given time step $\Delta t$ and interest rate $r$) that prices the market option value exactly lies in the interval $\sigma_1 < \sigma < \sigma_2$.

To price American options, the method is similar, except an adjustment is made to account for the possibility of early exercise. Namely, the risk-neutral valuation formula is now (see Section 1.14.1 on dynamic programming):

$$f_n^m = \max\left(\phi(S_n^m), e^{-r\Delta t}(pf_{n+1}^{m+1} + (1-p)f_n^{m+1})\right). \tag{9.22}$$

In the lattice calibration step, one has to adjust the lattice volaility to match the price of the single at-the-money option used as calibration target. Figure 9.2 shows a schematic representation of the lattice calibration procedure. Notice that the optimal value for the lattice volatility $\sigma$ does not necessarily coincide with the Black–Scholes implied volatility $\sigma^I$ of the option, but it converges to this value in the limit of time steps of vanishing length. The calibration procedure requires the use of a root-finding algorithm. The existence of a root is guaranteed with both choices $p = \frac{1}{2}$ and $u = \frac{1}{d}$, for in both cases the resulting families of binomial models allow for arbitrarily large or small values of the volatility. The worksheet **bin** contains an at-the-money European call as the calibration target or reference call option contract. The option is quoted in terms of a Black–Scholes implied volatility $\sigma^I$. The market price results from the Black–Scholes formula

$$C_{\text{ref}} = C(S_0, K_{\text{ref}}, r, \sigma^I, T_{\text{ref}} - t_0). \tag{9.23}$$

The current time is denoted by $t_0$ and the spot is $S_0$. To determine the lattice volatility $\sigma$, one has to find a root of the equation

$$f_0^0 = f_0^0(\sigma, r, \Delta t) = C_{\text{ref}} \tag{9.24}$$

for a given choice of $r$ and lattice geometry. Here we have explicitly written the dependence of the binomial approximation to the price, i.e., $f_0^0$, in terms of the lattice parameters. The value of $f_0^0$ is found iteratively using equation (9.21) or equation (9.22), depending on whether the option is European or American, respectively. The final time condition is given by equation (9.19) for calls and equation (9.20) for puts. The value for the strike is set as $K = K_{\text{ref}}$. Having finally obtained a value for $\sigma$, the model can be used to price other American or European options. $\partial$

CHAPTER • 10

# Project: The Trinomial Lattice Model

The main task in this project is to build a trinomial lattice model to price European and American claims within an explicit finite-difference scheme. Both drifted and nondrifted types of lattice geometries are considered. For the drifted lattice model, the drift is adjusted to account for the prevailing interest rate so as to maintain risk neutrality. As with binomial models, the model is parameterized by means of a suitably defined lattice volatility, which is then calibrated to match the price of a given at-the-money European option. Option prices are obtained for all single-barrier and plain-vanilla European as well as American-style claims. Extensions to Derman–Kani (i.e., local volatility) trinomial trees are left to the interested reader.

*Worksheets:* **pded1, pded2**

*Required Libraries:* **MFioxl, MFBlas, MFFuncs, MFZero, MFStat**

## 10.1 Building the Lattice

Trinomial lattices are normally based on lattices of fixed geometry and parameterized by the nodal transition probabilities. Consider a recombining two-dimensional tree with a total number of time steps $M \geq 1$. The nodes of the tree are placed along the time lines $t_m$, $m = 0, 1, \ldots, M$, where the initial (e.g., present) calendar time is denoted by $t_0$. We will denote the time to expiry by $T$, which defines a time step of size $\Delta t = (T - t_0)/M$ (i.e., $t_M = T$) for the lattice. At the $m$th time step, there are $(2m+1)$ nodes in a standard trinomial lattice.

The nodes are chosen on a log-rectangular grid and can be generally expressed as follows:

$$S_n^m = S_0^0 e^{m\mu\Delta t + n\Delta x}, \tag{10.1}$$

for $n = -m, -m+1, \ldots, 0, \ldots, m-1, m$. The spot (i.e., initial stock price) is $S_0 = S_0^0$. The choice of the parameters $\mu$ and $\Delta x$ is discussed shortly. Note that by taking logarithms of equation 10.1, $\Delta x$ gives a measure for the change in $\log S$ within a given time slice. Namely, $\Delta_n \log S_n^m \equiv \log S_{n+1}^m - \log S_n^m = \Delta x$ gives the node spacing for a fixed value of time. Changes due to a possible drift can arise from the difference $\Delta_m \log S_n^m \equiv \log S_n^{m+1} - \log S_n^m = \mu\ \Delta t$. In the stochastic process underlying the trinomial tree model, stock prices can jump from a node $S_n^m$ to the nodes $S_{n'}^{m+1}$, with $n' = n, n \pm 1$. There are three transition probabilities, $p_+$, $p_0$, and $p_-$, that correspond to an upward move, middle move (i.e., no move for zero drift), and downward move, respectively, for any trinomial tree. These risk-neutral probabilities are subject to two constraints; the first is that of probability conservation,

$$p_+ + p_0 + p_- = 1. \tag{10.2}$$

A trinomial tree is recombining, and the nodes span a cone within a rectangular grid arrangement in log-stock and time space (see Figure 10.1). Notice, though, that, as we discuss later, to price several options of different strikes at once, it is useful to extend the trinomial lattice to cover the complete rectangular grid of $(2M+1)(M+1)$ points, so at every time step $m$ we have $(2M+1)$ points $S_n^m$, where $n = -M, \ldots, 0, \ldots, M$.

In what follows we present three different geometric constructions of trinomial lattices. The first two, Cases 1 and 2, assume $\mu = 0$, while the third asks for an additional constraint on the probability amplitudes and adjusts the drift $\mu$ in such a way as to achieve risk neutrality.

### 10.1.1 Case 1 ($\mu = 0$)

Since $\mu = 0$, the risk-neutrality constraint $E[S_{t_{m+1}} | S_{t_m} = S] = e^{r\Delta t} S$ gives:

$$p_+ e^{\Delta x} + p_0 + p_- e^{-\Delta x} = e^{r\Delta t}. \tag{10.3}$$

Probability conservation (10.2) allows one to eliminate the variable $p_0 = 1 - (p_+ + p_-)$, and gives

$$p_+(e^{\Delta x} - 1) + p_-(e^{-\Delta x} - 1) = e^{r\Delta t} - 1. \tag{10.4}$$

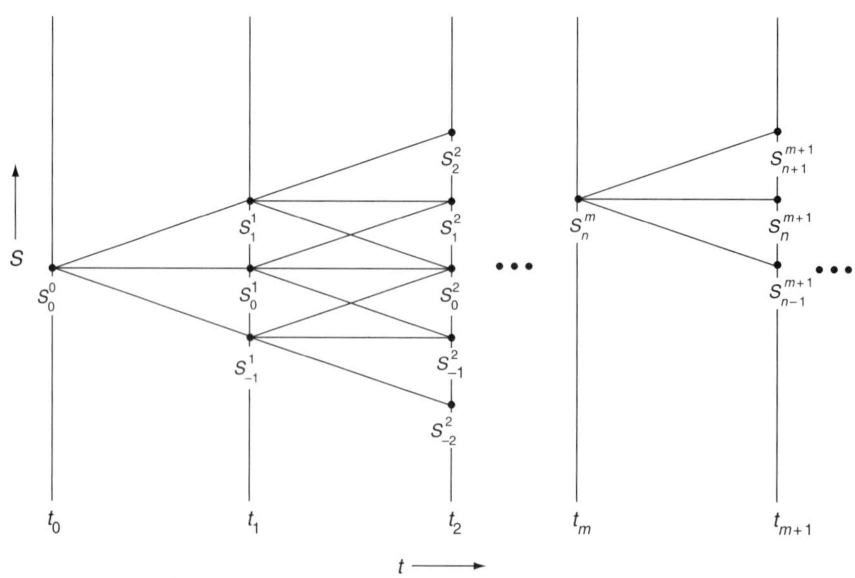


FIGURE 10.1 A schematic of the nondrifted ($\mu = 0$) trinomial lattice originating at current time $t = t_0$ with stock level $S_0^0$. At every time slice $t_m$, a stock at level $S_n^m$ can change to $S_{n'}^{m+1}$, with $n' = n, n \pm 1$. The drifted lattice has a similar geometry, except all nodes are shifted by an amount $\exp(\mu\ \Delta t)$ after every time step.

Let us introduce a lattice volatility parameter $\sigma$ by means of the following equation, similar to equation (9.4):

$$p_+(e^{2\Delta x}-1)+p_-(e^{-2\Delta x}-1)=e^{(2r+\sigma^2)\Delta t}-1. \tag{10.5}$$

Equations (10.4) and (10.5) are a linear system in the two unknowns $p_+$, $p_-$. By solving them we find the transition probabilities as a function of $\sigma$ and $r$:

$$p_+=\frac{(e^{-2\Delta x}-1)(e^{r\Delta t}-1)-(e^{-\Delta x}-1)(e^{(2r+\sigma^2)\Delta t}-1)}{(e^{\Delta x}-1)(e^{-2\Delta x}-1)-(e^{-\Delta x}-1)(e^{2\Delta x}-1)}, \tag{10.6}$$

$$p_-=\frac{(e^{r\Delta t}-1)-(e^{\Delta x}-1)p_+}{(e^{-\Delta x}-1)}. \tag{10.7}$$

### 10.1.2 Case 2 (Another Geometry with $\mu=0$)

An alternative definition for the lattice volatility is

$$p_+(\Delta x)^2+p_-(-\Delta x)^2+p_0 0^2=\sigma^2\Delta t. \tag{10.8}$$

This is also an acceptable definition because in the continuous-time limit it also converges to the Black–Scholes volatility. With this equation, we have

$$p_++p_-=\frac{\sigma^2\ \Delta t}{(\Delta x)^2}. \tag{10.9}$$

Equations (10.4) and (10.9) is a linear system of two equations in the two unknowns $p_+$, $p_-$. Solving gives

$$p_-=\frac{(e^{\Delta x}-1)\sigma^2\Delta t/(\Delta x)^2-(e^{r\Delta t}-1)}{(e^{\Delta x}-e^{-\Delta x})}, \tag{10.10}$$

and

$$p_+=\frac{\sigma^2\ \Delta t}{(\Delta x)^2}-p_-. \tag{10.11}$$

The expressions for the probabilities are in this case slightly simpler than in Case 1.

For both choices of the lattice volatility, one has to select appropriate values for $\Delta x$, given a time step $\Delta t$, so as to obtain acceptable probabilities, i.e., $p_+>0$, $p_->0$, and $p_++p_-<1$. For the simpler Case 2, we see immediately from equation (10.9) that the constraint $\sigma^2\ \Delta t/(\Delta x)^2<1$ must be obeyed. This is related to the usual stability constraint that arises in the direct, or explicit, PDE method for solving the Black–Scholes equation.

### 10.1.3 Case 3 (Geometry with $p_+=p_-$: Drifted Lattice)

If we ask for the symmetry condition $p_+=p_-\equiv p$, we need to adjust the lattice drift $\mu$ in such a way as to satisfy the risk-neutrality condition. Equation (10.2) gives $p_0=1-2p$, which is used to eliminate $p_0$. This is used in the risk-neutrality condition, which now includes the overall drift

$$p_+e^{\mu\Delta t+\Delta x}+p_-e^{\mu\Delta t-\Delta x}+p_0e^{\mu\Delta t}=e^{r\Delta t} \tag{10.12}$$

and, this, gives

$$p(e^{\Delta x} + e^{-\Delta x}) + (1 - 2p) = e^{(r-\mu)\Delta t}. \tag{10.13}$$

Taking logarithms gives the drift in terms of all other parameters:

$$\mu = r - \frac{1}{\Delta t} \log(2p(\cosh \Delta x - 1) + 1). \tag{10.14}$$

To define the lattice volatility parameter and express $p$ in terms of it, we set

$$(p_+ + p_-)(\Delta x)^2 = \sigma^2 \ \Delta t, \tag{10.15}$$

which reduces to

$$p = \frac{\sigma^2 \ \Delta t}{2(\Delta x)^2}. \tag{10.16}$$

As in Case 2, this equation gives the probability in terms of $\Delta x$, $\Delta t$, and $\sigma$. A possible strategy is to choose a sensible value for the probability $p$, given $\Delta t$ and $\sigma$, and then to arrive at the spacing in the logarithm of the stock price,

$$\Delta x = \sigma \sqrt{\Delta t / 2p}. \tag{10.17}$$

Note that the usual stability condition for the direct PDE solution of the Black–Scholes equation requires $p < 1/2$. $\Delta x$ is then given by equation (10.17) for a given value of the lattice volatility $\sigma$. The drift then follows from equation (10.14).

## 10.2 Pricing Procedure

Option prices are computed iteratively, starting from the maturity date $T$, at which point option prices are given by the payoff function $\phi(S)$. We denote the value of the option at node $S_n^m$ by $f_n^m = V(S = S_n^m, t = t_m)$. The final-time condition for a call struck at $K$ is

$$f_n^M = \phi(S_n^M) = \max(S_n^M - K, 0), \tag{10.18}$$

and for the put struck also at $K$

$$f_n^M = \phi(S_n^M) = \max(K - S_n^M, 0). \tag{10.19}$$

Hence, prices for European-style options at each node are computed recursively using the risk-neutral valuation formula:

$$f_n^m = e^{-r\Delta t}(p_+ f_{n+1}^{m+1} + p_- f_{n-1}^{m+1} + (1 - (p_+ + p_-)) f_n^{m+1}). \tag{10.20}$$

Figure 10.2 depicts the explicit scheme inherent in equation (10.20) for propagating prices at each time step. The nodes are placed according to equation (10.1), as explained earlier, and the probabilities $p_+$, $p_-$ are given as described within the respective cases. The iteration is best accomplished using a band matrix multiplication routine in the MFBlas numerical library. This is possible since the pricing equation (10.20) can be rewritten in matrix format, whereby

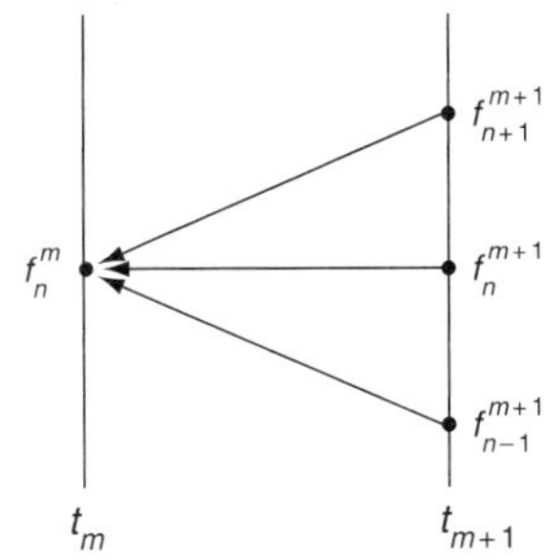


FIGURE 10.2 The explicit finite difference method makes use of prices at three adjacent nodes at a more later time step, $t = t_{m+1}$, for propagating the price to a given node at a more current time $t = t_m$.

the option price solution column vector denoted by $\mathbf{f}^m$ at time $t_m$ is $(2M+1)$-dimensional with components $f_{-M}^m, f_{-M+1}^m, \ldots, f_0^m, \ldots, f_{M-1}^m, f_M^m$:

$$\mathbf{f}^m = e^{-r\Delta t}\mathbf{T}\mathbf{f}^{m+1}. \tag{10.21}$$

This is a special linear system of equations with a tri-diagonal *transfer matrix* $\mathbf{T}$. This $(2M+1)$-dimensional matrix is given in terms of the transition probabilities for the upward and downward moves. Namely,

$$\mathbf{T} = \begin{pmatrix} 1-(p_+ + p_-) & p_+ & 0 \;.\;. & 0 \\ p_- & 1-(p_+ + p_-)\; p_+ & \cdot\;\cdot & \cdot \\ 0 & p_- & \cdot\;\cdot\;\cdot & \cdot \\ \cdot & 0 & \cdot\;\cdot\;\cdot & \cdot \\ \cdot & \cdot & \cdot\;\cdot\;\cdot & 0 \\ \cdot & \cdot & \cdot\;\cdot\;\cdot & p_+ \\ 0 & \cdot & \cdot\;\cdot\; p_- & 1-(p_+ + p_-) \end{pmatrix}. \tag{10.22}$$

Note that for the drifted lattice geometry $p_+ = p_-$, hence giving a symmetric banded matrix in this particular case.

To price American options, the iteration proceeds similarly, except an adjustment is made at every time step to account for the possibility of early exercise. Namely, the risk-neutral valuation formula is now

$$f_n^m = \max\left(\phi(S_n^m), e^{-r\Delta t}[p_+ f_{n+1}^{m+1} + p_- f_{n-1}^{m+1} + (1-(p_+ + p_-))f_n^{m+1}]\right). \tag{10.23}$$

The price of the option at current time $t_0$ and spot $S_0$ is given by the last iterate, $V(S_0, t_0) = f_0^0$.

The risk-neutral condition is exactly satisfied at all nodes if we restrict ourselves to the grid points belonging to the interior of the cone with $n = -m, \ldots, m$ at the $m$th time slice. If the foregoing equations are used to price options across all grid points with $n = -M, \ldots, M$, risk neutrality fails outside the boundaries of the cone and numerical errors arise. This is the case for implementing the strictly trinomial model, unless proper boundary conditions are imposed on the extreme nodes of the rectangular grid.

For the case of American put options we can compute the exercise boundary as a function of the time to maturity $T - t$. The exercise-boundary value $S_t^*$ at time $t$ corresponds to the highest value of $S$ for which it is optimal to exercise the option rather than holding it. The value $S_t^*$, for each $t = t_m$, is the largest node value $S_n^m$ for which $\phi(S_n^m) \geq \tilde{f}_n^m$, with the latter quantity given by the right-hand side of equation (10.20), i.e., the continuation value at $t_m$ .

## 10.3 Calibration

As in the binomial model, we determine the lattice volatility in such a way as to match the price of an at-the-money option chosen as calibration target. The resulting optimal implied lattice volatility computed again does not coincide with the implied Black–Scholes volatility $\sigma_I$, but it converges to this value in the limit of infinitesimal time steps. The lattice volatility compensates for the systematic errors in the discrete-time approximation scheme inherent in the trinomial method.

Calibration requires the use of a root-finding algorithm. The procedure is similar for all three lattice cases, as now discussed. The **pded1** spreadsheet contains a European at-the-money call with given maturity $T_{\text{ref}}$ and strike $K_{\text{ref}}$ as the calibration (reference) target. The price of the calibration target is provided as a Black–Scholes implied volatility $\sigma^I$. The market price of this call is then given by the Black–Scholes formula

$$C_{\text{ref}} = C(S_0, K_{\text{ref}}, r, \sigma^I, T_{\text{ref}} - t_0). \tag{10.24}$$

$t_0$ is the time at which we seek the price, and the corresponding spot price is assumed to be $S_0$. The implied lattice volatility $\sigma$ is obtained by inverting the following equation with a root finder in the MFZero library:

$$f_0^0 = f_0^0(\sigma, r, \Delta t, \Delta x) = C_{\text{ref}}. \tag{10.25}$$

Here we have explicitly written the dependence of the trinomial lattice option price, i.e., $f_0^0$, in terms of the lattice parameters. The value of $f_0^0$ is found iteratively using the earlier pricing equations for a European call option. Note that the interest rate $r$ is held fixed and $\Delta t$ is also fixed by the chosen number of time steps in the lattice. The value for the strike is set as $K = K_{\text{ref}}$, i.e. the reference strike.

## 10.4 Pricing Barrier Options

The procedure to price a barrier option is a modification of the method for plain-vanilla options, except we have to account for a boundary condition at the barrier $H$. We discuss, in detail, the case of single-barrier down-and-out options. The case of up-and-out options is similar, while the case of knock-in options reduces to that of knockouts thanks to the in-out symmetry relation

$$\text{Knock In} + \text{Knockout} = \text{Vanilla}. \tag{10.26}$$

To price a down-and-out, we can assume that the spot is above the lower barrier, i.e., $S_0 > H$; otherwise the option would be worthless. There is an important distinction between the cases $\mu = 0$ and $\mu \neq 0$. In case $\mu \neq 0$, it is not possible to adjust the lattice so that a horizontal line of lattice nodes lies exactly on the barrier. While iterating equation (10.20), one verifies the node positions with respect to the barrier. For all values of $n$ for which $S_n^m < H$ one sets $f_n^m = 0$ in the pricing equations. This limitation in approximating the real location of the barrier at each time slice gives rise to systematic numerical errors (see also Section 11.4 and Figure 11.1).

In the case $\mu = 0$, it is possible to adjust the lattice so that a subset of the nodes lies exactly on the barrier. Figure 10.3 shows such a description of a nondrifted lattice, where the choice of geometry is such that a set of horizontal nodes corresponds exactly to either a

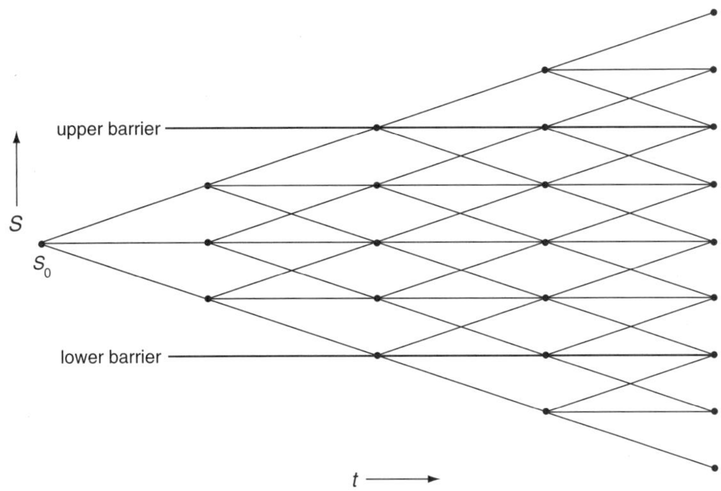


FIGURE 10.3 The spacing of a nondrifted ($\mu = 0$) trinomial lattice can be chosen to exactly match an upper- or lower-barrier level along a horizontal line of nodes.

lower or upper barrier level. Namely, one can select a spacing $\Delta x$ so that for a given positive integer $n_H$ we have $H = S_0 \exp(-n_H \,\Delta x)$, or, expressed otherwise:

$$\Delta x = \frac{1}{n_H} \log(S_0/H).$$

The price of the down-and-out option is then obtained by iterating the pricing equations, whereby one considers only the nodes lying at and above the barrier, i.e., $n \geq -n_H$, with the condition that the option prices $f_n^m = 0,\ n \leq -n_H$, for all $m$. Note that the approach can be used to value European as well as American-style barrier options.

## 0.5 Put-Call Parity in Trinomial Lattices

One of the consequences of risk neutrality is the put-call parity relation for European prices

$$S + P(S, t) - C(S, t) = Ke^{-r(T-t)} \tag{10.27}$$

across all nodes $(S, t) = (S_n^m, t_m)$. It is instructive to verify it directly. Consider prices at the final time line with $m = M$. Call and put prices $C_n^M$ and $P_n^M$ satisfy, by construction, the put-call parity relation at the terminal nodes,

$$S_n^M + P_n^M - C_n^M = K. \tag{10.28}$$

At the internal nodes, we can proceed by induction. Hence, begin by assuming that the put-call parity relation is satisfied at the $m$th time step,

$$S_n^m + P_n^m - C_n^m = Ke^{-r(T-t_m)}. \tag{10.29}$$

By applying equation (10.20) for the put and call we find that:

$$\begin{aligned} P_n^{m-1} - C_n^{m-1} = e^{-r\Delta t}(&p_+(P_{n+1}^m - C_{n+1}^m) + p_-(P_{n-1}^m - C_{n-1}^m) \\ &+(1-(p_+ + p_-))(P_n^m - C_n^m)). \end{aligned} \tag{10.30}$$

Using probability conservation and the induction hypothesis, $P_k^m - C_k^m = Ke^{-r(T-t_m)} - S_k^m, k = n, n \pm 1$. Equation (10.30) yields

$$P_n^{m-1} - C_n^{m-1} = e^{-r\Delta t}[Ke^{-r(T-t_m)} - (p_+ S_{n+1}^m + p_- S_{n-1}^m + (1-(p_+ + p_-))S_n^m]. \tag{10.31}$$

The second term on the right-hand side of this equation simplifies to $S_n^{m-1}$ as a consequence of the risk-neutrality condition. Multiplying out the discount term while using $t_{m-1} = t_m - \Delta t$ then gives

$$S_n^{m-1} + P_n^{m-1} - C_n^{m-1} = Ke^{-r(T-t_{m-1})}. \tag{10.32}$$

Put-call parity is therefore recovered.

## 10.6 Computing the Sensitivities

The sensitivities $\Delta = \partial V/\partial S$, $\Gamma = \partial^2 V/\partial S^2$ and the vega $\partial V/\partial \sigma^I$ of the option value $V$ at present time $t_0$ and spot $S = S_0$ can be approximated by finite differences:

$$\Delta \approx \frac{V(S_0 + dS, t_0) - V(S_0 - dS, t_0)}{2dS}, \tag{10.33}$$

$$\Gamma \approx \frac{V(S_0 + dS, t_0) + V(S_0 - dS, t_0) - 2V(S_0, t_0)}{(dS)^2}, \tag{10.34}$$

$$\frac{\partial V}{\partial \sigma^I} \approx \frac{V(\sigma^I + d\sigma, t_0) - V(\sigma^I - d\sigma, t_0)}{2d\sigma}. \tag{10.35}$$

The quantity $dS$ can be chosen as a small change in spot price, e.g., $dS \sim 0.001 S_0$ and likewise $d\sigma$ is a small increment in the volatility, $d\sigma \sim 0.001\sigma^I$, where $\sigma^I$ is the implied volatility at $S_0$. Note that for clarity of notation we have explicitly written the dependence of $V$ on spot and volatility only, respectively.

CHAPTER • 11

# Project: Crank–Nicolson Option Pricer

The purpose of this project is to implement an implicit finite-difference solution scheme to price standard as well as barrier-type European and American options using the Crank–Nicolson (CN) method. The CN method is also calibrated against a reference European option. Possible put-call parity mismatches introduced by the CN approximation are then studied across a whole range of values in the moneyness parameter. The unique approach makes use of a drifted trinomial lattice. An implementation of the CN method within nondrifted lattices as well as other extensions are left as exercises for the interested reader.

*Worksheets:* **cranic1, cranic2**

*Required Libraries:* **MFioxl, MFStat, MFFuncs, MFBlas, MFLapack, MFZero**

## 11.1 The Lattice for the Crank–Nicolson Pricer

Crank–Nicolson methods are among the more commonly used implicit finite-difference solvers for the Black–Scholes PDE. Here we implement a rather unique CN approach that borrows partly from the methodology used in the direct PDE trinomial lattice solver covered in the previous project. The first step is to build a trinomial lattice. This part was already covered in the first section of the previous project on trinomial lattice modeling and hence will not be repeated here. There we discussed the use of three types of lattices, two of which are driftless. In this project we will focus specifically on the drifted lattice approach. The use of nondrifted lattices in CN (which were seen in the direct trinomial solver to introduce explicit differences in the nodal transition probabilities for upward versus downward moves) will be left as a future exercise. As well, this project focuses on the calibration and subsequent pricing of plain European and single-barrier European options. The extension to price double-barrier options as well as American barrier options is also obvious within the framework provided here, although we shall leave this as a separate implementation exercise for the interested reader.

As described in the previous trinomial project, the nodes are chosen on a log-rectangular grid as given by equation (10.1) with nonzero drift parameter $\mu$. For a full description of the lattice, see the first section of the trinomial project. Again, making use of the risk-neutrality condition and taking logarithms gives the drift in terms of all other parameters, as given in equation (10.14) and repeated here for clarity:

$$\mu = r - \frac{1}{\Delta t}\log(2p(\cosh\Delta x - 1) + 1). \tag{11.1}$$

The probability $p$ is again given in terms of the lattice volatility parameter $\sigma$, the spacing $\Delta x$, and $\Delta t$:

$$p = \frac{\sigma^2 \Delta t}{2(\Delta x)^2}. \tag{11.2}$$

As in the direct method, one chooses a sensible value for the probability $p$, given a $\Delta t$ and a $\sigma$, and then arrives at the spacing in the logarithm of the stock price:

$$\Delta x = \sigma\sqrt{\Delta t / 2p}. \tag{11.3}$$

Note that $p$ is normally chosen in the range $0 < p < \frac{1}{2}$, although the CN method can be shown to be stable and convergent for all $p > 0$. To reiterate, the $M+1$ time slices are chosen with time step $\Delta t = (T - t_0)/M$, where $T$ is the maturity time and $t_0$ denotes present time.

## 11.2 Pricing with Crank–Nicolson

Here we shall explicitly discuss the pricing of European-style options; the extension to Americans is obvious and introduces the same extra step as discussed in the previous project. The pricing equations for the CN method differ significantly from the direct trinomial pricer, in that propagation of the solution takes into account both backward and forward motion. In particular, one can relate the option prices $f_n^m = V(S_n^m, t_m)$ at the nodes $S_n^m$ for time $t_m$ to the option prices $f_n^{m+1} = V(S_n^{m+1}, t_{m+1})$ at nodes $S_n^{m+1}$ for future time $t_{m+1} = t_m + \Delta t$, via the probability $p$ for forward-time upward and downward stock motion, as follows:

$$(1+p)f_n^m - \frac{p}{2}(f_{n-1}^m + f_{n+1}^m) = e^{-r\Delta t}\left[(1-p)f_n^{m+1} + \frac{p}{2}(f_{n-1}^{m+1} + f_{n+1}^{m+1})\right]. \tag{11.4}$$

Note the difference between this and the explicit finite-difference approach used in the trinomial lattice project (see Figure 10.2). In this implicit CN scheme, prices at three nodes at a later time step are related to prices at three nodes before a time step. Equation (11.4) can be rewritten in matrix format in which the option price solution column vector denoted by $\mathbf{f}^m$ at time $t_m$ is $(2M+1)$-dimensional, with components $f_{-M}^m, f_{-M+1}^m, \ldots, f_0^m, \ldots, f_{M-1}^m, f_M^m$:

$$\mathbf{Z}\mathbf{f}^m = e^{-r\Delta t}\mathbf{A}\mathbf{f}^{m+1}. \tag{11.5}$$

This is a linear system of equations with tridiagonal matrices $\mathbf{A}$, $\mathbf{Z}$ given in terms of the transfer matrix $\mathbf{T}$,

$$\mathbf{Z} = \tfrac{3}{2}\mathbf{1} - \tfrac{1}{2}\mathbf{T}, \tag{11.6}$$

$$\mathbf{A} = \tfrac{1}{2}\mathbf{1} + \tfrac{1}{2}\mathbf{T}, \tag{11.7}$$

where **1** is the $(2M+1)$-dimensional identity matrix and

$$\mathbf{T} = \begin{pmatrix} 1-2p & p & 0 & . & . & . & 0 \\ p & 1-2p & p & 0 & . & . & . \\ 0 & p & 1-2p & p & 0 & . & . \\ . & 0 & . & . & . & . & . \\ . & . & . & . & . & . & 0 \\ . & . & . & . & . & . & p \\ 0 & . & . & . & 0 & p & 1-2p \end{pmatrix}. \tag{11.8}$$

To implement the CN method, equation (11.5) is solved at every time step beginning with the known terminal payoff vector whose components are given by

$$f_n^M = \max(S_n^M - K, 0) \tag{11.9}$$

for the case of a call struck at $K$ and

$$f_n^M = \max(K - S_n^M, 0) \tag{11.10}$$

for a put struck at $K$. Equation (11.5) constitutes a system of $2M+1$ equations in the $2M+1$ unknowns $f_n^m$ with band diagonal matrix $\mathbf{Z}$ and is hence readily solved by LU factorization. The routine GBSV in the MFLapack library within MFlibs is useful for this purpose once the matrices $\mathbf{A}$, $\mathbf{Z}$ have been transformed to band matrix format. The latter operation is easily accomplished using the routines ST2B and GT2B in MFBlas. Having solved for $\mathbf{f}^{M-1}$ by using the known payoff solution vector $\mathbf{f}^M$, the procedure is then iterated by solving equation (11.5) for $\mathbf{f}^{M-2}$. Iterating $M$ times in this fashion gives the option price vectors at all time slices, including the vector $\mathbf{f}^0$ at present time $t_0$.

As a final important note, we observe that the CN pricing equation (11.5) assumes that the lattice grid takes into account large enough values of $S_M^m$ and small enough values of $S_{-M}^m$ where the put and call are negligible, respectively. Moreover, we have purposely excluded the proper corrections from the boundaries into the matrix pricing equations. The reader can experiment with the inclusion of boundary conditions at the lower and upper extremities of the rectangular grid . Without such inclusions the present CN approach will fail to correctly price options at nodes outside of the proper trinomial lattice (i.e., for nodes lying above or below the outer cone of the tree).

## 11.3 Calibration

As in the direct trinomial model, the lattice volatility is determined in such a way as to match the price of an at-the-money European option chosen as a calibration target. The resulting optimal implied lattice volatility computed again does not coincide with the implied Black–Scholes volatility $\sigma_I$, but it converges to this value in the limit of infinitesimal time steps. As in the other lattice methods, the lattice volatility compensates partly for the systematic errors in the discrete-time approximation scheme inherent in the trinomial method.

Calibration requires the use of a root-finding algorithm. The **cranic1** spreadsheet contains a European at-the-money call with given maturity $T_{\text{ref}}$ and strike $K_{\text{ref}}$ as the calibration target or reference. The price of the calibration target is provided as a Black–Scholes implied volatility $\sigma^I$. The market price of this call is then given by the Black–Scholes formula:

$$C_{\text{ref}} = C(S_0, K_{\text{ref}}, r, \sigma^I, T_{\text{ref}} - t_0). \tag{11.11}$$

$t_0$ is the time at which we seek the price, and the corresponding spot price is assumed to be $S_0$. The implied lattice volatility $\sigma$ is obtained by inverting the following equation with a root finder in the MFZero library:

$$f_0^0 = f_0^0(\sigma, r, \Delta t, \Delta x) = C_{\text{ref}}. \tag{11.12}$$

Here we have explicitly written the dependence of the CN approximation on the price, i.e., $f_0^0$, in terms of the lattice parameters. The value of $f_0^0$ is found iteratively using the earlier pricing equations for a European call option. Note that the interest rate $r$ is held fixed and that $\Delta t$ is also fixed by the chosen number of time steps in the lattice. The value for the strike is set as $K = K_{\text{ref}}$.

## 11.4 Pricing Barrier Options

The procedure for pricing barrier options is almost identical to what is formulated in Section 10.4 of the previous project. One important distinction arises, however, when using a drifted lattice (as is the case in the current CN approach) versus a nondrifted lattice. The differences that arise between the use of drifted and nondrifted lattices were also briefly discussed in the previous project, where the nondrifted lattice was favored over the use of drifted lattices when pricing options with a constant barrier level. Within the CN drifted-lattice approach, the price of a up-and-out barrier call, for example, with barrier at $S = H$, requires one to employ the pricing procedure as given in Section 11.2. At each time $t_m$, however, the price components $f_n^m$ must be reset to zero for all $n \geq n_H$ (i.e., all nodes at and above the barrier level $H$) before the next propagation time step. The integer $n_H$ can be taken to be the least integer value of $n$ such that $S_n^m \geq H$. Figure 11.1 demonstrates a possible source of inaccuracy arising from the use of a drifted-lattice geometry when pricing a barrier option, with barrier level at a fixed height. The zero-boundary conditions imposed on the "boundary" nodes creates only a coarse approximation to the actual horizontal straight-line barrier. Note

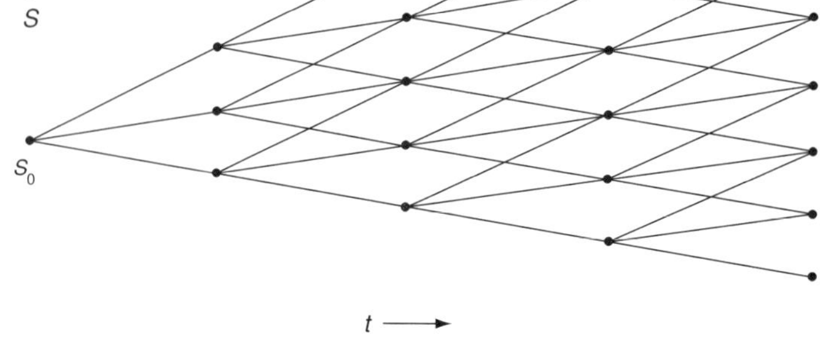


FIGURE 11.1 A drifted trinomial lattice used to price a barrier option. The barrier level lies along a horizontal line, which is inaccurately approximated by the zero-boundary nodes.

that as one makes the time step smaller and smaller, this approximation becomes more and more accurate. In the limit $\Delta t \to 0$, this approximation becomes exact.

The pricing of down-and-out options is similar, while the pricing of knock-in options reduces to that of knockouts, thanks to the in-out symmetry relation of equation (10.26). The reader may note that the spreadsheet can also be readily extended to include the pricing of American barrier options, if desired.

# CHAPTER · 12

# Project: Static Hedging of Barrier Options

The objective of this study is to hedge European barrier options by means of a static replication strategy involving a market-restricted set of available plain-vanilla European call and put options. The hedge trade occurs at the initial time and is unwound either at maturity or when the barrier is crossed.

*Worksheets:* **bhedge**

*Required Libraries:* **MFioxl, MFBlas, MFFuncs, MFLapack, MFStat, MFCollection**

## 12.1 Analytical Pricing Formulas for Barrier Options

We consider four flavors of single-barrier options: (1) down-and-out, (2) up-and-out, (3) down-and-in, (4) up-and-in. Each option can be either a call or a put, for a total of eight different types of contracts.

### 12.1.1 *Exact Formulas for Barrier Calls for the Case $H \leq K$*

Let us recall from Section 3.3 the pricing formulas for barrier options in the geometric Brownian motion model. The European down-and-out call option has nonzero value only for $S > H$:

$$C^{\mathrm{DO}}(S, K, T-t) = C(S, K, T-t) - (S/H)^{(1-k)} C(H^2/S, K, T-t), \tag{12.1}$$

with $k \equiv r/(\frac{1}{2}\sigma^2)$. This shows that the barrier option value at spot $S > H$ can be expressed in terms of the plain-vanilla call evaluated at effective spot values of $S$ and $H^2/S$. The corresponding down-and-in call option value is then

$$C^{\mathrm{DI}}(S, K, T-t) = C(S, K, T-t) - C^{\mathrm{DO}}(S, K, T-t). \tag{12.2}$$

The formula for the value of the call $C(S, K, T-t)$ is given by the plain Black–Scholes formula. Using it gives [i.e., equation (3.52)]:

$$C^{DO}(S, K, \tau) = SN(d_1(S/K)) - Ke^{-r\tau}N(d_2(S/K)) - S(H/S)^{k+1}N(d_1(H^2/SK)) + Ke^{-r\tau}(H/S)^{k-1}N(d_2(H^2/SK)), \tag{12.3}$$

where $d_1(x)$ and $d_2(x)$ are defined as

$$d_1(x) = \frac{\log x + (r + \frac{1}{2}\sigma^2)\tau}{\sigma\sqrt{\tau}} = \frac{\log x}{\sigma\sqrt{\tau}} + \tfrac{1}{2}(k+1)\sigma\sqrt{\tau}, \tag{12.4}$$

$$d_2(x) = d_1(x) - \sigma\sqrt{\tau}. \tag{12.5}$$

Note that we have reexpressed the formulas in terms of the time to maturity $\tau \equiv T - t$. As well, for clarity the obvious dependence on $k$ and $\sigma\sqrt{\tau}$ within the functions $d_1$ and $d_2$ is not written explicitly. Note that the down-and-in call option $C^{DI}(S, K, \tau)$ expressed in terms of cumulative normal distribution functions is just the negative of the sum of the last two terms in equation (12.3).

In contrast to the down-and-out call, the up-and-out call option is defined to have nonzero value for values $S < H$ and also has the same pay-off, namely, that of the plain call struck at $K$. The European up-and-out call option is zero for all spot values in the case $H \leq K$, i.e., $C^{UO} = 0$. This follows since the asset price $S$ must be below the barrier, $S < H$, for nonzero values of the option. However, the pay-off is that of a call struck at $K$, where $K \geq H$, which always gives a pay-off of zero, hence $C^{UO} = 0$. Then from in-out symmetry we immediately have $C^{UI}(S, K, \tau) = C(S, K, \tau)$.

### 12.1.2 Exact Formulas for Barrier Calls for the Case $H \geq K$

For a European down-and-out call option value we have [i.e., equation (3.51)]:

$$C^{DO}(S, K, \tau) = SN(d_1(S/H)) - Ke^{-r\tau}N(d_2(S/H)) - S(H/S)^{k+1}N(d_1(H/S)) + Ke^{-r\tau}(H/S)^{k-1}N(d_2(H/S))], \tag{12.6}$$

and from symmetry the corresponding down-and-in call has value

$$C^{DI}(S, K, \tau) = C(S, K, \tau) - C^{DO}(S, K, \tau). \tag{12.7}$$

The European up-and-in call option now has value [i.e., equation (3.62)]:

$$C^{UI}(S, K, \tau) = SN(d_1(S/H)) - Ke^{-r\tau}N(d_2(S/H)) - S(H/S)^{k+1}[N(-d_1(H^2/SK)) - N(-d_1(H/S))] + Ke^{-r\tau}(H/S)^{k-1}[N(-d_2(H^2/SK)) - N(-d_2(H/S))], \tag{12.8}$$

and from symmetry the corresponding up-and-out call has value

$$C^{UO}(S, K, \tau) = C(S, K, \tau) - C^{UI}(S, K, \tau). \tag{12.9}$$

### 12.1.3 Exact Formulas for Barrier Puts for the Case $H \leq K$

All four cases of put barrier options are defined in the same fashion as their corresponding call barrier options, except the payoff function is that of the plain-vanilla put rather than the call.

A European up-and-out put option with $H \leq K$ has price [i.e., equation (3.57)]:

$$\begin{aligned} P^{\mathrm{UO}}(S, K, \tau) = & -SN(-d_1(S/H)) + Ke^{-r\tau}N(-d_2(S/H)) \\ & + S(H/S)^{k+1}N(-d_1(H/S)) \\ & - Ke^{-r\tau}(H/S)^{k-1}N(-d_2(H/S)), \end{aligned} \tag{12.10}$$

and symmetry gives the up-and-in put in terms of the plain-vanilla put price $P(S, K, \tau)$,

$$P^{\mathrm{UI}}(S, K, \tau) = P(S, K, \tau) - P^{\mathrm{UO}}(S, K, \tau), \tag{12.11}$$

where

$$P(S, K, \tau) = -SN(-d_1(S/K)) + Ke^{-r\tau}N(-d_2(S/K)). \tag{12.12}$$

The down-and-in put price takes the form [i.e., equation (3.56)]:

$$\begin{aligned} P^{\mathrm{DI}}(S, K, \tau) = & -SN(-d_1(S/H)) + Ke^{-r\tau}N(-d_2(S/H)) \\ & + S(H/S)^{k+1}[N(d_1(H^2/SK)) - N(d_1(H/S))] \\ & - Ke^{-r\tau}(H/S)^{k-1}[N(d_2(H^2/SK)) - N(d_2(H/S))], \end{aligned} \tag{12.13}$$

and symmetry gives the down-and-out put,

$$P^{\mathrm{DO}}(S, K, \tau) = P(S, K, \tau) - P^{\mathrm{DI}}(S, K, \tau). \tag{12.14}$$

### 12.1.4 Exact Formulas for Barrier Puts for the Case $H \geq K$

The European down-and-out put $P^{\mathrm{DO}}(S, K, \tau) = 0$. This result is obtained since for any $S$ value below the barrier $H$ we have $P^{\mathrm{DO}} = 0$. The pay-off is also zero unless $S < K$, in which case $S < H$, giving $P^{\mathrm{DO}} = 0$. Hence $P^{\mathrm{DO}} = 0$ for all spot values in the allowed range $S \geq H$. The down-and-in put follows from usual symmetry,

$$P^{\mathrm{DI}}(S, K, \tau) = P(S, K, \tau). \tag{12.15}$$

The up-and-in put price for $H \geq K$ is given by [from equation (3.58)]

$$\begin{aligned} P^{\mathrm{UI}}(S, K, \tau) = & -S(H/S)^{k+1}N(-d_1(H^2/SK)) \\ & + Ke^{-r\tau}(H/S)^{k-1}N(-d_2(H^2/SK)), \end{aligned} \tag{12.16}$$

with symmetry giving

$$P^{\mathrm{UO}}(S, K, \tau) = P(S, K, \tau) - P^{\mathrm{UI}}(S, K, \tau). \tag{12.17}$$

## 12.2 Replication of Up-and-Out Barrier Options

Let us consider in detail the problem of replicating a European up-and-out call option $C^{UO}$ struck at $K$ with barrier $H$ and maturing at time $T$. The underlying theory of replication tells us that if two portfolios have equal value along the boundaries in $S$, $t$ space, then their worth is also the same at all points that are interior to the boundaries. For the call barrier in question, the payoff function gives us a choice of boundary, with $S = H$ defining a line of constant $S$ joining another line at the point $(H, T)$ given by $t = T$. The $t = T$ boundary gives

$$C^{UO}(S, K, \tau = 0) = \max(S - K, 0), \qquad S < H, \tag{12.18}$$

while the $S = H$ boundary gives

$$C^{UO}(S = H, K, \tau) = 0, \qquad \text{for all } \tau. \tag{12.19}$$

The other obvious boundary is at $S = 0$, where all is zero and any portfolio consisting of a linear combination of vanilla calls will automatically match this value.

We now consider replicating the up-and-out call with a portfolio consisting of a linear combination of plain calls:

$$\Pi(S, t) = C(S, K, T - t) + \sum_{i=0}^{N-1} a_i \Theta(T_i - t) C(S, K_i, T_i - t). \tag{12.20}$$

Note that we are using a notation that makes explicit the maturities $T_i$ of the various options. The first term is just a call struck at $K$ with the same maturity $T$ as the barrier option. The second term involves a linear combination of positions $a_i$ in the plain calls struck at $T_i$, where $t < T_i \leq T$. Note also the use of the Heaviside step function defined by $\Theta(x) = 1$ for $x \geq 0$ and $\Theta(x) = 0$ for $x < 0$. This function is used explicitly to emphasize that any option is set to zero value for any negative value of time to maturity $T_i - t$ i.e., the particular option becomes excluded from the hedge portfolio. We also make the choice $T_0 = T$, so one of the calls with position $a_0$ in the sum also has the same maturity as the barrier option. This call, as well as all other calls in the sum, is meant to have strike $K_i \approx H$. That is, the strikes $K_i$ are not necessarily set exactly equal to the barrier level $H$. This is meant to simulate a more realistic situation in which a trader does not have all strikes with given maturity always available. This situation arises in the **bhedge** spreadsheet application and is captured by use of the input field corresponding to the "precentage away from barrier." Let us denote this quantity by $p^H$, where $0\% \leq p^H \leq 100\%$. Then for every $i$th available maturity date $T_i$, the hedge should proceed to include the strike $K_i$ closest to $H$ for which $|K_i - H|/H \leq p^H$, if any such strikes are available. If none are available, then such a term is eliminated from the sum, giving fewer contract terms available for the replication and, hence, for the hedge.

The problem is then reduced to finding the $N$ constants $a_i$ that give us the hedge positions to be shorted (i.e., the $a_i$ are actually the negative positions for the hedge portfolio). For a replication at today's time we interpret $t$ as current calendar time and $S$ as the spot, with $\Pi(S, t)$ in equation (12.20) being the value of our approximate replicating portfolio consisting of $a_i$ positions in each call maturing at $T_i$. This problem is formulated in a precise manner as follows. We consider a sequence of $M$ increasing times: $t_0 < t_1 < \cdots < t_{M-1}$ with $t_0 \geq t$ and $t_{M-1} < T$. For convenience we pick these $t_\alpha$ to be equally spaced. Moreover, we have the extra freedom of generally using more time slices than maturities $T_i$; i.e., $M \geq N$. Note that the times $t_\alpha$ will not necessarily coincide with the maturities $T_i$. See Figure 12.1 for a schematic of the replication strategy.

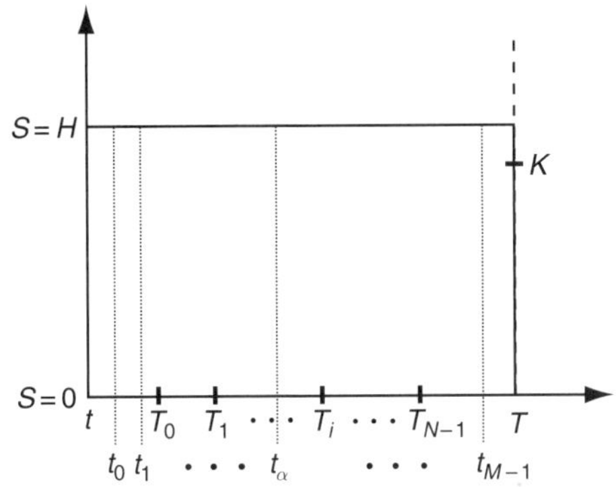


FIGURE 12.1 The replication strategy for an up-and-out European option with upper barrier at stock level $S = H$.

We then match $\Pi(H, t) = C^{\mathrm{UO}}(H, K, T - t) = 0$ at each $t = t_\alpha$ according to equation (12.19), which upon using equation (12.20) leads to the generally overdetermined linear system of $M$ equations in the $N$ unknowns $a_i$. In matrix form,

$$\sum_{i=0}^{N-1} A_{\alpha i} a_i = b_\alpha, \qquad \alpha = 0, 1, \ldots, M-1, \tag{12.21}$$

where the $M \times N$ matrix with elements $A_{\alpha i}$ is given by

$$A_{\alpha i} = \Theta(T_i - t_\alpha) C(H, K_i, T_i - t_\alpha) \tag{12.22}$$

and

$$b_\alpha = -C(H, K, T - t_\alpha). \tag{12.23}$$

This system is solved numerically by finding the minimum-norm solution via a singular value decomposition approach. The linear algebra library MFLapack is useful for this purpose. The solution vector is $\mathbf{a} \equiv (a_0, a_1, \ldots, a_{M-1})$. The replicating portfolio has value $\eta\Pi(S, t)$, where $\eta$ denotes the total position (number of purchased contracts) in the barrier option $C^{\mathrm{UO}}$. The total position that is shorted in each replicating $i$th call is then $\eta a_i$. The exact (i.e., target) portfolio value $\eta C^{\mathrm{UO}}$ is plotted against the result $\eta\Pi$, and one observes the mismatch between the two smooth curves as a function of $S$ at today's time $t$. The range of $S$ should be chosen judiciously within $S_{\min} \le S \le S_{\max}$, where $S_{\max} = H$ and $S_{\min}$ is considerably less than $K$ but greater than zero. Note that for the case $H \ge K$ one uses equation (12.9) for $C^{\mathrm{UO}}$, and for $H < K$ one simply has $C^{\mathrm{UO}} = 0$.

Typical results for the **bhedge** spreadsheet should indicate two smooth curves for the exact and approximate values of the portfolios as a function of spot $S$. Agreement should be overall quite good, within less than 5% for most points, and with maximum observed deviations of about only 10%. One can also experiment with increasing the number of equations (i.e., the number $M$ of time slices). As one increases this number past $N$, the results should not change in any noticeable way. Hence, the solution is provided for any $M \ge N$ choice. Figure 12.2 shows a replication of an up-and-out European call while allowing the possible strikes $K_i$ to deviate within 10% at most from the barrier;

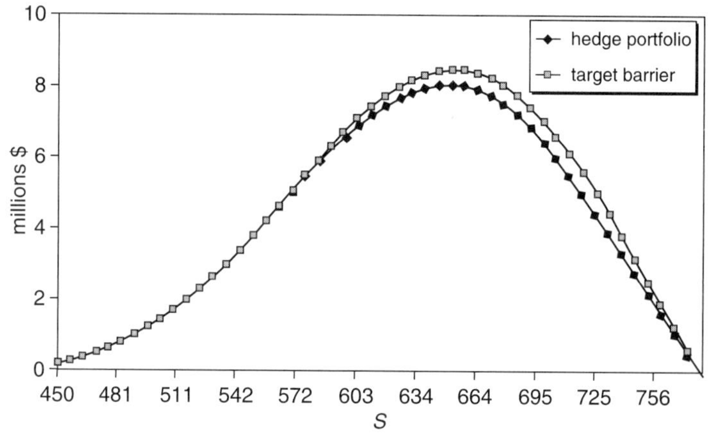


FIGURE 12.2 Comparison of an actual replication against the exact value for an up-and-out call option with upper barrier at $H = 780$, strike $K = 550$, interest rate $r = 7\%$, current date $t =$ September 1, 1999, and maturity date $T =$ April 15, 2000.

i.e., $p^H = 10\%$. The number of contracts is $\eta = 100{,}000$. The replicating portfolio consists of twelve plain-vanilla calls with positions: 100,000; 15,026.21; 11,953; 26,377.41; 42,179.29; 27,038.67; 105,598.27; 160,870.81; 67,371.38; 412,469.94; 1,495,103.62; −2,595,565.28, strikes: 550; 780; 775; 780; 780; 775; 780; 780; 780; 780; 780; 780, and maturities: 15-Apr-2000; 1-Oct-1999; 15-Oct-1999; 1-Nov-1999; 1-Dec-1999; 15-Dec-1999; 15-Jan-2000; 15-Feb-2000; 1-Mar-2000; 15-Mar-2000; 1-Apr-2000; 15-Apr-2000, respectively. The number of time slices was chosen to be $M = 20$. In practice, the computed positions can be rounded to the nearest integer multiple of 100. For example, 15,026.21 may be rounded to 15,000 (i.e., 150 lots of contract size 100).

This completes the discussion and static hedge implementation for up-and-out barrier calls. For the case of up-and-out barrier puts, the implementation is very similar, except for an important difference. The replication is again accomplished via a linear combination in vanilla calls, except the first term is a position in a put (rather than a call) struck at $K$. From put-call parity one also sees that the put can also be replaced by a call plus a position in cash (or bond) and a stock position. For the up-and-out put option with barrier at $H$, the analogous replication formula as in equation (12.20) is now

$$\Pi(S, t) = P(S, K, T - t) + \sum_{i=0}^{N-1} a_i \Theta(T_i - t) C(S, K_i, T_i - t). \tag{12.24}$$

One then solves a matrix equation of the same form as equation (12.21), where the $M \times N$ matrix with elements $A_{\alpha i}$ is again given by equation (12.22), but now

$$b_\alpha = -P(H, K, T - t_\alpha). \tag{12.25}$$

Note that the boundary condition along the line $S = 0$, all $t \leq T$, is automatically satisfied for any choice of the $a_i$ since the calls all have zero value at $S = 0$. This gives

$$\Pi(S = 0, t) = P(S = 0, K, T - t) = Ke^{-r(T-t)}, \tag{12.26}$$

which must be the case for the up-and-out put option value when $S = 0$.

## 12.3 Replication of Down-and-Out Barrier Options

The preceding section treats up-and-out type of knockout options. To replicate down-and-out options, the treatment is similar, except the expansion term in the $a_i$ is now done in puts rather than calls. This is dictated by the difference in the boundary conditions. That is, the zero-boundary conditions at $S = H$ and at the $t = T$ lines are the same, but the boundary at the $S = 0$ line is now replaced by the boundary condition at $S \to \infty$. Any linear combination of puts will give a zero-boundary condition. This latter boundary condition is convenient when expanding in puts as used here. Figure 12.3 gives a schematic of the barrier replication for down-and-out European options.

In particular, for a down-and-out call option we can consider replication using

$$\Pi(S, t) = C(S, K, T - t) + \sum_{i=0}^{N-1} a_i \Theta(T_i - t) P(S, K_i, T_i - t). \tag{12.27}$$

One then solves a matrix equation of the same form as equation (12.21), where the $M \times N$ matrix with elements $A_{\alpha i}$ is now given by

$$A_{\alpha i} = \Theta(T_i - t_\alpha) P(H, K_i, T_i - t_\alpha) \tag{12.28}$$

and

$$b_\alpha = -C(H, K, T - t_\alpha). \tag{12.29}$$

Note that the boundary condition $\Pi(S, t) \to C(S, K, T - t)$ as $S \to \infty$ is then automatically satisfied, as required.

The case of the down-and-out put option is then handled via the portfolio in puts:

$$\Pi(S, t) = P(S, K, T - t) + \sum_{i=0}^{N-1} a_i \Theta(T_i - t) P(S, K_i, T_i - t). \tag{12.30}$$

One then solves a matrix equation of the same form, where the $M \times N$ matrix with elements $A_{\alpha i}$ is exactly as in equation (12.28), and the coefficients $b_\alpha$ are now

$$b_\alpha = -P(H, K, T - t_\alpha). \tag{12.31}$$

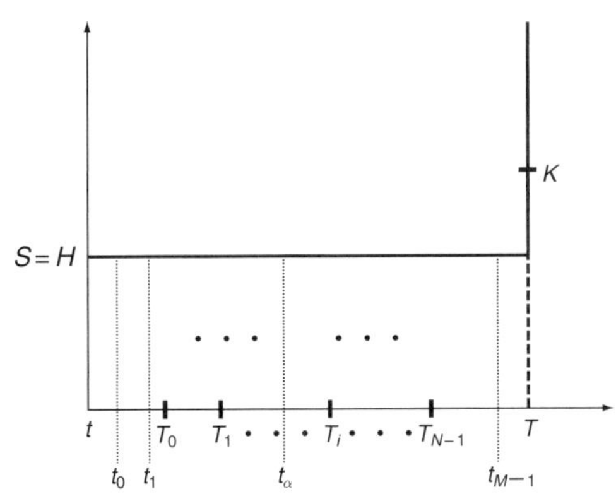


FIGURE 12.3 The replication strategy for a down-and-out European option with lower barrier at stock level $S = H$.

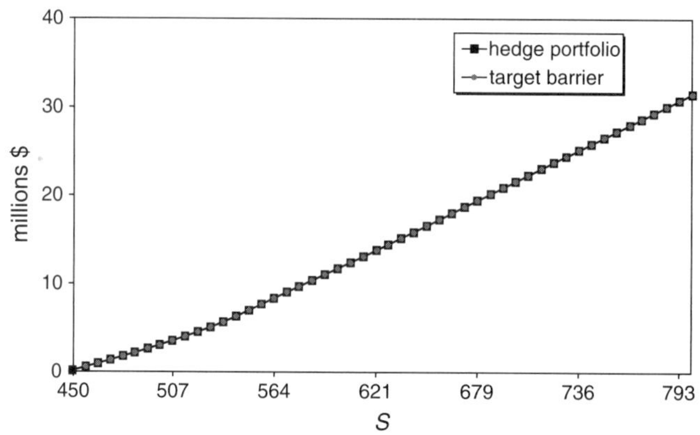


FIGURE 12.4 Comparison of an actual replication against the exact value for a down-and-out call option with upper barrier at $H = 450$, strike $K = 500$, interest rate $r = 7\%$, current date $t =$ September 1, 1999, and maturity date $T =$ March 1, 2000.

The correct boundary condition $\Pi(S, t) \to P(S, K, T - t) \to 0$, as $S \to \infty$ is also automatically satisfied.

Figure 12.4 shows a replication of a down-and-out European call when only allowing for strikes $K_i$ to match exactly the barrier level; i.e., $p^H = 0\%$. The nominal amount of such barrier contracts is $\eta = 100{,}000$. The replicating portfolio consists of five plain-vanilla puts with positions: 100,000; −39,151.60; −36,764.17; −2,047.41; −129.29, strikes: 500; 450; 450; 450; 450, and maturities: 1-Mar-2000; 15-Oct-1999; 1-Jan-2000; 1-Feb-2000; 1-Mar-2000, respectively. The number of time slices was again chosen to be $M = 20$. Note that the replication is relatively more accurate for the down-and-out versus the up-and-out call.

The preceding four replication strategies take care of all possible single-barrier European calls or puts, since the corresponding knock-in option values follow in a trivial manner from the aforementioned knock-in–knockout symmetry relationship.

CHAPTER • 13

# Project: Variance Swaps

Variance swaps are hedged by a combination of a dynamic and a static hedging strategy. The static part involves a replication of a logarithmic payoff function. The objective of this study is to construct the logarithmic payoff replication and hence to find hedge ratios for the static part of the strategy. This can be achieved by combining positions in calls (or puts), stocks, and bonds.

*Worksheets:* **varswaps**

*Required Libraries:* **MFioxl, MFBlas, MFFuncs, MFLapack, MFCollection**

## 13.1 The Logarithmic Pay-Off

A variance swap is a forward contract on an annualized variance or the square of the realized volatility. The payoff $\phi$ at final expiry time $T$ is given by

$$\phi = \mathcal{N} \times (\sigma_R^2 - K_{\text{var}}), \tag{13.1}$$

where $K_{\text{var}}$ is a fixed swap rate (i.e., the variance swap rate), $\mathcal{N}$ is the notional amount of the swap in dollars per annualized volatility point squared, and $\sigma_R^2$ is the realized variance (in annual terms) of an underlying market observable over the life of the contract. The underlying can be a stock price, a futures price, an index, etc. In Chapter 1, we discussed such a contract in detail where the underlying was chosen as a futures price and we showed how to (i) derive a fair value for the swap rate and (ii) replicate the realized variance in terms of a trading strategy involving a dynamic and a static component. In this project the goal is only to replicate the static component of the contract. Moreover, we shall assume that the underlying asset is a stock with price $S_t$ at time $t$. Two definitions of the historically realized variance are possible, depending on whether we use log-returns, in which case it is defined by

$$\frac{1}{n}\sum_{i=1}^{n} \log\left(\frac{S_i}{S_{i-1}}\right), \tag{13.2}$$

or arithmetic returns, in which case it is defined by

$$\frac{1}{n}\sum_{i=1}^{n}\frac{S_i - S_{i-1}}{S_{i-1}}. \tag{13.3}$$

The $S_i$ are quoted stock prices at interval times indexed by $i$. Note that if these are taken as daily closing prices, then one must convert the variance into a per annum basis (i.e., in terms of annualized variance). An example of a variance swap contract is a notional amount of $\mathcal{N} = \$100{,}000/(\text{one volatility point})^2$, with delivery swap rate of $K_{\text{var}} = (15\%)^2$ per annum on the S&P 500 daily closing index and maturity of 1 year.

A simple mathematical model can be constructed on the assumption that the stock price follows a diffusion process with stochastic time-dependent volatility $\sigma_t$ and constant drift (given by the interest rate $r$ within an assumed risk-neutral measure):

$$\frac{dS_t}{S_t} = r\,dt + \sigma_t\,dW_t. \tag{13.4}$$

The realized variance defining the pay-off is assumed to be given by the stochastic integral

$$\sigma_R^2 = \frac{1}{T}\int_0^T \sigma_t^2\,dt. \tag{13.5}$$

Let $F_t = e^{r(T-t)}S_t$ be the forward-price process. It follows from equation (13.4) that

$$\frac{dF_t}{F_t} = \sigma_t\,dW_t. \tag{13.6}$$

Ito's lemma gives

$$d\log F_t = \frac{dF_t}{F_t} - \frac{\sigma_t^2}{2}dt. \tag{13.7}$$

Hence, integrating gives

$$\sigma_R^2 = \frac{2}{T}\int_0^T \left(\frac{dF_t}{F_t} - d\log F_t\right) = -\frac{2}{T}\log\frac{F_T}{F_0} + \frac{2}{T}\int_0^T \frac{dF_t}{F_t}. \tag{13.8}$$

The first term of the realized variance (i.e., the logarithmic function) can be replicated statically, while the second term can be replicated dynamically by means of a self-financing strategy. In this project we shall only concern ourselves with replication of the static component.

## 13.2 Static Hedging: Replication of a Logarithmic Pay-Off

Logarithmic contracts are synthetic and, as such, are not traded directly, but they can be approximately replicated by means of portfolios in standard call or put options. Consider the following logarithmic payoff function:

$$f(S_T) = -\frac{2}{T}\log\left(\frac{S_T}{S_0}\right). \tag{13.9}$$

In practice, only a limited set of strikes is available for trading. In the **varswaps** worksheet, all available call options are represented in a table. Each column corresponds to a given maturity date and contains all the possible strikes assumed available for trading (for that given maturity date). The problem is to approximately replicate the pay-off in equation (13.9) by weighting positions in calls corresponding to only the available strikes as well as allowing for a variable cash (or bond) and stock position.

Consider the finite expansion

$$f(S_T) \sim w_{-1} + w_0 S_T + \sum_{i=1}^{N} w_i \max(S_T - K_i, 0). \tag{13.10}$$

The coefficient $w_{-1}$ gives the (dollar) cash position, while the coefficient $w_0$ gives the stock position, and the coefficients $w_i$ give the positions in the calls struck at values $K_i$ (i.e., the select set of calls with such strikes that are actually available for trading with given maturity). The goal is to find the positions $w_i$ providing the best fit in the least squares sense for the log-payoff on the left-hand side. More precisely, we determine the $N+2$ weights $w_i$ by matching approximate payoff function (13.10) with the exact logarithmic payoff function at $M$ number of points in the final stock price $S_T$: $S_T^1, S_T^2, \ldots, S_T^M$, where $M \geq N+2$. This leads to the linear system of $M$ equations in the $N+2$ unknown weights $w_i$:

$$f(S_T^j) \sim w_{-1} + w_0 S_T^j + \sum_{i=1}^{N} w_i \max(S_T^j - K_i, 0), \; j = 1, \ldots, M. \tag{13.11}$$

In matrix form this system is

$$\sum_{i=-1}^{N} A_{j,i} w_i = b_j, \qquad j = 1, \ldots, M, \tag{13.12}$$

where

$$A_{j,-1} = 1(\text{cash}) \qquad \text{or} \qquad A_{j,-1} = e^{rT}(\text{bond}), \tag{13.13}$$

$$A_{j,0} = S_T^j, \tag{13.14}$$

$$A_{j,i} = \max(S_T^j - K_i, 0); \qquad i \geq 1, \tag{13.15}$$

and

$$b_j = f(S_T^j) = -\frac{2}{T} \log\left(\frac{S_T^j}{S_0}\right). \tag{13.16}$$

The points $S_T^j$ are chosen so they are equally spaced (although they can also be unequally spaced) and the spot $S_0$ lies near the middle of the price range $[S_T^1, S_T^M]$. The system of equations (13.12) is solved numerically by finding the minimum-norm solution via, for example, a singular value decomposition of the matrix of elements $A_{i,j}$. The linear algebra numerical library MFLapack is useful for this purpose. This gives the required solution vector of all weights: $\mathbf{w} = (w_{-1}, w_0, w_1, \ldots, w_N)$. Note that when these weights are multiplied by the nominal position, we refer to them as the *hedge ratios*. This gives us the approximate replicating portfolio, with the pay-off approximating the target pay-off in equation (13.9). One can plot the exact target function $f(S_T)$ alongside the approximate function given by equation (13.10) as functions of final stock price $S_T$ in an appropriate range $[S_{\min}, S_{\max}]$, with $S_{\min} = S_T^1$ and $S_{\max} = S_T^M$. Typically, when using on the order of only five different strikes, one should observe fairly good agreement across all stock prices (i.e., 1–5% relative error). Also, the results should display the approximate payoff function as being greater than or equal to the target payoff function for all points in $S_T$. An example of an actual calculation is displayed in Figure 13.1. There the comparison is for a logarithmic pay-off with

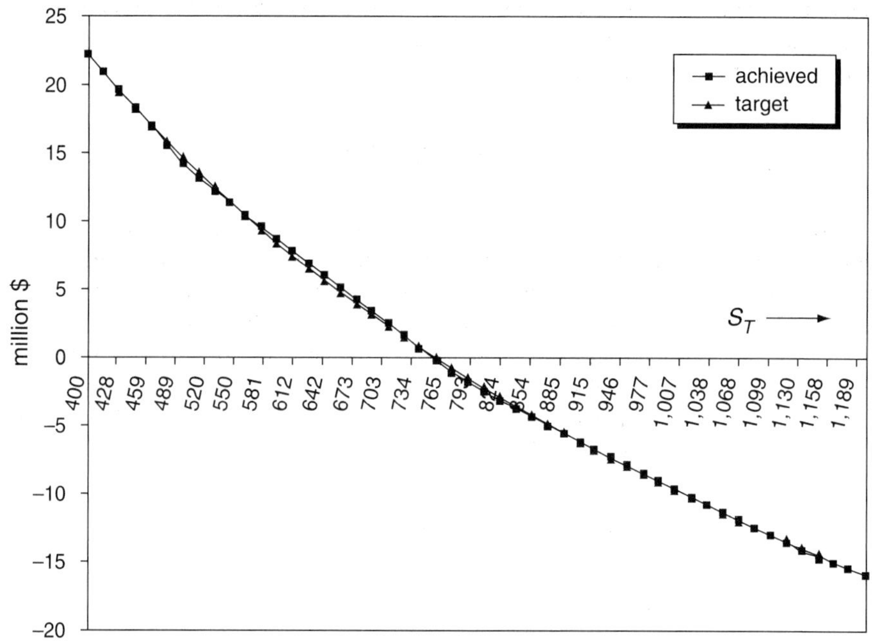


FIGURE 13.1 A comparison of an actual logarithmic pay-off with a replicating portfolio achieved using a cash and stock position as well as positions in only five available calls at strikes $K_1 = 343.04$, $K_2 = 505.10$, $K_3 = 783.52$, $K_4 = 1{,}137.93$, $K_5 = 876.90$.

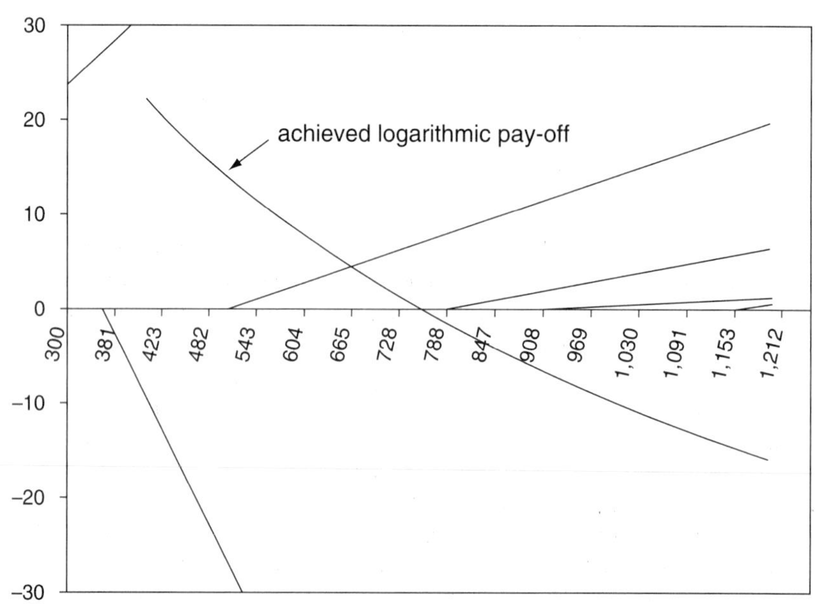


FIGURE 13.2 Decomposition of the achieved logarithmic payoff function of Figure 13.1 as a sum of pay-offs corresponding to positions in cash, stock, and one short and four long positions in calls at strikes $K_1 = 343.04$, $K_2 = 505.10$, $K_3 = 783.52$, $K_4 = 1{,}137.93$, $K_5 = 876.90$.

spot $S_0 = 758$, where the set of five strikes used are the ones corresponding to the available call contracts with time to maturity of $T = 21$ days. The pay-off has been rescaled by a notional amount of \$1 million per volatility point squared. Rapid convergence is observed with the use of only $M = 7$ stock price slices, although the plot shown is with $M = 50$. Figure 13.2 shows the decomposition of the replicating portfolio for the achieved payoff curve of Figure 13.1. The positions are: $w_{-1} = 699.84$, $w_0 = 78{,}673.12$, $w_1 = -161{,}399.18$ (short position), $w_2 = 28{,}390.46$, $w_3 = 15{,}552.98$, $w_4 = 9{,}429.72$, $w_5 = 3{,}717.31$. Examples of other detailed replications are found on the varswaps spreadsheet.

:HAPTER • 14

# ʾroject: Monte Carlo Value-at-Risk ·or Delta-Gamma Portfolios

The objective of this project is to develop multivariate Monte Carlo simulation procedures for computing the probability density function for the delta-gamma change in portfolio as well as to estimate portfolio value-at-risk (VaR) for a given confidence level. Two different probability distributions for the risk-factor returns are considered: multivariate normal and multivariate Student t-distributions. This project allows the reader to explore some of the differences between the use of a normal distribution and a heavy-tailed distribution model when computing VaR. This project also constitutes a template for future analysis of delta-gamma portfolio VaR under more general heavy-tailed distributions via the implementation of t-Copula methods. Such methods allow one to compute VaR for distributions where the returns can possess different degrees of freedom for different risk factors.

*Worksheet:* **var**

*Required Libraries:* **MFioxl, MFBlas, MFLapack, MFRangen, MFStat, MFSort**

## 4.1 Multivariate Normal Distribution

Let $\Phi(V)$ be the cumulative distribution function for the P&L of a portfolio. Let $P(V) =$ the probability that the change in portfolio value, denoted by $\Delta V$, is less than or equal to a value $V$. Postulating a multivariate distribution $p(\mathbf{r})$ for the returns, we have [see equation (4.72) of Chapter 4]:

$$\Phi(V) = \mathrm{P}(\Delta V \leq V) = \int_{\mathbb{R}^n} p(\mathbf{r})\Theta(V - \Delta V(\mathbf{r}))d\mathbf{r}. \tag{14.1}$$

Here $\mathbf{r}$ denotes the vector of returns $\mathbf{r}^T = (r_1, \ldots, r_n)$ and the integration is over the complete $n$-dimensional space of all risk-factor returns. [Note: Here we use superscript $T$ for the transpose of a matrix (or vector); e.g., $\mathbf{r}$ is $n \times 1$ and $\mathbf{r}^T$ is $1 \times n$.] The function $\Theta(x)$ is the Heaviside step function. Let us consider the case of a multivariate (Gaussian) normal probability distribution function in the space of the risk-factor returns

$$p^G(\mathbf{r}) = \frac{1}{\sqrt{(2\pi)^n|\mathbf{C}|}} \exp\left(-\tfrac{1}{2}\mathbf{r}^T\mathbf{C}^{-1}\mathbf{r}\right), \tag{14.2}$$

where $\mathbf{C}$ is the $n \times n$ covariance matrix and $|\mathbf{C}|$ is its determinant. The covariance matrix is assumed already given in this project (i.e., generated at random), although it can be readily computed from the returns corresponding to a risk-factor time series.

The value of the portfolio is assumed to be a function of $n$ risk factors denoted by $x_1, \ldots, x_n$. We shall denote a change in risk factors by the vector $d\mathbf{x}$, hence giving the return for the $i$th factor as $r_i = dx_i/x_i$. Given a change in risk factors, the change in value of the portfolio within the delta-gamma approximation [neglecting the $\Theta$ term in equation (4.27) which is trivial to include] is assumed to be given by the second-order Taylor expansion:

$$\Delta V(\mathbf{r}) = \mathbf{r}^T\mathbf{\Delta} + \tfrac{1}{2}\mathbf{r}^T\mathbf{\Gamma}\mathbf{r}. \tag{14.3}$$

The $n$-dimensional delta vector $\mathbf{\Delta}$ has components

$$\Delta_i = x_i \frac{\partial V}{\partial x_i}, \tag{14.4}$$

and the $n \times n$ gamma matrix has elements

$$\Gamma_{ij} = x_i x_j \frac{\partial^2 V}{\partial x_i \partial x_j}. \tag{14.5}$$

The first step is to generate a random delta vector, gamma matrix, and covariance matrix. This is the functionality of the randomize button on the **var** spreadsheet. In essence, one is fabricating sensitivities for a fictitious portfolio.[1] The gamma matrix must have the property that it is symmetric, $\Gamma^T = \Gamma$. The covariance matrix must be symmetric positive-definite. Based on these greeks and the covariance matrix, one then computes VaR and P&L using a plain Monte Carlo technique as follows.

To implement a plain Monte Carlo algorithm without invoking any additional variance reduction approaches, one begins by performing a Cholesky factorization of the covariance matrix

$$\mathbf{C} = \mathbf{U}^T\mathbf{U}. \tag{14.6}$$

This factorization is done only once, at the beginning of a simulation. The numerical library class MFLapack is useful for this purpose. Scenarios can then be generated in a two-step procedure. First, one samples vectors of independent standard normals $\mathbf{y}^{(i)}$, with components drawn independently from the standard normal distribution, $y_k^{(i)} \sim N(0, 1), k = 1, \ldots, n$. The vectors $\mathbf{y}^{(i)}$, $i = 1, \ldots, N_s$, represent intermediate $i$th scenarios. The random-number library class MFRangen is useful for this purpose. In the second step, the vectors $\mathbf{y}^{(i)}$ are transformed into actual scenario vectors for the correlated returns using

$$\mathbf{r}^{(i)} = \mathbf{U}^T\mathbf{y}^{(i)}. \tag{14.7}$$

[1] However, the user can also run a VaR simulation by inputting the values $\Delta_i$ and $\Gamma_{ij}$ precomputed for an actual portfolio with position $(\theta_1, \ldots, \theta_N)$ in $N$ assets or subportfolios (see Sections 4.2.1 and 4.2.2).

These two steps are repeated $N_s$ times, where $N_s$ is the total number of scenarios in the simulation. The scenarios are then distributed according to $\mathbf{r}^{(i)} \sim N(\mathbf{0}, \mathbf{C})$. For each scenario vector $\mathbf{r}^{(i)}$ we evaluate the portfolio variation, $\Delta V^{(i)} = \Delta V(\mathbf{r} = \mathbf{r}^{(i)})$, as given by equation (14.3). Once the portfolio variations under all scenarios are obtained, the outcomes are sorted in increasing order, where $\Delta V^{(i+1)} \geq \Delta V^{(i)}$. The MFSort library is useful for this purpose.

Given the sorted portfolio variations $\Delta V^{(i)}$ that were generated from the scenarios, the value-at-risk (VaR), defined by the probability

$$P\big(\Delta V \leq -\text{VaR}\big) = p, \tag{14.8}$$

where $p = 1 - \alpha$ and $\alpha$ is the confidence level (typically $\alpha = 95\%$ to $99\%$), is then estimated as $\text{VaR} = -\Delta V^{([[pN_s]])}$. Here $[[x]]$ denotes the integer part of a number $x$.

## 14.2 Multivariate Student t-Distributions

A popular model that introduces fat tails in the returns is the multivariate Student t-distribution with pdf,

$$p_\nu(\mathbf{r}; \mathbf{C}) = \frac{\Gamma((\nu+n)/2)}{\Gamma(\nu/2)(\nu\pi)^{n/2}|\mathbf{C}|^{1/2}} \left(1 + \frac{\mathbf{r}^T\mathbf{C}^{-1}\mathbf{r}}{\nu}\right)^{-\frac{\nu+n}{2}}, \tag{14.9}$$

where $\Gamma(\cdot)$ is the gamma function. In Chapter 4, this distribution was discussed for the univariate case $n = 1$. In contrast to the multivariate normal density given by equation (14.2), this density allows for an additional parameter $\nu$, i.e., the degrees of freedom parameter. Small values of $\nu \sim 3$ are not uncommon in historical time series and lead to fat-tailed distributions. The value of $\nu$ is an input to the calculations of VaR and P&L. It is interesting to point out a few special properties of the multivariate t-distribution. For values $\nu > 2$, t-distributed random variables with density [with density given by equation (14.9)] can be shown to have covariance matrix $(\frac{\nu}{\nu-2})\mathbf{C}$. In the special case that $\mathbf{C}$ has all unit diagonals, it follows that $\mathbf{C}$ corresponds to the correlation matrix of the distribution, with each marginal being a univariate $t$ with common degrees of freedom $\nu > 2$. More generally each variable has the distribution of a scaled t random variable with $\nu$ degrees of freedom. Another important property that can be numerically investigated in this project is that the multivariate t-density converges to the multivariate normal density, i.e., equation (14.9) becomes equation (14.2) in the limit $\nu \to \infty$.

A useful relationship between random variables of a multivariate t-distribution and those drawn from a multivariate normal is as follows. Assume the random vector $\mathbf{R}^T \equiv (R_1, \ldots, R_n) \sim t_\nu(\mathbf{0}, \mathbf{C})$; i.e., this is shorthand notation for a random vector whose components are jointly distributed according to the multivariate t-density in equation (14.9). Then $\mathbf{R}$ has the *same* distribution as the vector given by $\mathbf{X}/\sqrt{Y/\nu}$, where $\mathbf{X}^T = (X_1, \ldots, X_n) \sim N(\mathbf{0}, \mathbf{C})$ and $Y \sim \chi^2_\nu$ (is a chi-squared random variable with $\nu$ degrees of freedom) *independent* of $(X_1, \ldots, X_n)$. A chi-squared random number with assumed integer $\nu$ is generated simply by summing up $\nu$ independent and identically distributed standard normals $z_j \sim N(0, 1)$: $Y = \sum_{j=1}^{\nu} z_j^2$. From this property we conclude that a multivariate t random vector $\mathbf{R}$ is *generated* by a multivariate normal vector with an independent randomly scaled covariance matrix, i.e., using

$$\mathbf{R} = \frac{\mathbf{X}}{\sqrt{Y/\nu}} = \frac{\mathbf{U}^T\mathbf{Z}}{\sqrt{Y/\nu}} = \mathbf{U}^T\hat{\mathbf{R}}. \tag{14.10}$$

This result obtains from equation (14.6) with $\hat{\mathbf{R}}$ being a t random vector of uncorrelated (yet not independent, since they share a common $Y$ random variable) components: $\hat{\mathbf{R}} \sim t_\nu(\mathbf{0}, \mathbf{I})$, $\mathbf{I}$

is the $n \times n$ identity matrix. The random vector $\mathbf{Z} \sim N(\mathbf{0}, \mathbf{I})$ is an $n \times 1$ vector of independent standard normal components.

Hence, given an integer degree of freedom $\nu > 2$, the simulation procedure for generating multivariate t random vectors is similar to the procedure for generating multivariate normals, with only slight modifications. In fact, relation (14.10) points to the specific recipe. The covariance matrix is Cholesky factorized only once at the beginning of the simulation. Then for each return scenario $\mathbf{r}^{(i)}$, a vector $\mathbf{z}^{(i)} \sim N(\mathbf{0}, \mathbf{I})$ is generated and independently a random chi-squared $y_i$ value is generated. Then using equation (14.10): $\mathbf{r}^{(i)} = \mathbf{U}^T\mathbf{z}^{(i)}/\sqrt{y_i/\nu}$, $i = 1, \ldots, N_s$. Each $i$th scenario can therefore be obtained by generating $n + \nu$ independent standard normal random numbers. The random-number library routine *gennor* within the MFRangen class is useful for this purpose.

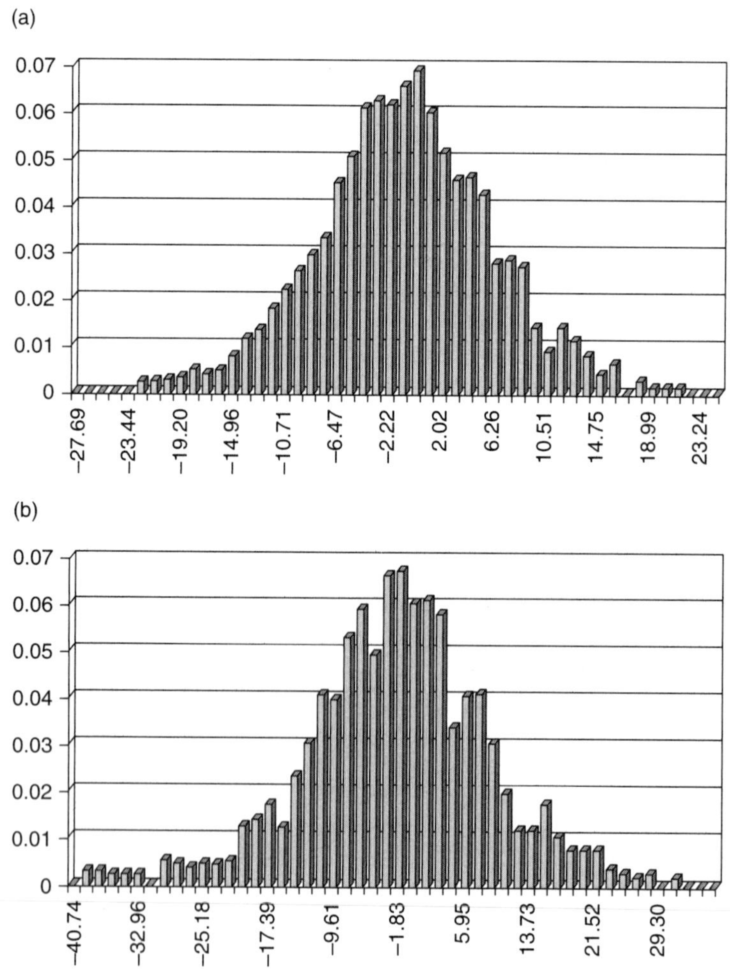


FIGURE 14.1 Histograms of the P&L, within the delta-gamma approximation, for a generic portfolio of 10 risk factors using the multivariate (a) normal versus (b) Student t-($\nu = 3$ degrees of freedom) distributions, respectively. The number of scenarios is $N_s = 2000$. Random gamma and delta sensitivites were chosen identically for both distributions, while, for precentile $p = 1\%$, the computed values for VaR were 17.53 versus 28.03 for distributions (a) and (b), respectively.

For both multivariate normal and Student t-distributions, one should observe 1–5% statistical error when using a number of scenarios on the order of $N_s \sim 10{,}000$. Moreover, the results of the simulations should demonstrate fatter tails for the P&L corresponding to the Student t-distribution as well as a respectively larger value for VaR at a given percentile. As well, one should observe a much more pronounced effect as the degrees of freedom $\nu$ is decreased. Figure 14.1 gives a comparison of the P&L and VaR for an actual Monte Carlo simulation on a portfolio of 10 risk factors.

CHAPTER • 15

# Project: Covariance Estimation and Scenario Generation in Value-at-Risk

This project investigates the covariance properties of return time series generated by a multivariate Monte Carlo simulation.

In particular, one generates a random, symmetric positive-definite matrix with a specific set of eigenvalues. The matrix is then interpreted as the covariance matrix of a multivariate normal distribution, and multivariate normal scenarios are subsequently generated. The covariance matrix is finally reestimated with one of the methods typically used in VaR implementations. The reestimated covariance is then analyzed in terms of eigenvalue spectral concentration as compared with the original covariance matrix.

*Worksheet:* **recov**

*Required Libraries:* **MFioxl, MFBlas, MFLapack, MFFuncs, MFRangen, MFSort**

## 15.1 Generating Covariance Matrices of a Given Spectrum

In this section we discribe a technique for generating a random positive-definite symmetric $N$-dimensional matrix with a specific preassigned set of eigenvalues. The first step is to generate a symmetric positive-definite (SPD) matrix. Two alternatives are possible. The first is to simply use the MATP routine in the random-number (and random-matrix) library MFRangen. This will generate an SPD type of matrix of a given dimension $N$ as specified by the user. The other alternative is to generate an upper triangular random matrix (using the normal random-number generator of MFRangen). This preliminary matrix, $\mathbf{A}$, can then be used to form the matrix $\mathbf{B} = \mathbf{A}^T\mathbf{A}$ (superscript $T$ stands for matrix transpose). Now, $\mathbf{B}$ is of type SPD as long as one makes sure that all of the generated diagonal elements of $\mathbf{A}$ are nonzero.

Note that the matrix $\mathbf{B}$ can, in principle, represent a covariance matrix. Most probably, however, this matrix will largely be dominated by only one or a very small number of principal components. It is of interest in the present study to consider covariance matrices whose eigenvalues are more equally spaced. In particular, one can enforce a set of preassigned eigenvalues. This leads to the next step, namely, creating a covariance matrix of given eigenvalues. To accomplish this, one makes use of the MFLapack library and performs a singular value decomposition (SVD) on the matrix **B**,

$$\mathbf{B} = \mathbf{O}\boldsymbol{\beta}\mathbf{O}^T. \tag{15.1}$$

The matrix **O** is now an orthonormal matrix whose columns $\mathbf{O}_i$ are normalized eigenvectors of **B**. $\boldsymbol{\beta}$ is the diagonal matrix of eigenvalues $\beta_i$ of **B**. That is, $\mathbf{B}\mathbf{O}_i = \beta_i \mathbf{O}_i$, with $\mathbf{O}_i \cdot \mathbf{O}_j = \delta_{ij}$, $i, j = 1, \ldots, N$. The $\mathbf{O}_i$ are essentially the randomly generated principal components of the covariance matrix. After having performed this SVD, the eigenvalues $\beta_i$ are readjusted (i.e., specifically reassigned) by making the change $\beta_i \to \alpha_i$ for a chosen set of $\alpha_i$, $i = 1, \ldots, N$. The new diagonal matrix $\boldsymbol{\alpha}$ of eigenvalues $\alpha_i$ is then used to give the desired covariance matrix:

$$\mathbf{C} = \mathbf{O}\boldsymbol{\alpha}\mathbf{O}^T. \tag{15.2}$$

One is now at liberty to choose an eigenvalue set. For example, by setting $\alpha_i = \kappa(\sqrt{N+1} - \sqrt{i})$ for some positive constant $\kappa$, one has a slowly decaying spectral density (i.e., eigenvalue density) as one moves away from the origin of zero eigenvalue.

## 15.2 Reestimating the Covariance Matrix and the Spectral Shift

As in the previous VaR project, we assume a multivariate normal distribution given by equation (14.2) for the returns. Scenarios are then generated for returns **r** using the same procedure described in detail in the plain Monte Carlo approach of the VaR project. Namely, one generates a vector $\mathbf{y}^{(k)}$ of independent standard normals and multiplies this vector by the Cholesky factored form of the foregoing covariance matrix **C** of equation (15.2). This gives a scenario $\mathbf{r}^{(k)}$. Each $\mathbf{r}^{(k)}$ is then used to form the exponentially weighted sum over $N_s$ scenarios:

$$\hat{C}^{\lambda}_{ij} \approx (1-\lambda) \sum_{k=1}^{N_s} \lambda^{k-1} r_i^{(k)} r_j^{(k)} \tag{15.3}$$

for all $i, j = 1, \ldots, N$ and where $\lambda$ is a damping parameter or decay factor strictly less than unity, $0 < \lambda < 1$. In particular, the value for $\lambda$ is typically chosen between 0.94 to 0.97. The choice of $\lambda = 0.97$ roughly corresponds to assuming a 1-year time window of trading days. This parameter, therefore, determines the relative weights given to past observations (i.e., the return scenarios) and hence the amount of data that is actually used to estimate the variance-covariance of the return time series. The factor $(1-\lambda)$ is a normalization since $\sum_{k=0}^{n} \lambda^k \approx (1-\lambda)^{-1}$ for large $n$. Note that equation (15.3) is not applicable for the special case $\lambda = 1$. Hence, for zero damping ($\lambda = 1$) one must replace equation (15.3) by

$$\hat{C}_{ij} \approx \frac{1}{N_s} \sum_{k=1}^{N_s} r_i^{(k)} r_j^{(k)}. \tag{15.4}$$

Note that the time series considered here are scenario sets, which are quite lengthy, typically of order 10,000.

Having estimated the covariance matrix using equation (15.3) or (15.4), one can then compare the $\hat{C}^{\lambda}_{ij}$ elements with the original matrix elements $C_{ij}$. A more interesting comparison, however, is obtained by computing the eigenvalues for both **C** and $\hat{\mathbf{C}}^{\lambda}$ matrices. Earlier

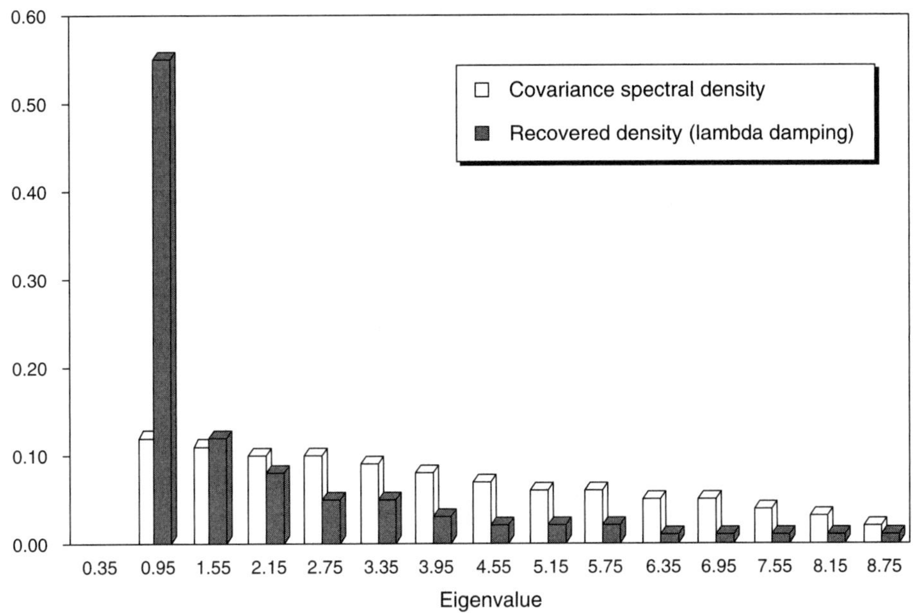


FIGURE 15.1 Eigenvalue distributions for a 100-dimensional covariance matrix using 10,000 scenarios. The recovered distribution is computed with damping factor $\lambda = 0.97$.

we denoted the eigenvalues of $\mathbf{C}$ by $\alpha_1, \alpha_2, \ldots, \alpha_N$. Correspondingly, we will denote the eigenvalues of $\hat{\mathbf{C}}^\lambda$ by $\alpha_1^\lambda, \alpha_2^\lambda, \ldots, \alpha_N^\lambda$. Note that these eigenvalues can be obtained in a variety of ways, one of which is the singular value decomposition, as given earlier, of the respective covariance matrices. The so-called vectors of singular values give the sets of eigenvalues. The objective is to compare eigenvalues in terms of the density (or distribution) for the $\alpha_i$ versus the distribution in the $\alpha_i^\lambda$. The eigenvalue density $f(\alpha)$ at the point $\alpha$ is defined as the number of eigenvalues lying between $\alpha$ and $\alpha + d\alpha$ for infinitesimal $d\alpha$. The densities are actually estimated by considering histogram plots of the respective eigenvalue sets. The density plots should demonstrate a probability increase or shift of distribution toward the origin in the spectrum of eigenvalues as the decay factor $\lambda$ is decreased from 1.0 to 0.94, the latter case corresponding to more damping of past observations. Figure 15.1 gives a histogram comparison of actual versus recovered eigenvalue distributions for a covariance matrix with 100 risk factors, as generated in the recov spreadsheet. A simple extension to this project is to include an analysis of the differences in the principal components of $\mathbf{C}$ and $\hat{\mathbf{C}}^\lambda$.

CHAPTER • 16

# Project: Interest Rate Trees: Calibration and Pricing

The purpose of this project is essentially twofold: to calibrate interest rate trees against market discount (zero-coupon) curves and to subsequently use the calibrated lattices to price interest rate products, such as bonds, bond options, caplets, floorlets, and swaptions. The theory and implementation allow for four different stochastic interest rate models: Black–Derman–Toy and Ho–Lee (within a binomial lattice approach) and the Hull–White model and Black–Karasinski (within a trinomial lattice approach).

*Worksheets:* **ir and ycc**

*Required Libraries:* **MFioxl, MFFuncs, MFBlas, MFLapack, MFFit**

## 16.1 Background Theory

In developing interest rate trees we consider a subdivision of calendar time $t \in [0, T]$ into $M$ subintervals $[T_0 = 0, T_1, T_2, \ldots, T_M]$ with time spacing $\Delta t = T_i - T_{i-1}$. Throughout we shall assume equal time steps, although the lattice methods we present can be extended to the more general case of unequal time steps. Discount bond prices at current calendar time $t$ maturing at calendar time $T$ are denoted by $Z_t(T)$ ($\equiv Z_t(r_t, T)$) (see Chapter 2). Consider, then, a generic (European-style) security with payoff function $\Lambda(r_T, T)$ depending only on the value attained at maturity time $T$ for the short rate $r_T$. If one assumes market completeness, the arbitrage-free price of such a security, at current time $t = 0$, is given by the expectation

$$P_0(r_0, T) = P_0(T) = E_0\left[\exp\left(-\int_0^T r_s ds\right)\Lambda(r_T, T)\right]. \tag{16.1}$$

Here, the numeraire is chosen as the rolled-over money market account $B_t \equiv e^{\int_0^t r_s ds}$, as discussed in Chapter 2. In more explicit terms, this expectation (which is conditional on the short rate's having value $r_0$ at time $t = 0$) has the form of an infinite product of conditional integrations for every incremental time $\Delta t \to 0$. In particular, if we denote the risk-neutral conditional probability density that the short rate will attain a value $r_i$ at time $T_i$, given $r_{i-1}$ at time $T_{i-1}$, by $p(r_i, r_{i-1}; \Delta t)$, then (for a generally stochastic $r_t$ diffusion process) the price can be accurately approximated by an $M$-dimensional integral,

$$P_0(r_0, T) = \int_0^\infty \cdots \int_0^\infty \prod_{i=1}^{M} p(r_i, r_{i-1}; \Delta t) e^{-r_{i-1}\Delta t} \Lambda(r_M, T) dr_M \ldots dr_1, \tag{16.2}$$

where $T_M = T$ is the terminal time. In the limit $\Delta t \to 0$ (or $M \to \infty$ since $T = M\ \Delta t$) this gives an exact path integral representation of the price. Lattice methods arise by choosing a finite number $M$ of time slices and evaluating equation (16.2) by using efficiently recombining lattice point integral approximations. For zero-coupon bonds we have a pay-off of one dollar with certainty ($\Lambda(r_T, T) = 1$); hence

$$Z_0(T) = Z_0(r_0, T) = E_0\left[\exp\left(-\int_0^T r_s ds\right)\right]. \tag{16.3}$$

Of interest are the Arrow–Debreu prices, denoted by $G(r_0, 0; r, T)$ and given by

$$G(r_0, 0; r, T) = E_0\left[\exp\left(-\int_0^T r_s ds\right)\delta(r_T - r)\middle| r_{t=0} = r_0\right], \tag{16.4}$$

which is the expectation of an infinitely narrow butterfly spread pay-off (i.e., the Dirac delta function) conditional on the short rate's starting at $r_0$ at time $t = 0$. These correspond to the worth at time $t = 0$, given (i.e., conditional on) current state $r_0$, of a riskless security that pays one dollar if state $r_T = r$ is attained at any later time $T > 0$. The zero-coupon bonds are expressed in terms of the Arrow–Debreu values as follows:

$$Z_0(T) = \int_0^\infty G(r_0, 0; r, T) dr. \tag{16.5}$$

An important consistency requirement is the continuity relation

$$G(r_0, 0; r_i, T_i) = \int_0^\infty G(r_0, 0; r_{i-1}, T_{i-1}) G(r_{i-1}, T_{i-1}; r_i, T_i) dr_{i-1}. \tag{16.6}$$

This formula is the basis for a discrete version that is used in the sections that follow to generate a forward induction procedure for propagating the Arrow–Debreu prices. The function $G(r_{i-1}, T_{i-1}; r_i, T_i)$ is the Arrow–Debreu value conditional on the short rate's having value $r_{i-1}$ at time $T_{i-1}$ and attaining a value of $r_i$ at a later time $T_i > T_{i-1}$. We conclude this section by noting that the quantity $Z(r, t; t + \Delta t) = Z_t(r, t + \Delta t)$ defined by the conditional expectation

$$\begin{aligned} Z(r, t; t + \Delta t) &= E_t\left[\exp\left(-\int_t^{t+\Delta t} r_s\ ds\right)\middle| r_t = r\right] \\ &= \int_0^\infty dr_T\ G(r, t; r_T, T = t + \Delta t) \end{aligned} \tag{16.7}$$

gives the price of a discount bond at time $t \geq 0$ (any time later than current time), with time to maturity of $\Delta t$, conditional on the short rate's having value $r$ at time $t$. Note that here we have explicitly denoted the conditional nature of the expectation. This formula, in conjunction with concatenating equation (16.6) for every time step $T_i - T_{i-1}$, forms the basis for producing lattice pricing formulas of derivatives, such as caplets, floorlets, and swaptions dealt with later.

## 16.2 Binomial Lattice Calibration for Discount Bonds

In developing binomial interest rate trees we subdivide calendar time $t \in [0, T]$ into $M$ subintervals $[T_0 = 0, T_1, T_2, \ldots, T_M]$ with time spacing $\Delta t = T_i - T_{i-1}$. At each time $t = T_i$ there are $i+1$ nodes corresponding to the attainable values of the short rate $r(j, i) \equiv r(j, T_i)$, $j = 0, 1, \ldots, i$. Note that throughout we use a notation for the short rate whereby $r_t$ denotes the continuous short rate variable at calendar time $t$, whereas $r(j, i) \equiv r(j, T_i)$ corresponds to the discretized short rate value at the node with state $j$ and time $T_i$. Also, note that the indexing of the nodes in binomial models is such that the index has nonnegative value: $j \geq 0$. Figure 16.1 gives a schematic of the binomial interest rate tree. The two binomial models considered in this project are the Ho–Lee (HL) and Black–Derman–Toy (BDT) models. The HL model is the simplest, with no mean reversion. The HL model follows a normal stochastic process

$$dr_t = \lambda(t)dt + \sigma(t)dW_t, \tag{16.8}$$

where $\lambda(t)$ and $\sigma(t)$ are deterministic drift and volatility functions, respectively. One obvious shortcoming of this model is the admittance of negative interest rates. The BDT model removes these deficiencies by considering the logarithm of the short rate, which is assumed to follow a stochastic process of the form

$$d\log r_t = \left[\theta(t) + \frac{d}{dt}(\log \sigma(t)) \log r_t\right] dt + \sigma(t)dW_t, \tag{16.9}$$

where $\sigma(t)$ is the lognormal volatility and the drift function allows for a drift component as well as a mean-reversion component for the variable $\log r_t$. Note that the speed of the mean reversion is zero for the case of constant volatility. Note that throughout this study we shall assume a constant volatility. Hence, mean reversion shall remain zero in the current implementation of the BDT model. In contrast, mean reversion is introduced in later sections where we implement the Hull–White and Black–Karasinski models using trinomial lattices.

The HL lattice model can be defined by a set of nodes placed according to

$$r(j, i) = r(j-1, i) + 2\sigma\sqrt{\Delta t}, \tag{16.10}$$

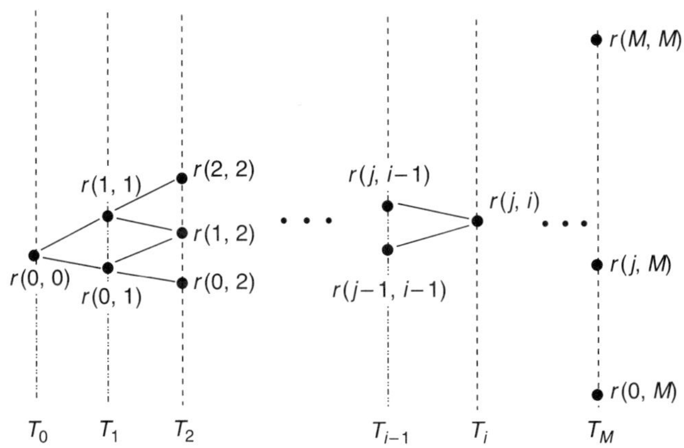


FIGURE 16.1 A binomial lattice originating at the current short-rate node $r(0, 0)$.

whereas the BDT lattice can be taken as

$$r(j,i) = r(j-1,i)\exp(2\sigma\sqrt{\Delta t}), \tag{16.11}$$

for given time slice $T_i = i\ \Delta t$, $i = 0, 1, \ldots, M$. Here $\sigma$ is a (lattice) volatility parameter for the short rate. Throughout the calibration to market zero-coupon bonds, as discussed in this section, the volatility shall be preset to a fixed value, independent of time. It should be noted that there exist a whole range of fixed $\sigma$ values that can produce identical matches to the same set of market zero-coupon bond prices. Different values for this parameter have the effect of shifting the overall drift of the tree so as to still keep it risk neutral. Note that the assumption of a fixed volatility eliminates the reversion component in the BDT model. By allowing the volatility $\sigma$ to be time dependent, one can further calibrate to a larger set of market instruments besides zero-coupon bonds. The first step to consider in the calibration of discount bond prices is the interpolation of yields from given treasury yield data. Consider the set of maturities $T = T_i$, $i = 1, 2, \ldots, M$, and set the current time $t = T_0 = 0$. The discount curve for the calibration consists of the set of prices $\{Z_0(T_1), Z_0(T_2), \ldots, Z_0(T_M)\}$ derived from the set of corresponding yields $y_t(T_i) = \{y_0(T_1), y_0(T_2), \ldots, y_0(T_M)\}$. This set of yields does not, in practice, match the actual input market set of $N$ maturity yields given at a fixed set of times denoted by the set $\{y_0(\bar{T}_1), y_0(\bar{T}_2), \ldots, y_0(\bar{T}_N)\}$. The latter are the actual treasury yields at times $\bar{T}_1 = 3$ months, $\bar{T}_2 = 6$ months, etc. The foregoing discount prices are therefore obtained after having interpolated for the yields $y_t(T_i)$ at each $i$th time step. This must be done either by employing a simple linear interpolation or by using a spline-fitting algorithm of higher order, such as a cubic spline. The MFFit numerical library class is useful for this purpose.

Lattice methods correspond to fixing the number of integrations in all the equations of the previous section into some fixed integer, such as the number of time steps in the case of pricing, and, in turn, evaluating each integral using *only* two (for the case of a binomial lattice) or three (for trinomial lattices) points of integration. An important approximation underlying the binomial lattice methodology is to set the conditional transition density for every time step $\Delta t$ simply as a constant, $p(r_i, r_{i-1}; \Delta t) = \frac{1}{2}$. Moreover, the short rate is taken as locally constant $r_{i-1}$ within time intervals $[T_{i-1}, T_i]$, hence giving the *conditional* Arrow–Debreu values for $\Delta t$ maturity as the simple form $G(r_k, T_{i-1}; r_j, T_i) \equiv G(k, T_{i-1}; j, T_i) = \frac{1}{2}e^{-r_{i-1}\Delta t}$. By adopting the binomial short-rate lattices defined by equation (16.10) (for the HL model) or (16.11) (for the BDT model), we now are in a position to obtain the discrete-time versions of the equations in the previous section.

To begin with, equation (16.6) takes the discrete form

$$G(0,0;j,T_i) = \sum_{k=j-1,j;0\le k\le i-1} G(0,0;k,T_{i-1})G(k,T_{i-1};j,T_i), \tag{16.12}$$

where

$$G(k,T_{i-1};j,T_i) = \frac{1}{2}\exp[-r(k,i-1)\Delta t]. \tag{16.13}$$

Note that equation (16.12) describes a procedure that takes into account the Arrow–Debreu prices at intermediate nodes $r(k, i-1)$ for the previous time $T_{i-1}$, which are subsequently used for time stepping by an amount $\Delta t$ until a terminal node $r(j, i)$ is reached at the time slice $T_i = T_{i-1} + \Delta t$. The sum involves only two possible values for $k$: $k = j-1$ and $k = j$, with the restriction that $0 \le k \le i-1$. For the extreme (highest or lowest) node there is only one term in the sum. This is the forward induction equation that is used in practice to generate all Arrow–Debreu prices $G(0, 0; j, T_i)$ for each $j$th node at terminal time $T_i$. It is

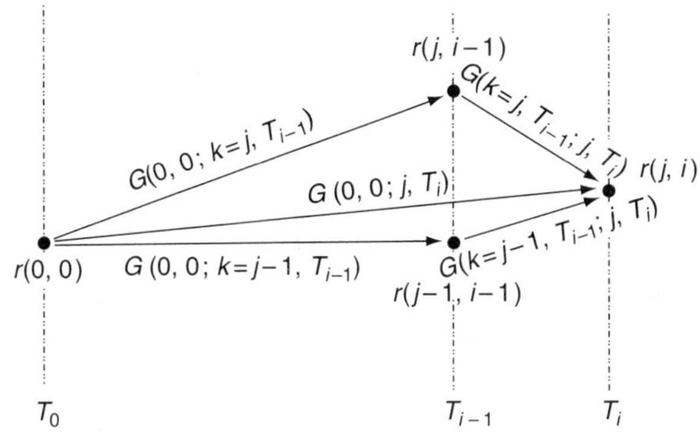


FIGURE 16.2 A pictorial representation of the forward propagation of Arrow–Debreu prices on a binomial lattice originating at the current short-rate node $r(0, 0)$. Their values, i.e., $G(0, 0; j, T_i)$, at nodes corresponding to a later time step $t_i$ are obtained as a sum of contributions from (at most) two intermediate time $T_{i-1}$ two-legged paths.

important to note that this forward induction equation can generally be used for any type of short-rate model (see Figure 16.2). The discrete-time version of the security price given by equation (16.2) takes the form

$$P_0(r_0, T_i) = \sum_{j=0}^{i} G(0, 0; j, T_i)\Lambda(r(j, T_i), T_i), \tag{16.14}$$

where $G(0, 0; j, T_i)$ are computed using the forward induction relation in equation (16.12). Specializing this formula to the case of zero-coupon bonds, which have a riskless pay-off of one dollar, we have the discrete-time lattice version of equation (16.5):

$$Z_0(T_i) = \sum_{j=0}^{i} G(0, 0; j, T_i). \tag{16.15}$$

Hence, the Arrow–Debreu prices at all the nodes of a given maturity $T$ are sufficient for determining the price of a discount bond of that maturity (see Figure 16.3). In the calibration procedure the market zero-coupon prices at times $T = T_i$ are used as input to the left-hand side of equation (16.15). By solving for the nodes at the $(i-1)$th time step, for every time slice $T_i$, we imply the whole lattice and hence obtain the market prices of all discount bonds correctly. In practice, the right-hand side, for each $T = T_i$, is computed by using the vector of Arrow–Debreu values $G(0, 0; k, T_{i-1})$, $k = 0, 1, \ldots, i-1$, which are assumed to be known from the previous time step, as well as a trial vector of nodes $r(k, i-1)$. These are plugged into forward induction equation (16.12) while using equation (16.13) and summing up all node contributions via equation (16.15). At the same time, one also makes use of the constraint among the $r(k, i-1)$, namely, equation (16.10) or (16.11), depending on whether one is calibrating the Ho–Lee or BDT model, respectively. Hence, this reduces the discount bond calibration problem to a succession of $M$ root-finding problems that make the left- and right-hand sides of equation (16.15) equal for each $T_i$. Note also the single-variable nature of the problem, since the expressions are reduced to finding just one node, i.e., the lowest one $r(0, i-1)$, and the rest follow for time slice $T_{i-1}$. Observe that at each maturity the nodes being computed are lagged by one time step. One can use the MFZero library class for the purpose of finding roots. To start the procedure off, one uses $G(0, 0; 0, 0) = 1$

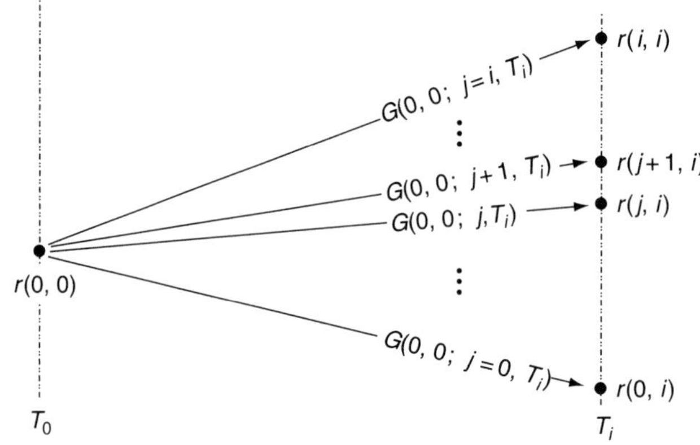


FIGURE 16.3 Lattice calibration of zero-coupon bonds of maturity $T_i$ uses a sum of Arrow–Debreu prices beginning from the current time $T_0$ and node $r(0,0)$ up to all nodes $r(j,i)$, $j=0,\ldots,i$, at time $T_i$. Note, however, that calibration up to maturity time $T_i$ determines the short-rate lattice points up to the previous time step with time $T_{i-1}$.

and solves for $r(0,0)$. Note that since there are only two branches in this case, giving $G(0,0;k=0,1,\Delta t)=\frac{1}{2}\exp[-r(0,0)\Delta t]$, one has

$$r(0,0)=-\frac{\log Z_0(T_1=\Delta t)}{\Delta t}, \tag{16.16}$$

where the first node, $r(0,0)$, is given by the smallest-maturity zero-coupon bond price (i.e., the initial term structure). If one assumes continuous compounding, then one can also avoid the numerical root-finding procedure in the case of the Ho–Lee model, which admits a simple analytical solution for the node positions $r(j,i-1)$ in terms of $Z_0(T_i)$ and the Arrow–Debreu prices for terminal time $T_{i-1}$.

## 16.3 Binomial Pricing of Forward Rate Agreements, Swaps, Caplets, Floorlets, Swaptions, and Other Derivatives

Recall from Chapter 2 the price of a plain-vanilla FRA of a given tenor $\tau=T_{i+1}-T_i$. Assuming continuous compounding, equation (2.8) can be used to give the net present value of an FRA (to the party receiving an initial nominal amount) with one dollar nominal:

$$\mathrm{PV(FRA)}_0=-Z_0(T_1)+e^{\tau f_0(T_1,T_2)}Z_0(T_2), \tag{16.17}$$

where the forward is given by

$$f_0(T_1,T_2)=\frac{1}{\tau}\log\left(\frac{Z_0(T_1)}{Z_0(T_2)}\right). \tag{16.18}$$

Since all expressions are completely determined by the prices of the zero-coupon bonds, it necessarily follows that all FRAs are also exactly priced by the binomial lattices obtained from the calibration procedure in the previous section. Moreover, as recalled from Chapter 2, a swap is just a collection of FRAs. Plain-vanilla swaps are, therefore, also priced exactly within the foregoing calibration framework.

The pricing of options such as caps (or caplets) is not as straightforward as that for FRAs. In particular, recall from Chapter 2, the pay-off of a caplet struck at fixed interest rate $R_K$, maturing at time $T$, on a floating reference rate $R^{\tau}(T)$ of tenor $\tau$ applied to the period $[T, T+\tau]$ in the future. The floating rate is typically the three- or six-month LIBOR. The pay-off of this caplet option is given by

$$\mathbf{Cpl}(R^{(\tau)}(T), T) = (R^{(\tau)}(T) - R_K)_+ \tau Z_T(r_T, T+\tau), \tag{16.19}$$

where $Z_T(r_T, T+\tau)$ is the discount function over that period, since the cash flow occurs at time $T+\tau$. Here we define $(x)_+ \equiv \max(x, 0)$ as usual. In order to obtain the price at current time $t = 0$ of this caplet one must take an expectation, or integral, of the pay-off with a risk-neutral distribution in the reference rate $R_t^{(\tau)}$, where $t = T$, i.e., the expiry or maturity time of the option on the (call-type) pay-off. The latter is, however, expressed in terms of a rate applied to the period of the tenor (i.e., the reference forward rate) and not the short rate used in the rate lattice calibration of the previous section. In particular, the short rate lattice gives the conditional distribution of the short rate.

To price the caplet, one must relate the short rate to this reference forward rate. In particular, the values of the short rate at the nodes $(j, i)$, $r(j, i)$ must be related to the values of the reference forward rates, denoted by $R^{(\tau)}(j, i)$, at these nodes. This is achieved by using the continuous-time relation for the forward rate, and this is where the conditional zero-coupon prices $Z(r, t; T)$ are useful. In particular, for continuous compounding,

$$R^{(\tau)}(t) = \frac{1}{\tau} \log\left(\frac{Z(r, t, t)}{Z(r, t, t+\tau)}\right), \tag{16.20}$$

and since $Z(r, t, t) = 1$,

$$R^{(\tau)}(t) = \frac{1}{\tau} \log\left(\frac{1}{Z(r, t, t+\tau)}\right). \tag{16.21}$$

Choosing $t = T_i$ and $T = T_i + n\,\Delta t$, where it is assumed that the tenor is exactly $n$ periods of the lattice time step, for some integer $n$, $\tau = n\,\Delta t$, we arrive at the discrete time value at the $j$th node:

$$R^{(\tau)}(j, i) = \frac{1}{\tau} \log\left(\frac{1}{Z(j, T_i, T_i + n\,\Delta t)}\right). \tag{16.22}$$

Here $Z(j, T_i, T_i + n\,\Delta t) \equiv Z(j, T_i, T_{i+n})$ is the zero-coupon value maturing at time $T_{i+n}$ ($n$ time steps in the future), conditional on the short rate's having value $r(j, i)$ at time $T_i$. Based on equations (16.19) and (16.22), we can write all components of the payoff vector of the caplet at each node $(j, i)$, denoted by $C^{(\tau)}(j, i)$, as

$$C^{(\tau)}(j, i) = (R^{(\tau)}(j, i) - R_K)_+ \tau Z(j, T_i, T_{i+n}), \tag{16.23}$$

where equation (16.22) is plugged in for $R^{(\tau)}(j, i)$. Note that the preceding equations assume continuous compounding, while a similar set of equations obtain for the case of discrete compounding, where the $\log(x)$ function is simply replaced by $x$. The foregoing payoff vector introduces an extra procedural step, requiring one to compute the quantities $Z(j, T_i, T_{i+n})$, which involve a separate forward induction starting from the nodes $r(j, i)$. In practice, these are computed using the discrete-time version of equation (16.7):

$$Z(j, T_i, T_{i+n}) = \sum_{k=j}^{j+n} G(j, T_i; k, T_{i+n}), \tag{16.24}$$

where the $(n+1)$ Arrow–Debreu values (conditional on beginning at a $j$th node at time $T_i$ and ending at node $k = j, j+1, \ldots, j+n$ at time $T_{i+n}$) on the right-hand side of this equation are computed by forward recursion using an adaptation of equation (16.12), rewritten here in a slightly different form:

$$G(j, T_i; k, T_{i+m}) = \sum_{s=k,k-1;\, j \le s \le j+m-1} \frac{1}{2} e^{-r(s,i+m-1)\Delta t} G(j, T_i; s, T_{i+m-1}). \qquad (16.25)$$

Here $m = 1, 2, \ldots, n$ and the iteration is readily carried out from time $T = T_i$ to final time $T_{i+n}$, where one initially has $G(j, T_i; j, T_i) = 1$ for any $j$ value. It is instructive to write out the Arrow–Debreu values explicitly for the first two time steps. For a single step (for $m = 1$) the terminal time is $T_{i+1}$, and we simply have

$$G(j, T_i; k, T_{i+1}) = \frac{1}{2} e^{-r(j,i)\Delta t}, \qquad (16.26)$$

where $k = j, j+1$ are the only two possible values for $k$. Not surprisingly, this is consistent with the relation in equation (16.13). Propagating out to the second step ($m = 2$), equation (16.25) gives

$$G(j, T_i; k, T_{i+2}) = \sum_{s=k,k-1;\, j \le s \le j+1} \frac{1}{2} e^{-r(j,i)\Delta t} \frac{1}{2} e^{-r(s,i+1)\Delta t}, \qquad (16.27)$$

where possible values for $k$ are $j, j+1, j+2$. Summing up the terms explicitly, these three Arrow–Debreu prices are

$$G(j, T_i; j, T_{i+2}) = \frac{1}{4} e^{-r(j,i)\Delta t} e^{-r(j,i+1)\Delta t},$$
$$G(j, T_i; j+1, T_{i+2}) = \frac{1}{4} e^{-r(j,i)\Delta t} \left[ e^{-r(j,i+1)\Delta t} + e^{-r(j+1,i+1)\Delta t} \right],$$
$$G(j, T_i; j+2, T_{i+2}) = \frac{1}{4} e^{-r(j,i)\Delta t} e^{-r(j+1,i+1)\Delta t}. \qquad (16.28)$$

Specializing equation (16.14) we therefore finally have the binomial lattice pricing formula for a caplet valued at current time $T_0 = 0$ and maturing at time $T_i$ of tenor $\tau = n\ \Delta t$:

$$\mathbf{Cpl}_0^{(\tau)}(R_K, T_i) = \sum_{j=0}^{i} G(0, 0; j, T_i) C^{(\tau)}(j, i). \qquad (16.29)$$

To summarize then, the application of this pricing formula contains two components. The first part is the computation of the $G(0, 0; j, T_i)$, which are already computed from the calibration step, as discussed in the previous section. The second part consists of computing the payoff components $C^{(\tau)}(j, i)$. These are obtained by first computing the conditional Arrow–Debreu prices $G(j, T_i; k, T_{i+n})$ by forward induction using equation (16.25). These quantities are then summed up to give the $Z(j, T_i, T_{i+n})$, as in equation (16.24). In turn, the latter are plugged into equation (16.22), giving the forward rates $R^{(\tau)}(j, i)$, and hence $C^{(\tau)}(j, i)$, using equation (16.23).

Figure 16.4 depicts, schematically, this procedure for pricing a caplet. For implementation considerations, note that the inputs within the **ir** sheet (for pricing a caplet) are the expiry time $T_i$, which for simplicity is assumed chosen as an integer number of time steps from current time $T_0$, and the tenor of the caplet is chosen as an integer number of time steps

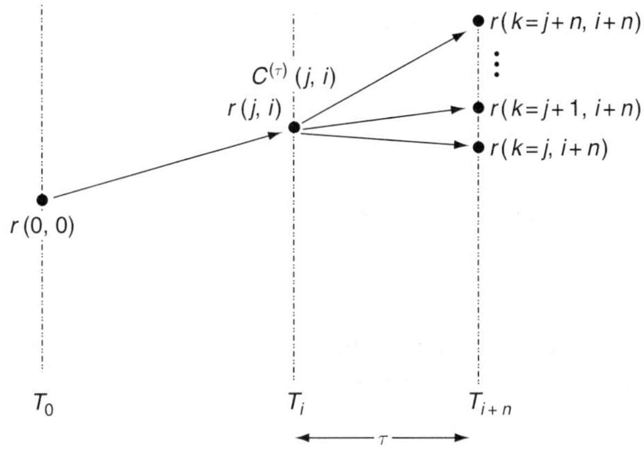


FIGURE 16.4 Schematic representation of the separate components used for the pricing of a caplet option of tenor $\tau$, expiring at time $T_i$. The initial leg starting from the current time node $r(0, 0)$ gives the Arrow–Debreu prices $G(0, 0; j, T_i)$ at each $j$th node $r(j, i)$ at time $T_i$. The payoff vector of the caplet with $j$th component $C^{(\tau)}(j, i)$ (for the $j$th node at time $T_i$) is obtained by summing all the Arrow–Debreu prices $G(j, T_i; k, T_{i+n})$ $(k = j, \ldots, j+n)$ that are conditional on starting at the node $r(j, i)$ at time $T_i$ and ending at nodes $r(k, i+n)$ at time $T_{i+n}$ for the period of the caplet.

past $T_i$, where the time step is the previously computed $\Delta t$. Note that, if needed, this apparent restriction can be readily lifted by using a different time-step value for the lattice past-maturity time $T_i$. The spreadsheet contains inputs for the number of time steps to reach the caplet (or floorlet) option expiry time from today, i.e., an integer $M$ with $M\ \Delta t = T_i$ and another integer for the number of steps defining the tenor.

The entire analysis for pricing a floorlet of the same maturity, struck at rate $R_K$, follows almost identically as in the case of the caplet, except the pay-off is now that of a put, $(R_K - R)_+$, instead of that of a call, $(R - R_K)_+$. Within this project one should allow for a computation of both types of options as well as the pricing of swaptions.

Next, we consider the pricing of European swaptions. Such options, as discussed in Chapter 2, come in two flavors: The payer swaption has pay-off given by equation (2.41), while the receiver swaption has the put type of pay-off. Let us consider a payer swaption, struck at rate $r_K$, on an underlying swap to start at time $T = T_{n_s}$ in the future and having a lifetime of $n$ periods of fixed tenor $\tau$:

$$\mathbf{PSO}_T = \tau(r_T^s - r_K)_+ \sum_{p=1}^{n} Z_T(T + p\tau). \qquad (16.30)$$

Here $r_t^s$ denotes the equilibrium swap rate at time $t$. Hence, the first reset time of the swap is assumed as $T = T_{n_s}$, with first payment time at $T + \tau$, the latter being the second reset time with second payments occurring at $T + 2\tau$, etc. Note that, as in the case of caplets, within the **ir** application spreadsheet the user enters both the option expiry time $T$ and the tenor $\tau$. In addition, the swaption contract is defined by entering the number of periods $n$, with each time period assumed constant and given by $\tau$. In particular, given a maturity $T$, we choose a number of time steps $n_s$ up to maturity with $n_s\ \Delta t = T$, thereby defining a fixed time step $\Delta t = T/n_s$. The contract is assumed to be specified as having tenor $\tau = m_s\ \Delta t$. The number of time steps within the swap is then $N_s = m_s n$, giving a swap lifetime of $N_s\ \Delta t$; i.e., the swap ends at calendar time given by the $(n_s + N_s)$th time slice: $T + n\tau = T_{n_s + N_s}$. Figure 16.5

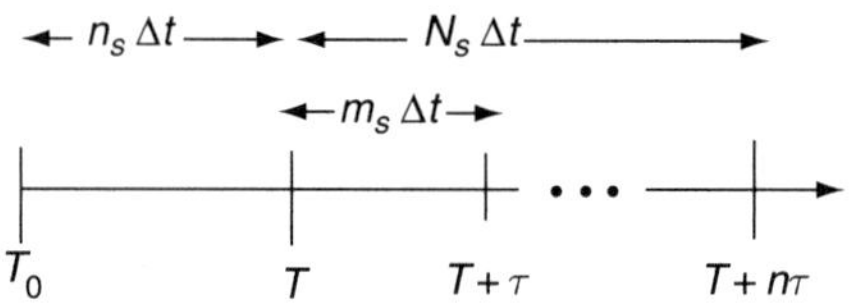


FIGURE 16.5 Time spacing for a swaption expiring at time $T$. The underlying swap has $n$ equal periods of tenor $\tau$.

shows a schematic of the time spacing for the swaption. Now recall from Chapter 2 that the equilibrum swap rate at time $T$ can be written as

$$r_T^s = \frac{Z_T(T) - Z_T(T+n\tau)}{\tau \sum_{p=1}^{n} Z_T(T+p\tau)} = \frac{1 - Z_T(T+n\tau)}{\tau \sum_{p=1}^{n} Z_T(T+p\tau)}. \tag{16.31}$$

The pay-off then takes the form

$$\mathbf{PSO}_T = (A - B)_+. \tag{16.32}$$

Here $A$ is a floating-rate bond

$$A = 1 - Z_T(T+n\tau), \tag{16.33}$$

and

$$B = \tau r_K \sum_{p=1}^{n} Z_T(T+p\tau), \tag{16.34}$$

is an annuity or fixed-coupon bond originating at time $T$ with fixed payments of amount $\tau r_K$ at $n$ periods of time $\tau$. The formula in equation (16.32) is directly suitable for implementation. Based on equations (16.33) and (16.34), the components of the payoff vector of the payer swaption at each $j$th node $r\,(j, i = n_s)$, denoted by $P^{(n\tau)}(j, i)$, are given by

$$P^{(n\tau)}(j, i = n_s) = \left(1 - Z(j, T_{n_s}, T_{n_s+N_s}) - \tau r_K \sum_{p=1}^{n} Z(j, T_{n_s}, T_{n_s+pm_s})\right)_+. \tag{16.35}$$

This pay-off therefore requires the evaluation of the zero-coupons $Z(j, T_{n_s}, T_{n_s+pm_s})$ conditional on the starting node $r(j, n_s)$ at time slice $T_{n_s}$ and maturing at times $T_{n_s+pm_s}$, $p = 1, \ldots, n$. These are computed in the same manner as described for the caplet case. Namely, equation (16.24) gives

$$Z(j, T_{n_s}, T_{n_s+pm_s}) = \sum_{k=j}^{j+pm_s} G(j, T_{n_s}; k, T_{n_s+pm_s}), \tag{16.36}$$

where conditional Arrow–Debreu prices now need to be calculated at every time slice $n_s + pm_s$, i.e., for $p = 1, \ldots, n$. The procedure for doing so is the same as in the caplet case, where forward recursion equation (16.25) is used repeatedly. This time the recursion is carried out for a total of $N_s = m_s n$ steps, and at each interval number $p$ of $m_s$ steps we extract a $(pm_s + 1)$-dimensional vector of Arrow–Debreu prices $G(j, T_{n_s}; k, T_{n_s+pm_s})$ with components

$k = j, j+1, \ldots, j+pm_s$. After having obtained the conditional zero coupons, the current price of the payer swaption is given by discounting the payoff vector:

$$\mathbf{PSO}_0^{(n\tau)}(r_K, T) = \sum_{j=0}^{n_s} G(0, 0; j, T) P^{(n\tau)}(j, n_s). \tag{16.37}$$

Lastly, note that the pricing of reciever swaptions follows in identical manner, except that the pay-off is simply replaced by the put type of expression $(B - A)_+$.

## 16.4 Trinomial Lattice Calibration and Pricing in the Hull–White Model

The implementation of trinomial lattices for interest rate trees shares some similarities with the case of stock price trees covered in the previous project on trinomial lattices for pricing equity options. There are, however, some important differences, stemming from the fact that the short rate is itself stochastic and, hence, discounting is inherently very different, as we have seen in the binomial lattice implementation. Before proceeding to implement a specific short-rate lattice, it is useful to note that there are various possible acceptable tree implementations. Namely, one could adapt the tree methodologies used in the previous trinomial lattice project, which deals with stock price processes, over into the case of a short-rate process. This requires appropriate modifications to account for the mean-reversion effect as well as calibration to discount bond prices across all time steps. The latter would require that the transition probabilites ($p_+$, $p_0$, and $p_-$) also depend on the nodal positions. One can, moreover, also incorporate a similar drift parameter (i.e., the $\mu$ parameter), which would now also depend on the $i$th time slice $T_i$. Such a viable lattice makes use of only *normal* branching. Here we shall deviate slightly and follow Hull and White's two-stage tree-building procedure [HW93, HW94, Hul00]. As shown later, this procedure has the added advantage of separating out the reversion term from the drift component. As well, the sampling of the short-rate nodes in the lattice is done in a more efficient manner by incorporating three types of possible branching modes.

### 16.4.1 *The First Stage: The Lattice with Zero Drift*

As discussed in Chapter 2, the Hull–White (HW) model is defined by the stochastic short-rate process, which can be written in the form

$$dr_t = [\theta(t) - a(t)r_t]dt + \sigma(t)dW_t, \tag{16.38}$$

where $\theta(t)$ is a time-dependent drift term. Throughout, we shall further restrict the mean reversion $a(t) = a$ and volatility $\sigma(t) = \sigma$ to be time-independent parameters. For present purposes this offers a reasonably good model that can be used to calibrate to zero-coupon bonds and subsequently to price interest rate options. Extensions that allow for the reversion speed and/or volatility functions to take on a time dependence (either numerically or analytically) can also be readily achieved. This would allow for exact calibration of the lattice model to a larger basket of instruments besides zero-coupon bonds. We leave this as an optional implementation exercise for the interested reader. The first step is to construct a tree for the related process with zero drift (and nonzero reversion) defined by

$$dr_t^* = -ar_t^*\, dt + \sigma\, dW_t. \tag{16.39}$$

Fixing $r_t^*$ within a time step, we compute the mean and variance of the random variable $r_{t+\Delta t}^* - r_t^*$ as given by the expectations

$$E[r_{t+\Delta t}^* - r_t^*] = -ar_t^* \, \Delta t, \tag{16.40}$$

$$E[(r_{t+\Delta t}^* - r_t^*)^2] = a^2(r_t^*)^2(\Delta t)^2 + \sigma^2 \, \Delta t, \tag{16.41}$$

where only terms up to order $(\Delta t)^2$ are included. The $r^*$ lattice has nodes defined by $r(j, i) = r^*(j, i)$, where

$$r^*(j, i) = r_0^* + j \, \Delta r, \tag{16.42}$$

with $r_0^* = 0$ and $j = -i, -i+1, \ldots, i-1, i$ for any time slice $T_i = i \, \Delta t$ (see Figure 16.6). Using equation (16.42) within equations (16.40) and (16.41) gives

$$E[(r_{t+\Delta t}^* - r_t^*)|r_t^* = j \, \Delta r] = -aj \, \Delta r \, \Delta t, \tag{16.43}$$

$$E[(r_{t+\Delta t}^* - r_t^*)^2|r_t^* = j \, \Delta r] = a^2 j^2 (\Delta r)^2 (\Delta t)^2 + \sigma^2 \, \Delta t. \tag{16.44}$$

At this point one finds explicit formulas for the transition probabilities $p_+$, $p_0$, and $p_-$ for, respectively, the higher, middle, and lower branches emanating from a given node $r(j, i)$. The three possible branching modes considered are depicted in Figure 16.7. Note the difference in convention with respect to the indexing of the nodes that was used in the binomial lattice. As in the previous project on trinomial lattice models, an up (down) move changes the $j$th index in $r(j, i)$ by $+1$ $(-1)$, while only for a middle move $j$ remains unchanged. For the case of normal branching we compute the expectations

$$\begin{aligned} E[(r_{t+\Delta t}^* - r_t^*)|r_t^* = j \, \Delta r] &= p_+(j+1)\Delta r + p_0 j \, \Delta r + p_-(j-1)\Delta r - j \, \Delta r \\ &= (p_+ - p_-)\Delta r \end{aligned} \tag{16.45}$$

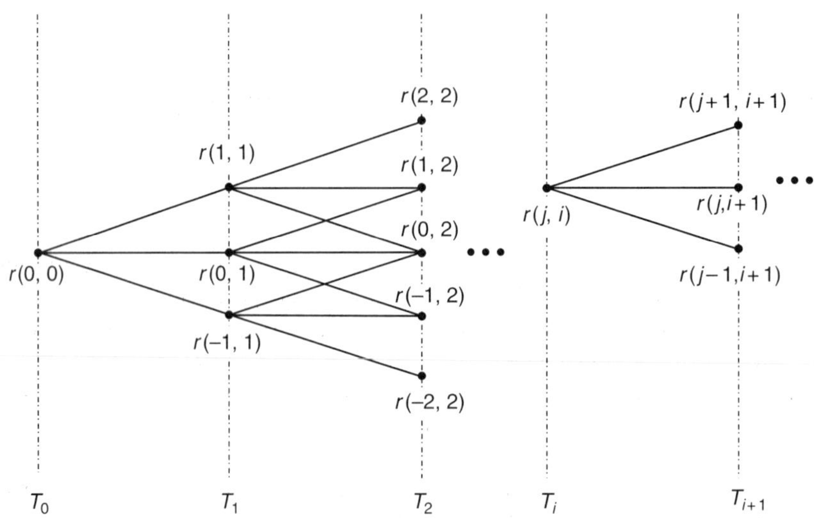


FIGURE 16.6 Schematic of the (driftless) symmetric trinomial $r^*$-lattice for the short-rate process with symmetric (normal) branching from all nodes.

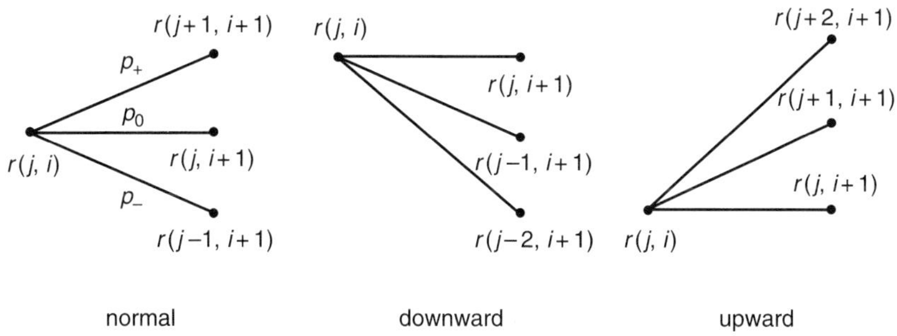


FIGURE 16.7 The three possible branching modes.

and

$$
\begin{aligned}
E[(r^*_{t+\Delta t} - r^*_t)^2 | r^*_t = j\ \Delta r] &= p_+(\Delta r)^2 + p_0 0^2 + p_-(\Delta r)^2 \\
&= (p_+ + p_-)(\Delta r)^2,
\end{aligned}
\tag{16.46}
$$

where we have used probability conservation $p_+ + p_0 + p_- = 1$. It has been observed in the past [HW94] that numerical efficiency is maximized by fixing the spacing to

$$
\Delta r = \sigma \sqrt{3\ \Delta t}. \tag{16.47}
$$

Using this value for the spacing and equating expectations in equations (16.45) and (16.43) and the expectation in equation (16.46) with that in equation (16.44) gives a linear system of two equations in $p_+$ and $p_-$ with unique solution

$$
p_{\pm}(j) = \frac{1}{6} + \frac{1}{2} aj\ \Delta t(aj\ \Delta t \mp 1). \tag{16.48}
$$

Probability conservation gives

$$
p_0(j) = \frac{2}{3} - (aj\ \Delta t)^2. \tag{16.49}
$$

Note that the argument $j$ is used to explicitly denote the dependence of the transition probabilities on the nodal $j$-position value.

A similar analysis gives the probabilities for downward branching:

$$
p^d_+(j) = \frac{7}{6} + \frac{1}{2} aj\ \Delta t(aj\ \Delta t - 3), \tag{16.50}
$$

$$
p^d_-(j) = \frac{1}{6} + \frac{1}{2} aj\ \Delta t(aj\ \Delta t - 1), \tag{16.51}
$$

$$
p^d_0(j) = -\frac{1}{3} - aj\ \Delta t(aj\ \Delta t - 2). \tag{16.52}
$$

The superscipt $d$ is used to denote the transition probabilities for downward branching.

Lastly, for upward branching we have

$$p_{+}^{u}(j) = \frac{1}{6} + \frac{1}{2} aj\,\Delta t(aj\,\Delta t + 1), \tag{16.53}$$

$$p_{-}^{u}(j) = \frac{7}{6} + \frac{1}{2} aj\,\Delta t(aj\,\Delta t + 3), \tag{16.54}$$

$$p_{0}^{u}(j) = -\frac{1}{3} - aj\,\Delta t(aj\,\Delta t + 2). \tag{16.55}$$

The superscipt $u$ denotes the transition probabilities for upward branching.

Note that the foregoing expressions are either concave or convex quadratic functions of $j$. One can readily derive conditions on $j$ for the transition probabilities to be strictly positive. Namely, for normal branching

$$-\frac{\sqrt{2/3}}{a\,\Delta t} < j < \frac{\sqrt{2/3}}{a\,\Delta t}, \tag{16.56}$$

for upward branching

$$\frac{-1-\sqrt{2/3}}{a\,\Delta t} < j < \frac{-1+\sqrt{2/3}}{a\,\Delta t}, \tag{16.57}$$

and for downward branching

$$\frac{1-\sqrt{2/3}}{a\,\Delta t} < j < \frac{1+\sqrt{2/3}}{a\,\Delta t}. \tag{16.58}$$

Throughout we assume $a > 0$. Let us define a maximum value $j_{\max}$ as the smallest integer greater than $(1-\sqrt{2/3})/(a\,\Delta t) \approx 0.1835/(a\,\Delta t)$, for the index $j$ at any time slice, and a minimum value as $j_{\min} = -j_{\max}$. This leads to the branching methodology for each node $(j, i)$, whereby normal branching is used for $j_{\min} < j < j_{\max}$, downward branching is used for extreme positive value $j = j_{\max}$, and upward branching is used for extreme negative value $j = j_{\min}$.

### 16.4.2 The Second Stage: Lattice Calibration with Drift and Reversion

The purpose of the first stage is to build the component of the $r$-tree (i.e., the $r^*$-tree) that encapsulates the mean-reversion and volatility aspect of the short-rate process. In the final tree implementation, considered in this section, one needs to incorporate a drift component. Namely, at each time slice the nodes will be drifted by an amount determined by the market prices of the zero-coupon bonds. The drift component is incorporated by considering the difference $\alpha_t = r_t - r_t^*$. This satisfies an ordinary differential equation where

$$d\alpha_t = [\theta(t) - a\alpha_t]dt, \tag{16.59}$$

with solution

$$\alpha_t = e^{-at}\left[\alpha_0 + \int_0^t e^{as}\theta(s)ds\right]. \tag{16.60}$$

Here $\alpha_0 = r_0 = r(0,0)$ since $r_0^* = 0$. Equation (16.59) provides an apparently trivial analytical link between the actual $r$-tree and the driftless $r^*$-tree since $\theta(t)$ can be obtained exactly from the initial-term structure [i.e., as function of the yield $y_0(t)$]. Indeed, the right-hand side of equation (16.60) can be computed explicitly by applying the formulas derived in Chapter 2. Namely, one can use equation (2.110) or (2.109) [note that there the drift function $\theta(t)$ is called $a(t)$ and the function $b$ is called $a$ here] into the integral of equation (16.60) to obtain $\alpha_t$. We will not adopt this methodology here since it leads to inaccurate results and, moreover, bypasses the importance of the pricing algorithm for the drifted trinomial lattice, which we now present.

To apply the trinomial lattice pricing methodology we simply extend the equations of Section 16.2 into the trinomial lattice case. In general, we must distinguish between the different possible branching. Let us first assume normal branching. In this case the Green's function propagation (i.e., the Arrow–Debreu forward recursion) equation (16.12) is modified to read

$$G(0,0;j,T_i) = \sum_{k=j\pm 1,j;|k|\le i-1} G(0,0;k,T_{i-1})G(k,T_{i-1};j,T_i), \tag{16.61}$$

where the Arrow–Debreu prices for a single time step are nonzero for $|k| \le i-1$ and $k = j,\ j\pm 1$, and given by

$$G(k,T_{i-1};j,T_i) = \begin{cases} p_+(k)e^{-r(k,i-1)\Delta t}, & k = j-1, \\ p_0(k)e^{-r(k,i-1)\Delta t}, & k = j, \\ p_-(k)e^{-r(k,i-1)\Delta t}, & k = j+1. \end{cases} \tag{16.62}$$

In contrast to the binomial case, the forward time propagation of Arrow–Debreu prices is now obtained by summing contributions up to three (as opposed to two) possible two-legged paths. Note that the probabilities for up/down and middle moves in equations (16.61) and (16.62) are the ones corresponding to normal branching. For terminal node values of $j$ close to either $j_{\min}$ or $j_{\max}$, equations (16.61) and (16.62) need to be slightly modified. Namely, for any given value of $j$, equation (16.61) must be modified to the more general case

$$G(0,0;j,T_i) = \sum_{k;|k|\le i-1} G(0,0;k,T_{i-1})p(j,k)e^{-r(k,i-1)\Delta t}. \tag{16.63}$$

This formula takes into account all (generally mixed) branching types. The quantities $p(j,k)$ denote the nodal transition probabilities for all possible nonzero contributions from intermediate nodes at positions $k$ for time $T_{i-1}$. The sum of the corresponding probability values to be used in equation (16.63) now depend on the terminal $j$ value. Assuming $j_{\max} > 2$, there are possibly seven distinct cases to consider after $j_{\max}$ time steps.

1. $j = j_{\max}$ gives two terms (one down branch and one normal branch) with $p(j,j) = p_+^d(j)$, $p(j,j-1) = p_+(j-1)$.
2. $j = j_{\max} - 1$ gives three terms (one down branch and two normal branches) with $p(j,j+1) = p_0^d(j+1)$, $p(j,j) = p_0(j)$, $p(j,j-1) = p_+(j-1)$.
3. $j_{\min} + 2 < j < j_{\max} - 2$ gives three terms (three normal branches) with $p(j,j+1) = p_-(j+1)$, $p(j,j) = p_0(j)$, $p(j,j-1) = p_+(j-1)$.
4. $j = j_{\min}$, gives two terms (one up branch and one normal branch) with $p(j,k=j+1) = p_-(j+1)$, $p(j,j) = p_-^u(j)$.
5. $j = j_{\min} + 1$ gives three terms (one up branch and two normal branches) with $p(j,k = j+1) = p_-(j+1)$, $p(j,j) = p_0(j)$, $p(j,j-1) = p_0^u(j-1)$.

6. $j = j_{\max} - 2$ gives four terms (one down branch and three normal branches) with $p(j, k = j+2) = p_-^d(j+2)$, $p(j, j+1) = p_-(j+1)$, $p(j, j) = p_0(j)$, $p(j, j-1) = p_+(j-1)$.
7. $j = j_{\min} + 2$ gives four terms (one up branch and three normal branches) with $p(j, k = j+1) = p_-(j+1)$, $p(j, j) = p_0(j)$, $p(j, j-1) = p_+(j-1)$, $p(j, j-2) = p_+^u(j-2)$.

The forward propagation of Arrow–Debreu prices therefore involves a sum of two, three or four terms in cases where the terminal node is close to $j_{\max}$ or $j_{\min}$. Most values of $j$, however, involve normal branching, with the use of a three-term sum.

The pricing of zero-coupon bonds is essentially similar to the binomial lattice case, in the sense that one iterates out to any given time slice $T_i$ to obtain the Arrow–Debreu prices $G(0, 0; j, T_i)$. The analogue of equation (16.14) takes the form

$$P_0(r_0, T_i) = \sum_{j=-i}^{i} G(0, 0; j, T_i)\Lambda(r(j, T_i), T_i). \tag{16.64}$$

Specializing to the case of zero-coupon bonds, the equation analogous to equation (16.15) for pricing zero-coupon bonds is

$$Z_0(T_i) = \sum_{j=-i}^{i} G(0, 0; j, T_i). \tag{16.65}$$

Inserting equation (16.63) into equation (16.65) gives

$$Z_0(T_i) = \sum_{j=-i}^{i} \sum_{|k| \leq i-1} G(0, 0; k, T_{i-1})p(j, k)e^{-r(k, i-1)\Delta t}. \tag{16.66}$$

Hence, in general, one finds that the trinomial lattice calibration for a short-rate model can be achieved using a numerical root-finding procedure in equation (16.66) analogous to the binomial lattice methodology. The HW model, however, offers extra flexibility since one can actually solve the calibration problem analytically in the case of continuous compounding. The calibration of the lattice nodes for the HW model proceeds as follows.

The preceding formulas are specialized to the case where the actual drifted lattice is represented by

$$r(j, i) = \alpha(i) + j\,\Delta r, \qquad -i \leq j \leq i, \tag{16.67}$$

with $\alpha(0) = r(0, 0)$ as the initial node and the spacing given by equation (16.47). The coefficients $\alpha(i)$ represent the central node $r(0, i)$ along each time slice $T_i$ and will therefore account for the drift of the lattice. Plugging this into equation (16.66) and taking logarithms we obtain the simple analytical form for the coefficients:

$$\alpha(i) = \frac{\log\left[\sum_{j=-i-1}^{i+1} \sum_{k;|k| \leq i} G(0, 0; k, T_i)p(j, k)e^{-k\Delta r \Delta t}\right] - \log Z_0(T_{i+1})}{\Delta t}. \tag{16.68}$$

Note that we have shifted the time slice index $i$ to $i+1$. This gives the central node at each time slice $T_i$, and hence from equation (16.67) all nodes $r(j, i)$ for time $T_i$ are obtained, based on the market price of the zero-coupon bond maturing at time $T_{i+1}$ and knowledge of the Arrow–Debreu prices out to time $T_i$. These Arrow–Debreu prices are in turn given by forward induction using equation (16.63) by using the already-known values for the node positions at time slice $i-1$.

One begins the calibration procedure with $G(0,0;0,0) = 1$, and the initial node $r(0,0) = \alpha(0)$ is given in terms of the interpolated zero-coupon price at the first maturity time $T_1 = \Delta t$ using equation (16.16). Based on this and the zero-coupon prices at further maturities, one obtains the rest of the lattice nodes using equations (16.68) and (16.63). For instance, after the first step we have normal branching with $G(0,0;0,T_1) = p_0 e^{-\alpha(0)\Delta t}$, $G(0,0;\pm 1,T_1) = p_{\pm} e^{-\alpha(0)\Delta t}$. Assuming normal branching, at the second time step we obtain $\alpha(1)$:

$$\alpha(1) = \frac{\log\left(\sum_{j=-1}^{1} e^{-j(\Delta r)(\Delta t)} G(0,0;j,T_1)\right) - \log Z_0(T_2)}{\Delta t}$$
$$= \frac{\log\left(p_- e^{(\Delta r)(\Delta t)} + p_0 + p_+ e^{-(\Delta r)(\Delta t)}\right) - \log Z_0(T_2)}{\Delta t} - \alpha(0). \tag{16.69}$$

This procedure is continued for the rest of the time steps, hence giving the calibrated lattice for as many time steps as needed.

For the calibration of short-rate models that do not admit a simple analytical solution, such as the Black–Karasinski model covered in Section 16.5, one can readily proceed to find the central nodes numerically via a root-finding routine similar to what was described earlier for the binomial lattice.

### 16.4.3 Pricing Options

Once the calibrated lattice is built, the procedure for pricing options (e.g., caplets, floorlets, swaptions) follows similar steps as described for the binomial lattices given in Section 16.3. The conditional zero-coupon bonds are now obtained using

$$Z(j, T_i, T_{i+n}) = \sum_{k=j-n}^{j+n} G(j, T_i; k, T_{i+n}), \tag{16.70}$$

where the $(2n+1)$ Arrow–Debreu values (conditional on beginning at a $j$th node at time $T_i$ and ending at node $k = j-n, \ldots, j+n$ at time $T_{i+n}$) are computed by a general extension of equation (16.63), i.e., using the forward recursion relation

$$G(j, T_i; k, T_{i+m}) = \sum_{s;|s|\leq i+m-1} p(k,s) e^{-r(s,i+m-1)\Delta t} G(j, T_i; s, T_{i+m-1}). \tag{16.71}$$

Just as in equation (16.63), this forward propagation formula takes into account all possible mixed branchings. Note that the starting node is denoted by index $j$, while the terminal node now has index $k$. The nodal transition probabilities $p(k,s)$ are again given as described just following equation (16.63). For instance, when $j_{\min} + 2 < k < j_{\max} - 2$, normal branching is used with three possible nonzero values for $p(k,s)$: $p(k, s = k \pm 1) = p_{\mp}(k \pm 1)$, $p(k, s = k) = p_0(k)$.

Based on knowledge of the conditional zero-coupon prices, all option-pricing formulas are indentical in form to those for the binomial lattice, except for the obvious modification in having to compute and sum up more terms due to $n$ extra terminal nodes for every $n$ steps. Hence, for example, the caplet price is obtained by modifying equation (16.29) slightly:

$$\mathbf{Cpl}_0^{(\tau)}(R_K, T_i) = \sum_{j=-i}^{i} G(0,0;j,T_i) C^{(\tau)}(j,i). \tag{16.72}$$

For evaluating swaptions, the conditional zero-coupon prices are now given by a formula similar to equation (16.36) (except for the summation involving more nodes):

$$Z(j, T_{n_s}, T_{n_s+pm_s}) = \sum_{k=j-pm_s}^{j+pm_s} G(j, T_{n_s}; k, T_{n_s+pm_s}), \tag{16.73}$$

This leads to the pricing formula analogous to equation (16.37) (with payoff vector containing $n_s$ more components) for the payer swaption:

$$\mathbf{PSO}_0^{(n\tau)}(r_K, T) = \sum_{j=-n_s}^{n_s} G(0, 0; j, T) P^{(n\tau)}(j, n_s), \tag{16.74}$$

where $P^{(n\tau)}(j, n_s)$ is again given by equation (16.35). Similar pricing formulas follow in the obvious manner for other instruments, such as floorlets and receiver swaptions. The interested reader can also apply the methodology presented here to interest rate derivatives involving more exotic payoff structures.

## 16.5 Calibration and Pricing within the Black–Karasinski Model

The Black–Karasinski (BK) model is described by the short-rate process

$$d \log r_t = [\theta(t) - a(t) \log r_t]dt + \sigma(t) dW_t, \tag{16.75}$$

where $a(t)$ is a time-dependent mean-reversion speed. This model is the lognormal version of the Hull–White model, with $r_t$ replaced by $\log r_t$. Hence, a nice feature of this model is that the occurrence of negative interest rates is not possible. Note also that the BK model is an extension of the BDT model. Throughout we shall again assume a constant reversion speed $a(t) = a$ and constant $\sigma(t) = \sigma$. In contrast to the BDT model, the BK model still incorporates mean reversion under such conditions. As mentioned for the HW model, extensions to time-dependent reversion and/or volatility can also be implemented with some modifications and are left as an optional exercise.

The tree-building procedure for the BK model follows in similar fashion to the HW model as described in Sections 16.4.1 and 16.4.2. The difference here is that the short-rate node values are now replaced by their logarithms. Namely, the spacing takes a similar form as equation (16.11) except that the nodes now also drift. In particular, we define a constant spacing for the logarithm of the short-rate nodes: $\Delta x = \log r\,(j, i) - \log r\,(j-1, i)$ for any time slice $T_i$, or equivalently

$$r\,(j, i) = r\,(j-1, i) \exp(\Delta x). \tag{16.76}$$

This leads to the geometry of the short-rate nodes defined by a modification of equation (16.67) to read

$$r\,(j, i) = \alpha(i) \exp(j\,\Delta x), \qquad -i \le j \le i, \tag{16.77}$$

with $\alpha(i) = r\,(0, i)$ corresponding to the central node at time $T_i$ with $\alpha(0) = r\,(0, 0)$. Although we are somewhat at liberty to choose a spacing for $\Delta x$ in terms of $\Delta t$, we shall, in analogy with the HW model, set the spacing as

$$\Delta x = \sigma \sqrt{3\,\Delta t}.$$

With this choice we can carry out algebra similar to that in the HW model (see Section 16.4.1). Now, however, we consider the random variables $x_t \equiv \log r_t$ (and equivalently $x_t^* \equiv \log r_t^*$) and compute conditional expectations $E[(x_{t+\Delta t}^* - x_t^*)^2 | x_t^* = j\,\Delta x]$ and $E[x_{t+\Delta t}^* - x_t^* | x_t^* = j\,\Delta x]$. From equation (16.75) it is evident that $x_t$ obeys the HW model. As can easily be verified, the end result is that one obtains exactly the same formulas for the $j$-nodal transition probabilities $p_0^{(j)}$ and $p_\pm^{(j)}$ for middle and up/down moves, respectively, as in the HW case. Note, however, that now the logarithmic spacing between the short-rate nodes is constant: $\Delta x = \Delta \log r = \sigma\sqrt{3\,\Delta t}$.

The propagation of the Arrow–Debreu prices follows the general recursion procedure described earlier. Namely, the Arrow–Debreu prices originating from the present node $r\,(0,0)$ are given recursively by equation (16.63), where the node positions $r\,(k, i-1)$ are given by equation (16.77). The zero-coupon bonds are again given by equation (16.66). By plugging equation (16.63) into equation (16.66) and the expression for the nodes given by equation (16.77) one observes that, in contrast to the HW model, one cannot analytically solve for the central nodes. This is due to the fact that the grid spacing is constant in the logarithm of the short rate rather than the short rate itself. More explicitly, for the BK model we have

$$Z_0(T_i) = \sum_{j=-i}^{i} \sum_{k;|k|\le i-1} G(0,0;k,T_{i-1})p(j,k)\exp\left[-\alpha(i-1)\Delta t e^{k\Delta x}\right]. \tag{16.78}$$

Given the market zero-coupon price at maturity $T_i$ and the vector of Arrow–Debreu prices that are determined by forward recursion up to a previous time $T_{i-1}$, the parameter $\alpha(i-1)$ in this last equation is determined numerically via a single variable root-finding procedure. One can use the function zeroin of the MFZero library for this purpose. Hence, by determining the set of parameters $\alpha(i)$, $i = 0, 1, \ldots, M$, one obtains the entire calibrated BK lattice out to time $T_M$. Option pricing within the BK model then follows the same trinomial methodology as in Section 16.4.3.

# Bibliography


[AA02] L. Andersen and J. Andreasen. Volatile volatilities. *Risk*, 15(12):163–168, December 2002.

[ACCL01] C. Albanese, G. Campolieti, P. Carr, and A. Lipton. Black–Scholes goes hypergeometric. *Risk*, 14:99–103, December 2001.

[ACCZ03] C. Albanese, G. Campolieti, O. Chen, and A. Zavidonov. Credit barrier models. *Risk*, 16:109–113, June 2003.

[ADEH99] P. Artzner, F. Delbaen, J.-M. Eber, and D. Heath. Coherent measures of risk. *Mathematical Finance*, 9(3):203–228, 1999.

[AJW02] C. Albanese, K. Jackson, and P. Wiberg. Dimension reduction in the computation of value-at-risk. *Journal of Risk Finance*, 3:41–53, 2002.

[AL00] M. Avellaneda and P. Laurence. *Quantitative Modeling of Derivative Securities: From Theory to Practice*. Chapman & Hall, New York, 2000.

[AS64] M. Abramowitz and I. Stegun. *Handbook of Mathematical Functions, with Formulas, Graphs and Mathematical Tables*. National Bureau of Standards, Washington, DC, 1964.

[Bac00] L. Bachelier. Theorie de la speculation. *Annales de l'Ecole Normale Superieure XVII*, 3:21–86, 1900.

[Bas88] Basel Committee on Banking Supervision. International convergence of capital measurement and capital standards. Technical report, Bank for International Settlements, 1988.

[Bas96a] Basel Committee on Banking Supervision. Amendment to the Capital Accord to incorporate market risks. Technical report, Bank for International Settlements, 1996.

[Bas96b] Basel Committee on Banking Supervision. Overview of the amendment to the Capital Accord to incorporate market risks. Technical report, Bank for International Settlements, 1996.

[Bat91] D.S. Bates. The crash of 87: Was it expected? The evidence from options markets. *Journal of Finance*, 46:1009–1044, 1991.

[BD96] M. Broadie and J. Detemple. American option valuation: New bounds, approximations, and a comparison of existing methods. *Review of Financial Studies*, 9:1211–1250, 1996.

[BDT90] F. Black, E. Derman, and W. Toy. A one-factor model of interest rates and its applications to treasury bond options. *Financial Analysts Journal*, pp. 33–39, Jan.–Feb. 1990.

[Bed95] T.S. Beder. VAR: Seductive but dangerous. *Financial Analysts Journal*, 51(5):12–24, 1995.

[BEG89] P. Boyle, J. Evnine, and S. Gibbs. Numerical evaluation of multivariate contingent claims. *Review of Financial Studies*, 2:241–250, 1989.

[BG97a] M. Broadie and P. Glasserman. Monte Carlo methods for pricing high-dimensional American options: An overview. *Net Exposure: The Electronic Journal of Financial Risk*, 3:15–37, 1997.

[BG97b] M. Broadie and P. Glasserman. Pricing American-style securities using simulation. *Journal of Economic Dynamics and Control*, 21:1323–1352, 1997.

[BGM97] A. Brace, D. Gatarek, and M. Musiela. The market model of interest rate dynamics. *Mathematical Finance*, 7:127–154, 1997.

[BH95] W.L. Briggs and V.E. Henson. *The DFT: An Owner's Manual for the Discrete Fourier Transform*. SIAM, Philadelphia, 1995.

[BK91] F. Black and P. Karasinski. Bond and option pricing when short rates are lognormal. *Financial Analysts Journal*, pp. 52–59, July–August 1991.

[BKT01] Phelim P. Boyle, Adam W. Kolkiewicz, and Ken Seng Tan. Pricing American derivatives using simulation: A biased-low approach. *XIth Annual International Afir Colloquium*, pp. 1323–1352, Sept. 2001.

[BM01] D. Brigo and F. Mercurio. *Interest Rate Models—Theory and Practice*. Springer-Finance, Heidelberg 2001.

[BM02] D. Brigo and F. Mercurio. Joint calibration of the LIBOR market model to caps and swaptions volatilities. *Risk*, 117–121, January 2002.

[Bol86] T. Bollerslev. Generalized autoregressive conditional heteroskedasticity. *Journal of Econometrics*, 31:307–327, 1986.

[Boy77] P. Boyle. Options: A Monte Carlo approach. *Journal of Financial Economics*, 4:323–338, 1977.

[BR96] M. Baxter and A. Rennie. *Financial Calculus: An Introduction to Derivative Pricing*. Cambridge University Press, Cambridge, UK, 1996.

[BS73] F. Black and M. Scholes. The pricing of options and corporate liabilities. *Journal of Political Economy*, 81:637–54, 1973.

[But80] E. Butkov. *Mathematical Physics*. Wiley, New York, 1980.

[Ca97] P. Carr and A. Chou. Breaking Barriers. *Risk*, 139–145, September 1997.

[CaM04a] G. Campolieti and R. Makarov. Monte Carlo path integral pricing of Asian options on state dependent volatility models using high performance computing. Submitted, Wilfrid Laurier University, 2004. <http://info.wlu.ca/~wwwmath/faculty/joe/research/asian.pdf>.

[CaM04b] G. Campolieti and R. Makarov. (2004) Pricing path-dependent options on state dependent volatility models with a Bessel bridge. Submitted, Wilfrid Laurier University, 2004. <http://info.wlu.ca/~wwwmath/faculty/joe/research/bridge.pdf>.

[CaM05] G. Campolieti and R. Makarov. Parallel lattice implementation for option pricing under mixed state-dependent volatility models. *Proceedings of HPCS 2005*. Forthcoming. <http://info.wlu.ca/~wwwmath/faculty/joe/research/lattice.pdf>.

[Car98] P. Carr. Randomization of the American put. *Review of Financial Studies*, 11:597–626, 1998.

[CH03] P. Carr and A. Hirsa. Why be backward? Forward equations for American options. *Risk*, 16(1):103–107, January 2003.

[CIR85] J.C. Cox, J.E. Ingersoll, and S.A. Ross. A theory of the term structure of interest rates. *Econometrica*, 53:385–407, 1985.

[CLM97] J.Y. Campbell, A.W. Lo, and A.C. MacKinlay. *The Econometrics of Financial Markets*. Princeton University Press, Princeton, NJ, 1997.

[CM00a] P. Carr and D.B. Madan. Determining volatility surfaces and option values from an implied volatility smile. Forthcoming in M. Avellaneda (editor) *Quantitative Analysis in Financial Markets: Collected Papers of the New York University Mathematical Finance Seminar*, Vol. II, World Scientific, NY, 2001.

[CN47] J. Crank and P. Nicolson. A practical method for numerical evaluation of solutions of partial differential equations of the heat conduction type. *Proceedings of the Cambridge Philosophical Society*, 43:50–67, 1947.

[CR76] J. Cox and S.A. Ross. The valuation of options for alternative stochastic processes. *Journal of Financial Economics*, 3:145–166, 1976.

[CR85] J. Cox and M. Rubinstein. *Option Markets*. Prentice Hall, Englewood Cliffs, NJ, 1985.

[Cra89] J.C. Crank. *Mathematics of Diffusion*. Oxford, UK, 1989.

[CRM79] J. Cox, S. Ross, and M. Rubinstein. Option pricing: A simplified approach. *Journal of Financial Economics*, 7:229–264, 1979.

[CT04] R. Cont and P. Tankov. *Financial Modelling with Jump Processes*. Chapman & Hall, New York, 2004.

[Dav02] B. Davies. *Integral Transforms and Their Applications*. Springer-Verlag, New York, 2002.

[DK94] E. Derman and I. Kani. Riding on a smile. *Risk*, 7:32–39, 1994.

[DK98] E. Derman and I. Kani. Stochastic implied trees: Arbitrage pricing with stochastic term and strike structure of volatility. *Journal of Derivatives*, 3:7–22, 1998.

[DKC96] E. Derman, I. Kani, and N. Chriss. Implied trinomial trees of the volatility smile. *Journal of Derivatives*, 3:7–22, 1996.

[DL03] D. Davydov and V. Linetsky. Pricing options on scalar diffusions: An eigenfunction expansion approach. *Operations Research*, 51:185–209, 2003.

[DM88] L.M. Delves and J.L. Mohamed. *Computational Methods for Integral Equations*. Cambridge University Press, New York, 1988.

[DP97] D. Duffie and J. Pan. An overview of value-at-risk. *Journal of Derivatives*, 4(3):7–49, 1997.

[DP01] D. Duffie and J. Pan. Analytic value-at-risk with jumps and credit risk. *Finance and Stochastics*, 5(2):155–180, 2001.

[DT02] Jerome Detemple and Weidong Tian. The valuation of American options for a class of diffusion processes. *Management Science*, 48:917–937, 2002.

[Duf01a] D. Duffie. *Dynamic Asset Pricing Theory*. Princeton University Press, Princeton, NJ, 2001.

[Duf01b] D.G. Duffy. *Green's Functions with Applications*. Chapman & Hall, New York, 2001.

[Dup94] B. Dupire. Pricing with a smile. *Risk*, 7:18–20, 1994.

[EG92] L.C. Evans and R.F. Gariepy. *Measure Theory and Fine Properties of Functions*. CRC Press, Boca Raton, FL, 1992.

[EK99] R.J. Elliott and P.E. Kopp. *Mathematics of Financial Markets*. Springer-Verlag, Heidelberg, 1999.

[EL98] N. El Karoui, M. Jeanblanc, and S. Shreve. Robustness of the Black–Scholes formula. *Mathmetical Finance*, 8:93–126, 1998.

[EM84] D. Emanuel and J. MacBeth. Further results on the constant elasticity of variance call option pricing model. *Journal of Financial and Quantitative Analysis*, 7:533–554, 1984.

[Eng82] R.F. Engle. Autoregressive conditional heteroskedasticity with estimates of the variance of United Kingdom inflation. *Econometrica*, 50(4):987–1007, July 1982.

[Fel71] W. Feller. *An Introduction to Probability Theory and Its Applications*, volume 2. Wiley, New York, 1971.

[Feu93] A. Feuerverger. A consistent test for bivariate dependence. *International Statistical Review*, 61(3):419–433, 1993.

[FH96] B. Flesaker and L. Hughston. Positive interest. *Risk*, 9:46–49, 1996.

[Fis03] George S. Fishman. *Monte Carlo Concepts, Algorithms and Applications*, Springer-Verlag, New York, 2003.

[GHS02] P. Glasserman, P. Heidelberger, and P. Shahabuddin. Portfolio value-at-risk with heavy-tailed risk factors. *Mathematical Finance*, 12(3):239–269, July 2002.

[Gir60] I. Girsanov. On transforming a certain class of stochastic processes by absolute substitution of measures. *Theory of Probability and Applications*, 5:285–301, 1960.

[GJ84] R. Geske and H. Johnson. The American put option valued analytically. *Journal of Finance*, 39:1511–1524, 1984.

[GKR95] H. Geman, N. El Karoui, and J.C. Rochet. Changes in numeraire, arbitrage and option prices. *Journal of Applied Probability*, 32:443–458, 1995.

[GL89] G.H. Golub and C.F. Van Loan. *Matrix Computations*, 2nd ed. Johns Hopkins University Press, Baltimore, 1989.

[Gla04] P. Glasserman. *Monte Carlo Methods in Financial Engineering*. Springer-Verlag, New York, 2004.

[GNRS86] V. Giorno, A.G. Nobile, L.M. Ricciardi, and L. Sacerdote. Some remarks on the Rayleigh process. *Journal of Applied Probability*, 23:398–408, 1986.

[Gol97] B. Goldys. A note on pricing interest rate derivatives when forward LIBOR rates are lognormal. *Finance and Stochastics*, 2:345–52, 1997.

[GRon] I.S. Gradsteyn and I.M. Ryzhik *Tables of Integrals, Series and Products*, 2nd ed. Academic Press, New York, 1989.

[GW98] C. Gasquet and P. Witomski. *Fourier Analysis and Applications: Filtering, Numerical Computation, Wavelets*. Springer-Verlag, New York, 1998.

[Hal99] W.G. Hallerbach. Decomposing portfolio value-at-risk: A general analysis. Technical Report TI99-034/2/ECFR Report 9907, Tinbergen Institute, *Journal of Risk*, 5(2):1–18, Winter 2003.

[Hau98] E.G. Haug. *The Complete Guide to Option Pricing Formulas*. McGraw-Hill, New York, 1998.

[HD68] G.W. Hill and A.W. Davis. Generalized asymptotic expansions of Cornish–Fisher type. *Annals of Mathematical Statistics*, 39:1264–1273, 1968.

[Hes93] S. Heston. A closed-form solution for options with stochastic volatility with applications to bond and currency options. *Review of Financial Studies*, 6(2):327–343, 1993.

[Hig96] N.J. Higham. *Accuracy and Stability of Numerical Algorithms*. SIAM, Philadelphia, 1996.

[HJM92] D. Heath, R. Jarrow, and A. Morton. Bond pricing and the term structure of interest rates: A new methodology for contingent claims valuation. *Econometrica*, 60(1):77–105, 1992.

[HK79] J.M. Harrison and D.M. Kreps. Martingales and arbitrage in multiperiod securities markets. *Journal of Economic Theory*, 20:381–408, 1979.

[HL86] T. Ho and S. Lee. Term structure movements and pricing interest rate contingent claims. *Journal of Finance*, 41:1011–1029, 1986.

[HP81] J.M. Harrison and S.R. Pliska. Martingales and stochastic integrals in the theory of continuous trading. *Stochastic Processes and Their Applications*, 11:215–260, 1981.

[Hul00] J.C. Hull. *Options, Futures and Other Derivatives*, 4th ed. Prentice Hall, New York, 2000.

[HW93] J. Hull and A. White. One-factor interest rate models and the valuation of interest rate derivative securities. *Journal of Financial and Quantitative Analysis*, 28:235–254, 1993.

[HW94] J. Hull and A. White. Numerical procedures for implementing term structure models I: Single-factor models. *Journal of Derivatives*, 2(1):7–16, 1994.

[HW98a] J. Hull and A. White. Binormal distributions and value-at-risk. *Net Exposure*, 1:1, 1998.

[HW98b] J. Hull and A. White. Value-at-risk when daily changes are not normally distributed. *Journal of Derivatives*, 5:9–19, 1998.

[HW00] J. Hull and A. White. Forward rate volatilities, swap rate volatilities, and the implementation of the LIBOR market model. *Journal of Fixed Income*, 10:46–62, 2000.

[HW01] J. Hull and A. White. The general Hull–White model and supercalibration. *Financial Analysts Journal*, 57:6, 2001.

[Ing87] J.E. Ingersoll, Jr. *Theory of Financial Decision Making*. Rowman and Littlefield, Lanham, MD, 1987.

[IW89] N. Ikeda and S. Watanabe. *Stochastic Differential Equations and Diffusion Processes.*, 2nd ed. North Holland-Kodansha, Amsterdam, 1989.

[Jam91] F. Jamshidian. Bond and option evaluation in the Gaussian interest rate model. *Research in Finance*, 9:131–170, 1991.

[Jor96] P. Jorion. Risk$^2$: Measuring the risk in value at risk. *Financial Analysts Journal*, 52(6):47–56, 1996.

[JR96] J.C. Jackwerth and M. Rubinstein. Recovering probability distributions from option prices. *Journal of Finance*, 51(5): 1611–1631, 1996.

[JS87] J. Jacod and A. N. Shiryaev. *Limit Theorems for Stochastic Processes*. Springer-Verlag, New York, 1987.

[KMRZ98] A. Kreinin, L. Merkoulovitch, D. Rosen, and M. Zerbs. Principal component analysis in quasi Monte Carlo simulation. *Algo Research Quarterly*, 1(2):21–30, December 1998.

[KS91] I. Karatzas and S. E. Shreve. *Brownian Motion and Stochastic Calculus*, 2nd ed. Springer-Verlag, New York, 1991.

[KS99] I. Karatzas and S.E. Shreve. *Methods of Mathematical Finance*. Springer-Verlag, New York, 1999.

[Kwo99] Yue-Kuen Kwok. *Mathematical Models of Financial Derivatives*. Springer-Verlag, New York, 1999.

[Lip03] Alexander Lipton (editor). *Exotic Options: The Cutting Edge Collection*. Risk Books, London, UK, 2003.

[LL96] D. Lamberton and B. Lapeyre. *Introduction to Stochastic Calculus Applied to Finance*. Chapman & Hall, London, 1996.

[LM86] R.J. Larsen and M.L. Marx. *An Introduction to Mathematical Statistics and Its Applications*, 2nd ed. Prentice Hall, New York, 1986.

[LS01] F. Longstaff and E. Schwartz. Valuing American options by simulation: A simple least-squares approach. *Review of Financial Studies*, 14:113–147, 2001.

[Mar78] W. Margrabe. The value of an option to exchange one asset for another. *Journal of Finance*, 33:177–186, 1978.

[Mer73] R. Merton. Theory of rational option pricing. *Bell Journal of Economics and Management Science*, 4:141–183, 1973.

[Mer92] R.C. Merton. *Continuous-Time Finance*. Blackwell Scientific, Oxford, UK, 1992.

[MF53] P.M. Morse and H. Feshbach. *Methods of Theoretical Physics, Part I*. McGraw-Hill, New York, 1953.

[MM03] P.K. Medina and S. Merino. *Mathematical Finance and Probability: A Discrete Introduction*. Birkhauser Verlag, Boston, 2003.

[Mor96a] J.P. Morgan/Reuters. *RiskMetrics — Technical Document*, 4th ed. J.P. Morgan, New York, 1996.

[Mor96b] J.P. Morgan/Reuters. *RiskMetrics — Technical Document*, 4th ed. Morgan Guaranty Trust Company, New York, 1996.

[MR97a] M. Musiela and M. Rutkowski. *Application of Martingale Methods in Financial Modelling, Applications of Mathematics*, volume 36. Springer-Verlag, New York, 1997.

[MR97b] M. Musiela and M. Rutkowski. *Arbitrage Pricing Theory of Derivative Securities: Theory and Applications*, volume 36. Springer-Verlag, New York, 1997.

[MR98] H. Mausser and D. Rosen. Beyond VaR: From measuring risk to managing risk. *Algorithmics Research Quarterly*, 1(2):5–20, December 1998.

[MSS97] K. Miltersen, K. Sandmann, and D. Sondermann. Closed-form solutions for term structure derivatives with lognormal interest rates. *Journal of Finance*, 52:409–430, 1997.

[MU99] J. Mina and A. Ulmer. Delta-Gamma four ways. Technical report, RiskMetrics Group, New York, 1999.

[Nef96] S. Neftci. *An Introduction to the Mathematics of Financial Derivatives*. Academic Press, New York, 1996.

[Nel91] D.B. Nelson. Conditional heteroskedasticity in asset returns: A new approach. *Econometrica*, 59(2):347–369, March 1991.

[Øks00] B. Øksendal. *Stochastic Differential Equations*, 5th ed., Springer-Verlag, New York, 2000.

[Par61] E. Parzen. On estimation of a probability density function and mode. *Annals of Mathematical Statistics*, 33(3):1065–1076, September 1961.

[Pău99] S. Păun. Sensitivities of value at risk for non-linear portfolios. Master's thesis, University of Toronto, 1999.

[PBM86] A.P. Prudnikov, Yu. A. Brychkov, and O.I. Marichev. *Integrals and Series*, volume 2. Gordon and Breach, Cambridge, MA, 1986.

[Pel00] A. Pelsser. *Efficient Methods for Valuing Interest Rate Derivatives*. Springer-Verlag, New York, 2000.

[Pli97] S.R. Pliska. *Introduction to Mathematical Finance: Discrete Time Models*. Blackwell, Oxford, UK, 1997.

[PTVF92] W.H. Press, S.A. Teukolsky, W.T. Vetterling, and B.P. Flannery. *Numerical Recipes in C: The Art of Scientific Computing*, 2nd ed. Cambridge University Press, New York, 1992.

[Reb98] R. Rebonato. *Interest-Rate Option Models: Understanding, Analyzing and Using Models for Exotic Interest-Rate Options*, 2nd ed. Wiley, New York, 1998.

[Ric66] J.R. Rice. A theory of condition. *SIAM Journal on Numerical Analysis*, 3(2):287–310, June 1966.

[Ros56] M. Rosenblatt. Remarks on some nonparametric estimates of a density function. *Annals of Mathematical Statistics*, 27(3):832–837, September 1956.

[Ros00] S.M. Ross. *Introduction to Probability Models*, 7th ed. Harcourt Academic Press, New York, 2000.

[Rub78] M. Rubinstein. Implied binomial trees. *Journal of Finance*, 49:455–480, 1978.

[Rub85] M. Rubinstein. Non-parametric tests of alternative option pricing models. *Journal of Finance*, 40:455–480, 1985.

[Rud76] W. Rudin. *Principles of Mathematical Analysis*, 3rd ed. McGraw-Hill, New York, 1976.

[RZ98] M. Reimers and M. Zerbs. Dimension reduction by asset blocks. *Algo Research Quarterly*, 1(2):43–58, December 1998.

[Shi99] A.N. Shiryaev. *Essentials of Stochastic Finance: Facts, Models, Theory*. World Scientific, Singapore, 1999.

[Sla60] L.J. Slater. *Confluent Hypergeometric Functions*. Cambridge University Press, New York, 1960.

[SS90] G.W. Stewart and J.-G. Sun. *Matrix Perturbation Theory*. Academic Press, London, 1990.

[Sta96] F. Stambaugh. Risk and value at risk. *European Management Journal*, 14(6):612–621, 1996.

[Stu96] M. Stutzer. A simple non-parameteric approach to derivative security valuation. *Journal of Finance*, 51:1633–1652, 1996.

[Sun92] J.-G. Sun. Rounding-error and perturbation bounds for the Cholesky and $LDL^T$ factorizations. *Linear Algebra and Its Applications*, 173:77–97, 1992.

[Til93] J.A. Tilley. Valuing American options in a path simulation model. *Transactions of the Society of Actuaries*, 45:499–520, 1993.

[Tit62] E.C. Titchmarsh. *Eigenfunction Expansions Associated with Second-Order Differential Equations*. Clarendon Press, Oxford, UK, 1962.

[TT90] J.R. Thompson and R.A. Tapia. *Nonparametric Function Estimation, Modeling, and Simulation*. SIAM, Philadelphia, 1990.

[Vas77] O.A. Vasicek. An equilibrium characterization of the term structure. *Journal of Financial Economics*, 5:177–188, 1977.

[WHD95] P. Wilmott, S. Howison, and J. Dewynne. *The Mathematics of Financial Derivatives: A Student Introduction*. Cambridge University Press, New York, 1995.

[Wil98] P. Wilmott. *Derivatives: The Theory and Practice of Financial Engineering*. Wiley, New York, 1998.

[Wil00] P. Wilmott. *Paul Wilmott on Quantitative Finance*. Wiley, New York, 2000.

# ndex

## C

**D**

E



F

## Q

## R



## S

**U**



**V**



**W**



**X**

## ELSEVIER LICENSE AGREEMENT

PLEASE READ THE FOLLOWING AGREEMENT CAREFULLY BEFORE USING THIS ELECTRONIC MEDIA PRODUCT. THIS ELECTRONIC MEDIA PRODUCT IS LICENSED UNDER THE TERMS CONTAINED IN THIS ELECTRONIC MEDIA LICENSE AGREEMENT ("Agreement"). BY USING THIS ELECTRONIC MEDIA PRODUCT, YOU, AN INDIVIDUAL OR ENTITY INCLUDING EMPLOYEES, AGENTS AND REPRESENTATIVES ("You" or "Your"), ACKNOWLEDGE THAT YOU HAVE READ THIS AGREEMENT, THAT YOU UNDERSTAND IT, AND THAT YOU AGREE TO BE BOUND BY THE TERMS AND CONDITIONS OF THIS AGREEMENT. ELSEVIER. ("Elsevier") EXPRESSLY DOES NOT AGREE TO LICENSE THIS ELECTRONIC MEDIA PRODUCT TO YOU UNLESS YOU ASSENT TO THIS AGREEMENT. IF YOU DO NOT AGREE WITH ANY OF THE FOLLOWING TERMS, YOU MAY, WITHIN THIRTY (30) DAYS AFTER YOUR RECEIPT OF THIS ELECTRONIC MEDIA PRODUCT RETURN THE UNUSED ELECTRONIC MEDIA PRODUCT AND ALL ACCOMPANYING DOCUMENTATION TO ELSEVIER FOR A FULL REFUND.

### DEFINITIONS

As used in this Agreement, these terms shall have the following meanings:

"Proprietary Material" means the valuable and proprietary information content of this Electronic Media Product including all indexes and graphic materials and software used to access, index, search and retrieve the information content from this Electronic Media Product developed or licensed by Elsevier and/or its affiliates, suppliers and licensors.

"Electronic Media Product" means the copy of the Proprietary Material and any other material delivered on Electronic Media and any other human-readable or machine-readable materials enclosed with this Agreement, including without limitation documentation relating to the same.

### OWNERSHIP

This Electronic Media Product has been supplied by and is proprietary to Elsevier and/or its affiliates, suppliers and licensors. The copyright in the Electronic Media Product belongs to Elsevier and/or its affiliates, suppliers and licensors and is protected by the national and state copyright, trademark, trade secret and other intellectual property laws of the United States and international treaty provisions, including without limitation the Universal Copyright Convention and the Berne Copyright Convention. You have no ownership rights in this Electronic Media Product. Except as expressly set forth herein, no part of this Electronic Media Product, including without limitation the Proprietary Material, may be modified, copied or distributed in hardcopy or machine-readable form without prior written consent from Elsevier. All rights not expressly granted to You herein are expressly reserved. Any other use of this Electronic Media Product by any person or entity is strictly prohibited and a violation of this Agreement.

### SCOPE OF RIGHTS LICENSED (PERMITTED USES)

Elsevier is granting to You a limited, non-exclusive, non-transferable license to use this Electronic Media Product in accordance with the terms of this Agreement. You may use or provide access to this Electronic Media Product on a single computer or terminal physically located at Your premises and in a secure network or move this Electronic Media Product to and use it on another single computer or terminal at the same location for personal use only, but under no circumstances may You use or provide access to any part or parts of this Electronic Media Product on more than one computer or terminal simultaneously.

You shall not (a) copy, download, or otherwise reproduce the Electronic Media Product in any medium, including, without limitation, online transmissions, local area networks, wide area networks, intranets, extranets and the Internet, or in any way, in whole or in part, except that You may print or download limited portions of the Proprietary Material that are the results of discrete searches; (b) alter, modify, or adapt the Electronic Media Product, including but not limited to decompiling, disassembling, reverse engineering, or creating derivative works, without the prior written approval of Elsevier; (c) sell, license or otherwise distribute to third parties the Electronic Media Product or any part or parts thereof; or (d) alter, remove, obscure or obstruct the display of any copyright, trademark or other proprietary notice on or in the Electronic Media Product or on any printout or download of portions of the Proprietary Materials.

### RESTRICTIONS ON TRANSFER

This License is personal to You, and neither Your rights hereunder nor the tangible embodiments of this Electronic Media Product, including without limitation the Proprietary Material, may be sold, assigned, transferred or sublicensed to any other person, including without limitation by operation of law, without the prior written consent of Elsevier. Any purported sale, assignment, transfer or sublicense without the prior written consent of Elsevier will be void and will automatically terminate the License granted hereunder.

NOTICE

This book cannot be returned for credit or refund if the media package has been opened or removed.

**TERM**

This Agreement will remain in effect until terminated pursuant to the terms of this Agreement. You may terminate this Agreement at any time by removing from Your system and destroying the Electronic Media Product. Unauthorized copying of the Electronic Media Product, including without limitation, the Proprietary Material and documentation, or otherwise failing to comply with the terms and conditions of this Agreement shall result in automatic termination of this license and will make available to Elsevier legal remedies. Upon termination of this Agreement, the license granted herein will terminate and You must immediately destroy the Electronic Media Product and accompanying documentation. All provisions relating to proprietary rights shall survive termination of this Agreement.

**LIMITED WARRANTY AND LIMITATION OF LIABILITY**

NEITHER ELSEVIER NOR ITS LICENSORS REPRESENT OR WARRANT THAT THE INFORMATION CONTAINED IN THE PROPRIETARY MATERIAL IS COMPLETE OR FREE FROM ERROR, AND NEITHER ASSUMES, AND BOTH EXPRESSLY DISCLAIM, ANY LIABILITY TO ANY PERSON FOR ANY LOSS OR DAMAGE CAUSED BY ERRORS OR OMISSIONS IN THE PROPRIETARY MATERIAL, WHETHER SUCH ERRORS OR OMISSIONS RESULT FROM NEGLIGENCE, ACCIDENT, OR ANY OTHER CAUSE. IN ADDITION, NEITHER ELSEVIER NOR ITS LICENSORS MAKE ANY REPRESENTATIONS OR WARRANTIES, EITHER EXPRESS OR IMPLIED, REGARDING THE PERFORMANCE OF YOUR NETWORK OR COMPUTER SYSTEM WHEN USED IN CONJUNCTION WITH THE ELECTRONIC MEDIA PRODUCT.

If this Electronic Media Product is defective, Elsevier will replace it at no charge if the defective Electronic Media Product is returned to Elsevier within sixty (60) days (or the greatest period allowable by applicable law) from the date of shipment.

Elsevier warrants that the software embodied in this Electronic Media Product will perform in substantial compliance with the documentation supplied in this Electronic Media Product. If You report a significant defect in performance in writing to Elsevier, and Elsevier is not able to correct same within sixty (60) days after its receipt of Your notification, You may return this Electronic Media Product, including all copies and documentation, to Elsevier and Elsevier will refund Your money.

**YOU UNDERSTAND THAT, EXCEPT FOR THE 60-DAY LIMITED WARRANTY RECITED ABOVE, ELSEVIER, ITS AFFILIATES, LICENSORS, SUPPLIERS AND AGENTS, MAKE NO WARRANTIES, EXPRESSED OR IMPLIED, WITH RESPECT TO THE ELECTRONIC MEDIA PRODUCT, INCLUDING, WITHOUT LIMITATION THE PROPRIETARY MATERIAL, AND SPECIFICALLY DISCLAIM ANY WARRANTY OF MERCHANTABILITY OR FITNESS FOR A PARTICULAR PURPOSE.**

If the information provided on this Electronic Media Product contains medical or health sciences information, it is intended for professional use within the medical field. Information about medical treatment or drug dosages is intended strictly for professional use, and because of rapid advances in the medical sciences, independent verification of diagnosis and drug dosages should be made.

IN NO EVENT WILL ELSEVIER, ITS AFFILIATES, LICENSORS, SUPPLIERS OR AGENTS, BE LIABLE TO YOU FOR ANY DAMAGES, INCLUDING, WITHOUT LIMITATION, ANY LOST PROFITS, LOST SAVINGS OR OTHER INCIDENTAL OR CONSEQUENTIAL DAMAGES, ARISING OUT OF YOUR USE OR INABILITY TO USE THE ELECTRONIC MEDIA PRODUCT REGARDLESS OF WHETHER SUCH DAMAGES ARE FORESEEABLE OR WHETHER SUCH DAMAGES ARE DEEMED TO RESULT FROM THE FAILURE OR INADEQUACY OF ANY EXCLUSIVE OR OTHER REMEDY.

**U.S. GOVERNMENT RESTRICTED RIGHTS**

The Electronic Media Product and documentation are provided with restricted rights. Use, duplication or disclosure by the U.S. Government is subject to restrictions as set forth in subparagraphs (a) through (d) of the Commercial Computer Restricted Rights clause at FAR 52.22719 or in subparagraph (c)(1)(ii) of the Rights in Technical Data and Computer Software clause at DFARS 252.2277013, or at 252.2117015, as applicable. Contractor/Manufacturer is Elsevier Inc., 360 Park Avenue South, New York, NY 10010-5107 USA.

**GOVERNING LAW**

This Agreement shall be governed by the laws of the State of New York, USA. In any dispute arising out of this Agreement, you and Elsevier each consent to the exclusive personal jurisdiction and venue in the state and federal courts within New York County, New York, USA.